Museums at the Forefront of the History and Philosophy of Geology: History Made, History in the Making

edited by

Gary D. Rosenberg
Milwaukee Public Museum
800 W. Wells Street
Milwaukee, Wisconsin 53233, USA
and Earth Sciences Department
Indiana University–Purdue University
Indianapolis, Indiana 46202, USA

Renee M. Clary
Department of Geosciences
Mississippi State University, Box 5448
Mississippi State, Mississippi 39762, USA

THE GEOLOGICAL SOCIETY OF AMERICA®

THE GEOLOGICAL SOCIETY OF AMERICA®
HISTORY AND PHILOSOPHY OF GEOLOGY DIVISION

Special Paper 535

3300 Penrose Place, P.O. Box 9140 ▪ Boulder, Colorado 80301-9140, USA

2018

Published by The Geological Society of America, Inc.
3300 Penrose Place, P.O. Box 9140, Boulder, Colorado 80301-9140, USA
www.geosociety.org

Printed in U.S.A.

GSA Books Science Editors: Richard A. Davis Jr. and Christian Koeberl

Library of Congress Cataloging-in-Publication Data

Names: Rosenberg, Gary D., editor. | Clary, R. M. (Renee M.), editor. | Geological Society of America.
Title: Museums at the forefront of the history and philosophy of geology : history made, history in the making / edited by Gary D. Rosenberg (Milwaukee Public Museum, Milwaukee, Wisconsin, and Earth Sciences Department, Indiana University-Purdue University, Indianapolis, Indiana), Renee M. Clary (Department of Geosciences, Mississippi State University, Mississippi).
Description: Boulder, Colorado : The Geological Society of America, 2018. | Series: Special paper ; 535 | Includes bibliographical references.
Identifiers: LCCN 2018021727 | ISBN 9780813725352 (pbk.)
Subjects: LCSH: Geological museums. | Geology—Exhibitions. | Geology—Social aspects. | Geology—Study and teaching.
Classification: LCC QE51 .M87 2018 | DDC 550.74/73—dc23
LC record available at https://lccn.loc.gov/2018021727

Cover: *Front: Charles Willson Peale: The Artist in His Museum,* 1822 self-portrait. Courtesy of the Pennsylvania Academy of the Fine Arts, Philadelphia. Gift of Mrs. Sarah Harrison (The Joseph Harrison, Jr., Collection). See Rosenberg and Clary, Chapter 1, and Newcomb, Chapter 11, in this volume for more information. *Background image:* Galerie de Paléontologie et d'Anatomie Comparée of the Muséum national d'Histoire naturelle, Paris, France. See Lipps, Chapter 2, this volume, for more information.

Contents

Preface and acknowledgments

Museums at the Forefront of the History and Philosophy of Geology: History Made, History in the Making is an outgrowth of a topical session of the same name held at the 2016 annual meeting of the Geological Society of America (GSA). We solicited additional submissions from colleagues in the United States and Europe for this volume.

Richard Conniff's essay "Our Natural History, Endangered," lamenting the declining status of natural history museums in Europe and America, appeared as an op-ed in the online 1 April 2016 edition of the *New York Times*[1] as we were organizing our topical session. It gave our presentations an immediacy that underscores the general cultural and specifically scientific importance of natural history museums.

Natural history museums have evolved over the past 500 years to become vanguards of science literacy and thus institutions of democracy. Curiosity about nature and distant cultures has proven to be a powerful lure, and museums have progressively improved public engagement through increasingly immersive exhibits, participation in field expeditions, and research using museum holdings, all facilitated by new technology. Natural history museums have dispersed across the globe and have demonstrated that fascination with ancient life, vanished environments, exotic animals in remote habitats, cultural diversity, and our place in the cosmos is universal. This volume samples the story of museum development and illustrates that the historical successes of natural history museums have positioned them to be preeminent facilitators of science literacy—hence contributors to the greater good—well into the future.

The topical session was sponsored by the GSA History and Philosophy of Geology Division, the Geoscience Education Division, and the Geology and Society Division; the Paleontological Society; the History of Earth Sciences Society; and the Paleontological Research Institution. The GSA History and Philosophy of Geology Division and the Paleontological Society generously provided financial assistance for invited speakers.

The History and Philosophy of Geology Division contributed funds from the Mary C. Rabbitt Bequest to underwrite publication of color images in this volume.

Numerous colleagues kindly reviewed our manuscripts. The GSA staff, notably April Leo, Managing Editor for Books, and Richard A. "Skip" Davis, Science Editor, devoted much time and effort to the production of this volume.

We thank all for their dedication that facilitated our efforts.

Gary D. Rosenberg, *Milwaukee Public Museum and Indiana University–Purdue University*
Renee M. Clary, *Mississippi State University*
January 2018

[1]Richard Conniff's online op-ed is available at http://nyti.ms/1ZRJC2d (last accessed 12 January 2018) and on page SR4 of the *New York Times* paper edition of 3 April 2016, with the headline: "Natural History, Endangered."

Dedication to Michele La Clergue Aldrich (1942–2016)

We dedicate this volume to the memory of Michele La Clergue Aldrich (1942–2016), a co-founder of the Geological Society of America's (GSA) History and Philosophy of Geology Division in 1976.

Michele was a steadfast participant, mentor, and advisor within the History and Philosophy of Geology Division. She served as chair, and then for eight years as secretary-treasurer and editor of the division's semi-annual newsletter. She held the division's institutional memory and unselfishly provided time and expertise to guide the division's growth.

Michele was a perfectionist. She produced exceptional historical research on topics that included the contributions of women in several fields of science, the accomplishments of the New York Natural History Survey, and geological discoveries of the pioneering surveys and other explorations of the western United States. As testimony to her expertise, William H. Goetzmann and Kay Sloan (1982, p. 10) wrote that "Michele Aldrich … ranks as one of this or any other generation's outstanding researchers, and over the years she has set a new standard for scholarly friendship."

Michele's longstanding collaboration with Alan Leviton in the history of geoscience encompassed significant research on museums. It is these accomplishments (Aldrich and Leviton, 2000a, 2000b; Leviton and Aldrich, 2003) that were the inspiration for this volume.

In 1992, Michele received the Mary C. Rabbitt Award from the GSA History and Philosophy of Geology Division for exceptional scholarly contributions. With Alan Leviton, Michele was further recognized by the division in 2007 with the Gerald M. and Sue T. Friedman Distinguished Service Award to the history and philosophy of geology.

In addition to her work for the History and Philosophy of Geology Division, Michele devoted years of service to the community of historians of the earth sciences through the History of Earth Sciences Society, the American Association for the Advancement of Science, the History of Science Society, the Forum for the History of Geology in America (founding member), and INHIGEO (International Commission on the History of Geological Sciences).

She was especially dedicated to facilitating the participation and recognition of women in the study of the history and philosophy of geology. Her research was an effort to tell the hidden history of women in geology as well.

Michele's colleagues fondly recall her attention to detail, exceptional editorial skills, and remarkable sense of humor. Michele excelled in pointing out "howlers"—those tragic, but humorous, literary faux pas that convey unintended, outrageous meanings.

We can only begin to outline the entirety of Michele's influence within the history and philosophy of geology. However, her publications remain as invaluable references; her service advanced the importance of the history and philosophy of the geosciences; and her impact endures through the many colleagues she

Clary, R.M., and Rosenberg, G.D., 2018, Dedication to Michele La Clergue Aldrich, *in* Rosenberg, G.D., and Clary, R.M., eds., Museums at the Forefront of the History and Philosophy of Geology: History Made, History in the Making: Geological Society of America Special Paper 535, p. vii–viii, https://doi.org/10.1130/2018.2535(00).

Michele La Clergue Aldrich's Enduring Legacy

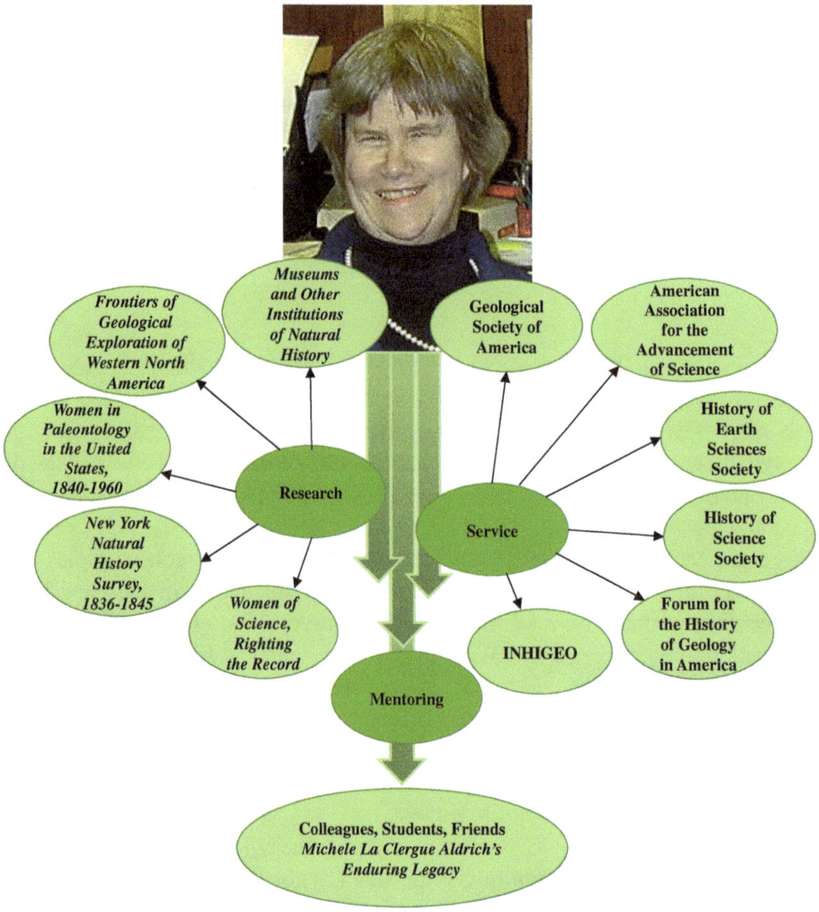

assisted, educated, and inspired. Her mentorship of young scholars continues as the Michele Aldrich History and Philosophy of Geology Student Research Award, which her husband Mark Aldrich established in her memory.

Renee M. Clary, *Mississippi State University*
Gary D. Rosenberg, *Milwaukee Public Museum and Indiana University–Purdue University*
February 2018

References Cited

Aldrich, M.L., and Leviton, A.E., 2000a, West and east: The California Academy of Sciences and the Smithsonian Institution, 1852–1906, *in* Ghiselin, M.T., and Leviton, A.E., eds., Cultures and Institutions of Natural History: Essays in the History and Philosophy of Science: Memoirs of the California Academy of Sciences No. 25, p. 183–202.

Aldrich, M.L., and Leviton, A.E., 2000b, India: A case study of natural history in a colonial setting, *in* Ghiselin, M.T., and Leviton, A.E., eds., Cultures and Institutions of Natural History: Essays in the History and Philosophy of Science: Memoirs of the California Academy of Sciences No. 25, p. 51–80.

Goetzmann, W.H., and Sloan, K., 1982, Looking Far North: The Harriman Expedition to Alaska, 1899: New York, Viking Press, 224 p.

Leviton, A.E., and Aldrich, M.L., eds., 2003, Museums and Other Institutions of Natural History: Past, Present, and Future: A Symposium Held on the Occasion of the 150th Anniversary of the California Academy of Sciences, June 16–17, 2003: Proceedings of the California Academy of Sciences, series 4, v. 55 (Suppl. 1), San Francisco, 329 p.

The Geological Society of America
Special Paper 535

Something to be said for natural history museums

Gary D. Rosenberg
Milwaukee Public Museum, 800 W. Wells Street, Milwaukee, Wisconsin 53233, USA, and
Earth Sciences Department, Indiana University–Purdue University, Indianapolis, Indiana 46202, USA

Renee M. Clary
Department of Geosciences, Mississippi State University, Box 5448, Mississippi State, Mississippi 39762, USA

MR. PEALE'S NATURAL HISTORY MUSEUM

Charles Willson Peale's self-portrait, *The Artist in His Museum* (book cover; Fig. 1; Newcomb[1]), represents a seminal period in the emergence of natural history museums. Peale (1741–1827) completed the painting in 1822 as the modern natural history museum began to take shape, first in Europe and then in the young American nation as an outcome of the liberalizing transformations of the Enlightenment in both Europe and America. Peale had founded his museum in 1784, the second natural history museum in America[2]. The Pennsylvania Academy of Fine Arts (PAFA), where the painting now resides, recounts that Peale was an artist, naturalist, showman, promoter of American ideals, and dedicated to public education[3].

Peale is all of those in his painting. He is a host on a stage, he looks directly at the viewer and, with a dramatic flourish and hint of a smile, pulls back a red drape emblematic of ignorance to invite the viewer into his museum to witness its wonders. One, on the floor at left, is the mount of a turkey. Others, including a bald eagle, which had been named the national symbol of American pride in 1782, sit along the walls in the gallery behind the drape. High on the rear wall are portraits of American luminaries whom Peale chose to honor. A family is exploring the exhibits in the background.

A femur and jaw, from what we now understand to have been a mastodon, lie in the foreground at right. It was then known as the "American incognitum" in reference to mysteries surrounding its identity[4]. Peale himself helped quarry the animal's remains in 1800, and the reassembled skeleton appears behind the drape as it is raised. Peale recorded the dig in 1806 in another painting that is called *Exhumation of the Mastodon.*

THEMES IN PEALE'S PAINTING AND IN THIS BOOK

This volume samples some 500 years of development of natural history museums from the Scientific Revolution to the present and Peale's painting captures its themes. Among them is the historical disengagement of exhibits of human-made objects (*artificialia*) from those produced by nature (*naturalia*). During the Scientific Revolution of the sixteenth and seventeenth centuries, both groups had been commingled in curiosity cabinets or "wünderkammern" (rooms of marvels or wonders). In the Philadelphia museum of Charles Willson Peale, nature takes precedence. His specimens are arranged according to Linnaean taxonomy[5], which had been formulated in the eighteenth century and quickly accepted. Taxonomic displays would be adopted by natural history museums and predominate until the end of

[1]All citations in this introduction are to papers in this volume unless stated otherwise.
[2]The Charleston Museum, South Carolina, was the first. See also Newcomb (this volume).
[3]PAFA website: https://www.pafa.org/collection/artist-his-museum (accessed 20 May 2018).
[4]See S.M. Rowland, 2009, Thomas Jefferson, extinction, and the evolving view of Earth history in the late eighteenth and early nineteenth centuries, *in* Rosenberg, G.D., ed., *The Revolution in Geology from the Renaissance to the Enlightenment*: Geological Society of America Memoir 203, p. 225–246, https://doi.org/10.1130/2009.1203(16).
[5]Website for *The Encyclopedia of Greater Philadelphia:* http://philadelphiaencyclopedia.org/archive/peales-philadelphia-museum/ (accessed 20 May 2018).

Rosenberg, G.D., and Clary, R.M., 2018, Something to be said for natural history museums, *in* Rosenberg, G.D., and Clary, R.M., eds., Museums at the Forefront of the History and Philosophy of Geology: History Made, History in the Making: Geological Society of America Special Paper 535, p. 1–7, https://doi.org/10.1130/2018.2535(01).

Figure 1. Charles Willson Peale, 1822, *The Artist in His Museum*, courtesy of the Pennsylvania Academy of the Fine Arts, Philadelphia. Gift of Mrs. Sarah Harrison (The Joseph Harrison, Jr., Collection).

the nineteenth century with the advent of Darwinian evolution (Cohen). Thus, Peale's painting stands for a pivotal transition from the archaic to modern perspective of natural history museums. Late in the nineteenth century, displays showing animals and plants in their natural habitats would emerge with the transformative taxidermy and dioramas that began in Milwaukee with Carl Akeley (Rosenberg).

Another theme is curiosity about the objects of nature and the acquisitive instinct for them, manifested as use of collections for research and for educational exhibits about nature. Research and education are both outcomes of curiosity and the urge to possess. Researchers strive for new knowledge. Teachers acquire the information that research makes available, and they impart it to students. Students learn best if they can integrate acquired knowledge into their existing cognitive frameworks, also an act of possession.

REORGANIZING NATURE: RECONTEXTUALIZATION AND RECONCEPTUALIZATION

Curiosity about nature and distant cultures has proven to be a powerful lure for museums over the past 500 years, but it has not always been that way. Long before the Scientific Revolution, curiosity was viewed in Western civilization with suspicion. During antiquity, in Greece, curiosity was regarded as a trait of busybodies and, later in Rome, as an expression of dangerous or useless knowledge. In the Middle Ages, curiosity was deemed a vice: It was a sin to be curious! Even during the Scientific Revolution, Francis Bacon and Galileo had their doubts about its value (Kaufmann, 1993[6]). Thus, the expression of curiosity has a history and cultural context that requires nurturing because it is vital to science literacy, and it is imperative because science literacy is an institution of democracy.

The focus in sixteenth- and seventeenth-century collections was on unusual, rare, and exquisite objects. Some of the most important were royal reserves that signified personal wealth and power (over nature as well as over the populace). Neither royal reserves nor collections in non-aristocratic hands were widely accessible to the populace (Kaufmann, 1993).

The commingling of naturalia and artificialia reflected esoteric correspondences with mystical or religious connotations that only the privileged were thought worthy of deciphering. Today those pretensions would seem strange. The context was meant to reinforce and signify the importance of the owner. For example, elegant, costly metalwork embellishing a cup made of rhinoceros horn or placed nearby undoubtedly cast an aura of sophistication upon the cup. "Philosophers' stones" in some collections were purported to hold the secret of immortality or of turning base metals into gold. The human body was considered a microcosm of the world (the macrocosm). In esoteric ways, the structure and workings of the microcosm and macrocosm were modeled similarly.

Remnants of these early collections can still be seen in cities such as Copenhagen, Stockholm, Vienna, Paris, and Florence, but most have been dispersed, sorted and consolidated, intermingled, or augmented with later collections altogether obscuring provenance. The disorganization opened the way for the modern museum because it made space for new organizations.

One of the remnants is in Florence, where a few samples of "all things formed in the earth" from Nicholas Steno's *Indice,* which he accumulated on his European travels and during his employment in the Medici Court, have been identified (Dominici and Cioppi). Another is on display at the Museum of Natural History, University of Copenhagen. It is the only object confirmed to have once been in Danish polymath Ole Worm's wünderkammern[7], a collection that is said to have impressed Nicholas Steno, a founder of modern geology, in the seventeenth century. It is a tree trunk with an ingrown horse jaw whose juxtaposition was thought to illustrate the merger of plant and animal principles through the creative force of nature (Tarp).

However, as the flow of wealth and power shifted in Europe, the objectives of collecting changed, and whether by dispersal and loss or by consolidation and augmentation, the organization of objects in collections, their place in society, and the conception of their place in nature began to change (Inglehart; Koeberl et al.; Klemun; Cohen; Dominici and Cioppi; Tarp). It became clear that neither naturalia nor artificialia held mystical powers and that naturalia alone could be used to decipher nature's mysteries, without the help of artificialia (Kaufmann, 1993). Interest in the extraordinary waned as interest in common objects gained (Kaufmann, 1993; Inglehart), though of course interest in exceptional objects would never disappear; remains of extraordinarily large dinosaurs will always be a draw, for example.

Collections became "recontextualized" and consequently "reconceptualized," two words that indicate an implicit theme in this volume[8]. Objects gain meaning (form concepts) in reference to others with which they are associated (their context) (Klemun; Cohen; Foucault, 1970[9]). Collections were reorganized and old associations faded. New associations and new conceptions, ideas, or models of the objects' place in nature arose. As naturalia and artificialia were separated, fancy metalwork did nothing to enhance the meaning of cups made of rhinoceros horn. Those same cups take on quite a different connotation today in view of the threatened status of the species from which they are far removed. Now, many would conclude that the only place for a rhinoceros horn is on the animal that grew it.

The recontextualization that occurred during the Scientific Revolution was an outcome of the rediscovery of Greek and Arabic geometry during the Renaissance and use of it to describe the organization of nature. It provided a standardized way to record the characteristics of objects that anyone with a modest education could comprehend (Edgerton, 1991[10]). It was an intuitive as well as mathematical approach to recording "scientific detail," e.g., the structure, shape, size, color, surface features, of an object and its relation to others in three-dimensional space. What Renaissance artists began, naturalists in the Scientific Revolution proceeded to formalize as modern science (Alpers, 1983[11]; Edgerton, 1991).

Description facilitated reconceptualization: comparison, new taxonomies, and modern renditions of natural history. Taxonomies based on value judgments faded. For example, the proximity of entities to God (i.e., the "Great Chain of Being" that ranked angels closest to God, followed by royalty, then successively commoners, animals, plants, and minerals and rocks) gave way to systematics based on "scientific detail."

Nicholas Steno provides an early example of a transition from speculative "poetry" about nature (Dominici and Cioppi)

[6]See T.D. Kaufmann, 1993, *The Mastery of Nature: Aspects of Art, Science, and Humanism in the Renaissance*: Princeton, New Jersey, Princeton University Press. See especially p. 174–194.

[7]See "Dr. Worm's Strange Objects and Curiosities," Jan. 2014, University Post, University of Copenhagen, https://uniavisen.dk/en/dr-worms-strange-objects-and-curiousities/ (accessed 5 March 2018).

[8]This also helps to explain why interdisciplinary collaborations are encouraged; they facilitate new associations of facts and generate new conceptions.

[9]See M. Foucault, 1970, *The Order of Things: An Archaeology of the Human Sciences*: New York, Vintage Books, 387 p.

[10]See S.Y. Edgerton, Jr., 1991, *The Heritage of Giotto's Geometry*: Ithaca, New York, Cornell University Press, 313 p.

[11]See S. Alpers, 1983, *The Art of Describing: Dutch Art in the Seventeenth Century*: Chicago, University of Chicago Press, 267 p.

to inferences based on observation. His comparison of puzzling "tongue stones" dug from the earth with the teeth of living sharks prompted him to realize that generative forces within the ground had nothing to do with the origin of the former. Rather, he concluded they were the remains of once-living things, a milestone in the history of paleontology.

Filippo Buonanni's augmentation of Athanasius Kircher's collection (Inglehart) offers another example from the seventeenth century. Kircher's esoteric fossil collection was in disarray when Buonanni supplemented it with common local fossils and shells. Buonanni thought that in numbers and variety of specimens, he could find exceptions to the experimentalist assertion of overarching "uniformity," that nature "conforms" to broad generalizations gleaned from few observations. In this case, experimentalists including Steno had generalized that all mollusks reproduce by eggs. For Buonanni, who held to Aristotelian spontaneous generation for the origin of fossils, that was an oversimplification. Inglehart concludes that, despite Buonanni's embrace of an archaic theory, his accumulation of common specimens in order to challenge a hypothesis situated Buonanni as a founder of modern museology.

Further, description of "scientific detail" led to a revolution in understanding the spatial relationships of fossils and their enclosing strata at geographic scales. Comparison and correlation of assemblages of fossils from different geographic areas facilitated mapping the sedimentary strata from which they were obtained. Diemer presents the important example of Scandinavian collections that would lead to worldwide correlations of Silurian fossil sequences (cf. Coorough Burke and Sheehan).

Linnaean taxonomy, based on the categorization of biologic form and published in the eighteenth century, was a logical precursor of Darwin's theory, formulated a century later, which explained how body plans evolve by natural selection[12]. These developments raised debates about natural laws and theology[13], rates of biologic change, and the place of extinctions in evolution (Cohen). A progression of exhibits still on display in the Paris Museum presents these contextual changes in dramatic fashion, from Linnaeus at one end, Darwin at the other, and Lamarck and Cuvier in between (Cohen; Lipps).

As they separated conceptually from fossils, other elements of geology such as minerals and rocks also acquired modern identities and classifications (Dominici and Cioppi; Tarp; Inglehart; Koeberl et al.; Klemun). This made it possible to conceive of their evolution in terms of the physical-chemical factors of which they were constructed, to determine their geographic distribution (Diemer), and to facilitate locating economically valuable ores (Klemun; Koeberl et al.).

Another spatial reorganization occurred with the accumulation in Vienna of what is now one of the world's largest and most important meteorite collections. And again, scientific detail made it possible to see that meteorites were different from terrestrial stones and eventually differentiated into stony and iron-nickel meteorites. The distinctions led not only to a better understanding of the organization of the solar system, but also of the structure and evolution of the Earth's interior (Koeberl et al.).

RECONTEXTUALIZATION IN THE MAKING

Great works of art extend beyond their frames: they leave something to the imagination. So too, great museums. They extend beyond their exhibits by stimulating the imagination as well. Both make space for visitors to participate in discovery. History attests to the progress that natural history museums have made inviting visitors in to discover natural wonders[14], just as Charles Willson Peale depicts in his painting. In defining visitors, we now include researchers, students, and educators together with the general public, all engaged in recontextualization and reconceptualization, finding new relationships and patterns among fossils, rocks, and minerals. And new relationships between the elements of geology and the visitors as well.

Augmentation and curation of geoscience collections, not a rehash of previously published data, will be the source of new insight. This is especially true for paleontology, which, unlike sciences such as astronomy, chemistry, physics, and ethology, is reliant on collections (Allmon et al.). Yet, if we remember that sequence stratigraphy, a relatively recent breakthrough, involved using major unconformities to redefine and correlate geographically distant stratigraphic packages in new ways, then it is clear that other subdisciplines of geoscience progress via recontextualization as well.

There are millions of fossils in collections worldwide whose contexts have not been thoroughly defined, let alone interpreted: curation of collections has never been a priority (Allmon et al.). For example, Liggett et al. relate the long history of what is a vast legacy collection of the U.S. Department of the Interior. The collections have been dispersed to numerous museums across the country (and some around the world). With progress in paleontology, awareness has grown that early collectors gave little thought to recording important scientific detail and that context is deficient without it. For example, the location of collecting sites, lithology of enclosing sediments and facies relationships, stratigraphic ranges of specimens, population structure, geographic distribution—"big data" (Allmon et al.) vital for defining context—may not be retrievable as

[12]Regarding the importance of Darwin's collections, see D. Bresson, 2018 (12 Feb.), How Darwin's fossil collection helped him to come up with evolution: *Forbes*, https://www.forbes.com/sites/davidbressan/2018/02/12/how-darwins-fossil-collection-helped-him-to-come-up-with-evolution/#2b6a88321df3 (accessed 20 May 2018).

[13]See M.T. Ghiselin, 2009, Natural theology, design and law, *in* Rosenberg, G.D., ed., *The Revolution in Geology from the Renaissance to the Enlightenment*: Geological Society of America Memoir 203, p. 277–283, https://doi.org/10.1130/2009.1203(20).

[14]Unfortunately, creation museums (Lipps) leave nothing to the imagination; they stipulate what the visitor is to learn with little leeway for uncertainty, debate, or alternative explanations—the stuff that makes science a process of discovery in which everyone can participate.

collecting localities have disappeared with population growth and development.

However, Antell is one of the first to assess this problem as she itemizes how several types of biases affect estimates of diversity of Cenozoic insect specimens housed at the Florissant, Colorado, museum with that of Florissant material distributed to other museums. Her objective is to test whether all groups are equally representative of the Florissant biota. Her comparisons suggest that bias can be accounted for, but residuals (i.e., data that do not fall within statistical patterns) should motivate further collection and research.

Whatever the outcome of efforts to rescue legacy collections, it is certain that "millions" of fossils in museums around the world (Allmon et al.) together with the known extant biota, are likely only a tiny fraction of the species that occupied our planet in the past and occupy it now[15]. This has big implications for biodiversity studies, which are utterly dependent on collections of fossils as well as extant specimens (Lieberman and Kimmig). The study of variations in diversity of organisms through time is one of the most important recontextualizations of specimens that contemporary museums have accomplished. Museum studies of biodiversity have influenced the conceptualization of extinction rates and patterns, the effects of climate change, and loss of habitats, for example (Antell; Lieberman and Kimmig).

Although one might argue that a census of living organisms alone is sufficient for evaluations of biodiversity, such a survey would be but a snapshot of a moment in time; collections in the context of the stratigraphic record are vital for conceptualizing long-term trends through deep time and across geographic space, as well as for comprehending the changing environmental context of biota from the past to the present (Antell; Lieberman and Kimmig).

NATURAL HISTORY MUSEUMS AT THE FOREFRONT OF SCIENCE LITERACY

The use of collections for education naturally expanded as their social context changed from private to shared ownership (Koeberl et al.; Klemun; Dominici and Cioppi; Cohen). Although Ole Worm in Denmark held to ideas we now consider archaic, in the preface to *Museum Wormianum* (1655), he urged students to study his collections to learn about nature directly rather than to accept without evidence either his word or "sophistry" and "superficial and scholastic [academic] philosophers" (Tarp).

Even now, there is something to be said for wünderkammer (Rosenberg); individual specimens without context still inspire curiosity. And even though life-sized replicas of dinosaurs, photographs of rocks and crystals, virtual reality, and computer-generated restorations of ancient life can provoke interest, the public wants to see real fossils (Evans; Lipps; Clary and Moe-Hoffman; Neitzke-Adamo et al.; Leviton and Aldrich), real crystals, real meteorites (Koeberl et al.; Aubele and Crumpler), real rocks, and real fragments of the Moon (Aubele and Crumpler). This should give pause to those who advocate downsizing collections.

Especially during the Enlightenment, collections became expressions of civic pride and springboards for public outreach, an explicit theme of Peale's painting which documents one of the many public museums that proliferated in Philadelphia during America's Enlightenment (Newcomb). More than the other three estates of geoscience progress (i.e., universities, governments, and the private sector), natural history museums have evolved to offer the most immediate impact on geoscience literacy. Research suggests that more people learn science informally than in classrooms[16]. Thus, in partnership with primary and secondary schools, and using the "Big Ideas" of the U.S. National Science Foundation's Earth Science Literacy Initiative[17] as a guide to planning exhibits and assessing their effectiveness, museums can facilitate science literacy for all citizens and thus their ability to adapt and live in harmony with the planet (Clary and Moe-Hoffman).

One approach is to engage the public with immersive exhibits that involve them in collections-based research, field expeditions in search of fossils, as well as lectures and other activities facilitated by new technology. In fact, one might say that many museums are "opening the windows" that have traditionally separated the public from exhibits, as well as "opening the doors" and expanding the museum experience outdoors. All in all, this is a recontextualization of the public vis-à-vis the museum.

It is not surprising that natural history museums are now dispersed across the globe. Curiosity about ancient life, vanished environments, exotic animals in temporally and geographically remote habitats, human origins, cultural evolution, and our place in the cosmos has proven to be universal (Lipps). While the social context of museums has become international, it is safe to predict that tension, again an outcome of recontextualization, could arise between some regional and international perspectives of natural history.

Small, regional museums can be as effective as encyclopedic museums with exhibits highlighting unique local geology and paleontology. They are always a draw and they convey a sense of local pride and shared custodianship. The New Mexico Museum of Natural History & Science in Albuquerque is situated among some of the most beautiful and accessible geology in the nation. The museum has recognized that the New Mexican terrain offers an entry to teach about geological processes

[15]See E.O. Wilson, 2018 (3 Mar.), The 8 million species we don't know, New York Times, p. SR10. Online at https://www.nytimes.com/2018/03/03/opinion/sunday/species-conservation-extinction.html (accessed 20 May 2018).

[16]See J. Falk and L.D. Dierking, 2002, *Lessons without Limit: How Free-Choice Learning Is Transforming Education*: Walnut Creek, California, AltaMira Press, 189 p.

[17]See Earth Science Literacy Initiative, 2010, Earth Science Literacy Principles: The Big Ideas and Supporting Concepts of Earth Science, National Science Foundation: http://www.earthscienceliteracy.org/ (accessed 20 May 2018).

operating on other planets that have features such as sand dunes and volcanoes comparable to those seen in New Mexico (Aubele and Crumpler). For the New Mexico Museum of Natural History & Science, context now includes other planets. The Internet, cell phones, and other aspects of new technology facilitate sharing these experiences globally (Hankla et al.).

Many encyclopedic museums started with an emphasis on the local geology. Such is the case with the California Academy of Sciences in San Francisco (Leviton and Aldrich). The initial collections were of local material, including specimens reflecting California's vast economic resources. The holdings quickly became internationalized and subsequently the museum moved to the forefront by "blurring the boundaries of the sciences" with the inclusion of a living rainforest, aquaria, and a planetarium (Leviton and Aldrich), as have other museums.

Liggett et al. recommend that public lands take on a role as outdoor museums of fossils and outcrops, "treasures of the nation." Summer dinosaur digs in the Hell Creek Formation, which the Indianapolis Children's Museum leads, have proven the value of expanding the museum experience outdoors (Evans). And Dinosaur National Monument has an exemplary history of attracting the public to fossil-laden outcrops. The enjoyment of nature can generate a sense of shared ownership and an interest in its conservation.

In fact, there is evidence that fossil hunts to prosaic, nearby localities will be equally meaningful for people who have never held a fossil. While one laments the loss of access to local fossil localities in many urban areas including Milwaukee, the city in collaboration with the Milwaukee Public Museum nevertheless has engaged its residents with larger-than-life reconstructions of Silurian fossils that have proven to attract visitors of all ages to its local Hartung Park, a former Silurian quarry (Coorough Burke and Sheehan).

The Milwaukee Public Museum has parlayed the local Schoonmaker Reef, the first fossil reef identified in North America (Coorough Burke and Sheehan), to one of the museum's most popular exhibits. Its "Silurian Reef" diorama brings local Silurian fossils "back to life" in a convincing Silurian sea (Coorough Burke and Sheehan; Rosenberg). Further, its holdings were the foundation of Silurian paleo-ecological research and the museum's innovative website, the Virtual Silurian Reef. In collaboration with the Chicago Field Museum, the museum's Silurian collection has been digitized and posted on the website, making it accessible to interested researchers and fossil aficionados worldwide.

Two university-affiliated museums are represented in this volume—the Rutgers Geology Museum in New Jersey and the Dunn-Seiler Museum at Mississippi State. Both museums foster geoscience literacy at a time when financial pressures have required universities and even some large, public museums to downsize their collections, research, and educational outreach. The Rutgers Museum was the first American museum dedicated to geology (Neitzke-Adamo et al.). The Dunn-Seiler Museum is one of two museums in the state of Mississippi housing paleontology collections. It is more than 100 miles from other natural history museums so it has a large proportion of constituents, many of whom are minority, and who are historically underserved in science education (Clary and Moe-Hoffman).

The Rutgers Museum was founded with a small private collection that Neitzke-Adamo et al. describe as a cabinet of curiosities. The Dunn-Seiler was similarly founded on a small collection, in its case one of local Mississippi fossils. Both Rutgers and Mississippi have expanded their holdings with specimens from beyond their neighborhoods, and have improved interpretive exhibits as well as their collections, which now facilitate research and education. Both museums engage visitors with collecting expeditions to campus or other local outcrops where youngsters can search for and keep the fossils they find. And they involve the public in activities in the museum dealing with environmental and other practical applications of geoscience that demonstrate its importance in their daily lives.

Finally, Hankla et al.'s paper analyzes their tour de force use of technology that has made it possible for people around the world to witness a distant fossil dig or mineral hunt as it is taking place and to interact in real time with geoscientists—all with little more than a smartphone. Their work makes it possible to share the experience with viewers who have little opportunity to participate in person. At a time when museum finances are under increasing strain and their mission to facilitate science literacy is at serious risk[18], Hankla et al.'s use of technology may prove to be an adaptation that can nurture the acquisitive instincts that youngsters and adults share for collecting and learning about fossils, rocks, and minerals. Just as museums have successfully evolved to do over some 500 years of history.

REDISCOVERING MR. PEALE'S MUSEUM

Great works of art and wonderful natural history museums deserve a second look. And then some. Both have content, they inform, challenge, please, trouble, mystify, soothe, engage, and do many other things that advance learning—a modification of thought and/ or behavior in response to stimuli. In other words, both facilitate adaptation to a changing world[19].

Charles Willson Peale's invitation to his museum is an invitation to participate in that progress. He offers witness to the remains of a strange and troubling skeleton, a great incognitum that apparently has no living counterpart. It seems out of context with the remains of living animals in his museum. The contrast creates a tension that one strives to grasp. But we know in retrospect that it would be joined by other fossils and its context would change. It would become one of many

[18]See R. Conniff, 2016 (1 Apr.), Our natural history, endangered: online op-ed available at http://nyti.ms/1ZRJC2d (accessed 20 May 2018) and on p. SR4 of the 3 April *New York Times* paper edition with the headline: "Natural History, Endangered."

[19]Unfortunately, creation museums strive for eternal stasis.

extinct organisms that would excite people with visions of former worlds. Mr. Peale's museum was auspicious in a way he could not have imagined: It would help change our concept of the history of life and share that process of understanding with whoever cared to visit.

Natural history museums are agents of change, not simply repositories of the past. They have a history of putting the elements of nature together in new contexts that generate new concepts that they share with the public and accordingly enhance science literacy. Science literacy is a hallmark of democracy. And that is what this book is about.

Acknowledgments

We thank Dr. Gregory P. Dietl for his review of this manuscript.

Manuscript Accepted by the Society 21 May 2018
Manuscript Published Online 6 July 2018

Printed in the USA

The Geological Society of America
Special Paper 535

Natural history museums: Facilitating science literacy across the globe

Jere H. Lipps
Museum of Paleontology, University of California, Berkeley, California 94720, USA

ABSTRACT

Natural history museums' (NHMs) primary missions are to collect, curate, and research natural history objects (life, earth, human cultures, and other specimens), and to use them for public education and outreach. The museums have the potential to enhance lifelong science literacy in unique, direct ways based on the collections they house. Ever since 1683, NHMs have exhibited specimens and educated visitors. Now, thousands of NHMs operate across the globe in ~100 countries, but no two of them are alike. Each resembles the others in the primary missions but differs significantly in collection size and diversity, research efforts, staff size and tasks, styles, public displays, outreach, and education. NHMs are thus complicated businesses due to the wide variety of tasks, objectives, and audiences.

Collections are the heart of a NHM, for everything depends on them. These collections are all biased for a number of reasons, but none of them could contain an example of every kind of natural history object. The big museums have the oldest and largest collections, while smaller NHMs have mostly local collections. Collections are further biased because only a small part of any of them can be exhibited; hence, specimens with certain attractive characteristics are selected for display and use in education and outreach. Many NHMs use replicas of specimens in occasional displays for a variety of reasons to enhance the visitor experience, chiefly to bring rare or fragile specimens to them. This is all normal and to be expected.

The overall outreach aim of NHMs should be to encourage and provide lifelong learning for everyone. People who attend NHMs are mostly educated, and, in Europe and America, chiefly white and middle to upper class. Ethnic or economically disadvantaged groups commonly find NHMs unwelcoming, alienating, and largely irrelevant to their own lives; hence, they make up only a small portion of attendees. In addition, people with physical and mental limitations of mobility, size, sight, hearing, and understanding must be accommodated in NHMs. Museums need to engage these people and to develop programs and exhibits that they will find attractive because these populations will increase in the future.

Exciting, stimulating, and engaging exhibits built around the collections of the NHMs can welcome all groups, if the culture and experiences of these people are understood. Sight, touch, sound, and smell are part of a more realistic exhibit and can

---

Erratum: This is a revised version of the chapter that was published online on 6 July 2018. Figures 12-14 from that version, and all references to them, have been removed; the paragraph describing them (current p. 16, 18) has been rewritten. Subsequent figures and their references have been renumbered.

Lipps, J.H., 2018, Natural history museums: Facilitating science literacy across the globe, *in* Rosenberg, G.D., and Clary, R.M., eds., Museums at the Forefront of the History and Philosophy of Geology: History Made, History in the Making: Geological Society of America Special Paper 535, p. 9–33, https://doi .org/10.1130/2018.2535(02).

reinforce the attractiveness of an exhibit. Real objects from the collections, displayed with imagination and creativity focused on the entire population served by the museum, can captivate and welcome people back again and encourage new visitors to attend.

Technology should be adopted to complement, not replace, exhibits of actual specimens from the NHM. Perhaps the most important computer technology will be artificial intelligence (AI). This bodes well for the future in planning, organizing, and integrating all aspects of the complicated functioning of a NHM.

INTRODUCING NATURAL HISTORY MUSEUMS

Approximately 100 countries in the world host natural history museums (NHMs) that serve people of many cultures, educations, and interests (Wikipedia List of World NHMs, 2017). Even within a single country, many NHMs may exist; for example, the United States has more than 325 NHMs in its 50 states (Wikipedia NHMs in the U.S., 2017). Fundamentally, each of these museums is dedicated to the preservation, curation, research, and display of natural history objects and, commonly, to the education of and outreach to the populations, especially children, they serve, including tourists from other areas. NHMs utilize a wide variety of collections, emphases, styles, presentations, and activities to achieve their missions and goals. Publicly, NHMs function as places of entertainment and, increasingly so, social events, and of learning about the natural world, even as they explore that world with superlative research in other parts of the museums. NHMs are multifunctional, serving both the scientific community and educational formal and informal endeavors for segments of societies.

Naturally, different opinions exist on many issues from collection development, research goals, and exhibition to visitor services and education. Most museums operate in substantially different ways in order to meet those aims. What works in one area, with one population, or with one size of museum is unlikely to work the same as operations of a museum under different conditions, cultures, and policies.

Importantly, times are changing as populations grow and age, as populations in some countries become more diverse, and as natural science itself changes. In addition, as cultural, financial, political, interests, globalization, and events in various regions develop, NHMs will need new ways to cope with them. Museum education too is rapidly changing to provide lifelong learning to people of different ages, different physical conditions, and variable understanding and experience. NHMs and their science education must therefore anticipate and plan for these changes because they are happening quickly (Merritt and Gangopadhyay, 2014). Thus, the NHMs and their service in science and education are complicated now and will become even more complex in the future.

A PERSONAL VIEW OF NATURAL HISTORY MUSEUMS

This chapter is a brief, personal overview of the fascinating world of natural history museums and how they work to provide science literacy through lifelong learning. It is primarily based on my experience in NHMs across the globe, including visits to many of them in North and South America, Asia, Europe, Australia, and Africa, in evaluating, planning, or advising on museums and their exhibitions, directing two specialized NHMs, and chairing the Berkeley Natural History Museums, a consortium of seven university museums. In addition, I have reviewed a small portion of the immense literature and huge numbers of websites (over 3,330,000 in a Google search, November 2017) on natural history museums. These experiences provide a background to understand how NHMs operate, how they pursue research, how they interact with their audiences and potential visitors, and which kinds of exhibits might enhance science literacy and interest. My intent is to stimulate the work of NHMs to increase science literacy through direct activities and exhibits. Of course, all NHMs strive to achieve these goals, and I applaud all of them. I hope this chapter may serve as a positive stimulant. In the references section of this chapter, I include internet links to a number of resources and to a few NHMs that are forging the way forward into a future of great potential for NHMs. I urge interested readers to peruse these websites, for they all contain impressive and different materials around the theme of natural history—truly an outstanding resource.

WHAT ARE NATURAL HISTORY MUSEUMS?

Natural history museums (NHMs) preserve examples of our natural and cultural world through the collection and use of organisms (animals, including humans, plants, and rarely microbes), geologic materials (fossils, minerals, rocks and their structures, and gems), archeological artifacts including mummies, astronomy, certain kinds of art, and various other items. Natural objects have long intrigued human beings, who, in response, have built collections of various sizes, probably ever since modern *Homo sapiens* first appeared 100,000 years ago. At first, these collections were the personal property of individuals or of some churches and, later, the first public museums developed (Farrington, 1915).

The first museum that accumulated collections and displayed them was the Ashmolian Museum, opened to the public in 1683 at Oxford University. It did two things, at least in the beginning, that have largely remained with NHMs everywhere: (1) The collection was chiefly in natural history but was mixed with items that are not natural history. (2) It charged admission (Asma, 2001; Fortey, 2008). Other NHMs opened later—the

Kunstkamera Peter The Great Museum of Anthropology and Ethnography in St. Petersburg, the oldest museum in Russia (1727); the Charleston Museum in South Carolina, USA, with one gallery of natural history specimens (including some remarkable fossils) among others on local history, America's first and oldest museum (1773); the Zoological Museum, Moscow, founded (1791) as the Cabinet of Natural History at the Moscow Imperial University; and the Muséum National d'Histoire Naturelle (Figs. 1 and 2) in Paris (1793). Other large museums that still exist today began in the nineteenth century, such as the natural history museums in Berlin (1810), Buenos Aires (1826), Sydney (1827), Brussels (1846), Washington, D.C. (1846), New York (1869), Tokyo (1877), London (1881), and Vienna (1889). More museums opened all over the world in the twentieth and twenty-first centuries, including the two youngest major museums: The Danish Natural History Museum, established in Copenhagen in 2004 by merger of the Botanic Garden, the Botanical Museum and Library, the Geological Museum, and the Zoological Museum, and North Korea's Natural History Museum and Zoo (2017/11) in Pyongyang in 2016 (Fig. 3).

In any NHM, the exhibits always represent a small part of the total collection, which is stored in cabinets in research areas that are not usually open to the public (Fig. 4) or in off-site facilities. The collections, in some ways, are like libraries that document nature, except that the knowledge of the objects is always increasing; hence, they become more valuable in scientific research and education as time and interests pass. They are invaluable, irreplaceable resources for research, display, and education.

The collections are always naturally biased by geography, geologic age, size of objects, means of preservation, history of the museum, who collected or studied materials, among other factors. Thus, if a museum had no one interested in certain groups of objects, those groups may not have been collected, or if expeditions went to special places, the collections naturally reflect those localities (Ashby, 2017). For example, microbes, which constitute the major part of life on Earth are seldom preserved in museums. Bacteria, archaeans, viruses, and single-celled eukaryotes without skeletons are not easily preserved, although single-celled organisms with skeletons—foraminifera, radiolarians, tintinnids,

Figure 1. Galerie de Paléontologie et d'Anatomie Comparée of the Muséum national d'Histoire naturelle, Paris, France. The Galerie was opened in 1896 with specimens collected by expeditions in the eighteenth and nineteenth centuries and with skeletons from animals that died in the nearby Ménagerie (zoo), all carefully arranged as a single collection. Jere H. Lipps photograph, 2005.

Figure 2. Galerie de Paléontologie et d'Anatomie Comparée. Whale (left) and primate (right) modern skeletons are displayed with many examples together. Jere H. Lipps photographs, 2006.

Figure 3. Entrance to the Natural History Museum and Zoo in Pyongyang, North Korea; opened in 2016. The museum displays mostly models and pictures of the Solar System; the Earth; the Paleozoic, Mesozoic and Cenozoic Eras; evolution; various animals including several dinosaurs; and plants in ~35,000 m^2. NK News image, 2016; used with permission of NK News.

Figure 4. Collections are stored in rows of cabinets at the Copenhagen Museum of Zoology, Danish Natural History Museum NHM (A), and with open draws in the Division of Birds, Smithsonian NHM (B), in the collection and research areas of these institutions. Most NHMs have rows and rows of cabinets in similar or specialized storage units. The public seldom sees the collections and research rooms of museums. Credits: Jere H. Lipps photograph, 2010 (A); Bird Collection, National Museum of Natural History, Smithsonian Institution, photo by Chip Clark, used with permission (B).

diatoms, coccolithophorids—and bacterial formations from the geologic record are collected mostly in larger museums. Soft-bodied animals are commonly preserved in glass bottles with fluid preservatives. Because of these complications, the organisms are not usually exhibited.

The exhibits are also biased toward the more intriguing larger, colorful animals, minerals, and rocks. Minerals, for example, are usually displayed as outstanding crystal or colorful forms. For animals, this means that males are more commonly exhibited because they are usually larger, more decorated with fur, horns, or other male attributes and are more colorful than females, and small animals such as mice, voles, and other burrowing creatures and certain scary or creepy kinds of animals are not displayed at all (Ashby, 2017). Some museums are now trying to display these specimens in their collections—for example, in Berlin, the NHM features a stunning display of soft-bodied animals preserved in glass bottles arranged in a large glass display case perhaps >3 m tall, >9 m long and >6 m wide in the Wet Collection Wing of the museum (Museum für Naturkunde, 2017/11). In displays, the animals and plants are commonly arranged systematically but sometimes arranged in ecological dioramas, evolutionary progressions, size, nature of preservation, reconstructions, or even as live specimens (Fig. 5). In the Grande Galerie de l'Évolution, one of the buildings in the Paris museum, a parade of vertebrate animals stands in the large, central area of the museum. While definitely biased toward outstanding specimens, the impact on the visitor of diversity, enormity, morphology, colors, sizes, and other attributes of vertebrates is impressive. Mineral and rock displays are also arranged according to classifications, and archeological materials are commonly arranged chronologically and geographically.

Many NHMs were originally or became centers for science research focused chiefly on organisms with additional spaces and staff devoted to other fields. All NHMs welcomed researchers even if they had no research staff. In museums with larger staffs, research is a particularly prized part of the work they do, because the research informs the collection, makes discoveries, adds to knowledge of our world, and commonly contributes to the solution of problems facing humanity. For a long time, research concentrated on the collection and classification of objects from localities all over the world for the largest museums and locally for smaller ones. Even the largest of these museums cannot possibly hold all the biodiversity on Earth. Collecting still continues and is a necessary task for the documentation and understanding of Earth's past and present, from which we can infer and plan for the future as well. The research has expanded to explore biodiversity, how that diversity evolved and is maintained in nature, how various habitats have been impacted by human influences, the

Figure 5. Living exhibits of plants on the roof (left) and of African penguins, *Sphenicus demersus,* in the gallery of African dioramas (right) at the California Academy of Sciences in San Francisco. The Academy also displays marine and freshwater life in its aquarium, coral reef animals and plants in an artificial reef, and plants, butterflies, frogs, and other animals in its glassed-in rainforest. It hosts a planetarium as well. Jere H. Lipps photographs, 2013.

sustainability of the biotas in the face of habitat degradation, environmental studies including massive predicted climate changes, history of humans, earth science studies that have importance in earthquake, tsunami, climate change, landslides, and floods, among many other earth characteristics that impact our societies each year, and others specific to certain museums. This research is documented in scientific publications, sometimes issued by the museums themselves, museum and other magazines and newsletters, and, in some cases today, on museum websites. A large museum with hundreds of researchers may contribute well over 500 scientific publications each year, and smaller museums without research staff still contribute numbers of such publications done by visiting scientists.

How these uses of collections develop in the future cannot be predicted. Thus, we never know when a new technique, instrument, data, or perspective will add to the value of these collections through new revelations (Fig. 6) or applications. The collections then constitute an invaluable resource of natural history objects to be used for research, outreach, and education (Winker, 2004), and the research produced by all these people working on NHM collections has enriched our knowledge of our world and its natural wealth.

Figure 6. Collection of polar bear skulls that provide evidence of climate change in the Arctic Ocean over the past century or so in the Copenhagen Museum of Zoology, Danish Natural History Museum (Sonne et al., 2013). This research was completely unanticipated when these specimens were collected and curated. Such uses of NHM collections are quite common as new problems, techniques, or instruments develop and require long-term or wide geographic collections of animals, plants, and microorganisms. Jere H. Lipps photograph, 2010.

As natural history museums progressed through the eighteenth to twenty-first centuries, most NHMs developed outreach enterprises based on their collections to children, adults, and families with educational goals and as entertainment or edutainment venues (Delicado, 2010; Berkowitz and Lightman, 2017). The numbers of visitors to most museums (at least in the United States) have been declining (Miller, 2010), although attendance at the major NHMs has been increasing somewhat or holding its own. As attendance declines, so do NHMs financial resources—fewer paying visitors and declining interest by philanthropists. This is a chief concern of NHMs, because attendance fees are a significant source of income needed to operate most museums. It is a trap too, because with decreasing funds comes decreasing staff in exhibits, research, and other essential operations that make the museums the attractive and important resources that they are.

Along with all these objectives, manifested fully in the late twentieth century, came increasing costs, addition of staff, collection growth, and a need to find other financial sources. While modern museums receive their incomes from a variety of sources—government, philanthropy, grants, and admission fees—admissions continue to be a major source of income for most. Only a few major museums, for example, the NHM London and the Smithsonian NHM, are free. Admission fees of others range from a few dollars to $35 for an adult visit. Specific research may also be funded by federal, state, and foundation grants. In the United States, only the largest museums have significant income, some generated from large endowments, ranging from >$2 billion to many that have no income at all (Grimes et al., 2015). The larger incomes include sources other than revenue, but clearly many smaller museums are dependent solely on revenues chiefly from visitors.

Modern NHMs have been built in many countries as well as states, provinces, counties, cities, universities, colleges, institutes, and private organizations. They range in size from large museums (Figs. 1 and 2) with large incomes employing hundreds of researchers, exhibit designers, other staff, and volunteers to smaller local museums (Figs. 7 and 8) operating chiefly on visitor revenue and run by a few employees and volunteers. Each is different in many aspects, including the choice of materials collected and exhibited, but all are dedicated to exhibiting natural history objects with varying aims to educate students as well as the general public. Some NHMs display living specimens in aquaria, rainforest, reef reconstructions, and gardens (Fig. 5) or have planetariums with their ever-popular star shows to attract audiences. Many even sponsor evening parties in their spaces for adults, sleep-overs for kids, special openings for teenagers, birthday parties, weddings, special events on holidays, and, of course, school and group tours.

Among the best NHMs are smaller ones that are part of a larger natural occurrence of the materials exhibited in the museum (Figs. 9–11), such as World Heritage Sites, and national or regional parks. These many NHMs are also visitor centers with exhibits or museums themselves built on field sites that allow visitors to see the actual materials, living or preserved, in natural settings nearby (Lipps, 2009). The museum and these field occurrences work in tandem to explain and educate visitors through direct involvement, sometimes even in collecting materials, and in the coordination of active research on the materials both in the museum and the nearby field sites. Most of these NHMs also use volunteers to help visitors in a sustained effort in the field as well as in the museum. In the United States, 331 million people visited the National Parks in 2016 (United States National Park

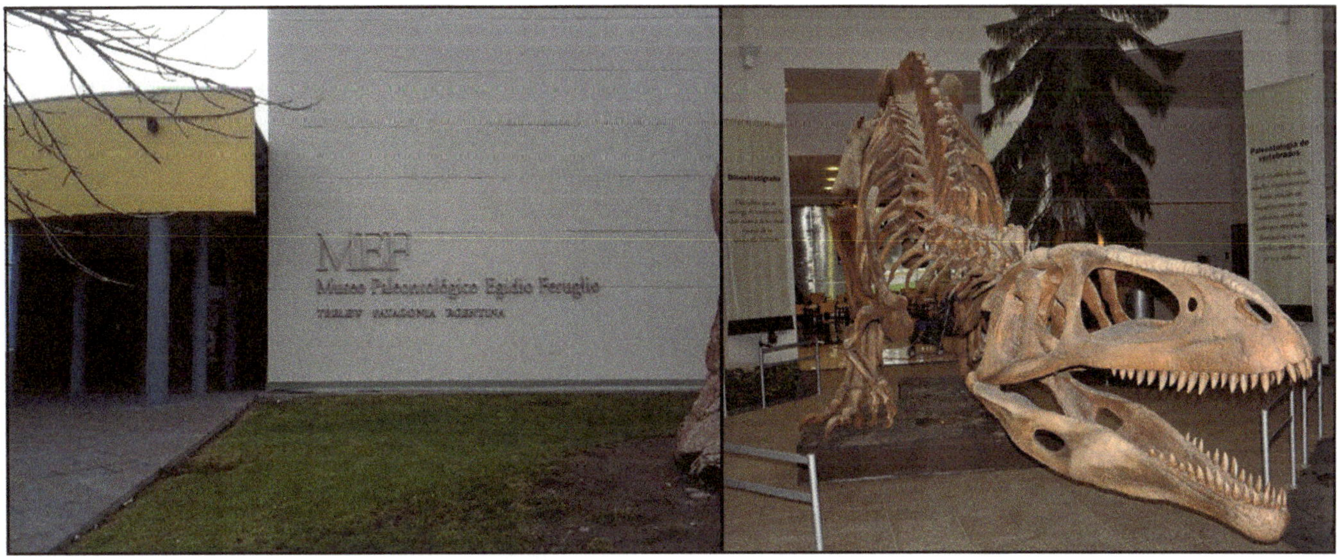

Figure 7. Museo Paleontológico Egidio Feruglio (left), Trelew, Patagonia, Argentina (Museo Paleontológico Egidio Feruglio, 2017/12), contains fossils including large dinosaurs (right), mammals, and invertebrates found in Patagonia. These include the largest known dinosaur, collections from Darwin's 1833 Pleistocene fauna (Darwin, 1838, 1846; Lipps, 2010), and many others. Jere H. Lipps photographs, 2014.

Figure 8. The Buena Vista Museum of Natural History and Science, a small museum, was established in a store in Bakersfield, California, to exhibit local fossils as well as artifacts and dioramas of Native Americans (Buena Vista Museum of Natural History, 2017/12). The fossils were collected from a bone bed (Pyenson, et al., 2009) in the nearby Miocene Round Mountain Silt. The bone bed outcrops over many acres east of Bakersfield and has yielded thousands of bones, many articulated, of marine and fewer terrestrial vertebrates. The Museum operates chiefly with the help of volunteers and holds fossil digs and other events for families throughout the year. Jere H. Lipps photograph, 2017.

Service, 2017/10), and many of those viewed the NHMs at visitor centers. These are highly valued educational sites with excellent displays and, of course, natural features in biology, geology, rivers, oceans, and other features.

NHMs may intergrade with other kinds of organizations; for example, science discovery centers, children's science museums, aquaria, botanical gardens, planetariums, and religious museums, among others. Science discovery centers and children's museums are favorite destinations for families because they provide a variety of activities for young people, commonly including natural history subjects but not necessarily real specimens. These organizations usually do not engage in object collection or research, but instead, they concentrate on fun education through hands-on activities. Children's science museums focus on fun and education using science materials including natural history. Others, such as the Exploratorium in San Francisco, are not museums but rather places where the public can interact personally with science (physics, chemistry, earth, and biology) activities. Botanical gardens, aquaria, and planetariums provide observation or interaction with science subjects. All of these play an important role in promoting science literacy, and each serves a distinct role in it.

Religious museums commonly display natural history objects to support their religious beliefs. Since these beliefs are not testable with science methods, the museums are not science museums even if they appear to be or state that they are. Their educational goals aim to direct visitors away from the science of natural history to their beliefs. Although their exhibits are commonly simple, clear, and to the point, hence done well, they do not explain science. A primary goal of natural history museums is to provide informal (non-classroom) education for all people about the sciences of biology, geology, anthropology, chemistry, and physics. This cannot be done in museums where explanations are

Figure 9. The Miguasha World Heritage Site and Provincial Park, Quebec, Canada (Miguasha World Heritage Site, 2017/11), has a museum (left) to show the specimens recovered from the cliffs below the museum (right). The cliffs yield an important and diverse fish fauna 370 million years old (Devonian). Jere H. Lipps photographs, 2005.

Figure 10. Peking Man World Heritage Site (inscribed 1987) and Museum at Zhoukoudian near Beijing, China. The museum, marked by a huge reconstructed head of Peking Man (left), has magnificent sculptures of Peking men and women (right) throughout the museum. Jere H. Lipps photographs, 2016.

Figure 11. The field collecting area surrounding the Peking Man Museum has paved trails to the caves (left) where Peking Man fossils were discovered, statues of animals, and Peking men, women, and families in reconstructed life situations (right). Jere H. Lipps photographs, 2016.

not testable using scientific evidence and data. Scientific organizations, like the National Center for Science Education, the Geological Society of America, the National Academy of Sciences, and the American Association for the Advancement of Science among many others, recommend that religion and science be kept separate because they operate in different ways. Religion is not science because "it invokes supernatural phenomena that cannot be tested, and therefore has no place in a science curriculum" (Geological Society of America, 2018), including the science education programs and exhibits of natural history museums.

AUDIENCES OF NATURAL HISTORY MUSEUMS

Collectively, museums are well attended. In America, museums of all kinds receive ~850 million visits a year, which is more than the 483 million in 2011 for all major sporting events and theme parks combined (Museum Facts, 2017). An additional 524 million adults visited online museum websites each year by 2006; that number has surely grown considerably since then. NHMs, of course, likely have smaller percentages of these numbers. The total attendance at U.S. NHMs is not recorded; the largest NHMs have estimated attendances of up to 7–8 million visits per year (U.S. NHM and Paris), while the larger provincial, state, county, and city museums may have hundreds of thousands to over a million visitors, and smaller, local NHMs may receive thousands of people each year. Presumably, museums in general have much larger attendances in other countries as well, and NHMs must make up a smaller proportion there as well.

NHMs serve local populations and tourists alike. In North America and Europe, museums attract audiences mostly from middle- and upper-class white populations that have an appreciation of museums already and can afford the costs of travel, parking, food, and admission. In other countries, museum goers may also be self-selected for the better educated and middle or upper class. While measures of economic class use are not generally available, the participation of ethnic minorities has been low (Bell et al., 2009; Farrell and Medvedeva, 2010; Dawson and Jensen, 2011; Dawson, 2014). In the United States, minority visitors (African American, Asian, and Hispanic) constitute ~9% of attendees at all museums, although they are 34% of the general population. The percentage of minorities in the U.S. population is expected to increase to 45% by 2035, and that may translate to more minorities visiting NHMs. However, many other factors influence whether people go to NHMs; therefore, projections are uncertain (Farrell and Medvedeva, 2010).

Museums that reach out to low-income people, usually students in schools, commonly provide transportation and free admission. Admission fees are thought to be an impediment to low-income and minority ethnic participation at NHMs. However, when the NHM London, eliminated admission fees, these groups did not visit more, but rather middle- and upper-class white visitors came more often (Dawson, 2014). Ethnic groups commonly found the NHMs unwelcoming and alienating due to language barriers, expressed authority, irrelevancy (to them), and costs such as food and gift shops in addition to admission fees (Garibay, 2009).

Even among people who could go to NHMs, attitudes and interests are varied. While some visitors are very positive about museums, others are negative or may have little interest. At an exhibit opening I witnessed, a man walked quickly around with a scowl on his face, then suggested to me that we should sell all the materials in the exhibit. Then we would have money, he said. I noted that also, then we have no fossils, to which he replied that he was a businessman; hence money is the important objective. A local politician also suggested that we make kits of all those fossils for each school in the area—a nice sentiment but a major task with 386 elementary schools in the area, each teaching a couple of classes where fossils could be used. Making kits would be an expensive and time-consuming endeavor that would deplete the collection. Better ways to intrigue businesspeople and politicians must be found.

Museum management may be populated by people who know little or nothing about museum values, collections, aims, research, or history; yet they make decisions about museum operations including scientific ones. These management boards or committees seldom assess their own work, so have no independent input as to how they are doing; overall, they rate about a B– (BoardSource, Museum Board Leadership, 2017). Good suggestions and good management require good knowledge.

Against these views are people who love museums and those values, who ask questions about the exhibits and how to get even more or better ones, who simply want to see interesting objects seldom encountered in everyday life, who might even wish to volunteer, teachers and students who treasure the educational goals of museums, and, of course, scientists who know well that the collections provide resources to solve problems that even now impact our societies. Richard Fortey (2008, p. 4) of the Natural History Museum, London, wrote: "You can judge a society by the quality of its museums." True enough, but societies now include far more people that do not or cannot take part in museum activities because of cultural, educational, and economic reasons, and their values and interests are not represented in NHMs. So Fortey's statement is becoming more of a worthy goal than what museums really do now. Dealing with these diverse attitudes, cultures, and actions is tricky (Ferilli et al., 2017), but the best way to counter the negative and enhance the positive is to achieve the NHMs' goals and aims, and to let the public know about them, chiefly by creatively increasing knowledge with innovation, about the museums, their collections, research, and educational opportunities, and attracting all people to visit.

SCIENCE LITERACY

Museums are considered an important part of informal science learning, which includes a wide variety of venues outside of regular classrooms (Fenichel and Schweingruber, 2010). NHMs are considered "designed environments" where learners are led through the learning process as opposed to "everyday environments" (television, walks in the parks, field trips, gardening, and others) that are self-motivated. Museums therefore play an important role here because they are valued for their resources and as venues for learning. But they do overlap with the everyday environments as well, if the learners are allowed to interact personally with at least some exhibits and displays. This should certainly become a goal of NHMs in general. Within either of these designed or everyday environments, little is constant across NHMs. Those with dedicated resources or staffs successfully build these environments, but not necessarily with input from learning experts, audience and population assessments, or teacher surveys.

Appreciation of NHMs requires some scientific interest, and ~70% of Americans are very interested in science (National Science Board, 2016). However, in recent years, science has been attacked and replaced by belief systems that are underlain by personal ideologies, economic considerations, and policy support causing the acceptance of science to be selective. These ideas are encouraged by corporations, agencies, and non-government organizations that fear an economic or policy impact on them if certain scientific facts and conclusions are accepted by the general public or government (Oreskes and Conway, 2010). For example, climate change, which will impact everyone, is often attacked, in spite of vast amounts of data and that nearly all scientists support it. On the other hand, an eclipse of the sun, also supported by science, is welcomed with glee. When scientists warned people not to look directly at the sun in 2017 across the United States because it would cause eye damage, most people took that advice seriously. But science doesn't matter now to an increasing number of people worldwide. These people are not necessarily scientifically illiterate but rather are commonly smart, politically pledged, or deeply dedicated to their own ideas. And this is the problem. Facts will not be enough to change their viewpoint or even interest them in particular sciences. They are already committed. Since NHMs are mostly about facts, these people may be less interested in visiting and more emotionally remote from the offerings of NHMs. Such people can be turned to science when evidence of science supports their own lives, ideology, or politics, but this seldom happens in a museum. Instead, museums must learn from psychology how to communicate meaningfully with them. Engaging them in that conversation by asking them questions is a good approach rather than questioning their views or values with facts. This will be self-correcting, at least for some people, as more of them suffer from erroneous opinions and turn back to science for real answers. NHMs surely can provide the communication and guidance to do this, because they have significant resources to deal with real-world problems and to encourage understanding of scientific processes and knowledge for society's sake. In NHMs, we do not need to dwell on facts—those are obvious in the collections and exhibits—we only need to engage people in them by focusing people on their own problems with questions and commentary about the museum's objects. This can be done with the design of exhibits so that the conversations are self-driven by the visitor—thus, moving away from exhibits that tell the facts and toward those that ask the questions leading to the facts. Such an approach can make for exciting and relevant exhibits that utilize most materials from the museums.

"Scientific literacy is the capacity to use scientific knowledge, to identify questions and to draw evidence-based conclusions to understand and help make decisions about the natural world and the changes made to it through human activity" (National Science Board, 2016). While no one can know all the facts, the public also should know how the process of science works—the development of alternative ideas or hypotheses (Chamberlin, 1890; Elliott and Brook, 2007; Understanding Science, 2017/12), testing those with evidence, and reexamining them with new data. Science assesses each alternative and disproves some by collecting evidence that corrects and hones the others (Platt, 1964). This is not a difficult process because most people use these methods unwittingly in their ordinary lives (Lipps, 1999); they are emblazoned in medicine—"get a second opinion."

Because of the large percentage of Americans who are interested in science and because scientists are held in high esteem second only to the military (National Science Board, 2016), NHMs have a huge, untapped potential audience, at least in America. The public can be engaged in science in many ways (Bergeron and Tuttle, 2013) from television to the internet, but science should ultimately be experienced directly. NHMs serve this purpose well as do science centers, field sites, science tours, schools, and other direct activities. Children are often given priority in these learning environments, but science literacy should be pursued in childhood, the teenage years, as adults, and as seniors. Science education is a lifelong endeavor. It is never ending, but that is part of the joy and excitement of science. NHMs can be an important part of that process.

WHAT CAN NATURAL HISTORY MUSEUMS DO TO INCREASE SCIENCE LITERACY?

With diverse and changing populations, NHMs cannot break their old molds easily. Too many alternatives, too many suggestions, and not enough actual data exist to guide changes. All museums should do marketing research or assemble focus groups of the populations they serve, examining not just the people who visit the museums but also the people who never go. Some museums already use these methods, but all would benefit from research and focus groups. Why, for example, do African Americans, Hispanic Americans, Asians, and economically disadvantaged people seldom go to museums? Is it cultural, are museums intimidating, are they economically difficult, or are museums simply unknown to these groups, hence not appreciated? Once all issues are discussed and understood, museums then can look at population projections for their own regions and plan for that future. In most American and European cities, minority populations will soon grow to be majority populations that NHMs will need to engage.

Some experts believe that NHMs should simply do what they have been doing very well for hundreds of years (Fortey, 2008; Benton, 2009). Others see a future of innovative new ways that museums can become educational partners of schools and families (Merritt and Gangopadhyay, 2014). NHMs did education from their very beginning, but that was mostly focused on the collections' objects as learning tools. Students could draw or photograph specimens, label their pictures, and write an essay about them, but interaction with the materials themselves was usually not included. But now, in the technology era, so much more is possible (Edson, 2014). Learning is immediate and as needed (Miller, 2010). Thus, museums are urged to move into new and expanding technologies as those develop. And that is a good idea that has been under way increasingly for the past three decades. Technology must play important and creative roles, with special device applications designed to illustrate exciting points in exhibits, interact with viewers with special "tours," games, or activities that can be done in person. New applications integrating new devices with all the exhibits can be created, but NHMs must not forget that their specimens are the most important items for visitors to encounter, not games or virtual tours. Technology also has an impact on various cultures that may not have these devices, and that will further alienate them. Ways to provide that technology to groups who do not have it or care to use it must be developed too.

NHMs commonly sponsor traveling exhibits in special spaces that may be relevant to the museum, or in some cases unrelated, even inappropriate, to a NHM. For example, the NHM, London had an exhibit of mechanical dinosaur models that packed so many people into the halls that getting from the front door to other parts of the museum was troublesome. Then one day, no more crowds. In the hall that had held the mechanical dinosaurs were insects, and no visitors. As an exhibit staffer told me, "Only dinosaurs and human sexuality bring in big crowds." That response seemed true enough, as I recalled seeing the human exhibit too, but no real objects from the collections to speak of were in either exhibit. The model dinosaurs were entertainment, while the human exhibit was curiosity driven and truly educational but unsupported by specimens from the collection. Hopefully, all those visitors had a chance to see the rest of the museum too. Nevertheless, because museums can collect both an entrance fee and a separate fee for special exhibits, these are welcomed, and they do add interest in the museum. In a market for entertainment, museums compete with cinema, theme parks, and social events. Indeed, surveys indicated that many people in the potential audiences would rather go to a social event than to a museum. Museums must therefore keep new material coming in, so people will return time after time, or the NHMs will saturate their markets. These exhibits are seldom constructed to appeal to the groups who do not come to NHMs. To do so, with advertising, might attract new audience participation from other local groups of people.

NHMs discovered that social activities will bring in new people—young people's nights, family sleepovers, birthday parties, teenage get-togethers (Fig. 12), seniors' night, holiday parties, and a host of other events. These events should feature exhibits or activities that connect the attendees with the museum itself rather than simply providing space for the social event. These novel ways of connecting additional people with science in the museums are a good start on the goals of NHMs.

In many museums, reproductions of actual objects exist side by side with real ones. Reproductions by casting, molding, 3D printing, or other means are done to bring rare objects to a museum (for example, a *Tyrannosaurus rex* skeleton; Fig. 13), to obviate mounting problems, or to make difficult exhibits possible. These modern methods of reproduction can be so exact, even microscopically, that most visitors cannot tell the reproductions from the originals. Indeed, some visitors even pilfer the last few tailbones from the lumbering dinosaur casts so often that the NHMs London and Paris keep spares to replace the misappropriated ones (Marie-The Venet-Peyre, 2006, personal commun.; Fortey, 2008). And some people object to having the casts on display (Benton, 2010). Yet, without those casts, dinosaurs, for example, could not be exhibited at all in most museums. So, casts are a trade-off—dinosaurs or no dinosaurs. Replications are not done in museums to fool the people, but to enhance their experience. Perhaps NHMs could explain these replications and tell the crowds what they are, and show how, why, and where they were created, and why they are valued in research and exhibits. That might be interesting to visitors. The Academy of Natural Sciences (2017/10) in Philadelphia has done exactly that with a nice exhibit on making dioramas, and the Burke Museum in Seattle does likewise for dinosaur casts. "Dinosaur Secrets" could be a similar exhibit focused on reproductions in most museums.

THE ESSENCE OF NATURAL HISTORY MUSEUMS SHOULD BE FEATURED IN ITS EXHIBITS

Collections of objects are the essence of NHMs. These museums have long done collection-based exhibits, so that the exhibits of 2017 frequently resemble those of the eighteenth and nineteenth centuries—a cataloging of objects set in locked cabinets. This old NHM model is still a valuable aspect of natural history learning because of the diversity held by NHMs. People want to see real things and the closer the better. The specimens from the collections are real and will be forever. How then can all kids, families, adults, and seniors benefit from designing exhibits in other ways (Bamberger and Tal, 2007)? Those collections and specimens should figure prominently in museums' displays by recognizing and building unique, attractive, and innovative exhibits that develop passion in the viewers. Such exhibits and specimens should challenge the interest, amazement, imagination, curiosity, and creativity of the visitors, including ethnic and economically disadvantaged groups. Exhibits should stimulate the visitors' self-exploration and imagination—get them involved and thinking about actual material.

All the impediments for groups of people, no matter who they are, may be overcome by letting individuals take charge of

Figure 12. Teen night at the California Academy of Sciences, San Francisco (California Academy of Sciences, 2017/12). This well-attended event introduces teenagers to the Natural History Museum. Photograph courtesy of Dr. Jonathan Foley, Executive Director of the California Academy of Sciences, San Francisco, California, 2017; used with permission.

Figure 13. Reproduction (cast) of a *Tyrannosaurus rex* mounted outside of the Museum of Paleontology, University of California, Berkeley (*Tyrannosaur rex* replica, 2017/11). The cast is made of polyurethane resin pulled from silicone rubber molds taken from the original fossils (over 300 bones), other *T. rex* bones where applicable, or reconstructed or sculptured of missing pieces. The original specimen (MOR 555) is at the Museum of the Rockies, Bozeman, Montana (https://museumoftherockies.org/). The assembled cast is held up by an armature of hollow steel pipe inserted in holes drilled through each bone and fastened together. It has been viewed by thousands of university students and visitors since its opening in 1995. Jack Penkethman photograph, 2000.

their own visit—clear and concise language, few staff instructions, displays relevant to people's level of experience, more interactive exhibits, more social interactions, and more wonder and interest. High-tech exhibits or games should not be used unless they relate directly to the exhibits and specimens. Parents can use the exhibits as well by reading and discussing the exhibits with their children. Families that do this spend more time with an exhibit, and the children learn more (To et al., 2016).

An ideal museum exhibit would bring most of the senses into play—sight, touch, smell, and sound—which reinforce memories and add to new experiences so that they are embedded in the minds of visitors. These senses, of course, must be used with care because different people will react to them in different ways. Odors, for example, are chemicals released from a source that can be particularly difficult because people sense them differently, and some people may find some odors very unpleasant, making them ill, while other people may find them pleasant (Oregon.gov/oha, 2018/2). Perceived pleasant odors may even be beneficial to health and well-being. Some museums have incorporated smells, and NHMs can carefully do likewise. All life has odors, some rocks and minerals smell, forests smell, fossil dig sites smell, and old collections themselves smell of musty specimens. When used appropriately, smells may enhance an exhibit experience in positive ways. Imagine viewing fossils from a tar pit and, at the same time, smelling that distinctive petroleum odor, or of looking at an exhibit of specimens in a cabinet and smelling that musty odor. These odors add dimensions, potentially lasting, to the NHM experience and along with bones or rocks to touch; abundant fossils, insects, and plants to see; and perhaps some likely animal and environmental sounds, make for a more lasting and meaningful visitor and learning experience.

Exhibits must also be designed to take into account the physical characteristics of viewers. These characteristics also are variable, hence, many visitors will be restrained by them. People who are short or tall may see an exhibit in a different way than intended. Color too presents difficulties that exhibit designers must consider. Color-blind people (10%–12% of males, fewer females) see colors in different degrees (American Academy of Ophthalmology, 2017/12). Some can see colors normally in good light but have difficulty in dim light, while others cannot distinguish certain colors in any light, or, uncommonly, some see everything in shades of gray. Color-blind people have difficulty with certain color combinations that impact museum exhibit and web design; for example, a red bird in a green tree may not be visible, or colored lettering on labels may make reading difficult or impossible. Simple changes in color or texture may make a color-blind person's viewing experience much more positive (How to Design for Color Blindness, 2017/12). Can NHMs be welcoming to blind people, say with touch and feel exhibits? The Natural History Museum Paris, for example, provides ways for people with a wide variety of limitations to visit and enjoy its museums (Muséum national d'Histoire naturelle, 2017/11). People who are limited in any way, physically or mentally, need consideration in exhibit design and museum layouts.

In my comments on displays and exhibits, I will focus on specimen exhibits I have seen and think promote imagination, exploration, self-involvement, innovation, appreciation, and creativity in at least some ways. They are not for every museum nor are they unique in any particular museum. They can be improved or, hopefully, will trigger other, more imaginative ideas for truly stimulating exhibits of the real objects. New exhibits should fascinate viewers with intriguing and captivating displays of objects that inspire, engage, and allow the imagination to run free and creatively.

Minimal Labels

Exhibit labels are important because they carry the significance of the display and engagement of the museum with its visitors. Labels are critical in setting the tone and connections to the audiences. Label designers thus require significant experience, good information about the objects in the display, interaction with experts, understanding of the visitors and their backgrounds and cultures, the educational or other goals of the exhibit and museum, the best ways and materials to make the labels, and knowledge of the exhibit construction and materials (Head Miller et al., 2013). Exhibit labels are complex but critical.

Too many labels are long, complicated (to the visitor) explanations that take away the joy (Academy of Natural Sciences, 2017/10) and excitement of specimens (Lipps, 2007) and may intimidate visitors, especially those who are new to museums. A label, poster, or diagram full of text and details may be more appropriate for a technical paper or lecture than for museum visitors. A trip to a museum can be discouraging for this reason, although most visitors ignore such explanations anyway. Only ~20%–40% of visitors read labels and then only for a few seconds (Rounds, 2004). For most visitors, detailed scientific information that includes technical names, locality, and ecology, for example, have little meaning and are mostly irrelevant to visitors; thus, much of the information and the exhibit itself are soon forgotten. For me, these kinds of explanations detract from the joy of imagination, discovery, and creativity by providing answers that seem to be final (too many facts), and nothing is left to the imagination.

Labels and explanations could use simple and concise terms that can be quickly understood by the average person. Labels should also be in languages in common use by its major audiences. A simple label might include short statements and/or simple diagrams that are easily and quickly grasped by laypeople, perhaps in distinctive layers and large bold format that can be seen and read easily from a distance. Fewer words, no technical words, and more images make for better labels. The American Alliance of Museums (2016) presents annual awards for excellence in design and labels; the winners of the 2016 awards represent some of the best in new designs and labels. Given the short attention span of most viewers, even some of these labels had too much text. Detailed information, of course, is important to some visitors and could be found by use of Quick Response (QR) codes used on some smart phones that link the user for more information to the museum's or other websites. Perhaps a way can be found to accommodate all visitors by moving longer, more informative materials to the side or making it available to those who want it and removing the readers from blocking the exhibit that others merely want to peruse. In general, labels should grab the visitor's attention and provoke thought about the specimens themselves rather than explaining everything to them. The labels should generate wonder and creativity, stimulate the imagination, and challenge the visitor of any age to get involved with the materials. They should be inventive, fun, and otherwise engaging to all visitors.

An Abundance of Specimens

Large displays of the same item (skulls, beetles, shark teeth, bones, gems, rocks, or almost anything) are impressive, enthralling, and engaging. They create an interest in these and related specimens. NHMs have done these kinds of exhibits, starting perhaps with the 1896 Gallerie de Paléontologie d'Anatomie Comparée, NHM, Paris (Figs. 1 and 2), in which hundreds of specimens of a single species' skulls, skeletons, or other bones were arranged on shelves side by side. That entire gallery is a tribute to sheer numbers as fascination.

More than 400 Dire wolf skulls (Fig. 14) are displayed on a wall at the George C. Page Museum at Rancho La Brea in Los

Figure 14. A wall of Dire wolf skulls in the George C. Page Museum at Rancho La Brea (La Brea Tar Pits and Museum, 2017/10), located on Wilshire Boulevard, in Los Angeles, California. Jere H. Lipps photograph, 2009.

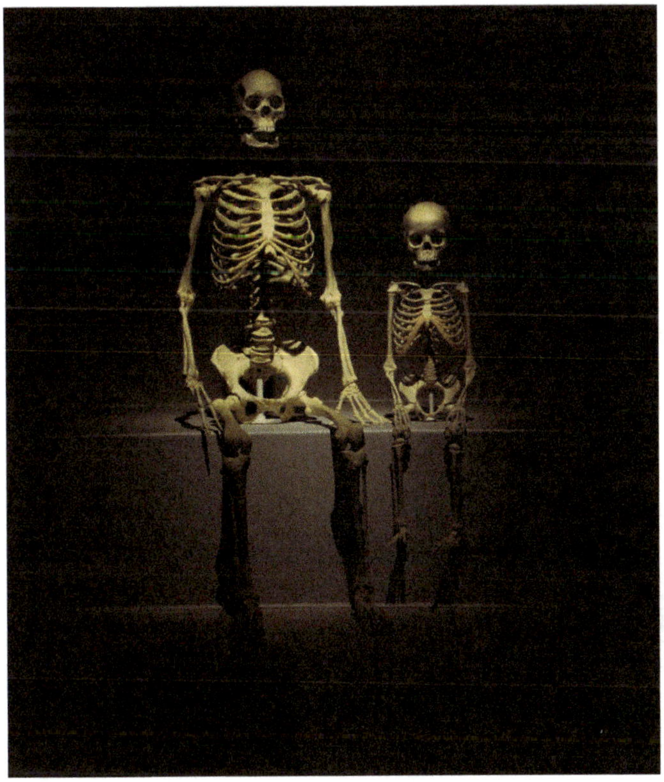

Angeles. Dire wolves are perhaps the most common vertebrate fossil found in the Late Pleistocene (<40,000 years old) tar seep deposits, and the wall of skulls emphasizes that. The skulls are displayed behind a piece of leucite but very close to the viewer. A similar wall of modern mammal skulls was displayed by the California Academy of Sciences; a bright, colorful wall of butterflies arranged in square patterns much like a quilt is on exhibit at the Carnegie Museum of Natural History in Pittsburgh; and primitive human stone tools fill a wall at the Peking Man Museum just outside of Beijing, China. Such displays arouse a sense of astonishment and curiosity that propels viewers to wonder and imagine even more about the objects. None of these walls has extensive labels—just many objects to admire.

Single or group specimens can be awe-inspiring as well, if displayed fittingly. Two human skeletons, an adult and a child, sit side by side on a bench in isolation from nearby exhibits (Fig. 15) in the Danish NHM; these skeletons are effective in drawing attention to the exhibit without any labels at all. The arrangement

Figure 15. Human skeletons displayed sitting side by side at the Copenhagen Zoological Museum, Natural History Museum (NHM) of Denmark. The pose and lighting evoke both interest and emotions. Jere H. Lipps photograph, 2010.

Figure 16. Black banners highlight specimens at a fossil exhibition at the Old County Courthouse, Santa Ana, California. Bruce Rubin photograph, 2012.

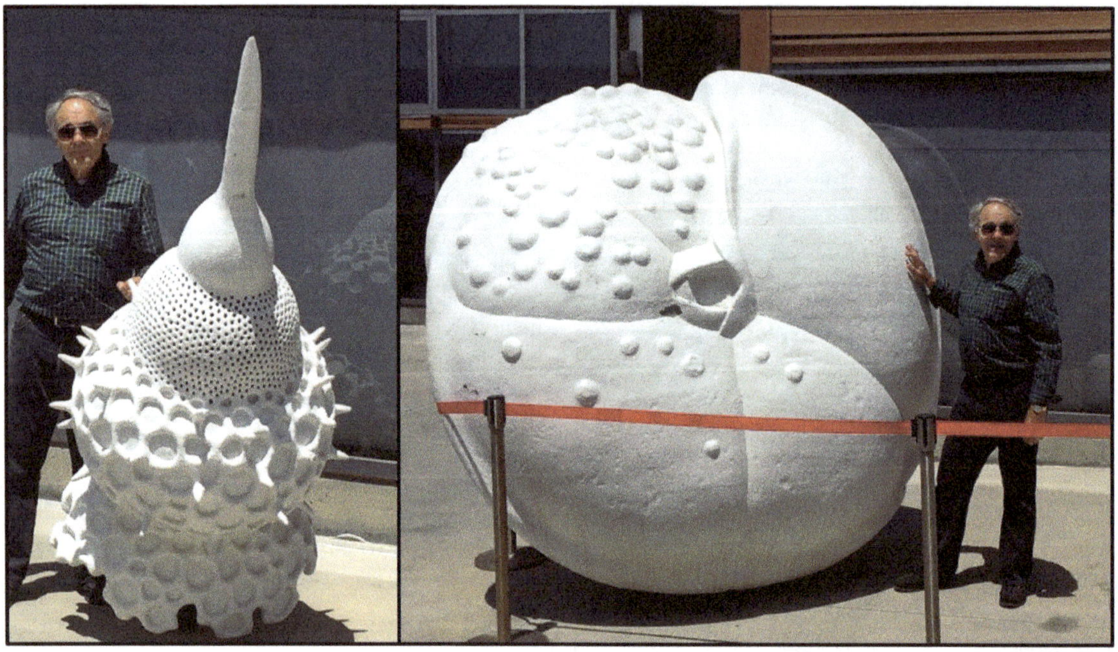

Figure 17. Larger-than-life models of a radiolarian (left) and foraminifera (right), both microscopic, single-celled protists, showing in great detail the morphologic structures of the skeletons. In life, they are smaller than 1 mm in diameter and have pseudopodia extending chiefly from the apertures. Taken by Meredith Rivin for Jere H. Lipps, photographs, 2013.

and on-off lighting of the two skeletons evoke memorable emotions and interest.

Photographs of specimens can also make effective displays. At the NHM, London, some time ago, huge photographs showing the faces of abused apes (on a black background) hung from near the ceiling to near the floor. A single ape face stared out at visitors from each one with a short description of the ape and its abuse. The black background and low lights added to the solemnity of the room. They were not real specimens, but the impact was special. I can't forget them. Black banners with single or grouped fossils were used effectively in a special exhibition at the Old County Courthouse in Santa Ana (Fig. 16). In displays such as these, special items or educational objectives can be emphasized.

Small or microscopic materials require special techniques. Microscopes themselves do not work well with general crowds because of the complications of their use and limited viewing opportunities and times. Another effective way to exhibit these materials is by making accurate models of them at a larger scale, even human size, to emphasize their complexity (Fig. 17). In this way, bacteria, protozoans, algae, and a host of other organisms that make up the majority of life on Earth, can be displayed and used to educate viewers about a completely different set of organisms than they normally see. Many of these organisms are critically important in human health and in environmental issues, and linking an exhibit of them to these problems would be exciting.

Touch Specimens

Everyone likes to touch and feel specimens, and some museums make that possible. One example is the Rock Pile in the Geology Museum at the University of Lucknow, India. The Pile consists of rocks and fossils all in a wide and tall pile without any order (Fig. 18). Kids and adults, I was told, love to grub through that pile trying to find rocks or fossils that excite them. It is, to them, exploration. Their imaginations run wild. While the educational value can be questioned, the Rock Pile sharpens the imagination, curiosity, and creativity of those kids and adults who join them. Of course, a pile of rocks or loose specimens in a museum is probably not a good idea unless they are too large to move or fixed in cement or plastic. Perhaps a plastic rock wall would be better, with rocks and fossils held firmly but protruding from both sides so that visitors can touch and feel them.

A special area where specimens of various sorts sit on tables and shelves so children can handle them was created at The Copenhagen Museum of Zoology, Danish NHM, and the San Diego NHM, among others. Kids and adults are excited by these (Fig. 19), and they can share the experience with each other. Sorting specimens seems to attract kids and adults alike. For example, shark teeth can easily be sorted by size, tooth differences, whole or broken, and color (Fig. 20). Any kid can do it.

A complete, 22-m-long (72-ft-long), blue whale skeleton is mounted in a dive position just outside the Santa Barbara Museum of Natural History (Fig. 21)—but not in the usual place high above the visitors. Instead, visitors can walk right into the rib cage of the animal and get a direct impression of its size and complexity. Staring from the rib cage forward through the skeleton to the skull is an amazing confirmation of evolution, because viewers can see the same bones that we have but modified for life in the ocean as a huge filter feeder.

Fishbowls

NHMs everywhere have glass windows into enclosures where scientists and preparators can be viewed working on specimens. Such so-called "fishbowls" (Figs. 22 and 23) are especially common in paleontology galleries but are also used to show other scientific work. While harboring interesting activities behind a glass window or wall provides a chance to see professionals in action, fishbowls also separate people from the activity. A few fishbowls have an audio system so that questions can be asked. That is not enough. People feel intimidated and reticent because the glass is a barrier, the people working behind the glass seem busy, objects may be difficult to see, and discussion is inhibited. Perhaps moving the activity, say preparation of a fossil in a block of sedimentary rock, cleaning artifacts, or sorting minerals, out into the room so that visitors can crowd around it would work better. The fishbowls could also have open windows, allowing easier conversation, closer examination, and eliminating the glass barrier. The people immediately feel welcomed, involved, and ask questions. They can even be allowed to do some of the work.

Involve the Visitor

Visitors can be involved in exhibits. They like it. For example, encourage them to make their own exhibits, as the Mace Brown Natural History Museum (2017/11) has done with their amateur exhibit program (Fig. 24). The San Diego (California) Natural History Museum has a permanent exhibit dedicated to Citizens Science called "Extraordinary Ideas from Ordinary People," demonstrating what these people have done and are doing at the museum (San Diego Natural History Museum, 2017/12). This is total involvement with rewards of accomplishment, audience appreciation, and learning about exhibit design and specimens in detail. Individual specimens can be used to connect with visitors as well. A skeleton cast of Lucy (Fig. 25), the early hominin from Ethiopia, is mounted standing at floor level with a measuring tape alongside her so that visitors can see how tall she is and compare themselves to her, thus promoting interaction. Or a durable cast of early human footprints could be mounted on the floor so that visitors can put their own feet into them. Exhibits should be personalized and invite the visitor to actively engage in them.

Visitors like interaction with exhibits, and many ways to do this without high tech are possible. For example, museum staff or volunteers can ask visitors in conversation which exhibits they liked best and why, and then post that list where other visitors can see it. Curators could do it too, but that puts up the "expert" barrier. Encourage visitors to use smart phones to take pictures

Figure 18. The Rock Pile with fossils in the patio of the Geology Museum at the University of Lucknow, India. The rocks and fossils are piled high and wide for visitors (kids, but parents get involved too) to grub for specimens that they like. It is a learning experience in the hands of an explorer. Jere H. Lipps photographs, 2017.

Figure 19. Loose skulls for touching and hand examination by kids (upper) and adults (lower) at the Copenhagen Zoological Museum, Natural History Museum of Denmark. Jere H. Lipps photographs, 2010.

of exhibits and to include themselves in pictures of exhibits they like, as a personalized souvenir. Perhaps provide "science strolls," where scientists, staff, or docents lead a group of visitors, but instead of giving out information, they just answer questions as the group walks. Many such activities can involve visitors more completely.

Science or curiosity laboratories, such as Q?rious at the U.S. NHM, the Santa Barbara NHM, and the Danish NHM (Fig. 19) among many others, provide separate areas with special displays, exhibits of working drawers full of specimens, microscopic examination of tiny specimens, and other activities related to museum science that are usually aimed at children. These displays require instruction to kids on how to use them, what to see, and how to manipulate the specimens; they are quite unlike a table full of skulls that visitors can rummage through themselves. These kinds of activities should be expanded to embrace lifelong learning for everyone.

Websites and Social Media

Websites and social media (Facebook, Instagram, Twitter, and others) have become an important means of communicating

Figure 20. An assemblage of fossil shark teeth from the Miocene of California. These can be sorted by kids and adults into various classes (size, color, completeness, and species). Bruce Rubin photograph, 2013.

with the public. Indeed, in the late 1990s, so many museums, universities, science organizations, and government agencies created websites loaded with information that people, including teachers, were overwhelmed. They needed a website to sort out the websites!

Even so, websites and social media are important functions for NHMs, but perhaps in different ways. Now NHMs mostly advertise their visiting and permanent exhibits, detail the visitor experience and how that might best be accomplished, introduce research, layout educational opportunities, and provide directions and parking instructions—all of which are important. However, NHMs could also include broader educational aspects of the collections by placing objects into the larger framework of science and by building strong links between their websites and actual specimens (Scotchmoor and Lipps, 1995). Websites designed to excite creativity and used to educate will benefit the public, schools, and the museum. Museums contain fantastic collections that generate interest in Web users all over the world. NHM websites have the power to reach very large numbers of

Figure 21. "Chad," "the most anatomically correct Blue Whale skeleton in the world," mounted at ground level outside the Santa Barbara, California, Museum of Natural History. Visitors can view most of the specimen at eye level, walk around the skeleton, and enter the rib cage of the whale—and what an experience that is! Jere H. Lipps photographs, 2017.

Figure 22. A "fishbowl" laboratory exhibit designed so that visitors can see workers preparing specimens at the George C. Page Museum of La Brea Tar Pits, Los Angeles. While interesting, the glass acts as a barrier between the visitors and the scientific staff. If the glass were removed or specimens moved to rolling tables where visitors could crowd around to see, talk, and even to touch the specimens, visitors would become more involved and intrigued. Jere H. Lipps photograph, 2013.

people, far more than those who come through the doors. Learning can begin before the visitor even gets to the NHM and last for a long time afterwards, if the visitor is encouraged or desires to learn more, if exhibits attract visitor attention and create wonder in their minds, and if visitors wish to follow up on items they saw in the museum. Follow-up on NHMs websites with coordinated and engaging material would help keep that attention and wonder alive and could encourage return visits. For these reasons, museums should develop general interest websites that connect with people everywhere (Edson, 2014). This kind of outreach may be costly to develop, but the service it provides and the interest it garners are well within the scope of NHMs.

Figure 23. Fishbowl exhibit made with windows into the permanent preparation laboratory at the Museo Paleontológico Egidio Feruglio, Trelew, Argentina. These windows could be sliding so that the laboratory could be opened to visitors without a glass barrier. Jere H. Lipps photograph, 2014.

Figure 24. Amateur exhibit at the Mace Brown Natural History Museum, Charleston, South Carolina. Amateurs are selected to assemble rotating exhibits based on their own interests, thus providing interaction, education, and connection with the museum. Robert Boessenecker photograph, 2017.

Figure 25. A cast of Lucy, the australopithecine found in Ethiopia, displayed at the California Academy of Sciences, San Francisco. The cast or replica is all the bones of Lucy (~40% of her skeleton) that were discovered, laid out on a flat surface in the position but not the orientation they would have had when Lucy was alive. The mount is placed upright at floor level, allowing for comparisons by visitors to themselves. Jere H. Lipps photograph, 2013.

Websites provide for long-term public learning with good, up-to-date information presented in a friendly and encouraging way. The larger public museums generally do these well because they have specialized staff to dedicate to the development, updating, and maintenance of their websites. The University of California Museum of Paleontology (UCMP), with one of the first 50 websites online but minimal public exhibits, developed several thousand pages of educational material for teachers, children, university students, and the general public on biology, paleontology, and geology (Scotchmoor and Lipps, 1995). This website was built by committed voluntary graduate students initially and later by designated staff. In addition to the UCMP website, the museum, through National Science Foundation funding, developed two extensive, useful, and popular websites: Understanding Science (2017) and Understanding Evolution (2017). These are an extension beyond the museum's primary aims in paleontology but are regarded as an important outreach activity. Website designing, updating, and development is a continuous process, as the science, museum and audience change. Those goals require a strong commitment and collaboration of staff, scientists, audience members, and funding sources.

Social media, such as Facebook, Instagram, and others, are popular and provide another means to attract visitors to museums. So far, these media have been used for announcements of exhibits, events, and staff, but they can be used effectively to provide information and encouragement to visit a museum (Baker, 2017). Posts of short educational matters that feature museum exhibits or activities are effective attractants. Users of these media also like to post their own pictures and stories that will interest their friends, families, and followers. Thus, a museum strategy might be to develop some exhibits catering to these users and other visitors—for example, an especially attractive specimen with a large, well-positioned font label with a colorful or appropriate background so that it is most photogenic and that will allow the photographer to take a position in the image as well. With several of these placed around the museum, the visitors would delight in taking images and selfies (images showing both the photographer and subject in the same picture). The social media provide a host of opportunities to directly involve users in museum visits and activities. To take advantage of these opportunities will require creativity and knowledge of the technology and how it is used psychologically.

Lastly, artificial intelligence, or AI, is advancing rapidly. AI is the use of computers to examine a wide array of interconnected information and make sense of it in a way similar to the workings of the human mind. Museums with broad-based and narrowly focused functions across many endeavors—research, collection management, visitor pleasure and satisfaction, lifelong education, and science literacy—are prime candidates to utilize AI (Merritt, 2017). The NHM should be closely integrated in all its seemingly dissimilar activities, which when taken case by case lead to a difficult analysis for planning and implementation. AI can help; but it will require investment in a full-time staff member or occasional use of computer experts. This should likely be a permanent investment because museums will evolve in the future in ways we do not yet conceive.

CONCLUSIONS

Natural history museums (NHMs) have a long history and unlimited potential to attract and maintain visitors to their exhibits, websites, and social media. Their collections, curated and preserved to show life, earth, human cultures, and other specimens, are used in research, education, and outreach. Thousands of NHMs have been built in the past two centuries in nearly every country in the world. Each one is different in style and objectives, but all seem to be focused on education, especially of children. This diversity of mission, age, and audience size and diversity make NHMs complicated businesses. Museum products are preservation of our heritage, education, joy, and well-being of our people, and scientific understanding and advancement of our world.

Only through public appreciation will museums thrive, but it must include all the people. Populations served by NHMs are mostly self-selected, generally well educated, and, in Europe and America, chiefly white and middle class. Other potential ethnic or economically disadvantaged groups seldom, if ever, visit museums. Museums will need to survey non-attending populations to figure out how best to engage these populations. As those populations increase in coming decades, so will audience diversity. NHMs need to plan now to develop close interactions with this changing audience.

NHMs need exciting and engaging exhibits that include the senses of sight, sound, touch, and smell. The combination of these senses can reinforce the pleasure or interest in an exhibit and the memories of it. Exhibits should emphasize the collections of the museum in ways that grab the attention of visitors rather than rely on traveling exhibits and replicated objects. Real objects from the collections, displayed with imagination, creativity, and innovation, focused on the entire population served by the museum, can captivate and welcome people back repeatedly.

Technology should not replace exhibits or specimens in museums but enhance the appreciation of them. NHMs should always ask: does this technology really add to our specimens or does it detract from them? On the other hand, artificial intelligence (AI) can enhance planning and understanding of the best ways to integrate the innumerable complexities of the museums.

Be flexible. Be innovative. Be encouraging. Be adventuresome. Be creative. Nothing will be lost, and much may be gained.

ACKNOWLEDGMENTS

I thank Professor Gary D. Rosenberg, professor emeritus, Department of Earth Sciences, Indiana University, Purdue University, Indianapolis, Indiana, and adjunct curator of geology, Milwaukee Public Museum, and Professor Renee Clary, director of the Dunn-Seiler Museum in the Department of Geosciences, Mississippi State University, Starkville, Mississippi, for organizing both the symposium on natural history museums at the 2016 Geological Society of America Annual Meeting and this book. They have been very patient and helpful editors. Professor William Pannapacker of Hope College, Holland, Michigan, has been very helpful and stimulating in discussions about NHMs since we first met in 2010 at a symposium in Copenhagen on museums. Drs. Mark Goodwin, Lisa White, Ken Finger, Pat Holroyd, and Diane Erwin at the Museum of Paleontology, Berkeley; Dr. Marie-The Venet-Peyre of the NHM Paris; Drs. Alexie Rozanov and M.A. Fedonkin of the Paleontological Institute in Moscow; Jeannine Pedersen-Guzman of Colorado State University; Meredith Rivin at the Burke Museum, Seattle; Professor Vibhuti Rai of University of Lucknow; Dr. Zhonghe Zhou at the Institute of Vertebrate Paleontology, Beijing; and the late Professor Adolf Seilacher of Tübingen University provided detailed information about their institutions that was particularly useful. Dr. M.P. Singh, Secretary, The Palaeontological Society of India, arranged for a tour of Indian institutions and museums in February and March 2017, during which I learned a great deal. I appreciate the help of the many other scientists and staff of museums I have visited in the United States, Europe, Russia, China, India, South America, Africa, Australia, and New Zealand, but there are too many to note individually. I especially benefitted as a visiting scientist at the Natural History Museum, London; Natural History Museums of Paris; the Institute and Museum for Geology and Paläontology, University of Tübingen; Oxford University; and as a research associate at The Natural History Museum, Los Angeles; as a fellow of the California Academy of Sciences, San Francisco; and as faculty curator at my home Museum of Paleontology, University of California, Berkeley. I thank these and other unnamed people and organizations for chances to see the exhibit galleries, research laboratories, and collection rooms of so many museums of such a wide variety. The people and institutions were kind and welcoming. I loved their museums, and I learned a lot.

REFERENCES CITED

Publications and websites referred to in this paper are listed here together. Websites are listed with the date of the last access by year and month and with notes on the contents.

Academy of Natural Sciences, 2017/10, www.ansp.org/visit/exhibits/secrets-of-the-diorama/: Philadelphia, Pennsylvania. [Collections, education, visitors, research.]

American Academy of Ophthalmology, 2017/12, www.aao.org/eye-health/diseases/what-is-color-blindness: San Francisco, California. [An explanation of color blindness and its effects on people.]

American Alliance of Museums, 2016, Excellence in exhibition label writing competition: Washington, D.C.: http://aam-us.org/docs/default-source/professional-networks/excellence-in-label-writing-2016.pdf?sfvrsn=6.

American Museum of Natural History, 2017/11, https://www.amnh.org/: New York, New York. [Collections, research, education, outreach, information, and educational materials.]

Ashby, J., 2017, Animal Kingdom: A Natural History in 100 Objects: Stroud, Gloucestershire, UK, The History Press, 328 p.

Asma, S.T., 2001, Stuffed Animals and Pickled Heads: The Culture and Evolution of Natural History Museums: Oxford, UK, Oxford University Press, 302 p.

Australian Museum, 2017/12, https://australianmuseum.net.au/: Sydney, Australia. [Exhibits, research, education for students, teachers, parents, image galleries of Australian animals with explanations.]

Baker, S., 2017, Identifying behaviors that generate positive interactions between science museums and people on Twitter: Museum Management and Curatorship, v. 32, p. 144–159, https://doi.org/10.1080/09647775.2016.1264882.

Bamberger, Y., and Tal, T., 2007, Learning in a personal context: Levels of choice in a free choice learning environment in science and natural history museums: Science Education, v. 91, p. 75–95, https://doi.org/10.1002/sce.20174.

Bell, P., Lewenstein, B., Shouse, A.W., and Feder, M.A., 2009, Learning Science in Informal Environments: People, Places, and Pursuits: Washington, D.C., The National Academies Press, 352 p.

Benton, T.H., 2009 (30 Oct.), Preserving the future of natural-history museums: The Chronicle of Higher Education, 4 p.

Benton, T.H., 2010 (1 July), Getting real at natural history museums: The Chronicle of Higher Education, 5 p.

Bergeron, A., and Tuttle, B., 2013, Magnetic: The Art and Science of Engagement: Washington, D.C., The American Alliance of Museums Press, 224 p., https://doi.org/10.1111/cura.12075.

Berkeley Natural History Museums, 2017/12, https://bnhm.berkeley.edu/about/origins/: Berkeley, University of California. [Gateway to the seven natural history museums of the University of California, Berkeley.]

Berkowitz, C., and Lightman, B., eds., 2017, Science Museums in Transition: Cultures of Display in Nineteenth-Century Britain and America: Pittsburgh, Pennsylvania, Pittsburgh University Press, 392 p., https://doi.org/10.2307/j.ctt1r6b0c8.

Best College Reviews of 30 US College Natural History Museums, 2017/10, www.bestcollegereviews.org/features/amazing-natural-history-museums/.

BoardSource, Museum Board Leadership, 2017, A National Report: Washington, D.C., BoardSource, www.aam-us.org/docs/default-source/press-releases/download-the-report.

Buena Vista Museum of Natural History, 2017/12, http://www.sharktoothhill.org/: Bakersfield, California. [Collection and exhibits of local Miocene fossils, Native American artifacts, children's activities, paleontological and geological field trips.]

Burke Museum of Natural History and Culture, 2017/10, www.burkemuseum.org/: University of Washington, Seattle. [Collections, preparation area, research, education in support of faculty and students at the University of Washington.]

California Academy of Sciences, 2017/12, www.calacademy.org/: San Francisco. [Large collections, research, Genomics Center, static and live exhibits, planetarium, aquaria, live rooftop, educational activities, information, hosts https://www.inaturalist.org, an online biological species identification application.]

Carnegie Museum of Natural History, 2017/12, https://carnegiemnh.org/: Pittsburgh, Pennsylvania. [Collections, education, research, outreach.]

Chamberlin, T.C., 1890, The method of multiple working hypotheses: Science, v. 15, p. 92–96. [Reprinted in Science, 1965, v. 148, p. 754–759.]

Charleston Museum, 2017/11, /www.charlestonmuseum.org/: Charleston, South Carolina. [Collections, images, activities.]

Danish Natural History Museum, 2017/10, http://snm.ku.dk/english/: Copenhagen, Denmark. [Museums Zoology, Botany, exhibits, research, children's activities.]

Darwin, C.R., ed., 1838, Fossil Mammalia Part 1 no. 1 of The Zoology of the Voyage of H.M.S. *Beagle* (by R. Owen; edited and superintended by C. Darwin): London, Smith Elder and Co.

Darwin, C.R., 1845, Journal of Researches into the Natural History and Geology of the Countries Visited during the Voyage of H.M.S. *Beagle* Round the World: Under the Command of Capt. Fitz Roy, R.N: London, J. Murray, 519 p.

Dawson, E., 2014, "Not designed for us": How science museums and science centers socially exclude low-income, minority ethnic groups: Science Education, v. 98, p. 981–1008, https://doi.org/10.1002/sce.21133.

Dawson, E., and Jensen, E., 2011, Toward a contextual turn in visitor studies: Evaluating visitor segmentation and identity-related motivations: Visitor Studies, v. 14, p. 127–140, https://doi.org/10.1080/10645578.2011.608001.

Delicado, A., 2010, For scientists, for students or for the public?: The shifting roles of natural history museums: Journal of History of Science and Technology, v. 4, Fall, http://johost.eu/print.asp?oid=96#_edn121.

Edson, M., 2014, The challenge of scaling up: Museums in an age of scale, *in* Merritt, E., and Gangopadhyay, P., convenors, Building the Future of Education Museums and the Learning Ecosystem: Washington, D.C., Center for the Future of Museums, American Alliance of Museums, p. 45–46, www.aam-us.org/docs/default-source/center-for-the-future-of-museums/building-the-future-of-education-museums-and-the-learning-ecosystem.pdf?sfvrsn=2.

Elliott, L.P., and Brook, B.W., 2007, Revisiting Chamberlin: Multiple working hypotheses for the 21st century: Bioscience, v. 57, p. 608–614, https://doi.org/10.1641/B570708.

Farrell, B., and Medvedeva, M., 2010, Demographic transformation and the future of museums: Washington, D.C., American Association of Museums Press: www.aam-us.org/docs/center-for-the-future-of-museums/demotransaam2010.pdf.

Farrington, O.C., 1915, The rise of natural history museums: Science. New Series, v. 42, p. 197–208, https://doi.org/10.1126/science.42.1076.197.

Fenichel, M., and Schweingruber, H.A., 2010, Surrounded by Science: Learning Science in Informal Environments: Washington, D.C., Board on Science Education, Center for Education, Division of Behavioral and Social Sciences and Education, The National Academies Press, 240 p.

Ferilli, G., Grossi, E., Sacco, P.L., and Blessi, G., 2017, Museum environments, visitors' behaviour, and well-being: Beyond the conventional wisdom: Museum Management and Curatorship, v. 32, p. 80–102, https://doi.org/10.1080/09647775.2016.1239125, https://doi.org/10.1080/09647775.2016.1239125.

Fortey, R., 2008, Dry Storeroom No. 1: The Secret Life of the Natural History Museum: New York, London, Alfred A. Knopf, 352 p.

Garibay, C., 2009, Latinos, leisure values, and decisions: Implications for informal science learning and engagement: The Informal Learning Review, v. 94, p. 10–13.

Geological Society of America, 2018/07, www.geosociety.org/GSA/Science_Policy/Position_Statements/Current_Statements/gsa/positions/position1.aspx. [Like many other scientific organizations, the GSA urges that science and religion be kept separate since they operate with different values. Science requires evidence and data while religion relies on beliefs for which evidence may be lacking.]

Grimes, J., Manjarrez, C.A., Miller, K.A., and Swan, D.W., 2015, Museum Universe Data File, FY 2015 Q3 Release: Documentation (IMLS-2015-MUDF-01): Washington, D.C., Institute of Museum and Library Services, Data file at https://data.imls.gov/Museum-Universe-Data-File/Museum-Universe-Data-File-FY-2015-Q3/ku5e-zr2b/data#column-menu.

Head Miller, J., Rawson, L.R., Serrell, B., and Bedford, L., 2013, Exhibition label writing at its best: Webinar, The American Alliance of Museums (https://aam-us.org/ProductCatalog/Product?ID=5164).

How to Design for Color Blindness, 2017/12, http://blog.usabilla.com/how-to-design-for-color-blindness/: Amsterdam, Usabilla. [A quick and simple guide to designing for color-blind people. A good introduction to the problem, although many sites provide even more detail.]

Institute for Museums and Library Services, 2015, https://data.imls.gov/Museum-Universe-Data-File/Museum-Universe-Data-File-FY-2015-Q3/ku5e-zr2b, Washington, D.C. [Contains descriptive information about museums in the 50 states and the District of Columbia based on public records and administrative data.]

Korea Biz Wire, 2017/11, http://koreabizwire.com/north-korea-opens-central-zoo-and-nature-museum-in-pyongyang/62529: Seoul, South Korea. [Description of natural history museum and zoo in North Korea.]

Kunstkamera, The Peter the Great Museum of Anthropology and Ethnography, Russia, 2017/11: http://www.kunstkamera.ru/en/museum/: St. Petersburg. [Exhibits and research on human anthropology and evolution, publishes two journals.]

La Brea Tar Pits and Museum, 2017/10, https://www.tarpits.org/: Los Angeles. [Tar pits, George C. Page Museum, exhibits, large collection of vertebrates and other organisms recovered from the Pleistocene tar pits, research, on-site activities.]

Lipps, J.H., 1999, This is science!: Paleontological Society Special Publication, v. 9, p. 3–16.

Lipps, J.H., 2007, Into focus: Museums and their exhibits: Palaeontologia Electronica, v. 10, no. 2, www.palaeo-electronica.org/paleo/2007_1/editorial/focus.htm.

Lipps, J.H., 2009, PaleoParks—The preservation and conservation of fossil sites worldwide: Carnets du Geologie, http://paleopolis.rediris.es/cg/CG2009_BOOK_03/Chapter1, 9 p.

Lipps, J.H., 2010, Charles Darwin and H.M.S. *Beagle*: Besides Galapagos: Proceedings of the California Academy of Sciences, Series 4, v. 61, Supplement II, no. 2, p. 7–30.

Mace Brown Museum of Natural History, Charleston University, 2017/11, http://geology.cofc.edu/natural-history-museum/index.php: Charleston, South Carolina. [Fossils, reports, images.]

Merritt, E., 2017, Artificial intelligence: The rise of the intelligent machine: TrendsWatch 2017: Washington, D.C., Center for the Future of Museums, American Alliance of Museums. p. 24–31: https://www.aam-us.org/2017/05/01/artificial-intelligence-the-rise-of-the-intelligent-machine/. [Forecasting reports issued annually since 2012 by the Center for the Future of Museums, which summarize emerging trends identified through CFM's research.]

Merritt, E., and Gangopadhyay, P., convenors, 2014, Building the Future of Education: Museums and the Learning Ecosystem: Washington, D.C., Center for the Future of Museums, American Alliance of Museums, 60 p., www.aam-us.org/docs/default-source/center-for-the-future-of-museums/building-the-future-of-education-museums-and-the-learning-ecosystem.pdf?sfvrsn=2.

Miguasha World Heritage Site, 2017/11, http://www.miguasha.ca/mig-en/a_world_heritage_site.php: Miguasha National Park, Quebec. [Museum with Devonian fossils, geology of the area, history of collecting, and associated field site.]

Miller, J.D., 2010, Adult science learning in the Internet era: Curator: The Museum Journal, v. 53, p. 191–208, https://doi.org/10.1111/j.2151-6952.2010.00019.x.

Museo Argentino de Ciencias Naturales, "Bernardino Rivadavia," 2017/12, www.macn.secyt.gov.ar/investigacion/homeinvest.php: Buenos Aires, Argentina. [Geology, paleontology, botany, ecology, zoology with images, research, learning center.]

Museum Facts, 2017/11, http://www.aam-us.org/about-museums/museum-facts: American Alliance of Museums, Washington, D.C. [Compilation of statistics demonstrating how museums are educational, trusted, beloved, and economic assets to communities everywhere.]

Museum für Naturkunde, 2017/11, www.naturkundemuseum.berlin/en: Berlin, Germany. [Exhibits of small portion of the 30 million specimens, many wet preservations of animals, visitor participation, education, research, images of animals.]

Muséum national d'Histoire naturelle, 2017/11, www.mnhn.fr/en/visit/: Paris. [Huge collections and exhibits in six buildings, botanical garden and zoo, extensive research, various public activities, outstanding exhibits, special section for blind or partially sighted visitors.]

Museum of the Rockies, 2017/10, www.museumoftherockies.org/: Bozeman, Montana. [Paleontology, history, photograph collections, research and exhibits, children's discovery center, planetarium, visitor services.]

Natural History Museum London, 2017/12, http://www.nhm.ac.uk/: London. [Full website, collections, exhibits, events, science of the museum's scientists.]

Natural History Museum Los Angeles, 2017/10, https://nhm.org/site/: Los Angeles, California. [Large collections, extensive research, exhibits, diverse education programs, teacher resources, citizen science and other activities, visitor services.]

Natural History Museum Smithsonian Institution, 2017/11, https://naturalhistory.si.edu/: Washington, D.C. [Full website on a large range of topics, large staff, world's largest collection, extensive research, Q'rius center for children and parents, teacher assistance.]

Natural History Museum Vienna, 2017/12, www.nhm-wien.ac.at/en/museum: Vienna, Austria. [Exhibits, research, education, planetarium, collections.]

National Museum of Nature and Science, 2017/11, www.kahaku.go.jp/english/index.php: Tokyo, Japan. [Exhibits, research, children's activities, scientific support.]

National Science Board, 2016, Science and Engineering Indicators 2016: Arlington, Virginia, National Science Foundation, NSB-2016-1.

North Korean Natural History Museum and Zoo, 2017/11, www.youtube.com/watch?v=l8T-HYtyoGo: Pyongyang, North Korea. [Variety of exhibits, mostly models and casts, linked to a zoo.]

Oregon.gov/oha, 2018/02, Odors and your health: http://www.oregon.gov/oha/ph/healthyenvironments/healthyneighborhoods/toxicsubstances/documents/OdorsAndYourHealth_Final.pdf. [A short summary about odors and people's health.]

Oreskes, N., and Conway, E.M., 2010, Merchants of Doubt: How a Handful of Scientists Obscured the Truth on Issues from Tobacco Smoke to Global Warming: New York, Bloomsbury Press, 355 p.

Orlov Paleontological Museum, 2017/10, https://www.paleo.ru/museum-en/: Moscow, Paleontological Institute, Russian Academy of Sciences. [Collections, research and exhibits on paleontology focused on Russian material.]

Paleontological Research Institute and Museum of the Earth, 2017/11, https://www.priweb.org/index.php: Ithaca, New York. [Collections, exhibits, teacher resources, children's activities, research.]

Paleozoological Museum of China, 2017/12, http://english.ivpp.cas.cn/rs/pm/: Institute for Vertebrate Paleontology and Paleoanthropology (IVPP), http://english.ivpp.cas.cn/: Beijing, Chinese Academy of Sciences. [Vertebrate collections, early humans, paleontology research.]

Peking Man World Heritage Site, 2017/11, http://whc.unesco.org/en/list/449/: Zhoukoudian, China. [History of site, images, maps, documents, collections, exhibits linked with trails on the discovery site.]

Platt, J.R., 1964, Strong inference: Science, v. 146, p. 347–353, https://doi.org/10.1126/science.146.3642.347.

Pyenson, N.D., Irmis, R.B., Lipps, J.H., Barnes, L.G., Mitchell, E.D., Jr., and McLeod, S.A., 2009, The origin of a widespread marine bonebed deposited during the middle Miocene Climatic Optimum: Geology, v. 37, p. 519–522, https://doi.org/10.1130/G25509A.1.

Q?rius, 2017/10, http://qrius.si.edu/: Washington, D.C., Smithsonian Museum of Natural History. [Q?rius is a special room at the museum where children's activities are laid out.]

Rounds, J., 2004, Strategies for the curiosity-driven museum visitor: New York, Curator, v. 47, p. 389–412.

Royal Belgian Museum of Natural History, 2017/12, www.naturalsciences.be/en/museum/home: Brussels, Royal Belgian Institute of Natural Sciences. [3D largest natural history museum in Europe, animals, earth science. Paleontology, images and text on these topics.]

San Diego Natural History Museum, 2017/12, www.sdnhm.org: San Diego, California. [Visitor information, exhibits, events, science, education for schools and others.]

Santa Barbara Museum of Natural History, 2017/09, www.sbnature.org/about/8.html: Santa Barbara, California. [Collections, research, education, outreach, children's room.]

Santa Barbara Museum of Natural History, The Blue Whale Skeketon, 2018/06, http://www.sbnature.org/visit/exhibitions/58/blue-whale-skeleton. Santa Barbara, California. [Story of the exhibited Blue whale skeleton.]

Scotchmoor, J., and Lipps, J.H., 1995, Where is the REAL Museum of Paleontology?: Geotimes, v. 40, p. 27–29.

Sonne, C., Bechshoft, T.O., Riget, F.F., Baagoe, H.J., Hedayat, A., Andersen, M., Bech-Jensen, J.-E., Hyldstrup, L., Letcher, R.J., and Dietz, R., 2013, Size and density of East Greenland polar bear (*Ursus maritimus*) skulls: Valuable bio-indicators of environmental changes?: Ecological Indicators, v. 34, p. 290–295, https://doi.org/10.1016/j.ecolind.2013.04.015.

State Darwin Museum, 2017/11, http://www.darwinmuseum.ru/: Moscow, Russia. [Exhibits and research on the evolution, ecology, biogeography, behavior of life, English version.]

To, C., Tenenbaum, H.R., and Wormald, D., 2016, What do parents and children talk about at a natural history museum?: Curator: The Museum Journal, v. 59, p. 369–385, https://doi.org/10.1111/cura.12174.

Tyrannosaurus rex replica, 2017/11, http://www.ucmp.berkeley.edu/trex/index.html: Berkeley, Museum of Paleontology, University of California. [The construction of this dinosaur replica (cast).]

Understanding Evolution, 2017/12, https://evolution.berkeley.edu/evolibrary/home.php: Berkeley, Museum of Paleontology, University of California. [A dedicated website to the science of evolution, useful to students, teachers, and the general public who wish to know and understand evolution.]

Understanding Science, 2017/12, https://undsci.berkeley.edu/article/whatisscience_01: Berkeley, Museum of Paleontology, University of California. [A primer on the nature and process of science, with examples from biology, definitions, and resources; useful to students, teachers, and the general public in understanding science.]

United States National Park Service, 2017/10, www.nps.gov/findapark/index.htm: Washington, D.C. [Searchable listing of National Parks in the United States.]

University of California Museum of Paleontology, 2017/11, www.ucmp.berkeley.edu/: Berkeley, California. [An academic research museum for faculty, researchers, students, and staff, large collections, educational activities for all, history of the museum, information on the biology, geology, and paleontology of life, educational material for lifelong learning and teachers.]

Wadia Institute of Himalayan Geology Museum, 2017/03, http://wihg.res.in/museum.htm: Dehradun, India. [History and description of the museum.]

Wikipedia List of Natural History Museums in the United States, 2017/11, https://en.wikipedia.org/wiki/List_of_natural_history_museums_in_the_United_States 2017/11.

Wikipedia List of Natural History Museums in the World, 2017/11, https://en.wikipedia.org/wiki/List_of_natural_history_museums/.

Winker, K., 2004, Natural history museums in a post-biodiversity era: Bioscience, v. 54, p. 455–459, https://doi.org/10.1641/0006-3568(2004)054[0455:NHMIAP]2.0.CO;2.

Zoological Museum of Moscow State University, 2017/11, http://zmmu.msu.ru/en: Moscow, Russia. [Extensive collections, active research, exhibits on animals worldwide, educational outreach.]

Manuscript Accepted by the Society 3 January 2018
Manuscript Published Online 6 July 2018
Revised Manuscript Published Online 3 August 2018

The Geological Society of America
Special Paper 535

Bridging the two fossil records: Paleontology's "big data" future resides in museum collections

Warren D. Allmon*,†
Gregory P. Dietl*
Jonathan R. Hendricks*
Robert M. Ross
Paleontological Research Institution, 1259 Trumansburg Road, Ithaca, New York 14850, USA, and
Department of Earth and Atmospheric Sciences, Cornell University, Ithaca, New York 14853, USA

ABSTRACT

There are two fossil records: the *physical fossil record,* which consists of specimens, and the *abstracted fossil record,* which is made up of data derived from those specimens. Museum collections are the conduit between these two fossil records. Over the past several decades, the abstracted fossil record has provided many important insights about the major features of life's history, but it has relied mostly on limited types of data (primarily taxonomic occurrence data) derived from ultimately finite literature sources. In contrast, specimen collections and modern tools for digitizing information about them present an opportunity to transform paleobiology into a "big data" science. Digitally capturing non-traditional (e.g., paleoecological, taphonomic, geochemical, and morphological) data from millions of specimens in museum collections and then integrating them with other unique big data resources has the potential to lead to the most important paleontological discoveries of the twenty-first century.

What we know about the past record relied heavily on museum collections—the cumulation of centuries of investigation of the fossil record. The sample of past biodiversity will accumulate only with continued exploration of the fossil record ... and restudy of existing collections....
—J. Sepkoski (1992, p. 80)

*Authors contributed equally and are listed alphabetically.
†Corresponding author: wda1@cornell.edu.

Allmon, W.D., Dietl, G.P., Hendricks, J.R., and Ross, R.M., 2018, Bridging the two fossil records: Paleontology's "big data" future resides in museum collections, *in* Rosenberg, G.D., and Clary, R.M., eds., Museums at the Forefront of the History and Philosophy of Geology: History Made, History in the Making: Geological Society of America Special Paper 535, p. 35–44, https://doi.org/10.1130/2018.2535(03).

INTRODUCTION: THE TWO FOSSIL RECORDS

There are two fossil records, and the data-driven science of paleontology balances upon both, like a beam that spans two pillars. The first fossil record—which we call the *physical fossil record*—consists of material objects: discovered specimens observed in the field or (more frequently) residing in collections, as well as specimens remaining undiscovered in nature. Yet fossils are not just things we can observe, they are also objects rich with biological, temporal, and geographic context. Because each fossil provides a record of the *consequences* of past biological and physical processes—rather than direct observation of past organisms, environments, and processes themselves—paleontology, like all historical sciences, is to some degree a science of inference and abstraction. Interpreting a fossil as anything other than a rock requires inference and abstraction based on context and comparison with modern observable analogs. This contextual and comparative information forms the second pillar of the fossil record, which we call the *abstracted fossil record*.[1]

Paleobiologists frequently divide themselves between two camps: those who tout their "specimen-based research," and those who identify as "quantitative" or "analytical" paleobiologists, often spending more time with computer databases than specimen collections. Both approaches, of course, rely upon abstracted information derived from the physical fossil record. A key difference, however, is the extent to which information from the abstracted fossil record is disconnected from the physical fossil record. As just one example, genus-level temporal duration is a common variable in many quantitative paleobiological analyses (e.g., Liow, 2007; Powell, 2007; Heim and Peters, 2011). The temporal duration of an individual genus is an amalgamation of the durations of individual species assigned to that genus. In turn, the duration of each species is based on known stratigraphic occurrences of individual specimens hypothesized to belong to that species. Thus, two levels of abstraction—specimens assigned to the species and species assigned to the genus—separate the genus-level duration from the material specimens that support it.

Such issues of the relationships between objects (the physical) and the data (the abstracted) drive much of philosophy in general—"what is the relation between concepts and things?" (Lord, 2006)—and have been discussed by a number of modern philosophers of science (e.g., Daston, 2000, 2004; Hein, 2000; Pinna, 2000). A related central theme in the philosophy of science, especially in recent decades, has been between "realist" views, in which "discoveries" are made about the physical world, and once found are found forever, and "constructionist" or "constructivist" views, which consider data and the interpretations derived from them (or, abstractions) as "inventions," created in the context of culture and biases and requiring occasional returns to the objects subjected to earlier interpretation (e.g., Daston, 2000; Haack, 2007). In the context of paleontology, interpretations based on fossils—such as the assignment of a specimen to a particular species or genus (i.e., the abstracted fossil record) are clearly much more like "constructions" than absolute or indisputable facts (such as the geographic position where a fossil was found or its longest dimension). It is, therefore, perhaps not surprising that the history of paleontology as a science is very much a history of shifting interpretations about the nature of fossils and their implications for the history of the Earth (Rudwick, 1976; Waterston, 1979, p. 10; Cutler, 2003).[2]

If we accept even a partially "constructionist" view of paleontology—in which our physical fossil record (objects) requires constant reinterpretation—then we must recognize specimens themselves as paleontology's only true source of "data,"[3] and the museum collections in which they are housed as the conduit between the two fossil records, allowing testing of past interpretations.[4] Yet paleontological collections are not just repositories for verification of past hypotheses and interpretations. In this essay, we argue that specimen collections and modern tools for digitizing information about them together present an opportunity to expand and transform quantitative paleobiology—which for decades has depended on highly abstracted data derived from an ultimately limited corpus of published literature—into what has come to be called a "big data" science.[5] In such a formulation, greater electronic access to a substantial portion of the millions of fossil specimens currently residing in museum drawers—the "unpublished fossil record" (Teichert et al., 1987)[6]—and the spectrum of potential data associated with them, present

[1]David Sepkoski (2017) has compellingly suggested a slightly different taxonomy: the "original" archive, he says, "is the earth itself," which he labels "archive$_0$." A collection of specimens is "archive$_1$," a pictorial atlas of fossils is "archive$_2$," a "text-only catalog" is "archive$_3$," and a compilation of numerical data—a database—is "archive$_4$." His first two "archives" (0 and 1) are equal to our "physical fossil record," and the next three (2–4) correspond to our "abstracted fossil record."

[2]As Daston noted, paraphrasing Cuvier, "the opposition between secrets of nature laid bare or 'discovered' and 'vast edifices [constructed] on imaginary bases' still haunts our discussions of scientific objects…." (Daston, 2000, p. 5).

[3]We follow Leonelli (2016, p. 77) in defining data as "any product of research activities … that is collected, stored, and disseminated *in order to be used as evidence for knowledge claims*" (emphasis in original). This definition views data as a relational category, which can be assigned to any objects, from physical specimens to the abstracted information about them.

[4]Fossil "collections" consist not just of specimens, but also of the associated human-created structures that make those specimens and their associated information available to researchers. It is these structures—which include physical ordering, safe storage, labeling, and access—that are the "conduit" through which the specimens "flow."

[5]There is no consensus about what "big data" means (Snijders et al., 2012; Mayer-Schönberger and Cukier, 2013). Initially, the idea focused on access to massive amounts of data and the manner in which it was stored and analyzed. The significance of "big data," however, lies not merely in the size of a given database, but in compiling data across previously unassociated databases. For the purposes of this paper, "big data" are of sufficient scale to provide opportunities to find new patterns, ask new questions, and generate new hypotheses that would not have been otherwise feasible or perhaps imaginable. "Big data" will thus likely ultimately be a relative term: what seems "big" today may seem small in the world of tomorrow as our capacity to store, retrieve, and analyze data continues to improve.

[6]The "unpublished" fossil record consists of specimens that have never been mentioned or used in any published scientific work, including systematics, biostratigraphy, or paleoenvironmental reconstructions. Teichert et al. (1987) included in this definition materials mentioned in unpublished theses as prominent examples of fossils that have been studied, but whose abstracted information is nevertheless inaccessible to other researchers.

an opportunity to undertake truly massive quantitative analyses. Such analyses may lead to new patterns, questions, and hypotheses about life's history, but only if we choose to re-embrace the physical fossil record as paleontology's largest and primary source of data. Before considering how paleontology might transform itself into a "big data" science—as well as the challenges that such a transformation presents—we first review how the science of paleontology is philosophically connected to specimens. We then discuss why fossil collections themselves have potentially inexhaustible evidential value, and consider how abstracted data have driven paleobiological research in recent decades.

FOSSILS AND DATA—OBJECTS AND ABSTRACTIONS

Fossils as a Primary Source of Data

Paleontologists study the physical fossil record and, from individual specimens, make empirical observations, which may be recorded as numbers, text, or images that are often called "data." Once these observations are made, the paleontologist usually only works with the derived, abstracted data, rather than the specimen itself. The fossil specimen (the object) is commonly incorporated into a museum collection and stored there long after the paleontologist has finished making her observations. This practice—which differentiates paleontology from sciences such as astronomy, physics, chemistry, and ethology, which do not involve collection of objects—has persisted for centuries, resulting in assembled specimen collections that last well beyond the careers and lifetimes of the paleontologists who first collected and studied them (e.g., Rudwick, 1976; Impey and MacGregor, 1985; Findlen, 1994; Knell, 2000). Paleontologists may also make representational images of the specimens, and these may be gathered together, forming "virtual" (or "paper") collections (e.g., Rudwick, 2000; Davidson, 2008; Hendricks et al., 2015). Both kinds of collections, however, have in common that they contain potentially important information that exists beyond the particular observations made on individual specimens. The process of organizing collections for the simple sake of keeping inventory and retrieving information—for example, assigning identifying numbers to specimens, recording those numbers in a specimen catalog, and, more recently, adding that information to a computerized database—has long been central to museum collections management. Thus, museum collections are the conduit between the physical and abstracted fossil records. Collections are in this way not only archives of previously discovered knowledge, but also reserves of potential future observations—that is, they have great "evidential value."

Evidential Value of Fossil Collections

What makes fossil collections (and the information abstracted from them) into powerful epistemic tools is their ability to act as evidence for a variety of knowledge claims about phenomena across research situations. There are four major reasons why paleontological collections—if they and the information they contain can be made widely available—have nearly inexhaustible evidential value (i.e., effectively unlimited potential for repurposing to serve new scientific goals; sensu Leonelli, 2016).

New Discoveries—New Collections

One of the most important justifications for maintaining large collections in paleontology (and other natural history disciplines) is the potential opportunity for future discoveries using previously collected specimens. There are many examples of older paleontological collections that have yielded new information and sometimes astonishing and unique discoveries (e.g., Allmon, 2005, and references therein; Callomon and Grădinaru, 2005; Boessenkool et al., 2009; Boersma and Pyenson, 2015; Kundrát and Ebbestad, 2015; Lindkvist, 2016). A number of recent studies have also not only discovered something new in existing fossil collections but used them to check previous results, sometimes confirming, supplementing, or disputing published literature data (e.g., Hunter and Donovan, 2005; Davis and Pyenson, 2007; Harnik, 2009; Dean et al., 2010). There are also countless examples of uses of existing fossil collections as sources of data for analyses that were never envisioned when they were first collected (e.g., Parham et al., 2012; Benton et al., 2014; Harnik et al., 2017).

In addition to existing collections, there is still clearly much to be discovered in the world's fossil-bearing rocks, many of which have never been carefully or extensively examined or collected; new collections are therefore clearly justified (e.g., Jackson and Johnson, 2001; Adrain and Westrop, 2003). In some cases, these new collections integrate new kinds of contextual data such as precise geographic coordinates and physicochemical data, and newly collected specimens may shed light on previously made collections.

Verification and/or Record of Observations and Conclusions

The scientific "crown jewels" of many museum fossil collections are "type specimens"—those that formally represent names assigned to new taxa based on the rules of biological nomenclature. This "authority of the specimen" (Waterston, 1979, p. 11) is treated as the central basis for all systematic biology and paleontology. Types are both an "archive of past research" (Knell, 1997) and a basis for future testing of the results of that research. The very concept of the type specimen is premised on the idea of permanently securing the potential for repeatedly comparing specimens with ideas derived from them (Mayr, 1989; Daston, 2004). More broadly, many authors have long argued that the practice of systematic biology itself depends absolutely on the presence of specimen collections, against which hypotheses can be tested (Mayr, 1969, p. 101; Waterston, 1979; Ellis, 2008, p. 173; Sunderland, 2016). "Systematic work that does not rely on museum specimens to verify or falsify the identities of the taxa studied," declared Winston (2007, p. 47) flatly, "is not science."

> Because the scholarship of biodiversity includes scrutinising earlier work, evaluating what was written before, and adding new information and insight, it should always be possible to return to those specimens.

They are the primary evidence for the information presented. The ability of researchers to re-examine the primary data and question the conclusions of previous work is a crucial part of what makes this a scientific activity. (Schilthuizen et al., 2015, p. 237)

Replication of Observations

Fundamental to science is the repeatability of results. Yet, unlike most phenomena studied by physicists and chemists, every object studied by paleontologists (and systematic biologists) is historically unique in that it is the product of events that occurred only once in time and space. This means an individual fossil cannot be replicated or easily re-obtained when more observations are needed later. If specimens are not accessible, and similar ones cannot be collected, additional data cannot be gathered. Similarly, it is frequently the case that if fossils are not collected in the field, they cannot be revisited easily later, due to their destruction by natural causes (such as erosion) or human causes (such as development). Many older museum collections contain specimens that simply cannot be collected again because the localities are gone (see examples in Allmon, 2005). Some new collections are strategically made in locations that will or are expected to become inaccessible in the future, even if there is no immediate or pressing scientific reason for doing so (i.e., "salvage paleontology"; e.g., Stone, 2013; Dooley et al., 2015).

Incomplete Nature of Observations

For a variety of reasons, we cannot make all the observations on a fossil at once. First, there may simply be no time to do so in the field. Much paleontological field work is done under conditions that are less than optimal for careful observation. Second, a paleontologist may be unsure of what is "important" to observe. This may be especially the case for unexpected or novel discoveries. Third, many techniques and technologies—from CT scans to isotope sclerochronology—may only be possible in laboratory settings, and many such techniques may become available only long after the fossils are first seen in the field. Fourth, the problems of paleontology change through time: new research questions beget the need for new kinds of observations of previously made collections.

Summary

The evidential value of specimens in collections is large and greatly underdetermined, highlighting the importance of their untapped potential in relation to existing and future research opportunities in paleontology. The nature of the relationship between fossils and the information derived from them has always been complex, but never static. We are always finding new ways to use old collections.

From Fossils to the Paleobiology Database

David Sepkoski (2013) recently argued persuasively that—although paleontology as a science has been "data-driven" almost since its beginnings in the early nineteenth century—there has been a major conceptual shift in the field over the past half-century, particularly "in the way paleontologists have understood the nature of their data and the kinds of knowledge derivable from it: a shift from imagining the record as a visual collection of objects that can be organized in physical or virtual space (the specimen cabinet or the illustrated catalog) to an abstract, randomly accessible collection of data points (a database) existing in information 'cyberspace'" (D. Sepkoski, 2013, p. 439). This recent shift has sharpened the distinction of what the fossil record "is," in both an epistemological sense, as the source of knowledge claims about the history of life, and in an ontological sense, as a collection not only of physical specimens but of "pure data" (D. Sepkoski, 2013, p. 435).[7]

A prime example of this changing view of data in paleontology is "quantitative paleobiology" (sensu Gilinsky and Signor, 1991), which arose in the 1970s and rapidly developed in the 1980s to address some of the largest questions about the history of life, many of which were tied to documentation of Phanerozoic-scale biodiversity patterns (e.g., J. Sepkoski et al., 1981; for historical review, see D. Sepkoski, 2012). The data for these studies—typically compendia of family- (J. Sepkoski, 1982) or genus-level (J. Sepkoski, 2002) taxon durations—were compiled largely from the published systematic literature, rather than directly from specimens. The tradition of using abstracted data from the literature to discover new insights about the history of life continues today. Its most important modern manifestation is the Paleobiology Database (hereafter, PaleoDB[8]), which has become an integral part of the landscape of paleontological research. Begun in the late 1990s, the PaleoDB is the largest single source of paleontological data ever compiled and has offered rich new insights about the history of life (to date, nearly 300 published papers have utilized the PaleoDB as a source of data). Although some of the PaleoDB information is derived from the original fieldwork and taxonomic studies of contributors, the majority of the data have been extracted from published literature (Peters et al., 2014). Various attempts are now under way to greatly expand the reach of the literature that is summarized in the PaleoDB, mainly by advanced information technology (see Peters et al., 2014, for discussion).

As valuable as the PaleoDB has become for discovering major features of life's history (e.g., Uhen et al., 2013), abstracted data compilations from the literature are ultimately finite and

[7]Recently, Tamborini (2015) has drawn attention to another basic philosophical division in paleontological "data." By way of a historical analysis of German paleontology in the late nineteenth century, Tamborini distinguishes between what would later be called the "ideographic" approach—in which paleontology's primary goal is to describe the morphological features of extinct organisms—and the "nomothetic" approach—using paleontological data to generate and test large general theories of evolutionary pattern and causation. We here take as given that the "nomothetic approach" is the appropriate goal of modern paleobiology but agree with Gould (1980), who promulgated these terms in the field, that "[i]deographic factors determine the parameters and then enter as boundary conditions into a nomothetic model" (Gould, 1980, p. 115, *in* Tamborini, 2015, p. 15).

[8]http://paleobiodb.org

dependent wholly on the research interests of the paleontologists who captured primary data through observation of fossil specimens. It is easy to criticize any compilation of information from the abstracted fossil record as incomplete or inadequate (just as it is easy to criticize the physical fossil record as incomplete; Kidwell and Flessa, 1996). Since at least the beginning of the "paleobiological revolution" of the 1970s (D. Sepkoski, 2012), there has been criticism of using large databases to make conclusions about large-scale patterns in the history of life. Skeptics have argued that such analyses make use of data that are not accurate or complete enough, or are otherwise inappropriate, for the purposes to which it has been put (see, for example, Ager, 1988; Boucot, 1990; Prothero and Liter, 2008; Gaskell, 2012, p. 88; Prothero, 2015; and other examples discussed in Raup, 1986, p. 104–105; J. Sepkoski, 1996; Benton, 1999; Miller, 2000, p. 55–56). Advocates of using such data compilations, however, have countered that the point is not the absolute completeness of the information but the shape of the patterns that large quantities of data reveal, even in the face of data error, as long as that error does not contain substantial directional bias; these patterns have been repeatedly tested internally and against supposedly independent data sources (e.g., J. Sepkoski, 1996; Benton, 1999; Adrain and Westrop, 2000; Miller, 2000; Alroy, 2003, 2010; Wagner et al., 2007; Mondal and Harries, 2015). These authors argue that, at least for some major patterns, large-scale patterns have become sufficiently well supported that active hypothesis testing of whether they are broadly correct is effectively finished.

The main argument in support of this conclusion is that additions or corrections to large paleontological databases usually do not significantly alter the large-scale patterns derived from those databases (e.g., J. Sepkoski, 1992, 1996; Benton, 1999; Miller 2000). Yet despite the apparently general consensus that "large-scale" (e.g., global, Phanerozoic) patterns derived from such literature-based databases are robust, questions continue about the utility of these large databases for producing unbiased results about patterns at *smaller* temporal, geographic, taxonomic, and ecological scales. For instance, Vermeij and Leighton (2003), citing examples from mollusks to mammals, argued that the blunt instrument of global diversity analyses unavoidably masks important variation among habitats, regions, and clades. Further, global, Phanerozoic-scale paleobiological analyses often employ higher taxa (e.g., genera) as units of analysis. Such higher taxa—which by their nature are arbitrarily circumscribed—may not always be suitable proxies for species, which in turn are fundamental to understanding many ecological and evolutionary processes (e.g., Hendricks et al., 2014). Regardless of their "reality" in nature, species in practice are defined by specimens (see above), and many of their most interesting ecological and evolutionary qualities are derived from the occurrences and distributions of those specimens in space and time. Even so, published descriptions and other accounts of many species are based upon only a small subset of the specimens that have been collected. New kinds of data that can be linked back to both these published specimens and those that have been collected, but not yet scientifically studied, are needed to better understand linkages—and identify potential disjunctions—between species-level patterns observed at regional scales (ranging from geographic ranges and temporal durations to ecological interactions) with those observed for higher taxa at larger spatiotemporal scales.

PALEONTOLOGY'S "BIG DATA" FUTURE

Which species interactions are most likely to escalate as the climate warms over the next century? How can paleontologists identify species at risk of extinction before they become extinct? When is an ecosystem likely to reach a potentially disastrous tipping point? When will evolution make an invading species spread even faster? Searching for answers to these questions, and many, many more (Seddon et al., 2014), across massively sized data sets (i.e., "big data") has the potential to reveal patterns, trends, and correlations previously unknown, inspiring questions and generating hypotheses (McCulloch, 2013)[9] that could lead to a more "predictive" paleontology.

Our main argument here is that the greatest potential source of new data in paleontology comes not from the corpus of existing literature, but from specimens themselves. Worldwide, paleontological collections hold many tens (probably hundreds) of millions of individual fossils (Allmon, 1997, 2000). Most of these are associated with geographic, stratigraphic, and taxonomic information, as well as data unique to biological individuals, including size, shape, morphological features associated with ontogenetic stage, indicators of taphonomic history, and sometimes records of environmental history (e.g., isotopic data) and past ecological interactions (e.g., predation scars). The data potentially available in these collections, therefore, exist at two levels: the "traditional" specimen-related data of geography, stratigraphy, and taxonomy, and more derivative data, such as chemistry, morphology, taphonomy, or ecology. Furthermore, new kinds of observations and data are invented as new questions and technologies emerge. Each specimen thereby presents an array of potential information for addressing a diversity of paleontological research questions.

Combining such data from millions of specimens in museum collections, and then integrating them with other unique "big data" resources (e.g., high-resolution paleoclimate and paleocean models, paleogeographic reconstructions, genomic data, etc.), has the potential to lead to the most important paleontological discoveries of the twenty-first century. Acquiring and digitizing such large quantities of specimen-based data, mobilizing them

[9]"Big data" research has been criticized for placing too much emphasis on data mining and statistical correlations (Leonelli, 2014). Such concerns reflect deeper debates about the proper role of hypotheses in research that have challenged the scientific community since the seventeenth century (Elliott et al., 2016). Here, we follow Elliott et al. (2016) in thinking of the scientific method as an iteration between both observationally driven and hypothesis-driven modes of investigation.

for dissemination (i.e., making data "travel"[10] so that they can be reused), and linking these data to other ecological and environmental data would be a massive undertaking, but we encourage paleontologists to think imaginatively about what new questions might be answered if—at least in theory—every aspect of every specimen sitting in a museum drawer right now could be captured, digitized, and made freely available for study.

FOUR CHALLENGES

Some steps toward a specimen-based "big data" science in paleontology have already been taken. The Integrated Digitized Biocollections database (iDigBio[11]; see overviews in Page et al., 2015; MacFadden and Guralnick, 2017) has been developed based on data derived directly from specimens themselves, rather than literature sources. Identified museum specimens of known stratigraphic provenance are georeferenced (i.e., their geographic coordinates are determined) and entered into the databases of individual museums before being served to iDigBio; photographs of the fossils themselves are sometimes also captured. These specimen data are then aggregated by iDigBio into a single, online database. New software tools are also being developed to connect the specimen-based neontological and paleontological data aggregated by iDigBio with the literature-based data available in the PaleoDB. For instance, the Enhancing Paleontological and Neontological Data Discovery[12] (ePANDDA) application programming interface (API) enables researchers to access and search these two databases simultaneously, resulting in a more comprehensive data set. Despite their great promise, however, these efforts are still in their infancy, and many challenges remain to realize the full potential of "big data" paleontology. These challenges boil down to the necessity of much larger quantities and greater diversity of high-quality data, as well as greater engagement of taxonomic experts, both of which obviously require much larger investment in collections and human resources.

1. Comprehensiveness

If paleontology is to become a "big data" science, greater attention and energy must be invested in specimens themselves. iDigBio currently contains over 4.6 million individual fossil specimen records, and more than 330,000 of these include additional media such as digital images. Yet this large data set is small in size in comparison to its potential scope, because the vast majority of fossil specimens in collections have not yet been digitized (see Challenge #4). Compounding the problem of small sample size is the reality that specimen data in iDigBio (or any large database) represent a highly selected fraction of what is potentially available in museum collections. All taxonomic groups, geographic regions, and stratigraphic intervals are not adequately represented. Furthermore, many informative aspects of individual specimens are also not included. For instance, the specimen-based data that have been most successfully included into large databases, such as iDigBio, are mainly taxonomic, geographic and/or stratigraphic in nature, using highly standardized vocabularies (Wieczorek et al., 2012). Heterogeneous data that document other morphological, taphonomic, ecological, or geochemical aspects of fossil specimens, however, are typically less standardized in their format, which has made their integration into large databases a challenge (Brewer et al., 2012; see Pauli et al., 2017, for an example with isotopic data). Such bias in large databases exists because what is included in them is not the result of "well-documented scientific choices.... Rather, it is the serendipitous result of social, political, economic, and technical factors, which determine which data get to travel" (Leonelli, 2016, p. 166). Unless paleontology finds a way to improve the comprehensiveness of its "big data," databases that are limited in scope to selected published data or individual museum collections will continue to provide what Leonelli (2016, p. 166) aptly described as a "privileged dissemination platform for a minority of irrationally selected datasets," which encourages an "inherently conservative and implicitly partial platform to discovery."

2. Quality and Reliability of Taxonomic Identifications

Deriving biological meaning from the fossil record—whether viewed as the world's collection of physical specimens or abstracted data that summarize particular aspects of those specimens—ultimately depends on high levels of trust (confidence) among the individual scientists involved. A concrete example is the issue of recognition (and differentiation) of discrete units of evolutionary lineages that many evolutionary biologists would recognize as species (Hendricks et al., 2014; Allmon, 2016). In terms of information content, taxonomic identifications are the metadata that bind specimens both within and across collections, just as a manufacturer's number binds the same product sold by different retailers. Put another way, a taxonomic identification attributed to a fossil specimen is a hypothesis about its phylogenetic position and evolutionary history. Specimen identifications are—along with spatiotemporal context—therefore essential for reconstructing the history of life. Conversely, although identified specimens lacking geographic and temporal context are typically deaccessioned and removed from research collections due to lack of any potential evidential value, those bearing spatiotemporal context—but lacking identifications—are usually stored and protected by museums for decades on end as they await experts to assign them names. Indeed, a digitized specimen record bearing detailed geographic, temporal, ecological, and taphonomic data—but lacking any level of identification—is nearly useless for analyses (even a fossil recognized as "problematic" or *incertae sedis* excludes other possibilities and therefore has some

[10]The dissemination of massive amounts of data is a defining characteristic of the epistemology of "big data" science (Leonelli, 2016). Therefore, the metaphor of data "journey" used by Leonelli (2016, p. 39) is adopted here to designate the "movement of scientific data from their production site to many other sites within or beyond the same field of research."

[11]www.idigbio.org

[12]https://epandda.org/

value). Although this point may seem obvious (just as it is obvious that we could not reconstruct the history of human civilization without the names of key individuals and information about when and where they lived), the problem of taxonomic identifications is not trivial. One of the greatest challenges for paleontology in the age of "big data" is the same challenge that has always vexed museum collections: attaining correct identifications for individual fossil specimens.

A large proportion of the fossil specimens currently residing in museum collections are not identified to the species level, and many that have been assigned names are misidentified. The effort required to collect specimens in the field is often much less than the effort needed to identify those specimens, especially at lower taxonomic levels. Generally, however, names are assigned to specimens—often long after they were first collected—by comparing them with images in printed literature. In this respect, specimen-based paleontology remains mired in the technology of the nineteenth century. Many fossil species have been figured only a small number of times, frequently in images of poor quality by modern standards, in publications that are out of print or otherwise difficult to attain (though digital scans made available through initiatives such as the Biodiversity Heritage Library[13] are now making some of these works more widely accessible). Comparisons between specimens and images in the printed literature available on hand may result in identifications that are correct at higher taxonomic levels but erroneous at lower levels. Paleontology needs new, free digital resources to help both professional and avocational paleontologists identify fossils; the new Digital Atlas of Ancient Life[14] project presents one example for how this might be accomplished (Hendricks et al., 2015).

Misidentifications of fossils and nomenclatural conflicts (e.g., synonymous species names or differences of opinion concerning generic assignments) often result in significant quantities of "messy" taxonomic data in large databases compiled from specimen-based data. Paleontology desperately needs resources such as the World Registry of Marine Species (WoRMS[15]), which seeks "to provide an authoritative and comprehensive list of names of marine organisms, including information on synonymy."[16] Although the PaleoDB allows for entry and accumulation of taxonomic opinion,[17] it was never intended to serve as a comprehensive nomenclatural platform for paleontology; rather, it offers a helpful but incomplete reflection of the taxonomic opinions contained within the literature. The further development of a single common listing could provide nomenclatural consistency across museum collections and tie into larger specimen-based databases, allowing universal updates to taxonomic names where necessary. With expanded active involvement of taxonomic experts, the PaleoDB could potentially evolve to meet this need, but development of a dedicated taxonomic database for paleontology is also worthy of consideration.[18]

Mayer-Schönberger and Cukier (2013) rejected the epistemological significance of such "messiness" of large databases, claiming that "big data" is self-correcting because of its diversity and variability. In a world of small data, reducing bias and errors and ensuring high quality of data are essential. Large data sets, however, make it likely that errors are automatically reduced by a process philosophers call "triangulation"—the tendency of reliable data to cluster together (Wylie, 2002; Leonelli, 2014). We agree, however, with Leonelli (2014) that if data sources share the same bias, it is also possible that bias will be amplified, rather than reduced, especially when a database is far from being comprehensive, which, as discussed above, is presently the case in paleontology—and will be for the foreseeable future.

3. Lack of Engagement by Researchers in Data Curation Activities

Realizing the full evidential value of museum collections in "big data" science requires the support and cooperation of the broader paleontological research community. Such support, however, is currently not encouraged because the scientific credit system typically does not place a high value on curatorial efforts by researchers to improve museum collections or the contribution by researchers to museum specimen collections and data outside of the publications stemming from research. For example, a 2002 National Research Council (NRC) report on the status and future of geoscience collections specifically recommended development of systems for rewarding care and use of collections, but little progress has been made to date (NRC, 2002). Many "curators" in large museums today perform little actual curation—now usually called "collections management"—activity; they are instead researchers who may or may not use, or even know much about, the collections under their nominal supervision. Incentivizing researchers to engage more with collections could help address this challenge.

4. Availability of Funding to "Digitize Everything"

Huge amounts of labor are needed to make data "travel" (i.e., to enable it to be reused and gain new evidential value) (Leonelli, 2016), starting with gathering the basic data in digital form. Funding for digital database creation and curation, however, remains scarce relative to the manifest need. Some crude calculations can provide a sense of the scope of this challenge. It is likely that the 20 or so largest museum collections of invertebrate fossils in the United States contain collectively *at least* 75 million specimens (Allmon and White, 2000). If we assume that ~5% of these specimens are currently represented by digital records, that ~50% are

[13]https://www.biodiversitylibrary.org/
[14]http://www.digitalatlasofancientlife.org
[15]http://www.marinespecies.org/
[16]http://www.marinespecies.org/about.php
[17]paleobiodb.org/data1.2/opinions_doc.html

[18]We recognize that there is a certain irony in the fact that a discipline sometimes pejoratively characterized as "stamp collecting" has no comprehensive listing of fossil species.

already identified, and that it costs approximately US$1 to digitize each specimen, we can then estimate that it would cost perhaps US$35 million to digitize all of the currently *identified* fossil specimens, and more than US$75 million to digitize all fossil specimens in U.S. collections.[19] By comparison, the U.S. National Science Foundation's 2017 budget for digitization of its Advancing Digitization of Biological Collections (ADBC) program for *all* natural history collections was US$10 million.[20]

These numbers make it obvious that digitizing all or even most fossils in museum collections is highly unlikely to occur in the near future. If this is true, then other questions arise, including how much digitization is sufficient (and for what purpose) and which specimens and associated data will be prioritized over others? These are vital questions for the future of museums, collections, and paleontological research.

CONCLUSIONS

Specimen collections are, have been, and shall forever be the epistemological engine of paleontology, offering effectively inexhaustible evidential value. Indeed, the physical fossil record is paleontology's only path toward a truly "big data" future, which may well present insights that we cannot currently predict and lead us to new questions that we cannot currently imagine asking. While adding more specimens with quality taxonomic and spatiotemporal data to databases such as iDigBio should always be a central digitization priority for museum collections, our discipline's "big data" future also resides in other aspects of the abstracted fossil record that generally go unrecorded in existing online databases (examples include morphological, ecological, taphonomic, and geochemical data). By its nature, the published literature is not designed to and cannot make available such "dark data." Excluding this kind of information risks a conceptual conservatism in the questions we can ask about the fossil record.

Moving our discipline in this direction will require considerable effort and resources. But, there are places where we can begin this transformation now. For example, many collections remain inadequately studied and represented in the large distributed databases that are increasingly important research tools in paleontology. Collections might therefore profitably focus their data curation efforts on areas that could be identified as less well represented in large databases, such as particular geologic ages, taxonomic groups, or geographic regions. Making taxonomy more reliable and consistent in order to increase the fitness of data is another major challenge for collections. Collections care might therefore be focused on taxa that are in particular need of clarification. For collections to maximize their contribution to databases in an age of "big data" science, they must also somehow be mobilized to release more of the information that they contain. This means more than just "putting them online" and should also include getting more researchers into the collections to study them, identify material, and sort what is important from what is not.

The "big data" future of paleontology is in many ways the next logical phase of the exploratory role that natural history museums and their collections have always played. "Data curators" (e.g., Freitas and Curry, 2016; Leonelli, 2016) are now becoming just as important as specimen curators for assembling big data and making it possible to integrate and analyze them. Such expanded roles for data and the people who curate them will not replace museum specimen collections. Instead, they will make these collections even more valuable and widely accessible.

Physical specimens have always formed the first pillar of the fossil record. We suggest that the untold quantities of data that are potentially abstractable from the physical fossil record could provide a twenty-first-century upgrade to the second pillar, which modern paleontology remains balanced near, albeit staidly.

ACKNOWLEDGMENTS

We thank Gary Rosenberg and Renee Clary for inviting us to participate in the symposium and volume; Amanda Schmitt for assistance with references; Erica Clites, Kirk Johnson, Bruce Lieberman, and Austin Hendy for discussion; and Mike Benton and an anonymous reviewer for comments on previous drafts.

REFERENCES CITED

Adrain, J.M., and Westrop, S.R., 2000, An empirical assessment of taxic paleobiology: Science, v. 289, p. 110–112, https://doi.org/10.1126/science.289.5476.110.

Adrain, J.M., and Westrop, S.R., 2003, Paleobiodiversity: We need new data: Paleobiology, v. 29, p. 22–25, https://doi.org/10.1666/0094-8373(2003)029<0022:PWNND>2.0.CO;2.

Ager, D.V., 1988, Extinctions and survivals in the Brachiopoda and the dangers of data bases, *in* Larwood, G.P., ed., Extinction and Survival in the Fossil Record: Oxford, UK, Oxford University Press, p. 89–97.

Allmon, W.D., 1997, Collections in paleontology, *in* Lane, H.R., Lipps, J., Steininger, F.F., and Ziegler, W., eds., Paleontology in the 21st Century Workshop: Kleine Senckenbergreihe, no. 25, p. 155–159.

Allmon, W.D., 2000, Collections, *in* Lane, H.R., Lipps, J., Steininger, F.F., Kaesler, R.L. Ziegler, W., and Lipps, J., eds., Fossils and the Future, Paleontology in the 21st Century: Frankfurt, Senckenberg-Buch, no. 74, p. 203–214.

Allmon, W.D., 2005, The importance of museum collections in paleobiology: Paleobiology, v. 31, p. 1–5.

Allmon, W.D., 2016, Studying species in the fossil record: A review and recommendations for a more unified approach, *in* Allmon, W.D., and Yacobucci, M.M., eds., Species and Speciation in the Fossil Record: Chicago, University of Chicago Press, p. 59–120, https://doi.org/10.7208/chicago/9780226377582.001.0001.

[19]These cost estimates are based on our participation in three paleo-themed "Thematic Collections Network" grants (PaleoNiches, Eastern Pacific Invertebrate Communities of the Cenozoic [EPICC], and Cretaceous World projects) funded by the U.S. National Science Foundation's Advancing Digitization of Biodiversity Collections (ADBC) program. Our experience is that digitization costs are extremely variable among institutions and are surprisingly poorly understood by both institutions and funding agencies. This situation appears to be the result of a combination of differences in work flow, staffing, institutional infrastructure, and details of methods and technology. The "$1 per specimen" is the estimated average cost of digitally capturing "basic label data" and georeferencing for one specimen or set of specimens. Importantly, this estimate does *not* include any imaging of specimens, which our experience suggests costs significantly more.

[20]https://www.nsf.gov/pubs/2015/nsf15576/nsf15576.pdf

Allmon, W.D., and White, R.D., 2000, Introduction, *in* White, R.D., and Allmon, W.D., eds., Guidelines for the Management and Curation of Invertebrate Fossil Collections: Paleontological Society Special Publications, v. 10, p. 1–4.

Alroy, J., 2003, Global databases will yield reliable measures of global biodiversity: Paleobiology, v. 29, p. 26–29, https://doi.org/10.1666/0094-8373(2003)029<0026:GDWYRM>2.0.CO;2.

Alroy, J., 2010, Fair sampling of taxonomic richness and unbiased estimation of origination and extinction rates, *in* Alroy, J., and Hunt, G., eds., Quantitative Methods in Paleobiology: The Paleontological Society Papers, v. 16, p. 55–80.

Benton, M., 1999, The history of life: Large databases in palaeontology, *in* Harper, D.A., ed., Numerical Palaeobiology: Computer-based Modelling and Analysis of Fossils and Their Distributions: New York, John Wiley, p. 249–283.

Benton, M.J., Forth, J., and Langer, M.C., 2014, Models for the rise of the dinosaurs: Current Biology, v. 24, p. R87–R95, https://doi.org/10.1016/j.cub.2013.11.063.

Boersma, A.T., and Pyenson, N.D., 2015, *Albicetus oxymycterus*, a new generic name and redescription of a basal physeteroid (Mammalia, Cetacea) from the Miocene of California, and the evolution of body size in sperm whales: PLoS One, v. 10, p. e0135551, https://doi.org/10.1371/journal.pone.0135551.

Boessenkool, S., Austin, J.J., Worthy, T.H., Scofield, P., Cooper, A., Seddon, P.J., and Waters, J.M., 2009, Relict or colonizer?: Extinction and range expansion of penguins in southern New Zealand: Proceedings of the Royal Society B: Biological Sciences, v. 276, p. 815–821, https://doi.org/10.1098/rspb.2008.1246.

Boucot, A.J., 1990, Phanerozoic extinctions: How similar are they to each other?, *in* Kauffman, E.G., and Walliser, O.H., eds., Extinction Events in Earth History: Berlin, Heidelberg, Lecture Notes in Earth Sciences, v. 30, Springer, p. 5–30, https://doi.org/10.1007/BFb0011131.

Brewer, S., Jackson, S.T., and Williams, J.W., 2012, Paleoecoinformatics: Applying geohistorical data to ecological questions: Trends in Ecology & Evolution, v. 27, p. 104–112, https://doi.org/10.1016/j.tree.2011.09.009.

Callomon, J.H., and Grădinaru, E., 2005, From the thesaurus of the museum collections. I. Liassic ammonites from Munteana (Svinita Zone, Southern Carpathians, Romania): Acta Palaeontologia Romaniae, v. 5, p. 49–65.

Cutler, A., 2003, The Seashell on the Mountaintop: New York, Dutton, 240 p.

Daston, L., 2000, Introduction: The coming into being of scientific objects, *in* Daston, L., ed., Biographies of Scientific Objects: Chicago, University of Chicago Press, p. 1–14.

Daston, L., 2004, Type specimens and scientific memory: Critical Inquiry, v. 31, p. 153–182, https://doi.org/10.1086/427306.

Davidson, J.P., 2008, A History of Paleontology Illustration: Bloomington, Indiana University Press, 217 p.

Davis, E.B., and Pyenson, N.D., 2007, Diversity biases in terrestrial mammalian assemblages and quantifying the differences between museum collections and published accounts: A case study from the Miocene of Nevada: Palaeogeography, Palaeoclimatology, Palaeoecology, v. 250, p. 139–149, https://doi.org/10.1016/j.palaeo.2007.03.006.

Dean, M.T., Owen, A.W., Bowdler-Hicks, A., and Akhurst, M.C., 2010, Discriminating faunal assemblages and their palaeoecology based on museum collections: The Carboniferous Hurlet and Index limestones of western Scotland: Scottish Journal of Geology, v. 46, p. 45–57, https://doi.org/10.1144/0036-9276/01-399.

Dooley, A.C., Jr., Heckert, A.B., Fraser, N.C., Byrd, C.J., Vodden, R., and Beard, J., 2015, Salvage paleontology at the Virginia Solite Quarry (Upper Triassic)—Rescuing a lagerstätte: Geological Society of America Abstracts with Programs, v. 47, no. 2, p. 24.

Elliott, K.C., Cheruvelil, K.S., Montgomery, G.M., and Soranno, P.A., 2016, Conceptions of good science in our data-rich world: Bioscience, v. 66, p. 880–889, https://doi.org/10.1093/biosci/biw115.

Ellis, R., 2008, Rethinking the value of biological specimens: Laboratories, museums and the Barcoding of Life Initiative: Museum & Society, v. 6, no. 2, p. 172–191.

Findlen, P., 1994, Possessing Nature: Museums, Collecting, and Scientific Culture in Early Modern Italy: Berkeley, University of California Press, 449 p.

Freitas, A., and Curry, E., 2016, Big data curation, *in* Cavanillas, J.M., et al., eds., New Horizons for a Data-Driven Economy: New York, Springer, p. 87–118, https://doi.org/10.1007/978-3-319-21569-3_6

Gaskell, I., 2012, Museums and philosophy—Of art, and many other things, Part II: Philosophy Compass, v. 7, no. 2, p. 85–102, https://doi.org/10.1111/j.1747-9991.2011.00469.x.

Gilinsky, N.L., and Signor, P.W., eds., 1991, Analytical Paleobiology: The Paleontological Society, Short Courses in Paleontology, no. 4, 216 p.

Gould, S.J., 1980, The promise of paleobiology as a nomothetic, evolutionary discipline: Paleobiology, v. 6, no. 1, p. 96–118.

Haack, S., 2007, Defending Science—Within Reason. Between Scientism and Cynicism: Amherst, New York, Prometheus Books, 411 p.

Harnik, P.G., 2009, Unveiling rare diversity by integrating museum, literature, and field data: Paleobiology, v. 35, no. 2, p. 190–208, https://doi.org/10.1666/07062.1.

Harnik, P.G., Serb, J.M., and Sherratt, E., 2017, Morphological evolution following the closure of the Central American Seaway: Geological Society of America Abstracts with Programs, v. 49, no. 6, https://doi.org/10.1130/abs/2017AM-301854.

Heim, N.A., and Peters, S.E., 2011, Regional environmental breadth predicts geographic range and longevity in fossil marine genera: PLoS One, v. 6, no. 5, p. e18946.

Hein, H.S., 2000, The Museum in Transition. A Philosophical Perspective: Washington, D.C., Smithsonian Institution Press, 203 p.

Hendricks, J.R., Saupe, E.E., Myers, C.E., Hermsen, E.J., and Allmon, W.D., 2014, The generification of the fossil record: Paleobiology, v. 40, no. 4, p. 511–528.

Hendricks, J.R., Stigall, A.L., and Lieberman, B.S., 2015, The Digital Atlas of Ancient Life: Delivering information on paleontology and biogeography via the web: Palaeontologia Electronica, v. 18, no. 2, p. 1–9.

Hunter, A., and Donovan, S., 2005, Field sampling bias, museum collections and completeness of the fossil record: Lethaia, v. 38, no. 4, p. 305–314, https://doi.org/10.1080/00241160500289559.

Impey, O., and MacGregor, A., 1985, The Origins of Museums: The Cabinet of Curiosities in Sixteenth- and Seventeenth-Century Europe: Oxford, UK, Oxford University Press, 335 p.

Jackson, J.B., and Johnson, K.G., 2001, Measuring past biodiversity: Science, v. 293, p. 2401–2404, https://doi.org/10.1126/science.1063789.

Kidwell, S.M., and Flessa, K.W., 1996, The quality of the fossil record: Populations, species, and communities: Annual Review of Earth and Planetary Sciences, v. 24, no. 1, p. 433–464, https://doi.org/10.1146/annurev.earth.24.1.433.

Knell, S.J., 1997, What's important?, *in* Pettitt, C.W., and Nudds, J.R., eds., The Value and Valuation of Natural Science Collections: London, the Geological Society, p. 11–16.

Knell, S.J., 2000, The Culture of English Geology 1815–1851: A Science Revealed through Its Collecting: Aldershot, UK, Ashgate, 377 p.

Kundrát, M.L.W., and Ebbestad, J.O.R., 2015, New tooth of Peking Man recognized in laboratory at Uppsala University: Acta Anthropologica Sinica, v. 34, no. 1, p. 1–14.

Leonelli, S., 2014, What difference does quantity make?: On the epistemology of Big Data in biology: Big Data & Society, April–June 2014, p. 1–11.

Leonelli, S., 2016, Data-Centric Biology: A Philosophical Study: Chicago, University of Chicago Press, 275 p.

Lindkvist, M., 2016, The importance of curation: A case-study of the subfossil lemur collection in the Museum of Evolution [Ph.D. thesis]: Uppsala, Sweden, Uppsala University, 84 p.

Liow, L.H., 2007, Does versatility as measured by geographic range, bathymetric range, and morphological variability contribute to taxon longevity?: Global Ecology and Biogeography, v. 16, p. 117–128, https://doi.org/10.1111/j.1466-8238.2006.00269.x.

Lord, B., 2006, Philosophy and the museum: An introduction to the special issue: Museum Management and Curatorship, v. 21, p. 79–87, https://doi.org/10.1080/09647770600102102.

MacFadden, B.J., and Guralnick, R.P., 2017, Horses in the cloud: Big data exploration and mining of fossil and extant *Equus* (Mammalia: Equidae): Paleobiology, v. 43, no. 1, p. 1–14, https://doi.org/10.1017/pab.2016.42.

Mayer-Schönberger, V., and Cukier, K., 2013, Big Data: A Revolution That Will Transform How We Live, Work and Think: New York, Houghton Mifflin Harcourt, 242 p.

Mayr, E., 1969, Principles of Systematic Zoology: New York, McGraw-Hill, 434 p.

Mayr, E., 1989, Attaching names to objects, *in* Ruse, M., ed., What the Philosophy of Biology Is: Essays for David Hull: Dordrecht, Netherlands, Kluwer Academic, p. 235–243, https://doi.org/10.1007/978-94-009-1169-7_12.

McCulloch, E.S., 2013, Harnessing the power of big data in biological research: Bioscience, v. 63, no. 9, p. 715–716, https://doi.org/10.1093/bioscience/63.9.715.

Miller, A.I., 2000, Conversations about Phanerozoic global diversity, *in* Erwin, D.H., and Wing, S.L., eds., Deep Time. Paleobiology's Perspective: Paleobiology 26 (Suppl. to No. 4), p. 53–73.

Mondal, S., and Harries, P.J., 2015, The effect of taxonomic corrections on Phanerozoic generic richness trends in marine bivalves with a discussion of the clade's overall history: Paleobiology, v. 42, no. 1, p. 157–171, https://doi.org/10.1017/pab.2015.35.

National Research Council (NRC) (Committee on the Preservation of Geoscience Data and Collections), 2002, Geoscience Data and Collections: National Resources in Peril: Washington, D.C., The National Academies Press, 107 p.

Page, L.M., MacFadden, B.J., Fortes, J.A., Soltis, P.S., and Riccardi, G., 2015, Digitization of biodiversity collections reveals biggest data on biodiversity: Bioscience, v. 65, no. 9, p. 841–842, https://doi.org/10.1093/biosci/biv104.

Parham, J.F., Donoghue, P.C.J., Bell, C.J., Calway, T.D., Head, J.J., Holroyd, P.A., Inoue, J.G., Irmis, R.B., Joyce, W.G., Ksepka, D.T., Patané, J.S.L., Smith, N.D., Tarver, J.E., van Tuinen, M., Yang, Z., Angielczyk, K.D., Greenwood, J.M., Hipsley, C.A., Jacobs, L., Makovicky, P.J., Müller, J., Smith, K.T., Theodor, J.M., Warnock, R.C.M., and Benton, M.J., 2012, Best practices for justifying fossil calibrations: Systematic Biology, v. 61, no. 2, p. 346–359, https://doi.org/10.1093/sysbio/syr107.

Pauli, J.N., Newsome, S.D., Cook, J.A., Harrod, C., Steffan, S.A., Backer, C.J.O., Ben-David, M., Bloom, D., Bowen, G.J., Cerling, T.E., Cicero, C., Cook, C., Dohm, M., Dharampal, P.S., Graves, G., Gropp, R., Hobson, K.A., Jordan, C., MacFadden, B., Pilaar Birch, S., Poelen, J., Ratnasingham, S., Russell, L., Stricker, C.A., Uhen, M.D., Yarnes, C.T., and Hayden, B., 2017, Why we need a centralized repository for isotopic data: Proceedings of the National Academy of Sciences of the United States of America, v. 114, no. 12, p. 2997–3001, https://doi.org/10.1073/pnas.1701742114.

Peters, S.E., Zhang, C., Livny, M., and Ré, C., 2014, A machine reading system for assembling synthetic paleontological databases: PLoS One, v. 9, p. e113523, https://doi.org/10.1371/journal.pone.0113523.

Pinna, G., 2000, A philosophy for natural history museums, *in* Ghiselin, M.T., and Leviton, A.E., eds., Cultures and Institutions of Natural History: San Francisco, California Academy of Sciences, p. 333–337.

Powell, M.G., 2007, Geographic range and genus longevity of late Paleozoic brachiopods: Paleobiology, v. 33, no. 4, p. 530–546, https://doi.org/10.1666/07011.1.

Prothero, D.R., 2015, Garbage in, garbage out: The effect of immature taxonomy on database compilations of North American fossil mammals, *in* Sullivan, R.M., and Lucas, S.G., eds., Fossil Record 4: New Mexico Museum of Natural History and Science Bulletin, v. 68, p. 257–264.

Prothero, D.R., and Liter, M.R., 2008, Systematics of the dromomerycines and aletomerycines (Artiodactyla: Palaeomerycidae) from the Miocene and Pliocene of North America, *in* Lucas, S.G., Morgan, G.S., Spielmann, J.A. and Prothero, D.R., eds., Neogene Mammals: New Mexico Museum of Natural History and Science Bulletin, v. 44, p. 273–298.

Raup, D.M., 1986, The Nemesis Affair: A Story of the Death of Dinosaurs and the Ways of Science: New York, W.W. Norton, 220 p.

Rudwick, M.J.S., 1976, The Meaning of Fossils: Episodes in the History of Palaeontology: 2nd edition: Chicago, University of Chicago Press, 287 p.

Rudwick, M.J.S., 2000, Georges Cuvier's paper museum of fossil bones: Archives of Natural History, v. 27, no. 1, p. 51–68, https://doi.org/10.3366/anh.2000.27.1.51.

Schilthuizen, M., Vairappan, C.S., Slade, E.M., Mann, D.J., and Miller, J.A., 2015, Specimens as primary data: Museums and 'open science': Trends in Ecology & Evolution, v. 30, no. 5, p. 237–238, https://doi.org/10.1016/j.tree.2015.03.002.

Seddon, A.W.R., Mackay, A.W., Baker, A.G., Birks, H.J.B., Breman, E., Buck, C.E., Ellis, E.C., Froyd, C.E., Gill, J.L., Gillson, L., Johnson, E.A., Jones, V.J., Juggins, S., Macias-Fauria, M., Mills, K., Morris, J.L., Nogués-Bravo, D., Punyasena, S.W., Roland, T.P., Tanentzap, A.J., Willis, K.J., Aberhan, M., van Asperen, E.N., Austin, W.E.N., Battarbee, R.W., Bhagwat, S., Belanger, C.L., Bennett, K.D., Birks, H.H., Bronk Ramsey, C., Brooks, S.J., de Bruyn, M., Butler, P.G., Chambers, F.M., Clarke, S.J., Davies, A.L., Dearing, J.A., Ezard, T.H.G., Feurdean, A., Flower, R.J., Gell, P., Hausmann, S., Hogan, E.J., Hopkins, M.J., Jeffers, E.S., Korhola, A.A., Marchant, R., Kiefer, T., Lamentowicz, M., Larocque-Tobler, I., López-Merino, L., Liow, L.H., McGowan, S., Miller, J.H., Montoya, E., Morton, O., Nogué, S., Onoufriou, C., Boush, L.P., Rodriguez-Sanchez, F., Rose, N.L., Sayer, C.D., Shaw, H.E., Payne, R., Simpson, G., Sohar, K., Whitehouse, N.J., Williams, J.W., and Witkowski, A., 2014, Looking forward through the past: Identification of 50 priority research questions in palaeoecology: Journal of Ecology, v. 102, p. 256–267, https://doi.org/10.1111/1365-2745.12195.

Sepkoski, D., 2012, Rereading the Fossil Record: The Growth of Paleobiology as an Evolutionary Discipline: Chicago, University of Chicago Press, 440 p., https://doi.org/10.7208/chicago/9780226748580.001.0001.

Sepkoski, D., 2013, Towards "a natural history of data": Evolving practices and epistemologies of data in paleontology 1800–2000: Journal of the History of Biology, v. 46, p. 401–444, https://doi.org/10.1007/s10739-012-9336-6.

Sepkoski, D., 2017, The Earth as archive: Contingency, narrative, and the history of Life, *in* Daston, L., ed., Science in the Archives: Pasts, Presents, Futures: Chicago, University of Chicago Press, p. 53–83.

Sepkoski, J.J., Jr., 1982, A compendium of marine animal families: Milwaukee Public Museum Contributions to Biology and Geology, v. 51, no. 1, p. 1–125.

Sepkoski, J.J., Jr., 1992, Phylogenetic and ecologic patterns in the Phanerozoic history of marine biodiversity, *in* Eldredge, N., ed., Systematics, Ecology, and the Biodiversity Crisis: New York, Columbia University Press, p. 77–100.

Sepkoski, J.J., Jr., 1996, Patterns of Phanerozoic extinction: A perspective from global data bases, *in* Walliser, O., ed., Global Events and Event Stratigraphy in the Phanerozoic: Berlin, Springer, p. 35–51, https://doi.org/10.1007/978-3-642-79634-0_4.

Sepkoski, J.J., Jr., 2002, A compendium of fossil marine animal genera: Bulletins of American Paleontology, no. 363, 560 p.

Sepkoski, J.J., Jr., Bambach, R.K., Raup, D.M., and Valentine, J.W., 1981, Phanerozoic marine diversity and the fossil record: Nature, v. 293, p. 435–437, https://doi.org/10.1038/293435a0.

Snijders, C., Matzat, U., and Reips, U.D., 2012, Big data: Big gaps of knowledge in the field of internet science: International Journal of Internet Science, v. 7, p. 1–5.

Stone, R., 2013, Salvage paleontology on the seaway: Science, v. 341, no. 6143, p. 232, https://doi.org/10.1126/science.341.6143.232.

Sunderland, M.E., 2016, Specimens and collections, *in* Lightman, B., ed., A Companion to the History of Science: West Sussex, UK, John Wiley & Sons Ltd., p. 488–499.

Tamborini, M., 2015, The constitution of paleobiological data [Ph.D. dissertation]: Germany, University of Heidelberg, 264 p.

Teichert, C., Sweet, W.C., and Boucot, A.J., 1987, The unpublished fossil record: Implications: Senckenbergiana Lethaea, v. 68, no. 1–4, p. 5–19.

Uhen, M.D., Barnosky, A.D., Bills, B., Blois, J., Carrano, M.T., Carrasco, M.A., Erickson, G.M., Eronen, J.T., Fortelius, M., Graham, R.W., and Grimm, E.C., 2013, From card catalogs to computers: Databases in vertebrate paleontology: Journal of Vertebrate Paleontology, v. 33, no. 1, p. 13–28.

Vermeij, G.J., and Leighton, L.R., 2003, Does global diversity mean anything?: Paleobiology, v. 29, p. 3–7, https://doi.org/10.1666/0094-8373(2003)029<0003:DGDMA>2.0.CO;2.

Wagner, P.J.M., Aberhan, M., Hendy, A., and Kiessling, W., 2007, The effects of taxonomic standardization on sampling-standardized estimates of historical diversity: Proceedings of the Royal Society B: Biological Sciences, v. 274, no. 1608, p. 439–444, https://doi.org/10.1098/rspb.2006.3742.

Waterston, C.D., 1979, The unique role of the curator in palaeontology, *in* Bassett, M.G., ed., Curation of Palaeontological Collections: Special Papers in Palaeontology, no. 22, p. 7–15.

Wieczorek, J., Bloom, D., Guralnick, R., Blum, S., Doring, M., Giovanni, R., Robertson, T., and Vieglais, D., 2012, Darwin Core: An evolving community-developed biodiversity data standard: PLoS One, v. 7, no. 1, p. e29715, https://doi.org/10.1371/journal.pone.0029715.

Winston, J.E., 2007, Archives of a small planet: The significance of museum collections and museum-based research in invertebrate taxonomy: Zootaxa, v. 1668, p. 47–54.

Wylie, A., 2002, Thinking from Things: Essays in the Philosophy of Archeology: Berkeley, University of California Press, 339 p.

Manuscript Accepted by the Society 30 November 2017
Manuscript Published Online 25 May 2018

Printed in the USA

The Geological Society of America
Special Paper 535

Filippo Buonanni and the Kircher Museum

Ashley J. Inglehart
*Department of History and Philosophy of Science, Indiana University,
Ballentine Hall 644, 1020 E. Kirkwood Avenue, Bloomington, Indiana 47405, USA*

ABSTRACT

This paper investigates the role played by Italian naturalist, Filippo Buonanni (1638–1723), in the reorganization of the Kircher Museum. Specifically, it considers Buonanni's unique philosophy of nature, which can be gleaned from his public debates about spontaneous generation and fossil formation. This paper argues that Buonanni rejected a conformity of causes in nature, which in turn informed his approach to natural history. His unique approach helped to transform Kircher's collection from a cabinet of wonders to one of the earliest examples of the modern museum.

INTRODUCTION

The Kircher Museum (Rome, 1651–1773) has remained the subject of fascination and inquiry for centuries, and in recent decades, there has been a renaissance of literature investigating its namesake's influence on chemistry, music, and natural philosophy. Less well known, however, is that Kircher's student—Filippo Buonanni—would go on to save the collection from obscurity following Kircher's death in 1680. With the notable exception of Paula Findlen's *Possessing Nature* (Findlen, 1996), Buonanni's stewardship of the Kircher Museum has been all but forgotten. Instead, Buonanni is generally relegated to a passing footnote in conversations on the generation of insects.

In between the death of Athanasius Kircher (1602–1680) and Buonanni's subsequent accession of the museum in 1698, Buonanni became involved in very public disputes concerning spontaneous generation. History has generally considered him to be on the losing end. This paper argues, however, that a close inspection of Buonanni's public controversies shows that he maintained a more sophisticated philosophy of nature than is typically ascribed to him. It further investigates how these debates about generation inform early modern understandings of fossil formation, methodological assumptions about science more generally, and the curation of museum collections.

FATHER KIRCHER AND HIS COLLECTION OF WONDERS

Father Athanasius Kircher was an eccentric, Jesuit polymath and among the more colorful characters of the seventeenth century. He seems to have had a rather polarizing effect upon those who met him: they either loved him or they hated him. The Royal Society thought him to be gullible and greeted him with immense skepticism, and the eminent German philosopher, Gottfried Leibniz, once lamented that "the man understands nothing" (Cobb, 2006, p. 75–76). And yet, he successfully charmed and impressed a sufficient number of benefactors (Queen Christina of Sweden among them) to procure one of the most successful cabinets of wonder in history.

Kircher was born to a large family in Geisa, Germany, where he was educated at the Jesuit Gymnasium in Fulda. Before his ordination as a Jesuit priest in 1628, Kircher had obtained an ideal Renaissance education. He was fluent in Greek and Latin in addition to studying the humanities, natural science, burgeoning seventeenth-century mathematics, theology, and philosophy. His curiosity of nature and the eagerness to disseminate the knowledge he had obtained of it would last throughout his life. The Thirty Years' War led Kircher out of Germany and into papal territory at Avignon, France, and, eventually, to Rome. There he

Inglehart, A.J., 2018, Filippo Buonanni and the Kircher Museum, *in* Rosenberg, G.D., and Clary, R.M., eds., Museums at the Forefront of the History and Philosophy of Geology: History Made, History in the Making: Geological Society of America Special Paper 535, p. 45–58, https://doi.org/10.1130/2018.2535(04).

would be appointed Professor of Mathematics at the Romanum Collegium in 1638 (Koertge, 2008, p. 374–375; Fletcher et al., 2011, Ch. 1). Here, he is depicted as such in a portrait of him printed in the *Musaeum Celeberrimum* (De Sepibus, 1678), the museum's first catalogue (see Fig. 1).

At the Collegium, Kircher became popular with students, whom he would invite to view his growing collection of artifacts and relics of nature. In 1651, Roman senator, Alfonso Doninno, bequeathed to the local Collegium Romanum, a large collection of fossils, "monuments of antiquity," and portraits. The generous bequest from so venerable a patron placed pressure upon the Collegium and Society of Jesus to create a public space worthy of the display. Kircher was soon after charged with the task of curating their joint collections. The result—Museum Kircherianum—was one of the very first public museums (Findlen, 2004, p. 29–30; Fletcher et al., 2011, p. 184).

Regarding the fossil collection with which he was tasked, Kircher had a good deal to say about their origin. Until recently, Kircher is generally remembered for having misidentified fossils and their origins (Parcell, 2009, p. 70). This characterization, however, is somewhat unfair. To begin, fossils, as understood in the early seventeenth century, were of an altogether different genus. The seventeenth-century conception of fossil included *any* object retrieved from below the Earth's surface, from gemstones to minerals, metals to what we would now call fossils, and virtually any stony object dug from the Earth (Rossi, 1984, p. 6).[1] Of these stony objects, the most problematic to explain were fossilized shells, which (unlike other fossils) had no living analogues. Even more strange was the petrification of the area surrounding them, leaving behind a mere ghostly imprint of their existence. Having no analogue among the living, they were often called *Naturae Ludibria, lusus naturae* or "sports of nature" (Rappaport, 1997, p. 119).

Figure 1. Portrait of Kircher, as printed in Georgio de Sepi's *Musaeum Celeberrimum* (1678).

Stephen J. Gould, in particular, has done an excellent job of evaluating Kircher's paleontological thought to show that Kircher—far from advocating for the inorganic origin of all fossils—has received the "short end of the stick" in the triumphalist historiography of geology. For his own part, Kircher divided fossils into two distinct categories: those that were the remains of actual organisms and those that were minerals that had been made to resemble other organisms (Gould, 2004, p. 215ff.). The former usually included things such as teeth and bones that had been made stony by petrifying waters. The latter, however, required a more complex explanation.

Kircher maintained the commonly held view of his time that a seminal or seed-like power manifested at Creation was responsible for the generation of plants, animals, and even minerals. According to Kircher, God endowed the world with *panspermia,* or the universal seeds of nature, at Creation (Kircher, [1678] r2004, Book 12, p. 347; cf. De Boodt, 1647, p. 22–23. For a detailed account of Kircher's *panspermia,* see Rowland, 2004, p. 191–206). A *seminal spiritus,* the generative power of these invisible seeds, was used by God to create life from chaotic matter and "play[ed] the role of the medium between the Creator and creatures" (Hirai, 2007, p. 78–79). According to this worldview, the invisible seeds have some kind of formative power that is responsible for plants and animals, minerals and crystals alike. Whereas living beings come only from matter that was once animated and are the result of the seminal power of these seeds, minerals are merely the result of a saline power that is merely "akin" to the formative power of seeds (Kircher, 1678, reprinted in 2004, Book 12, p. 346–356. See also Chang, 2011, p. 326; Smith, 2011, p. 224; Hirai, 2016, p. 265). This saline power alludes not to groundwater but to the salt of alchemists, following the tradition of Paracelsian *tria prima* (Chang, 2011, p. 326).

So-called "spontaneous generation" is the result of latent seeds, invisible to the naked eye, which lay hidden until activated by the ambient heat of putrefaction (Kircher, [1678] r2004, Book 12, p. 346ff.). Upon the decay of the previous creature, the emergent form of the seed is then responsible for the production of a different creature altogether. Through this notion of emergent forms, Kircher is able to explain why some minerals are made to resemble so closely other living creatures. Namely, the seed of an animal such as a fish can act upon the inanimate matter from

[1]The Latin *fossilia* is in fact derived from the verb *fodere* meaning "to dig." See *Oxford English Dictionary*: www.oed.com (accessed September 2016).

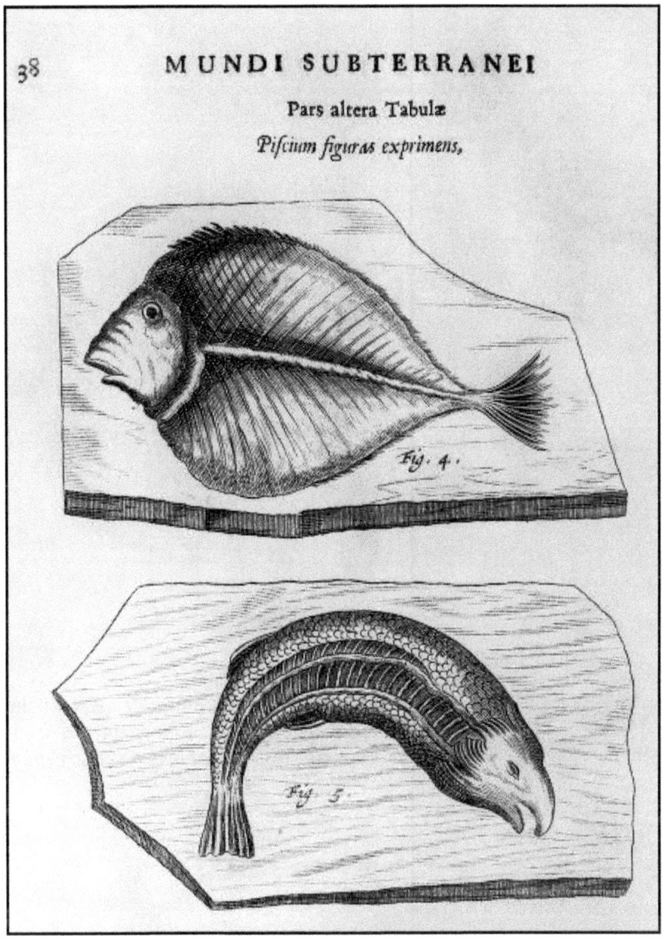

Figure 2. Fossilized fish, from Kircher's *Mundus Subterraneus* ([1665] 2004).

under the Earth's surface, generating a fish-like object in rock or stone (Kircher, [1678] 2004, Book 8, Ch. 2, p. 5–6; Rossi, p. 6–8). Kircher's *Mundus Subterraneus* ([1665] 2004) depicts several examples of these sorts of fossils, in which the form of a fish has been clearly imposed upon the inanimate matter of the rock or stone (see Fig. 2).

In addition to fossils of the sort mentioned above, the early Kircher Museum also boasted several specimens of fossils and artifacts from China, including books by leading Chinese converts (Hsia, 2004, p. 393). Kircher discusses them in his 1667 *China Illustrata* (*China Illustrated*) in which he describes the regional geography and natural history of the area based largely on information from the overseas missions of his fellow Jesuits. Notably, Kircher describes the fossils and stones collected from China as "sports of nature" (Yamanda, 2006, p. 68).[2] The museum of the Collegio Romano, which soon became known as "Kircher's Museum," became "a sort of philosophical gymnasium" under the stewardship of the legendary Jesuit (Sardo, 2004, p. 60). Kircher's Museum became well known for the fantastic elements of its collection throughout Rome and abroad. The space for the museum was a powerful arrangement of vaulted rooms decorated floor to ceiling with allegorical paintings and unusual objects retrieved from around the world. Most prominent of these were the Egyptian obelisks, placed alongside astronomical maps (Hagan, 2008, p. 43).

Just two years before Kircher's death, his disciple at the Collegium, Georgio de Sepi, printed the first account and Catalog of Kircher's Museum (1678). In terms of its organization, the treatise emphasized the more incredible elements of the collection. In particular, the frontispiece alone served as an advertisement for the museum and its marvels. As word of the collection spread, it soon was, in Kircher's own words, "visited by all the nations of the world and a prince cannot become better known in this theatre of the world than to have his likeness here" (Findlen, 1996, p. 641; Sardo, 2004, p. 53). We can catch a glimpse of the breathtaking wonder that the museum must have aroused within its visitors from de Sepi's frontispiece, which illustrates the foyer of the museum (see Fig. 3). When Kircher died in November of 1680, he was nearly 80 years old. His death would leave behind a vacancy that would remain unfilled for nearly 20 years.

FILIPPO BUONANNI AND *RECREATIO*

Filippo Buonanni, the devoted student of Kircher who would succeed him as Professor of Mathematics at the Collegium, differed from his teacher in a number of respects. Unlike Kircher, who hailed from Germany and traveled throughout Europe, Buonanni spent his life in Rome, where his proximity to the sea influenced his ideas about natural history (Findlen, 1996, p. 160). In a more striking contrast, Buonanni lacked the international notoriety and consequential patronage that Kircher had gained (Franceschini, 2008, p. 592). He nonetheless shared Kircher's devotion to Aristotle and commitment to the study of natural history. Shortly after Kircher's death, Buonanni became embroiled in a dispute over the subject of spontaneous generation following the publication of his *Ricreatione dell'occhio e della mente nell'osservazione delle chiocciole* (*Recreation of the Eye and the Mind in the Observation of Snails*) in 1681 (see Fig. 4). While this dispute may initially seem tangential to the nature of fossils, the curation of the museum, or geology more generally, a close analysis can provide a deep insight into the early curation of the Kircher Museum.

Because the treatise has never been translated into English, it has received relatively little attention in the secondary literature. It was nonetheless famed in its own time for its detailed illustrations of seashells and their methodical categorization (Rappaport, 1997, p. 127). Buonanni was truly unique in his devotion to the book's subject matter. He treasured his beloved mollusks. While several texts had already considered the origin of fossils, the problem of generation, or the constitution of the Earth, Buonanni presented the first reference book for mollusks and other testaceous

[2]Curiously, he also introduces the term *geologus* in this text as well, defining the term as "one who investigates stones." See Yamada (2006).

Figure 3. A contemporary view of the Kircher Museum as seen in the frontispiece for *Musaeum Celeberrimum* (1678).

animals.[3] *Ricreatione* boasts hundreds of detailed engravings, each one depicting a uniquely different specimen. The treatise gained an even larger audience in 1684, when a Latin edition was published in Rome by the name of *Recreatio mentis et oculi in observatione animalium testaceorum* (hereafter *Recreatio*).

[3]For earlier examples of publications that examine the origin of fossils, see Robert Hooke's *Micrographia* (1665); Antoine Goudin's *Philosophy, Following the Principles of St. Thomas* ([1668] 1864); Nicolaus Steno's *Prodromus to a Dissertation Concerning Solids Naturally Contained within Solids* (1671); Robert Boyle's *Origin and Virtues of Gems* ([1672] 2000); and Robert Plott's *A Natural History of Oxfordshire* (1677).

Figure 4. *Recreatio* frontispiece (Buonanni, 1684).

Recreatio is organized into four parts, with the majority of the written content in the first and third parts. The first part of the book contains 12 chapters devoted to the origin and causes of testaceous animals, i.e., those animals such as barnacles, mollusks, sponges, and shellfish whose correlative parts are like plants, causing them to lack the sensation and properties of motion shared by most other animals. The second part divides testaceous animals into three classes, and the third part is organized into "problems," following a traditional Scholastic style. Finally, the fourth part, which comprises more than half of the impressive tome, contains several hundred detailed illustrations of Buonanni's collection of seashells. Each class of testaceons has hundreds of detailed, beautiful engravings depicting unique examples of the animals discussed (see Fig. 5).

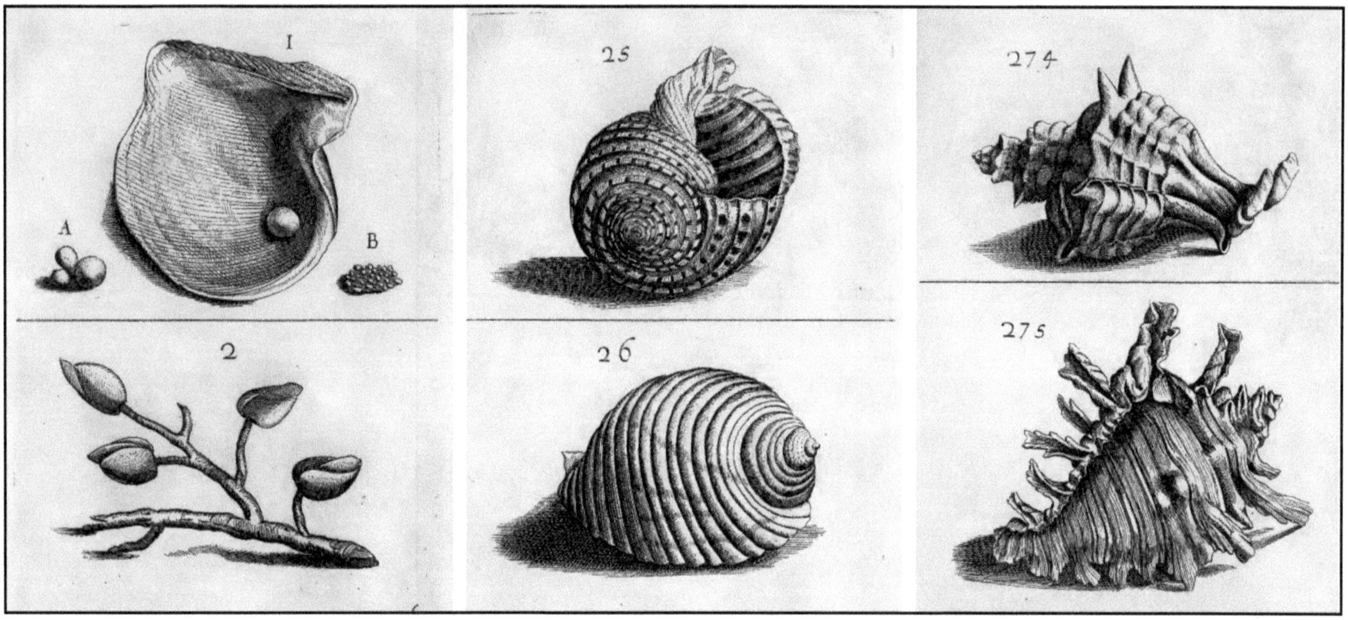

Figure 5. Testaceous seashells, as depicted in Buonanni's *Recreatio* (Buonanni, 1684).

In the first part of *Recreatio,* Buonanni provides in-depth discussion on the production or generation of a variety of mollusks. Like a good Aristotelian, Buonanni describes them in terms of spontaneous generation. The subject of spontaneous generation comes up early in the text with the subject of snails, "Whether snails are generated by propagation of a species, or rather are born spontaneously from themselves" [4] (Buonanni, 1684, p. 22). He begins the chapter by explaining that Aristotle flatly denies the propagation of the species in testaceous animals, claiming instead that they are born spontaneously from putrefied matter. (Although in some cases, this happens in the form of "shoots.")

Buonanni continues by citing Aristotle's metaphysics, where Aristotle has a term for creatures of this kind: *spontaneous.* Spontaneously generated animals, Buonanni explains, are those living beings whose "matter is able to be moved even by itself in just the same way that the seed usually moves it in the generation of other animals…"[5] (Buonanni, 1684, p. 23; cf. Aristotle's *Metaphysics,* Bk. VII: 1034b: 1–5 *in* Aristotle, 1995, p. 1633). He goes on to respond to Nicolaus Steno (1638–1686) citing the latter's *De solido intra solidum naturaliter contento dissertationis prodromus* (*The Prodromus to a Dissertation Concerning Solids Naturally Contained within Solids,* hereafter known as *Prodromus*), published 1669:

> Fr. Steno has taught in his most Illustrious *Prodromus* that snails are, in all actuality, born from eggs, [as he writes that] "From experience it is evident that oysters, as well as other testaceous animals, are born from eggs and not from putrefied matter." These words, however, are in direct opposition to Aristotle, who—after exhausting many fisherman with the use of his immense salary from Alexander of Macedon—gained a thorough knowledge of the secrets of nature about undersea creatures. Undoubtedly, Aristotle himself affirms that, "As a general rule, all testaceous animals are spontaneous in nature, [developing] in mud, and diverse according to the differences of the mud from which they arise: Oysters from slimy mud on the one hand, and snails from sandy mud on the other"[6] (Buonanni, 1684, p. 23–24; cf. Steno, 1671, p. 78; Aristotle's *History of Animals,* Bk. V: 547b: 18–24 *in* Aristotle, 1995, p. 864)

Worth noting is that when Buonanni appeals to the expertise of Aristotle over that of Steno, he does so by appealing to the general experiences of the fisherman with whom Aristotle conversed as a counter example to Steno's own claim. This is an interesting

[4] "Quaeritur an Cochleae propagatione Specierum generentur, aut potius sponte ex se nascantur."

[5] "… materia potest se ipsa moueri eo motu, quo semen mouet in generatione aliorum Animalium, qui motus dicitur spontaneus, …."

[6] "Nasci vero ex Cocleis oua docuit in suo Prodomo Illustrissimus Praeful Steno, ubi habit: Experientia constat, Ostrea, & alia Testacea ex ouis non ex putredine nasci. Haec autem sententia directe Aristoteli opponitur, qui postquam multos Picatores ingentibus Alexandri Macedonis expensis defatigasset, ut sub mari Naturae Aracna pernosceret, constanter affirmauit: Universim omnia Testacea sponte naturae in limo diversa pro differentia limi oriri, nam in caenoso Ostreae, in arenoso conchae."

example of giving epistemic priority to general folk experience. Missing from his retort, however, is that Steno actually has a response to this objection. Namely, that many shells found within the sandy beaches are as old as Noah's flood, or the "Great Deluge," thus predating the sand (Steno, 1671, p. 90). Ultimately, Buonanni maintains the view that all testaceous animals are the result of spontaneous generation and that many of their petrified shells may be similarly generated from stony matter.

The specifics of the flood and the origin of fossil shells became another point of contention between Steno and Buonanni. Buonanni considers the question of whether the fossil shells that we find from beneath the Earth's surface are made of stone and merely shaped with images of sea life imposed on them, or whether they were actually living sea creatures (Buonanni, 1684, p. 41–53). Here, again, he responds to Steno, who says that all shells once belonged to those of living testaceous animals—even in those areas where living sea creatures are no longer found. They were later calcined or petrified by means of the petrifying force of fluids (Buonanni, 1684, p. 42, cf. Steno, 1671, p. 83–95). But Buonanni found the notion of water transporting heavy shells over great distances—especially in mountainous terrain—to be problematic, suggesting instead that latent seeds could be deposited by the floods and thus capable of informing stony matter. On this point, he actually breaks away from Aristotle, who maintains that all shells are from marine life, and follows Kircher (Buonanni, 1684, p. 41–43; cf. Rappaport, 1997, p. 127). Figure 6 shows the sort of fossilized shells that were the subject of their debates. Much like the fossilized fish described by Kircher, one can see how the imprint of shells has been imposed upon inanimate rock.

Rhoda Rappaport has suggested that Buonanni would develop only much later in life a more sophisticated account of fossils in which "the wise Aristotelian," as he was called, divided them into (1) the remains of organisms and (2) the products of "natural powers" (Rappaport, 1997, p. 128). More likely, however, Buonanni's divisions were consistently similar to those of Kircher. Just as Kircher was far from advocating the inorganic origin of all fossils, so too was Buonanni. Ultimately, whatever contributions he may have tried to make regarding the constitutions of mollusks and fossils were soon to be overshadowed by his views on spontaneous generation.

GENERATION DISPUTES

Because of its insistence on spontaneous generation, *Recreatio* was not well received by experimentalists within Italy or abroad. This negative response was due in no small part to a previous controversy that had occurred more than a decade earlier between Kircher and Francesco Redi (1626–1697), a courtier from the Royal Court of Duke Ferdinando. Redi—assisted on occasion with the aid of his closest friend, Nicolaus Steno—conducted a series of controlled experiments utilizing a variety of substances that could putrefy, from raw meat and fish, milk and cheese, to fruits, such as pumpkin. What they discovered is that, when left out to rot, maggots and worms appeared only in the uncovered containers (Redi, 1909, p. 74–76; Wilson, 1997, p. 200; Cobb, 2006, p. 82ff.). In Figure 7, Redi has illustrated their reproduction cycle.

Redi ([1668] 1909) published a detailed account of his experiments in 1668, *Esperienze intorno alla generazione degl'insetti* (*Experiments on the Generation of Insects*), through which he aimed to demonstrate that creatures such as maggots—previously believed to be the result of some kind of generative process from corrupt and putrid matter—actually develop from material laid by adult flies. In doing so, he sought to refute Kircher's own claims about generation (Wilson, 1997, p. 193; Cobb, 2006, p. 85ff.; Bertoloni Meli, 2011, p. 180, 183). Indeed, a good deal of Redi's treatise could be seen as a direct response to Kircher. He frequently names Kircher specifically, and on one such occasion describes the Jesuit, "though a man worthy of esteem" as erroneous in his claim that one could breed flies from putrefied fly corpses (Redi, 1909, p. 34).

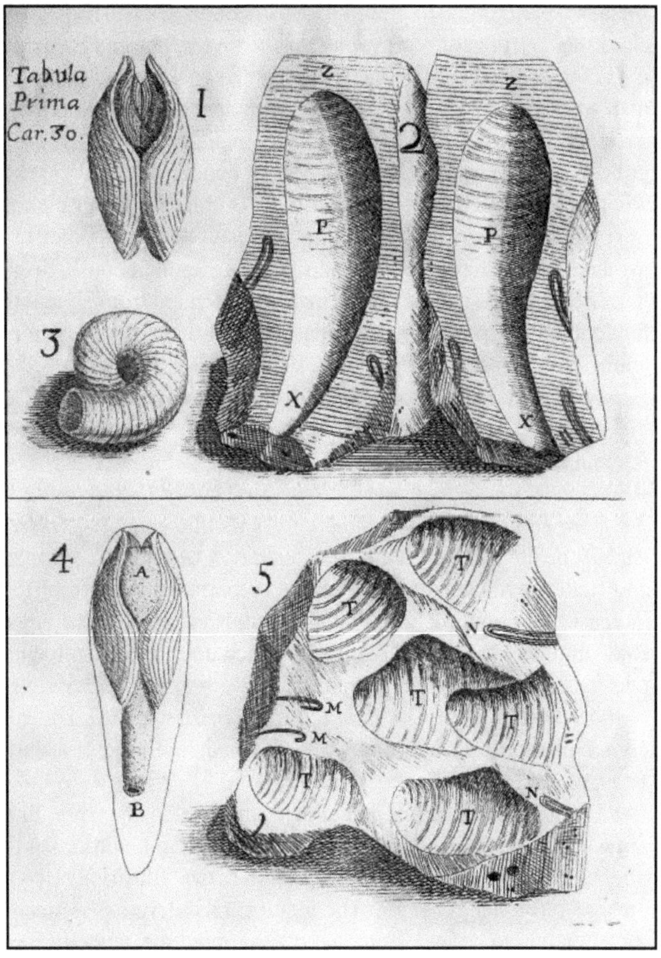

Figure 6. Mollusk fossils from Buonanni's *Recreatio* (Buonanni, 1684). In this chapter, Buonanni is discussing microscopic worms found within the fossil, the burrows of which are labeled "M."

Figure 7. The reproduction cycle of flies, as portrayed in Redi's *Esperienze* (Redi, 1909).

Given that Buonnani was both a loyal student of Kircher and a self-identified Aristotelian, it is hardly surprising that he would speak disparagingly of Redi's experiments. He explicitly criticizes the courtier for reaffirming that ordinary flies and gnats do not generate from rotting plants or meat. Buonanni replies that he may well find Redi's assertions that spontaneous generation from putrefied bodies *never* happens to be ridiculous. Buonanni even goes so far as to say that he really does not understand (*nescio*) by what means Redi thinks his experiment is able to counter claims of spontaneous generation. The Italian Jesuit then cites other living beings that have been said to generate spontaneously from putrefied matter, such as worms that are generated from Hyacinth flowers, before providing a similar account of mollusks (Buonanni, 1684, p. 42; see also Basile, 1987, p. 150ff.).

Francesco Redi responded to Buonanni in 1684, no doubt at the insistence from others within the community of Italian experimentalists, with *Osservazioni intorno agli animali viventi che si trovano negli animali viventi* (*Observations of Living Animals That Are Found in Living Animals*). As its title suggests, *Osservazioni* deals primarily with the generation of parasitic worms, and Redi provides within the treatise a continued defense of the increasingly popular claim that all animals originate from eggs. Buonanni eventually replied to the refined courtier in 1691 with *Observationes ca. viventia, quae in rebus non viventibus reperuntur* (*Observations Concerning Living Beings, Which Are Not Found in Other Living Things*), published in Rome (Fazzari, 1999, p. 99ff.).

Buonanni makes his intentions apparent in the very first sentence of his dedication to the volume wherein he proposes to offer evidence of spontaneous generation to the "Illustrious Leoni Strozze" (Buonanni, 1691). The treatise thus served as a means for Buonanni to defend his earlier claims about the generation of mollusks and attempted to prove that, contra Redi, spontaneous generation could in fact happen without seeds. In the massive tome of more than 600 pages, Buonanni frequently quotes Redi in order to respond to the celebrated courtier. On one such occasion, he provides a counter example to Redi in the chapter, "Ex Verminbus generatis in Floribus Hyacinthi" ("Worms That Are Generated from Hyacinth Flowers") (Buonanni, 1691, p. 145). At this point, one almost has to wonder if the two Italian scholars are talking past each other, as this is the exact same example provided by Buonanni in his much earlier *Recreatio*.

BUONANNI'S PHILOSOPHY OF NATURE

It is tempting for the twenty-first-century reader to view Buonanni as naive and simply too entrenched within an Aristotelian paradigm to appreciate or even fully understand Redi's experimental approach. Indeed, several of his own contemporaries seem to have done just this. For example, the famed Bolognese physician and anatomist, Marcello Malpighi (1628–1694)—respected friend and colleague of both Redi and Steno—notably complained in his private correspondence that the Jesuit "…corrupts the true method of philosophizing a posteriori, rendering everything uncertain and every possible extravagance as credible." Further, wrote Malpighi, this methodological approach "maltreats Redi and other Moderns on some matters … for Buonanni thinks that their flies are produced spontaneously and not from eggs deposited in them; and he proves all this, not by a series of observations of his own, but by the conjectures and assertions of others" (Adelmann, 1966, v. 1, p. 636).[7]

The grievance aired by Malpighi in this letter is Buonanni's rejection of an experimental epistemology. That Buonanni would give priority to Aristotelian text and natural history over the tried and true empirical results of those such as Redi, Malpighi, and Steno surely would seem nothing short of ridiculous to them. But Buonanni's program was additionally informed by another such principle that seems to be lost on Malpighi and his peers, thus missing the heart of Buonanni's response. Underlying their disagreements about generation, however, is Buonanni's explicit rejection that nature has uniformity. That is, Buonanni was quick to point out that just because an experiment produces results for one species, its conclusions could not, and *should* not, be generalized to others. In order to appreciate the significance of Buonanni's view of nature, it is important to see how closely this implicit assumption was linked with the experimental philosophy of the seventeenth century.

For example, consider one of the most celebrated experiments in natural philosophy, the discovery by William Harvey (1578–1657) of the circulation of the blood. A major part of Harvey's proof that blood circulates throughout the human body relied upon the large mass of blood pumped by the heart from the veins and to the arteries (Frank, 1980, p. 11; Bertoloni Meli, 2011, p. 146). His calculations, however, relied upon the

[7]Buonanni, sadly, did not reciprocate Malpighi's sentiments, but held Malpighi in rather high esteem and wished to smooth matters over with him. See Adelmann and Malpighi (1975, *The Correspondence of Marcello Malpighi*, p. 182, n. 9).

vivisection of a wide range of animals, including not just mammals, but also birds, lizards, frogs, eels, and fish (Frank, 1980, p. 17). (Cold blooded animals are, in fact, easier to observe due to their slower heart rate at low temperatures.) This particular method, what we might call comparative anatomy, is intended to provide a *demonstration* that begins with the particulars and uses induction to conclude about animals more generally. Harvey—like his Paduan anatomy teacher, Fabricius, before him—practiced this method of investigation following Aristotle, who assumes a similitude and unity across species, allowing for the induction from the specific to the general (Cunningham, 1997, Ch. 6). While this methodological approach did not go unnoticed or without criticism by Harvey's detractors, the underlying supposition that nature is uniform or generalizable became commonplace among experimentalists of the "New Science" by the second half of the seventeenth century (Frank, 1980, p. 17).

Vis-à-vis generation, Nicolaus Steno applies similar reasoning to the dissection of his famed dogfish. During his investigations with a viviparous dogfish shark, which he conducted while at Florence in 1667 with the aid of Francesco Redi, Steno noted similarities between the generative organs of the dogfish and that of the oviparous stingray that he had dissected several years earlier. His observations led him to the conclusion that, "the testicles of women are analogous to the ovary, whatever the manner the eggs themselves, or the matter that they contain, pass from the testicles to the uterus" (Cobb, 2006, p. 99). From this discovery, Steno surmised that that the egg was not produced by the uterus but came instead from those so-called female testicles, which he rightly identified as ovaries (Roger, 1997, p. 207; Bertoloni Meli, 2011, p. 209–217).

Here, too, lies the underlying assumption that the results from specific empirical investigations are generalizable across species. In truth, our privileged knowledge of the veracity of Steno's claims may prevent our fully recognizing just how giant a leap Steno and Redi were making to draw a conclusion about the entirety of avian and even mammalian reproduction based on his investigations of sharks and stingrays. This methodological assumption nevertheless played a vital role in the execution of these dissections and experiments and is ultimately what Buonanni finds so problematic and vehemently rejects. Thus, in this context, we can more fully appreciate Buonanni's claim that he has no idea by what means Redi claims to have proven spontaneous generation to be impossible. Namely, his criticism is Redi's fallacious application of the experiment to the whole of nature more generally. It is the methodological content of this complaint that makes Buonanni so unrelenting in each of his responses, regardless of whatever new discovery Redi, Steno, or others might report.

Domenico Bertoloni Meli discusses a related dispute between Buonanni and the future archdeacon of Bologna University, Antonio Felice Marsigli. Marsigli described the discovery of snails that reproduce sexually as a challenge to Buonanni's own claims of spontaneous generation. To this, Buonanni responded that just because some snails are generated from eggs does not mean that there is sufficient reason to assume that all do. Buonanni flat-out rejected "the uniformity of nature" and subsequently called "for greater caution in defending universal claims and questioned the method of induction, arguing that not all stars are fixed, since planets move, and not all birds fly, as the ostrich never leaves the ground" (Bertoloni Meli, 2011, p. 254–255).

The public disputes in which Buonanni engaged are distinct from later eighteenth-century debates between uniformitarianism (i.e., the view that geological processes have spatiotemporal uniformity and affect Earth gradually over time) and catastrophism (or the view that the formative geological processes are in fact large-scale, catastrophic events such as earthquakes or the Great Flood). Buonanni likewise does not address the functional morphology of the seashells and fossils he collected. Rather, he maintained a more general skepticism about the role of experimental induction. This skepticism put him at odds with contemporary experimentalists such as Steno, who used individual experiments to make universal claims such as "All animals come from eggs" and "No animals result from spontaneous generation."

Buonanni was hardly alone among their contemporaries to maintain this philosophical position. This philosophy of nature was shared by other traditional scholars within the Italian community, such as Bolognese naturalist, Giovanni Trionfetti, and his student, the physician, Giovanni Sbaraglia. Yet, Buonanni's access to—and later his position as curator for—the Kircher Museum put him in the unique position to apply that philosophy of nature more generally to both the benefit and the knowledge of a much larger audience. For Buonanni, the study of mollusks is inextricably linked to the museum. Buonanni's views on nature then would inform his curation of the Kircher Museum and play an important organizing principle under his stewardship. Namely, his rejecting conformity of causes in nature resulted in his collecting multiple specimens of natural artifacts such as shells and fossils, treating each as a unique object with a unique history worth recording. His doing so would have profound consequences for the future of the Museum Kircherianum.

BUONANNI AND MUSEUM CURATION

When Buonanni composed his response to Redi from the Collegium in 1691, Kircher's collection of wonders was in disarray and at the risk of perishing entirely. Georgio de Sepi, the disciple of Kircher responsible for publishing the Museum's first catalogue, openly lamented "more in sorrow than in anger" the neglect of Kircher's collection upon the death of its namesake (Fletcher et al., 2011, p. 184). In 1687, for example, the collection was described by one visitor as "formerly one of the most curious [cabinets] in Europe," but upon his visit was "very much mangl'd and dismembered." This inimical state of affairs was largely due to the Collegium not having yet found a suitable replacement for Kircher as curator of the museum (Findlen, 1996, p. 34). The fate of the collection would change, luckily, in 1698, when Buonanni finally succeeded Kircher.

Under his stewardship of the Kircher Museum, the cabinet was successfully restored to its glory days as a space of public wonder and learning. Buonanni moved the collection from a cramped, dark corridor near the infirmary to a more open space near the chapel of St. Louis. With the new space available to him, he decorated the ceilings, and Buonanni further added to the remarkable collection of wonders his own extensive collection of hundreds of mollusks shells and microscopes. In doing so, the "Wise Aristotelian" single-handedly transformed the museum into a space that would remain relevant and regularly visited until the dissolution of the Society of Jesus in 1773, nearly a century after Kircher's own death (Findlen, 1996, p. 126–127). (The Society would be subsequently restored by papal authority in 1814. See O'Malley SJ, 2014, p. 116.)

Paula Findlen has elegantly (and convincingly) argued that for naturalists such as Buonanni, the museum represented the reconstitution of the Aristotelian research program and a response to the experimental philosophy threatening them. In the case of Kircher, she explains how the collecting of artifacts of nature when conjoined with interrogation produced new attitudes toward nature and new methods of natural history (Findlen, 1996). Buonanni was certainly among the more recalcitrant Aristotelians of the late seventeenth century during a time when Aristotle was becoming increasingly unpopular. He consequently was vehemently opposed to the experimentalist epistemology of those such as Steno and Redi.

And yet, his refusal to generalize from induction is arguably at tension with an Aristotelian methodology, which assumes a unity across species. This break from Aristotle is even more striking given that Buonanni's method of collecting is distinguishable from that of his beloved professor, Kircher. The artifacts most treasured by Buonanni are not curious articles from afar, but example after example of shells and artifacts that could be found locally. Buonanni was creating a public record. He would become a unique figure in history because his conception of the museum as a space to preserve and actively study mollusk shells and their related fossils—each of which are to be treated as a uniquely significant specimen—greatly informed his research on mollusks and, later, his role as curator. As curator of the collection, he became one who served the public, expanding knowledge beyond small scientific communities.

Buonanni first articulates the role played by the museum in the study of seashells and fossils as early as *Recreatio.* In the last chapter of Part I, "The various museums to which snails are brought back and in which conch shells are preserved,"[8] Buonanni urges other scholars of museums to include exhibitions with snails, likewise urging them not to include any words or decorations on the shells (Buonanni, 1684, p. 81ff.). He makes mention of the Kircher Museum before concluding the chapter with an illusion to the book's own title. Buonanni presents illustrations of testaceous animals in the following pages so that they can be easily identified even without color. Thus, he explains that testaceous animals should be observed for pleasure to nothing less than the mind's eye (Buonanni, 1684, p. 85).

The year 1709 saw the publication of the second catalogue for the Kircher Museum, *Musaeum Kircherianum sive musaeum,* which was published by Buonanni. Whereas the original catalogue to the museum was 139 pages, including the illustrations and index, Buonanni's *Musaeum Kircherianum sive musaeum* totals 770 pages. The impressive catalogue divides the entirety of the museum's collection into a total of 12 classes, which include categories such as mathematical instruments, land and sea animals, artifacts from antiquity, and a collection of microscopes also donated by Buonanni. Most classes include numerous engravings depicting the items belonging to the museum. One surprising exception to his trend of including ample illustrations is Class VI, which discusses the Museum's collection of stones and fossils.

The testaceous animals and their remains added by Buonanni to Kircher's Museum are treated in Class XII. Buonanni, not surprisingly, has several chapters devoted to their generation and production. In describing the origin and causes of testaceous animals, the content conspicuously echoes that of Buonanni's earlier *Recreatio,* to the point that several chapters have the same title as earlier ones, such as "The qualities of other retrieved testaceous animals in which the Providence of God is manifest"[9] and "The various museums to which snails are brought back and in which conch shells are preserved,"[10] discussed earlier. Further inspection of those chapters shows that, by and large, they contain the same content verbatim.

In many ways, the overall organization of *Museum* also mirrors that of *Recreatio.* Much like his first publication, the treatise is divided into four parts. The first, described above, provided a meticulous and detailed account of the museum's collection. The second part provides a brief discussion on the classes of testaceous animals, and the third part is organized into "problems," following a traditional Scholastic style. And the fourth part, as with *Recreatio,* contains several detailed illustrations of Buonanni's collection of seashells. Another striking parallel between the 1709 Catalogue for the Kircher Museum and Buonanni's earlier *Recreatio* is his use of a copper engraving to introduce the numerous, detailed illustrations of seashells. At first glance, the images look indistinguishable. Closer inspection, however, shows that they are actually mirror images of each other. One possibility is that Buonanni used the back of the copper engraving, allowing an artist to engrave new text for the subsequent publication. Given that the written content is the same, however, his incentive remains unclear (see Figs. 8 and 9). In the Catalogue, as with *Recreatio,* the engraving is followed by Part IV

[8]"Referuntur varia Musea, in quibus Cochlearum, & Concarum Testae conferuantur."

[9]"Referuntur Proprietates aliquorumTestaceorum, in quibus elucet Providentia Divina."

[10]"Referuntur varia Musea, in quibus Cochlearum, & Concarum Testae conferuantur."

Figure 8. From *Recreatio* (Buonanni, 1684).

of the treatise, roughly a hundred pages, which are a series of detailed engravings of the seashell collection added by Buonanni. In examining the close relationship between Buonanni's mollusk research and the Kircher Museum, it is apparent that Buonanni's understanding of nature was not only influenced by his work as curator, but by applying the museum as a means of investigating natural history.

In this regard, Buonanni's methodological approach stands in stark contrast to that of his predecessor. For Father Kircher, wonder and marvel were among "the essential components in the study of nature and in the unravelling of its secrets" (Lugli, 1986, p. 123). Kircher's own presentation of the museum's collection reflected this outlook, as he embellished its corridors with optical chambers and marvelous curiosities from

Figure 9. From *Musaeum Kircerianum* (Buonanni, 1709).

around the world. Specimens of nature such as seashells, by contrast, were used to line the gardens and grottos—a common sixteenth-century decorating technique (Lugli, 1986). Whereas Kircher sought to possess one of every curiosity from around the world, Buonanni includes specimen after specimen after specimen of local seashells and their fossils. Buonanni thus helped to transform the museum from a mere cabinet of wonders and curiosities to a more robust public record, making his tenure at the Kircher Museum one of the very earliest examples of a modern museum.

Previous scholars have noted the many parallels between the impact of the printing press in Renaissance Europe and that of the Internet upon the present "Information Age" in which we reside. Both are responsible for the widespread dissemination of information at unprecedented speeds because of technological advancements, and both consequently had profound impacts upon culture and learning. A comparable relationship can be found between Buonanni's curation of the Kircher Museum and more recent efforts by scientific communities to make museum collections accessible through electronic databases, such as the digitization of fossil collections for open access. Both are responsible for the publication of unprecedented quantities of data regarding the many individual shells, fossils, and other natural artifacts unique to a specific location (for Buonanni, the shores of Rome). In finding these parallels, we can begin to see how Buonanni's story traverses through centuries, resonating throughout time and illustrating the important role played by museums to create a record of natural artifacts and their unique geological history; to preserve those artifacts and encourage their use in research; and to relay this information both to scientific communities and to the public more generally.

CONCLUSION

However laughable we may presently find Buonanni's views on spontaneous generation or the inorganic production of fossils, he was right to question the underlying assumptions of experimentalists such as Steno and Redi. In particular, there is perfectly good reason to question the conformity of causes in nature. While Buonanni's use of the Museum Catalogue to promote his own ideas on testaceous animals could be seen as somewhat opportunistic, in truth it was this aspect of his role as curator that helped to make the Kircher Museum actively engaged not only with the past, but with the present and future. In 1895, for example, Smithsonian Institution administrator, George Brown Goode, delivered a paper to the Museum Association on "The Relationship and Responsibilities of the Museum." In the address, he describes the responsibility of museums: to advance learning; to impart special information to the general public; and to preserve materials for record and for future comparative studies. But he also explains that museums must be dynamic and "constantly engaged in aggressive work" in order to be useful or reputable. Failure to do so means that the museum has finished its work, and "A FINISHED MUSEUM," he explains in all caps, "IS A DEAD MUSEUM, AND A DEAD MUSEUM IS A USELESS MUSEUM" (Goode, 2008, p. 111–117).

Buonanni made certain that the Kircher Museum was anything but finished. Under his stewardship, the Kircher Museum went on to serve as an early building block of a legacy for the geology museum—one in which each artifact, shell, or fossil is treated as having a unique and independent history, and one that serves to inform the general public rather than a smaller community of scientific scholars. He consequently helped to develop a tradition of the museum that is engaged not just with the past but with the future. It is, then, somewhat of a historical irony that it was his insistence upon the principle for which he has been ridiculed—his rejecting the conformity in nature—that allowed him to bring the museum to the forefront.

ACKNOWLEDGMENTS

I would like to thank Gary Rosenberg, Curtis Sommerlatte, two anonymous reviewers, and the GSA staff for insightful comments and suggestions that have much improved the quality of this paper.

Note: All figures are from copper engravings and in the public domain.

REFERENCES CITED

Adelmann, H.B., 1966, Marcello Malpighi and the Evolution of Embryology: Ithaca, New York, Cornell University Press, 5 vols., 551 p.

Adelmann, H.B., and Malpighi, M., 1975, The Correspondence of Marcello Malpighi: Ithaca, New York, Cornell University Press, 5 vols., 2227 p.

Aristotle, and Barnes, J., ed., 1995, Complete Works of Aristotle: The Revised Oxford Translation: Princeton, New Jersey, Princeton University Press, 2 vols., p. 774–993 and p. 1552–1728.

Basile, B., 1987, Polemiche sulla generazione spontanea: Redi, Buonanni, Malpighi, *in* Basile, B., ed., L'invenzione del vero: La letteratura scientifica da Galileo a Algarotti: Rome, Salerno, p.125–167.

Bertoloni Meli, D., 2011, Mechanism, Experiment, Disease: Marcello Malpighi and Seventeenth-century Anatomy: Baltimore, Johns Hopkins University Press, 439 p.

Buonanni, F., 1684, Recreatio mentis et oculi in observatione animalium testaceorum: Digitized copy of original from Harvard University Library of Dept. of Mollusks: Rome, Varesi, 569 p.

Buonanni, F., 1691, Observationes circa viventia, quae in rebus non viventibus reperuntur, Digitized by the Elektronische Bibliothek Schweiz project: Rome, Antonio Hercules, 585 p.

Buonanni, F., 1709, Musaeum Kicherianum Sive Musaeum AP Athanasio Kichero in Collegio Romano Societatis Jesu: Digitized by Google Books: Rome, Collegio Romano Societatis Jesu, 748 p.

Chang, K.-M.(K.), 2011, Alchemy as studies of life and matter: Reconsidering the place of vitalism in early modern chemistry: Isis, v. 102, no. 2, p. 322–329, https://doi.org/10.1086/660141.

Cobb, M., 2006, Generation: The Seventeenth-Century Scientists Who Unraveled the Secrets of Sex, Life, and Growth: New York, Bloomsbury Publisher, 333 p.

Cunningham, A., 1997, The Anatomical Renaissance: The Resurrection of the Anatomical Projects of the Ancients: Aldershot, UK, Scholar Press, 283 p.

De Boodt, A.B., 1647, Gemmarum et Lapidum Historia: Quam Olim Edidit, 602 p.

De Sepibus, G., 1678, Romani Collegii Societatis Jesu Musaeum celeberrimum, Cujus magnum Antiquariae rei, statuarium imaginum, picturarumque partem Ex Legato Alphonsi Domini, Senatus PopulusQue Romanus: Amstelo dami, Janssonio-Waesbergiana, frontispiece.

Fazzari, M., 1999, Redi, Buonanni e La Contraversia Sulla Generazione Spontanea, *in* Bernardi, W., and Guerrini, L., eds., Francesco Redi: Un Protagonista Della Scienza Moderna. Documenti, Esperimenti, Immagini: Florence, Olschki, p. 97–127.

Findlen, P., 1996, Possessing Nature: Museums, Collecting, and Scientific Culture in Early Modern Italy: Berkeley, University of California Press, 449 p.

Findlen, P., 2004, Introduction: The last man who knew everything … Or did he? Athanasius Kircher, S.J. (1602–1680) and his world, *in* Findlen, P., ed., Athanasius Kircher: The Last Man Who Knew Everything: New York and London, Routledge, p. 1–48.

Fletcher, J.E., Fletcher, E., and Kircher, A., 2011, A Study of the Life and Works of Athanasius Kircher, Germanus Incredibilis: With a Selection of His Unpublished Correspondence and an Annotated Translation of His Autobiography: Leiden, Brill, 656 p.

Frank, R.G., 1980, Harvey and the Oxford Physiologists: A Study of Scientific Ideas: Berkeley, University of California, 368 p.

Goode, G.B., 2008, The relationships and responsibilities of museums, *in* Andrei, M.A., and Genoways, H.H., eds., Museum Origins: Readings in Early Museum History and Philosophy: New York, Routledge, p. 111–124.

Goudin, A., [1668] 1864, Philosophie suivant les principes de saint Thomas, Tome 1: Vol. 1, Paris, Vve Poussielgue-Rusand. Digitized by the Bibliothèque nationale de France, département Philosophie, histoire, sciences de l'homme, 582 p.; http://gallica.bnf.fr/ark:/12148/bpt6k653059.

Gould, S.J., 2004, Father Athanasius on the isthmus of a middle state: Understanding Kircher's paleontology, *in* Findlen, P., ed., Athanasius Kircher: The Last Man Who Knew Everything: New York and London, Routledge, p. 207–238.

Hagan, H.A., 2008, The history of the origin and development of museums, *in* Andrei, M.A., and Genoways, H.H., eds., Museum Origins: Readings in Early Museum History and Philosophy: New York, Routledge, p. 39–48.

Hirai, H., 2007, Kircher's chymical interpretation of the creation and spontaneous generation, *in* Principe, L., ed., Chymists and Chymistry: Studies in the History of Alchemy and Early Modern Chemistry: Sagamore Beach, Massachusetts, Science History Publications, Watson Publishing International, p. 77–87.

Hirai, H., 2016, Mysteries of living corpuscles, *in* Distelzweig, P., Goldberg, B., and Ragland, E.R., eds., Early Modern Medicine and Natural Philosophy: Dordrecht, Netherlands, Springer, p. 255-269.

Hooke, R., 1665, Micrographia: Or Some Physiological Descriptions of Minute Bodies Made by Magnifying Glasses with Observations and Inquiries Thereupon: London, Royal Society. Digitized by Duke University Libraries, 246 p.; https://www.biodiversitylibrary.org/item/198549#page/5/mode/1up.

Hsia, F., 2004, Athanasius Kircher's *China Illustrated* (1667): An Apologia Pro Vita Sua, *in* Findlen, P., ed., Athanasius Kircher: The Last Man Who Knew Everything: New York and London, Routledge, p. 383–404.

Hunter, M., and Davis, E.B., eds., 2000, The Works of Robert Boyle: Volume 7, London, Routledge, p. 3–72

Franceschini, P., 2008, Buonanni, Filippo, *in* Koertge, N., ed., Complete Dictionary of Scientific Biography, vol. 2: Charles Scribner's Sons, Gale Virtual Reference Library, p. 591–592.

Kircher, A., [1678] 2004, Mundus Subterraneus, Facsimile Reproduction: Compiled by Gian Battista Vai: 3rd ed., Series 1678: Bologna, Italy, Museo Geologico Giovanni Capellini, 507 p.

Koertge, N., 2008, Complete Dictionary of Scientific Biography: Charles Scribner's Sons, Gale Virtual Reference Library.

Lugli, A., 1986, Inquiry as collection: The Athanasius Kircher Museum in Rome: RES: Anthropology and Aesthetics, v. 12, p. 109–124.

O'Malley, J.W., SJ, 2014, The Jesuits: A History from Ignatius to the Present: Lanham, Maryland, Rowman & Littlefield, 160 p.

Parcell, W.C., 2009, Signs and symbols in Kircher's *Mundus Subterraneus, in* Rosenberg, G.D., ed., The Revolution in Geology from the Renaissance to the Enlightenment: Geological Society of America Memoir 203, p. 63–74, https://doi.org/10.1130/2009.1203(04).

Plot, R., 1677, The Natural History of Oxford-shire: Being an Essay toward the Natural History of England: Oxford, UK, Printed at the Theater in Oxford: and are to be had there: And in London at Mr. S. Millers, at the Star near the West-end of St. Paul's Church-yard, 358 p.

Rappaport, R., 1997, When Geologists Were Historians: 1665–1750: Ithaca, New York, Cornell University Press, 336 p.

Redi, F., [1668] 1909, Experiments on the Generation of Insects: Chicago, Open Court Publishing, 174 p.

Roger, J., 1997, The Life Sciences in Eighteenth Century French Thought: Edited by Benson, K., translated by Ellrich, R.: Stanford, California, Stanford University Press, 804 p.

Rossi, P., 1984, The Dark Abyss of Time: The History of the Earth & the History of Nations from Hooke to Vico: Chicago, University of Chicago Press, 338 p.

Rowland, I.D., 2004, Athanasius Kircher, Giorando Bruno, and the panspermia of the infinite universe, *in* Findlen, P., ed., Athanasius Kircher: The Last Man Who Knew Everything: New York and London, Routledge, p. 191–206.

Sardo, E.L., 2004, Kircher's Rome, *in* Findlen, P., ed., Athanasius Kircher: The Last Man Who Knew Everything: New York and London, Routledge, p. 51–62.

Smith, J.E.H., 2011, Divine Machines: Leibniz and the Sciences of Life: Princeton, New Jersey, Princeton University Press, 380 p.

Steno, N., 1671, The Prodomus to a Dissertation Concerning Solids Naturally Contained Within Solids. Translated by Oldenburg, H.: London, Moses Pitt, 112 p.

Wilson, C., 1997, The Invisible World: Early Modern Philosophy and the Invention of the Microscope: Princeton, New Jersey, Princeton University Press, 280 p.

Yamanda, T., 2006, Kircher and Steno on the "geocosm," with a reassessment of the role of Gassendi's works, *in* Vai, G.B., and Caldwell, W.G.E., eds., The Origins of Geology in Italy: Geological Society of America Special Paper 411, p. 65–80, https://doi.org/10.1130/2006.2411(05).

Manuscript Accepted by the Society 30 November 2017
Manuscript Published Online 25 May 2018

Printed in the USA

The Geological Society of America
Special Paper 535

All is not lost: History from fossils and catalogues at the Museum of Natural History, University of Florence

Stefano Dominici
Elisabetta Cioppi
Museo di Storia Naturale, Università degli Studi, 50121 Firenze, Italy

ABSTRACT

Sedimentary strata and fossils of Tuscany have been the object of inquiry from the late Middle Ages into the onset of modern science, passing through the art and words of Leonardo da Vinci, and culminating in the work of Nicolas Steno on a Galilean foundation. In the Age of Enlightenment, the Florentine Giovanni Targioni Tozzetti perfected Steno's scheme for a history of Tuscany to be extended to a general theory of the Earth, corresponding with European savants, writing the oldest catalogue of fossils now hosted at the Museum of Natural History, University of Florence, and passing on Steno's taxonomy. A few decades later, founders of modern geology, Georges Cuvier, Giambattista Brocchi, and Charles Lyell walked Tuscan fossiliferous hills and studied public and private collections, focusing on the anatomy of Tertiary species as a means to track the making of the modern fauna. The international impact of Brocchi's *Subapennine Fossil Conchology* reached the young Charles Darwin, offering a theoretical background for the early development of modern evolutionary theory and fueling the modern taxonomic study of Tertiary marine shells. Under Igino Cocchi, in the year of national unity (1861), the Museum became the Italian Central Paleontological Collection, attracting collections from all over Italy and stirring an enduring international interest in Tertiary and Quaternary faunas, including fossil primates. With fossil specimens brought in by Steno, Targioni, and Cocchi, among many others, and with the organization of its catalogues reflecting the onset of modern taxonomy, the Museum of Natural History, University of Florence, is today an archive of the history of science as a whole and a means to bring environmental consciousness to future generations.

INTRODUCTION

In the age of Wikipedia, many narratives in the history of science follow a dualistic model where theories are opposed. Advocates of one theory eventually prevail, by way of stronger empirical evidence and better arguments, and knowledge progresses in a linear fashion. In the process, the position of protagonists is polarized in a tale of "good guys and bad guys" (so the argument runs): the good guy is armed with the power of observation; the bad guy is led by unscientific prejudices. The good guy wins in the end, eventually "fathering" (or perhaps "mothering") a new science. This simplistic model is persistently resurrected in much popular literature on the history of geology. Evolutionary biology, the other discipline that so much relied

Dominici, S., and Cioppi, E., 2018, All is not lost: History from fossils and catalogues at the Museum of Natural History, University of Florence, *in* Rosenberg, G.D., and Clary, R.M., eds., Museums at the Forefront of the History and Philosophy of Geology: History Made, History in the Making: Geological Society of America Special Paper 535, p. 59–80, https://doi.org/10.1130/2018.2535(05).

on geological evidence in its early days, is also involved. Even in manuals of geology and in dictionaries, the dualistic model persists in the choice of words, such as "natural-miraculous," "uniformitarian-catastrophic," "evolutionist-creationist," "long-short" (timescales), and "gradual-stepwise" (antinomies largely taken as equivalents by unlearned readers). As this case suggests, much still needs to be done to explain the history of science to the general public. Simplifying, the process comprises three steps: (1) historians analyze primary literature, place evidence in context, and write secondary literature (e.g., Rudwick, 2005); (2) secondary literature is studied by university teachers and professionals working at institutions that operate in education and outreach; and (3) knowledge is transferred to the public. Since fossils (the more spectacular, the better) attract a wide public and elicit the imagination, museums of paleontology are important connections between the work of historians of geology and the general visitor. When a museum hosts specimens and other evidence that directly forms part of that history, the role of the historian merges with that of the museologist. The personnel of these institutions of conservation, studying written information (metadata) connected to fossils, such as labels, inventories, catalogues, and other manuscripts, act in fact as historians. Even if not formally trained, they deal with primary data and seek for historical context, as historians do. This is the case with the Section of Geology and Palaeontology of the Museum of Natural History, University of Florence, where the need to explain not just geology, but also its history, is felt to be essential. In Tuscany, fossils have been dug up and commented upon since at least the late Middle Ages, then in the Renaissance and during the birth of modern science, until geology and evolutionary biology became what they are today. In this process, Florence has acted as the center of ferment in natural history. There may not be another place in the world where the main phases in the history of the science can be documented by books, manuscripts of many kinds, and actual fossils as they are in Florence, from the Renaissance to the present, without interruption of continuity. The authors of this paper are paleontologists with a degree in geology, compelled to become historians by the quantity of available evidence, and confronting themselves with the work of professionals of this field of humanistic sciences, trying to avoid misinterpretation and anachronism. This has led to a history of geology—a more intricate list of "good guys and bad guys," if you will, as seen from a peculiar perspective, briefly dealt with in this paper.

PREHISTORY OF GEOLOGY IN TUSCANY

In Tuscany, knowledge of the history of the Earth for the period that precedes civil history has been deeply related to the question: why are there sea shells inside rocks? There are documents, some textual and some visual, that suggest that more than one man attempted in Tuscany to understand the origin of these objects, long before the birth of modern science in Europe. Three such men were Restoro d'Arezzo (thirteenth century), Giovanni Boccaccio (1313–1375), and Leonardo da Vinci (1452–1519), spanning two centuries of history between the late Middle Ages and the Renaissance. What did sea shells in rocks mean to those three Italians?

Restoro's "Composition of the World with Its Causes," written in Italian in 1282 and widely read through the following century, is the oldest document available on this subject. Restoro found substantial agreement between his interpretation of nature, as seen in Tuscany, and the words of the Holy Scripture, as outlined in the following passages.

> And by then we had found and extracted, almost at the top of a very high mountain, great amounts of fish bones, which we call snails, and some call shells [...]. And that district, there where these mountains are found, there where the sand and fish bones are found, is a sign that there once was the sea in that district. [...] And it could also be that the mountain was caused by the water of the flood, which covered the ground and then moved it from one place to another. (Restoro, 1282, *in* Narducci, 1859, p. 245, translated by authors)

In the words of the first historiographer of fossil conchology, Giambattista Brocchi (1772–1826), the Renaissance humanist Giovanni Boccaccio, born in Certaldo, a small village in Tuscany, "must have been accustomed to seeing, starting in his childhood, the great mass of mollusc shells filling the hills around the village [Certaldo], where there is many a copy, [...] that in some places they render barren the soil" (Brocchi, 1814, p. III–IV). Boccaccio briefly expressed his opinion on dug-up marine shells in the novel *Filocolo,* written ca. 1336.

> In fertile Italy, there is a small part the ancients called Tuscia, in the middle of which, among very lovely plains, rises a small hill, left there by the waters, avengers of the ire of Jupiter when the sins of Lycaon resulted in the flooding of the world; this is according to the opinion of many, which I repute true, and as evidence of this truth there appears this small hill full of sea shells. (Boccaccio, 1339, translated *in* Dominici, 2011, p. 4)

The thinking of Leonardo on sea shells and the Biblical flood is mainly contained in a text written between 1506 and 1510, made public only in the nineteenth century and now known as Codex Leicester, but it is also apparent in his sketches, drawings, and paintings (Gould, 1998; Kemp, 2006; Cioppi, 2011; Dominici, 2017). Referring to the valley of river Arno downstream of Florence, Leonardo shunned the theory of the flood and expressed his own understanding of why sea shells can be found in rocks.

> Further on was deposited the mud in which the shells lived, which rose by degrees according to the levels of the Arno which flowed into the more or less turbid sea. And from time to time, the sea bottom rose, depositing these shells in layers, as can be seen in the cut at Gonzoli Hill, eroded by the Arno which is wearing away its base, in which cut the aforesaid layers of shells can be seen in bluish clay, along with

other marine objects [...] If the shells had instead been transported by the muddy deluge they would have been mixed, arranged separately in the mud, and not in ordered steps and layers, as we see them now. (Leonardo, ca. 1506–1510, translated *in* Gould, 1998, p. 40)

According to Kemp (2006), Leonardo had read "Composition of the World with Its Causes" by Restoro d'Arezzo and knew the classics of Medieval scholarship, including Avicenna's *Book of Healing* of 1027, where it is written that "it is possible that each time the land was exposed by the ebbing of the sea a layer was left, since we see some mountains that seem to have been piled up layer by layer" (Avicenna, translated in Cutler, 2003, p. 95). As a matter of fact, Leonardo considered the layer-cake structure of those hills in his youth, placing it on the forefront of his first dated work entitled, "On the day of the Madonna of the Snow, 5 August 1473." Here he gives a visual rendition of Avicenna's words, taking from a real place the horizontally stratified nature of so much Tuscan landscape (Rosenberg, 2001; but see Nova, 2015), particularly similar to the Volterra foothills (Dominici, 2017; Fig. 1).

Overall, there is ample evidence that, at the passage between the Middle and Modern Ages, the succession of strata that make up the lower reliefs of Tuscany, and the sea shells contained within, attracted learned men and was the subject of their enquiry, for purposes that transcended local history. In the two centuries that followed Leonardo's early landscape, additional objects dug up from the earth would amplify that interest, as testified by the building of natural history collections in Florence, when, during the Renaissance, this became the undisputed cultural capital of Tuscany.

RENAISSANCE TAXONOMIES

The collection of antiquities and rarities for private use hosted in the "treasure room" of the first Grand Duke of Tuscany, Cosimo de' Medici (1519–1574), possibly contained fossils and other "things of nature" ("rerum naturae," *in* Mercati, 1717, e.g., p. 1: to avoid anachronism, we adopt definitions used at the time). Antiquities were first kept in the Medici Palace in Via Larga, then in Palazzo della Signoria, after the Medici family moved there in 1540. A small *Cabinét,* or *Wunderkammer,* was built by Cosimo's son, Francis I (1541–1587), an alchemist and philosopher correspondent of Ulisse Aldrovandi (1522–1605), whom in his turn possessed in Bologna a museum rich with "petrifactions." Minerals owned by Francis I, and others dating back to Lorenzo de' Medici (1449–1492), are still curated in the Museum of Natural History, University of Florence (Cipriani et al., 2011). In 1596, Andrea Cesalpino (1519–1603), physician and botanist at the University of Pisa and in the service of Francis' successor, Ferdinand I (1549–1609), published *De metallicis,* reporting the recent discovery of elephant bones at San Giovanni Valdarno, including a femoral head two spans wide and some petrified sea shells, which again were interpreted in relationship with a landscape that differed from the modern ("as the sea retreated and the ground became salty, the remains present in it were transformed into stones": Cesalpino, 1596, *in* Cioppi and Dominici, 2011, p. 20). Cesalpino is an eminent example of the Late Renaissance

Figure 1. Volterra view on Pliocene hills, a possible template for Leonardo's 1473 sketch of the Tuscan landscape.

philosopher who, at the service of a patron of the arts, works at a taxonomy of the natural world on Aristotelian principles of matter and form (Rossi, 2001). In conjunction with *Metallotheca Vaticana* of his teacher Michele Mercati (1541–1593), Cesalpino's work on the mineral world was an answer to *De Re Metallica* of Georgius Agricola (1494–1555), widely read in the sixteenth century. Both born in Tuscany and educated at the Pisa University, Cesalpino and Mercati were the first to present the "natural things" of their region to a European public. This happened at a time when a network of apothecaries, botanists, and physicians acting as natural philosophers started the first Europe-wide correspondence, allowing to be known, through printed words and engravings which objects were hosted in their cabinets and which theories underlaid their arrangements. "Fossils" meant to them "any distinct object dug up from the earth, [...] including mineral ores, natural crystals, and useful rocks" (Rudwick, 1972, p. 1). The Grand Duke Ferdinand I had in Cesalpino an eminent adviser when in 1595 he moved to Pisa, where his court lived during the summer months and where the core of the natural history collections resided. How this "choice of what there was in his palaces belonging to Natural History" (Targioni Tozzetti, 1763b, p. 3) was arranged in the "Galleria de' Semplici" (established in 1591), and what their nature and size was, we can only imagine by comparison with similar collections known to occur elsewhere in Europe. A template of this arrangement is Mercati's *Metallotheca* itself. Interrupted in 1593 by the death of its author and published only in 1717, the book includes the description of the first nine cabinets out of 19 that formed the whole *Metallotheca* (the name given to the actual collection exhibited in the Vatican, no longer existing). The ninth cabinet included "stones with a shape" (*lapides idiomorphoi*). On the meaning of shapes resembling true animals or animal parts, Mercati believed that "forms cannot pass from objects to stones, nor do they represent objects whose matter has been changed: certainly they cannot ascribe their shape to any of the creatures that live among us" (Mercati, 1717, p. 219). Instead, he relied on the opinion of contemporary astronomers in attributing that shape to the action of "celestial radiations" coming from the eighth sky, where fixed stars resemble animals (Accordi, 1980). The arrangement in the gallery of "stones with a shape" mirrored that of any other Late Renaissance museum, notably those stones of Konrad Gesner (1516–1565), Johannes Kentman (1518–1574), Ulisse Aldrovandi, or Ferrante Imperato (1550–1631; see Rudwick, 1972). In Florence, in particular, the Uffizi Gallery, standing "as a public model of what it meant to transform a private passion for beautiful objects into a public statement about the value of culture" (Findlen, 1998, p. 113) and at the same time "a symbol of Medici power" (Findlen, 1994, p. 114), had become one of the most valued museums in Europe, while the flowering of scientific collections pushed their curators to share taxonomic strategies (Arnold, 2006). Next, in 1666, we know that Grand Duke Ferdinand II (1621–1670), grandson of Ferdinand I, invited to Florence Nicolas Steno (1638–1686), the Danish anatomist, who, in the years to come, would increase and give order to the natural history collections.

TAXONOMY DURING THE BIRTH OF MODERN SCIENCE

> *Why would it not be permitted to hope for great things, if anatomy was transformed so that experimental knowledge would rely only on well established facts, and reason accepted only what has been demonstrated; in other words, if anatomy used the language of mathematics?*
>
> —Nicolas Steno (1667, *in* Kardel and Maquet, 2013, p. 594)

Much has been written about Nicolas Steno (born Niels Stensen in Copenhagen, Niccolò Stenone for the Italians: Cutler, 2003; Vai, 2009; Di Orio et al., 2016), also in the form of the hagiography (Steno had been beatified by the Roman Catholic Church). In the final analysis, the greatest input Steno brought to the study of several natural objects, from the human body to the Earth, was methodological, based on geometry (Rotschuh, 1968; Rosenberg, 2006; Kardel, 2009), following the idea that mathematics is the language spoken by nature, as eminently stated before him by Galileo Galilei (1564–1642) and René Descartes (1596–1650; see Rossi, 2001). Experience and more inductive reasoning led him to criticize some of Descartes' assessments, after being a Cartesian himself (Olden-Jørgensen, 2009). Steno's empiricism was very effective; he was an exceptional observer, with an enviable ability to anatomize ("he possesses an inconceivable patience, and by exercise he has acquired an unusual skill. Neither butterfly nor fly can escape his attention": Graindorge, 1665, *in* Kardel and Maquet, 2013, p. 164). Steno was well experienced in public demonstrations and the first to apply the experimental method of enquiry outside the realm of physics (Rudwick, 1972). He arrived in Tuscany in the spring of 1666, after a two-year stay in Paris and a brief sojourn in Montpelier. In Florence, he became part of the circle of those disciples of Galilei who formed the Accademia del Cimento (active between 1657–1667), under the auspices of Leopoldo de' Medici (1617–1675), future cardinal and brother of Ferdinando II. The Galileian cultural milieu of the Florentine court, coupled with Steno's previous experiences in Holland and France, were conducive to the first application of geometric reasoning to facts concerning anatomy and the Earth as expressed in two essays, "The Head of a Shark Dissected" (Stensen, 1667) and the brief "Forerunner to a Dissertation Concerning a Solid Body Enclosed by Process of Nature within a Solid" (Stensen, 1669). The latter, translated in English as early as 1671, and often referred to as *De Solido,* contains the heritage left by Steno to all people concerned with the study of the Earth: a general theory on how all solids form, from shells to strata, to mountains, and its practical application to explain what he had seen in Tuscany. After its publication, Steno left Florence to make a long journey through Europe, "from Italy over the Alps to Austria and Hungary, further to Bohemia and through the whole of Germany and then from the Netherlands, Belgium and Germany back to his beloved Florence" (quotation from the late Gustav Scherz, Steno's main biographer, translated *in* Kardel and Maquet, 2013). During the journey, he presumably collected specimens useful for his natural philosophy concerning "all things formed in the earth," and, while back in Florence, during 1671–1672, he actively worked in Pisa at the cabinet of Ferdinand II's son, Grand Duke Cosimo III (1642–1723). The oldest document connecting the Museum of Natural History, University of Florence, with the collection he saw there is a handwritten note left by Steno in the "Inventory of the Gallery and the Botanical Garden" of the Pisa University. This note indicates that Steno was commissioned by Leopoldo de' Medici to "remove from the Gallery of Pisa certain curiosities for the Gallery that is being established in Florence," to have made a catalogue of the things "removed," and to have it delivered to the cardinal on 7 May 1672 (Savi, 1827, p. 38, translated by authors; De Rosa, 1986). The manuscript contains additional information on the nature of the specimens taken to Florence.

I've removed the following number of pieces, ["and these are in the number of two hundred seventy-eight, unspecified, but mentioned with general observations such as": Savi, 1827, p. 38]

No. 48 specimens, including crystals and bodies that can be reduced to crystals, for their figure and transparence.

No. 24 specimens, including mines of rare silver, tin, lead, iron, and cinnabar.

Plus other 22 specimens of petrifactions, salts, and zoophytes etc. (Stensen, 1672, *in* Savi, 1827, p. 38; but see De Rosa, 1986, p. 78–80, for a slightly different transcription of the manuscript)

According to one of the transcriptions (Savi, 1827), Steno took to Florence 278 specimens (252 specimens, counting the numbers transcribed by De Rosa, 1986), subdivided in groups, based on affinities between specimens.

The second document concerning Steno's ordering of natural history collections is a manuscript of February 1763 entitled, "Catalogue of Natural Things of the Royal Gallery of Florence," commissioned by Pietro Leopoldo to the naturalist Giovanni Targioni Tozzetti (1712–1783). It comprises an "Index of Natural Things Maybe Dictated by Niccolò Stenone, and Copied from the Original Existing in the Royal Gallery" (Fig. 2). This so-called *Indice* contains fully described item numbers 1–255, followed by entry numbers 256–304, the latter group preceded by the title "Catalogue of curious things that remain to be removed" (Targioni Tozzetti, 1763a, folio 61r). So we have transcriptions of two manuscripts by Steno, one by Savi and De Rosa, and one by Targioni Tozzetti, sharing the words "removing curiosities." A cross-check confirms the two report on nearly identical quantities: if we add items listed by letters in the *Indice* (item 44 is followed by letters A–Z; item 163 by letters A–N) to items listed by numbers, the total is 288, coming very close to the 278 counted by Savi (252 in De Rosa, 1986). Scholars rightly pointed out that the *Indice* contains specimens collected during the 1667–1670 European tour, such as those from Tirol and Bohemia, together with specimens coming from localities Steno had personally visited in Tuscany, such as Libbiano (Scherz, *in* Kardel and Maquet, 2013). It is possible that Steno's own specimens were added to the collection of the grand duke in Pisa, where he could study them side by side with the "curiosities" that the grand duke had gathered before his arrival in 1667 (Cioppi and Dominici, 2011). In this hypothesis, on 7 May 1672, Steno brought to Florence a selection of old and new specimens, selected and grouped according to his own original vision.

The first 114 numbers of the *Indice* refer to minerals. Crystalline habits are described as simple geometric figures (columns, pyramids, and cubes—habits that seem better developed in specimens 1–40; see Scherz, 1956, *in* Kardel and Maquet, 2013, for a further subdivision). Items 115–199 include mineralized remains of ancient organisms ("petrifactions" of the Pisa list), such as bivalve and gastropod shells, ammonites, crabs, fishes, and bones of terrestrial animals. Individual entries do not usually refer to isolated sea shells but to aggregates, such as one of the few specimens still recognizable today (Fig. 3;

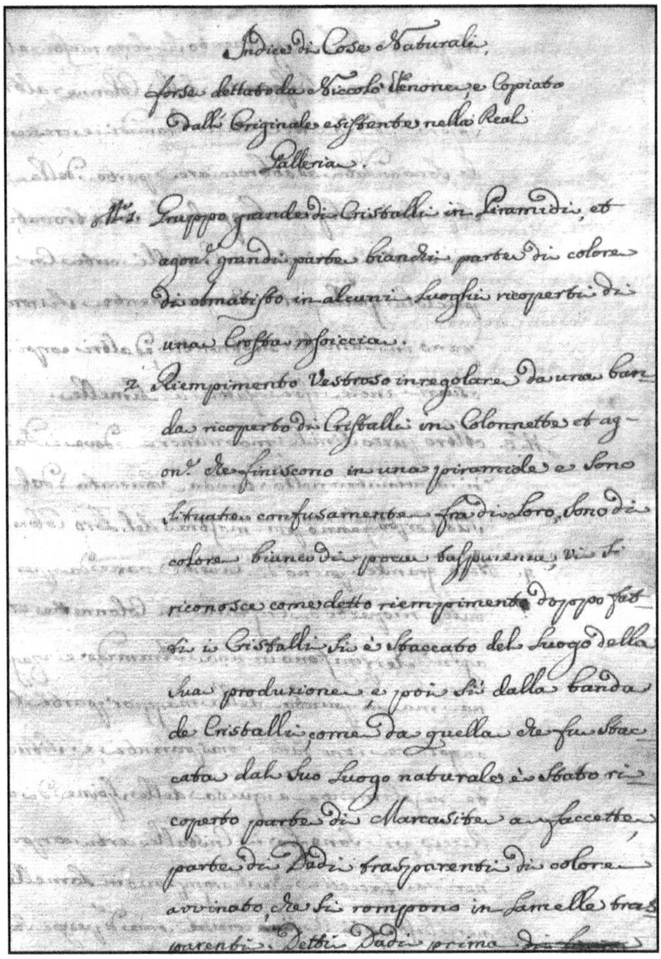
Indice di Cose Naturali
forse dettato da Niccolò Stenone, e Copiato
dall'Originale esistente nella Real
Galleria.

Figure 2. Title page of Steno's *Indice* contained in the "Catalogue of Natural Things" of 1763, a manuscript by Giovanni Targioni Tozzetti (folio 38v).

Table 1). Entries 200–214 are human artifacts, and 215–255 are mineralized parts of animals with plant-like shapes ("zoophytes" of the Pisa list). In separating inorganic forms from remains of ancient organisms, human artifacts, and zoophytes, the *Indice* shows an approach to taxonomy similar to that of *Metallotheca Vaticana*. Forms such as the "Islebian fish" (Fig. 4, Table 1), belonging also to Mercati's collection (Fig. 5), did probably already form part of the cabinet of curiosities of the grand duke. The attitude to name just some of the minerals and petrifactions suggests instead, unlike that of the late Renaissance collector, who would have named them all, a lack of interest for taxonomy and justifying Targioni Tozzetti's opinion, according to which, Steno's was "a very small, non methodic note on minerals and crystallizations" (Targioni Tozzetti, 1763c, "*Prefazione*," folio 5r). Moreover, many of the descriptions refer to natural aggregates of minerals or fossils bound together by a stony substance that would have been in itself worth of Steno's consideration, because it is the reciprocal contact between objects that reveals their order of formation ("principle of molding": Gould, 1981, p. 23; Cutler, 2003; Kardel, 2009). These specimens comprise the majority and would have perfectly served Steno's reasoning on solids: sea shells give shape to the rocks because rocks in the beginning were soft (Stensen, 1669, contrary to Mercati's statement that "forms cannot pass from objects to stones"). The second group of specimens, numbered 256–304, is even less homogeneous, containing again curiosities typical of a *wunderkammer* (such as item 257: "Human skull covered for the most part by a crust with a branch of red coral grown on its top," still in Pisa: Cipriani et al., 2011) and specimens from his travels both inside and outside of Tuscany (e.g., to Libbiano, Tirol, Bohemia, etc.).

Figure 3. Item no. 141 of Steno's *Indice,* described as: "two crusts made of stone around two wooden sticks, with the wood inside, united one to another, between which are attached certain small valves" (Targioni Tozzetti, 1763a, folio 50r/50v; see Table 1). This is now part of the mineralogical collections of the Museum of Natural History, University of Florence (Cipriani et al., 2011), catalogue number G54669. Dimensions of the specimen are 25 cm, 12 cm, and 3 cm.

In 1672, Steno left Florence and headed back to his hometown of Copenhagen for two years as anatomist to the Danish king. Leopoldo de' Medici entrusted the completion of the catalogue of the Grand-Ducal collection to Lorenzo Magalotti, also a member of the Accademia del Cimento, who was saddened "doubly, because in addition to losing a friend and may I say a spiritual father, I have been burdened with a very annoying occupation, which is that of the superintendence of the Museum of natural things" (Manni, 1775, p. 135). In 1691, the Sicilian botanist of the Giardino de' Semplici in Florence, Paolo Boccone (1633–1704), testified to the lack of order of the Natural History collection ("the gallery of natural things of the most illustrious Grand Duke Cosimo III [is] not exposed to the public, in Florence, for lack of order": Boccone, 1697, p. 178). Steno's order survived only in the *Indice,* whose meaning is still in part obscure.

Physical Geography of Tuscany

John Ray (1627–1705), the naturalist, traveler, and precursor to Carl Linnaeus (1707–1778) in proposing a taxonomy of animals and plants based on the reproductive system, testified to the fame soon reached in Europe by Steno's interpretation on the origin of fossils.

> I come now to give an account of the opinions of the best authors concerning the original and production of these stones. The first and to me most probable opinion is that they were originally the shells and bones of living fishes and other animals bred in the sea. This was the general opinion of the ancients, insomuch that Steno says, it was never made a question among them, whether such bodies came from any places but the sea. It has of late times and is now received and embraced by diverse and learned and ingenious philosophers, as in the precedent age by Fracostorius, and in the present by Nicolaus [*sic*] Steno and Mr. Robert Hook, after whom I need name no more to give it countenance and authority in the world. (Ray, 1673, p. 120)

The historical importance of Steno's *De Solido,* however, was not that it set principles that have guided those who came after him, but more importantly, that it was the first of a series of speculative theories of the Earth, whose authors appeared more and more as "world-makers," including Ray himself (Rudwick, 1972; Rappaport, 1997). Principles that are taught today in universities, such as the principle of superposition, were not passed on as such to Steno's contemporaries. His fame lasted for only a short time in England, overcome by the flourishing of more elaborated theories of the Earth until the end of the eighteenth century (Rudwick, 2005). *De Solido* quickly lost momentum and fell into oblivion nearly everywhere, except in Florence, thanks to the work of Giovanni Targioni Tozzetti himself. Not just an erudite, Targioni Tozzetti actively endeavored to put science to work for the benefit of his contemporaries—savants and laymen alike (Arrigoni, 1987). He learned to collect fossil shells, plants, and other objects of natural history under the tutelage of the botanist

TABLE 1. METADATA ASSOCIATED WITH TWO OF THE SPECIMENS OF STENO'S *INDICE* (FIGS. 3 AND 4): DESCRIPTIONS CONNECTING ENTRY IN STENO'S MANUSCRIPT OF 1672 TO THE CORRESPONDING ENTRIES IN SUBSEQUENT INVENTORIES AND CATALOGUES

	Steno's Pisa manuscript (De Rosa, 1986, p. 78–79)	Steno's *Indice* (Targioni Tozzetti, 1763a, 50r–50v)	Targioni's catalogue (Targioni Tozzetti, 1763b, p. 447–448)	1790–1793 museum catalogue	1820 museum catalogue	1844–1846 museum catalogue	Modern museum catalogue
Description	36 specimens, among which earth, stones with shells and Islebian stones.	N. 141. Two crusts of stone made around two small wooden sticks with wood inside, connected, among which some small shells are attached.	2324 - A petrifaction 7-and-a-half inches long and 21 lines large, looking like if in origin it was two sticks encrusted and bound together by sandy cement of the ancient sea. […] [follows Steno's description]	5996 - Brown, opaque fistular carbonate lime with shells impressed on its substance, shaped around plant branches, completely decomposed, from Germany. Room 2, shelf 18, lower part.	5003 - Brown, opaque fistular carbonate lime with shells impressed on its substance, shaped around plant branches, completely decomposed, from Germany. Room 12, shelf 5, lower part.	3417 - Brown, opaque fistular carbonate lime with shells impressed on its substance, shaped around plant branches, completely decomposed, from Germany. Room 2, shelf 18, lower part.	G54669 - Calcite. Brown, opaque fistular carbonate lime with shells impressed on its substance, shaped around plant branches, completely decomposed, from Germany.
Original text	36 p. tra terra, e pietre con nicchi, e pietre Islebiane.	N. 141. Due croste di pietra fatte intorno a due bastoncini di legno con legno dentro unite, fra di loro, fra le quali sono attaccati certi nicchietti.	2324. Una petrificazione lunga pollici 7 e mezzo e larga linee 21, che apparisce essere stata in origine due fuscelletti incrostati e riuniti insieme da tartaro renoso dell'antico mare. […] Lo Stenone nel suo catalogo così descrive questa petrificazione […]	N. 5996 - Calce carbonata fistolare bruna opaca che ha impressi dei tastacei sulla sua sostanza, e formata attorno a rami di piante, affatto decomposti, della Germania. Stanza II, scaffale 18 inf.	5003 - Calce carbonata fistolare bruna opaca che ha impresso dei testacei sulla sua sostanza, e formata attorno dei rami di piante affatto decomposti; della Germania. Stanza XII, scaff. 5 minor.	3417 - Calce carbonata fistolare bruna opaca che ha impresso dei testacei sulla sua sostanza, e formata attorno dei rami di piante affatto decomposti; della Germania. Stanza 2, scaffale 18 inf.	G54669 - Calcite. Calce carbonata fistolare bruna opaca che ha impresso dei testacei sulla sua sostanza, e formata attorno dei rami di piante affatto decomposti; della Germania.
Description	36 specimens, among which earth, stones with shells and Islebian stones.	N. 163N. A black Islebian stone with its fish of copper mine color.		N. 2301 - Impression of a fish on a slab made of cupriferous and micaceous shale; from Mansfeld in Saxony. Room 9, shelf 14, lower part.	9473 - Skeleton of a Gadus species; from Mansfeld [*sic*] in Saxony. Room 9, shelf 3, upper.	11793 - *Palaeoniscus Duvernoy* from the anthracitic shale, from Marsfeld in Rhine Bavaria. Scaffale 83 inf.	IGF12980 - *Paleoniscus vratislaviensis*, Agassiz 1833. Permian. Slaty bituminous shales. Mansfeld, Saxony.
Original text	36 p. tra terra, e pietre con nicchi, e pietre Islebiane.	N. 163N. Una pietra Islebiana nera col suo pesce di colore di Miniera di rame.		N. 2301 - Impronta di pesce sopra una lastra di scisto corneo, cuprifero, e micaceo; di Mansfeld in Sassonia. Stanza IX, scaffale 14 inf.	9473 - Scheletro di una specie di Gado; di Mansfeld [*sic*] in Sassonia. Stanza IX, scaffale 3 superiore.	11793 - *Palaeoniscus Duvernoy* dello scisto antracico, di Marsfeld nella Bavaria Renana. Scaffale 83 inf.	IGF12980 - *Paleoniscus vratislaviensis*, Agassiz 1833. Permiano. Scisti ardesiaci bituminosi. Mansfeld, Sassonia.

Figure 4. Specimen IGF12980 tentatively identified with item no. 163N of Steno's *Indice,* concisely described as: "A black Islebian stone with its fish with Copper Mine colour" (Targioni Tozzetti, 1763a, folio 53r; see Table 1). "Islebian" refers to the town of Eisleben, in Saxony (Germany), near the locality of Mansfeld, well known to Renaissance apothecaries and physicians for the fossil fishes that were dug up there, the so-called "*Lapis Islebianus*." Dimensions of each half of the specimen are 12.5 cm, 8.5 cm, and 1.0 cm.

Figure 6. Fossils and catalogues of Giovanni Targioni Tozzetti's private museum: a Tertiary shell named "Serpula," with its catalogue number (n. 51); relative page of the catalogue, where we learn that it belonged to Pier Antonio Micheli's museum acquired by Targioni Tozzetti in 1739; plate with "Serpula n. 51" hand-drawn by Giovanni's son, Ottaviano Targioni Tozzetti.

Pier Antonio Micheli (1679–1737). Targioni Tozzetti authored six volumes of his travels starting from 1749 and amassed in Florence a large private museum of natural history (Fig. 6), visited by savants from all over Europe (Cipriani and Scarpellini, 2007). His theoretical framework and the concept of "physical chorography," directly derived from Steno's geometric approach to the study of Tuscany (Targioni Tozzetti, 1754, containing also the taxonomy used to order the Granducal collections), was raised to international renown in 1795 through a review by Nicolas Desmarest (1725–1815), in the French *Encyclopédie Methodique.* When compiling the first volume of *Geographié-Physique* of 1795, Desmarest gave a full account of Targioni Tozzetti's input to a science that would later be incorporated into geology *tout court* (Laboulais-Lesage, 2004; Rudwick, 2005) and included his name among the great theorists of the seventeenth and eighteenth centuries, side by side with Steno, Buffon, and Hutton.

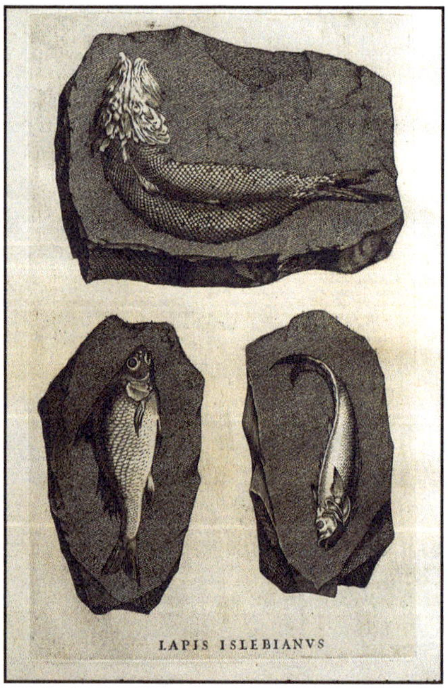

Figure 5. *Lapis Islebianus* or "Islebian stone" from Michele Mercati's *Metallotheca Vaticana* completed in 1593, unpublished, but accessible to Steno during his stay in Florence.

> Targioni had the good luck to have Tuscany as the theater of his observations, and to be guided at the same time by two skilled naturalists, Steno and Micheli. [...] Tuscany offers in a small extension a large variety of soils and terrains, that reunites under the same point of view the relative relationships, surprising for the contrast of their structure and for the different nature of the stony and earthy substances that compose them, allowing to easily understand their limits. These circumstances are essential to determine the order and the époques of nature. [...] Targioni has not ceased to meditate and reflect [on the primitive mountains and secondary hills] to shed light on the great system devised by Steno on the different deposits of the sea. (Desmarest, 1795, p. 525, 530)

Targioni Tozzetti corresponded on the fossils of Tuscany with Georges-Louis Leclerc de Buffon (1707–1788: Targioni Tozzetti, 1755), and there are reasons to believe that he passed on to the geognost Giovanni Arduino (1714–1795; see Vaccari, 2006) the idea that geometric relationships define the order of deposition of strata (Dominici, 2009). The first to testify to this debt is Arduino himself, in one of his famous letters, written in 1759.

In this way, from stratum to stratum, various kinds of rocks and other fossils succeed one another, as far as the loftiest summit of the Alp: and very high up one finds various beautiful seashells [...] All of the innumerable and quite diverse strata which make up this mountain, and indeed all of those vast ranges of mountains and Alps [...] can convince any mind capable of sound reasoning that at one time they were all ongoingly continuous: and that they were broken and disconnected simply by the force of flowing waters, which through immensely lengthy work created them, and which even now is creating so many bizarre formations and deep, tortuous valleys. This truth, extended to the whole terrestrial surface, has long been known to many men of great merit: and the aforementioned Signor Targioni provides its most convincing demonstration, with his observations concerning the mountains and hills of Tuscany (which I have in large part seen myself), as all may read in the [...] accounts of his travels, which especially concerning mineralogy and the theory of the Earth contain things of high regard. (Arduino, 1759, *in* Ell, 2011, p. 279)

Targioni Tozzetti made Tuscany internationally known and promoted Steno's conceptual framework, stimulating others to consider the importance of Tuscan "organic remains" in relationship with their stratigraphic occurrence. As he made explicit in the last volume of the 1763 catalogue of the Royal Gallery, any system of classification of things "dug up from the ground" is connected to a particular theory of the Earth (Targioni Tozzetti, 1863d).

Improved Taxonomies

Names convey the substance of things and are aids for the memory of the philosopher. Late Renaissance cabinets, with heterogeneous and idiosyncratic collections, enlarged and gradually gave way to seventeenth-century museums, "repositories of the collective imagination of their society" (Findlen, 1994, p. 9) and the context of scientific practice (Arnold, 2006). An eminent example of this was the museum of Athanasius Kircher (1602–1680) in Rome, with taxonomies perfected and published by his disciple, Paolo Bonanni (1638–1723), as a sound basis for contemporary knowledge on modern and fossil shells (Rappaport, 1997). In the first half of the eighteenth century, the Medicean natural history collection had been lost in part by the negligence of the son of Cosimo III and last grand duke of his dynasty, Gian Gastone de' Medici (1671–1737), who allowed the Florentine naturalist and collector, Niccolò Gualtieri (1688–1744; Manganelli and Benocci, 2011), and the naturalist, Giovanni de Baillou (1679–1758), appointed in 1735 director of the *Real Galleria,* to use the collections as they pleased. In fact, *chevalier* Baillou accumulated, during the years of his stay in Florence, a collection "rather for a prince than for a private person," in the words of a British contemporary, where "minerals and other fossils" had an important role. Baillou appropriated Medicean specimens (Targioni Tozzetti, 1763c, "Prefazione," folio 4r) and included them in the collection he sold in 1749 to Emperor Francis III of Lorraine (1708–1765) and sent to Vienna (Olmi, 2012). After the death of Micheli in 1732 and Gualtieri in 1744, Targioni Tozzetti remained the only taxonomist who could impose order of the collection that remained. In the 11 volumes of the handwritten "Catalogue of the Mineralogical Collection" of his own private museum, he had included the fossils among the minerals and left a dedicated volume for fossil shells, mostly coming from Tuscany (Cioppi and Dominici, 2011). Targioni Tozzetti's revision of the natural history collection of the grand duke resulted instead in five volumes comprising "2340 things of the Animal Kingdom, 375 of the Vegetal Kingdom, and 734 of the Fossil Kingdom" (Targioni Tozzetti, 1763a, p. 2v). Targioni Tozzetti never adopted the binomial system introduced by Linnaeus, and minerals were dealt with in the fifth volume of the catalogue (Targioni Tozzetti, 1763c) under the heading "Fossil Kingdom" (the category was informally used in Italy until the nineteenth century; e.g., Pedrini, 1791; Breislak, 1811). Fossils of organic bodies were included (1) at the end of the second volume of the Animal Kingdom, under the heading "Testacea fossilia" ("fossil shells"); (2) in the appendix "LAPIDES CONCHYLIATI vulgo LUMACHELLE" ("shelly stones"; Targioni Tozzetti, 1763b, p. 411, 442, respectively; see also Table 1); (3) in the third volume with the plants; and (4) in the fourth volume, under the heading "Steno's *Indice*" (Fig. 2). To Targioni Tozzetti, the classification of the "fossil kingdom" was subordinated to a theory of the Earth; but since theories are volatile, so are systems of classification.

I did not want to include in this third part [the Fossil Kingdom, i.e., volume 5] those many fragments of organic bodies—Animals and Plants, found underground, turned into stone or not, of which a worthy collection is seen in this Imperial Gallery, formed as it seems by the famous Nicolas Steno; I've distributed them instead in the respective natural classes [...] I feared that the novelty of my system [Targioni Tozzetti, 1754], not yet sufficiently digested, would have been little liked. [...] Not even the modern system of the celebrated Carl Linnaeus, notwithstanding in some parts it is better than the ancient ones, is a sufficient Ariane's thread. (Fig. 7: Targioni Tozzetti, 1763d, folios 1v–2v)

Grand Duke Peter Leopold (1747–1792), who ruled Tuscany in 1765–1790, entrusted Felice Fontana (1730–1805), physicist and physiologist from Trento, to establish a Royal Museum of Physics and Natural History, which was projected and realized during the 1760s and the early 1770s and opened in 1775 as one of the leading museums of Europe, the first public institution directed also to the general public (Fig. 8). The museum contained instruments that belonged to the Medici dynasty, including some of Galileo's and those of the Accademia del Cimento, and tens of machines and other devices that Fontana had built himself. Natural history collections, separated a few years before from the Galleria degli Uffizi and catalogued by Targioni, were included in the heritage, although in a very subsidiary position with respect to the instruments. In the *Saggio del Real Gabinetto di Fisica e di Storia Naturale,* in Fontana's words (Contardi, 2002), 29 pages were dedicated to the description of scientific

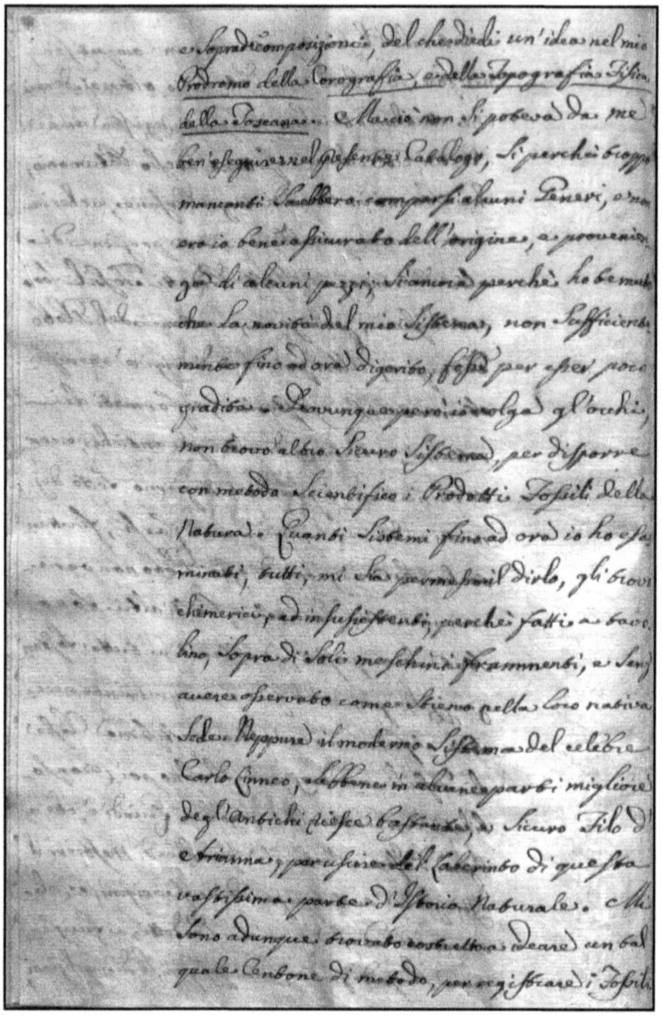

Figure 7. Remarks opening the fifth volume of the catalogue written by Targioni Tozzetti for the Royal Gallery, dedicated to the "fossil kingdom," where he mentions "things dug up from the ground" in relationship with his theory of the Earth (Targioni Tozzetti, 1754) and Linnaeus' system (Targioni Tozzetti, 1763d, folio 2v).

Col presente foglio Sarà permesso vedere il Real Museo alle ore 10. della Mattina del dì

Il Direttore

Si avverte che passati quattro, o cinque minuti dopo l'ora prefissa non sarà altrimenti permesso l'ingresso nella Sopraindicata mattina.

Figure 8. Ticket of 1775 to the Royal Museum of Physics and Natural History of Florence, one of the oldest museums to open to the general public.

instruments and their use, but only two pages for the whole natural history collections, with just a few words on the new system of classification that had been adopted.

> The symmetry that reigns in and out the shelves, the order and regularity of the specimens is such and so new that it enraptures the observer, and at the same time teaches him, and amuses him. Everything is seen ordered by class—Shells are ordered according to the method of Argenville [Antoine Joseph Dezallier d'Argenville, 1680–1765]. With the fossils [i.e., minerals] the system of Valerio [Johan Gottschalk Wallerius, 1709–1785] was adopted; with the animals that of Linnaeus. Minerals and earths, in the number of about ten thousand, are inside small crystal cups [...] everything is under the eye and at the same time protected by crystals from humidity and dust. (Fontana, 1775, p. 30)

When Fontana traveled throughout Europe for five years between 1775 and 1780, he made no substantial effort to increase the natural history collections of the Royal Museum, and among these, upon his return, he mentioned mainly zoological specimens, leaving very little room for true fossils (Cipriani et al., 2011). In presenting a plan for teaching in the New Accademia del Cimento that he imagined, natural history held a marginal place.

> Natural History, the way it is presently distributed in the Royal Museum, does not need to be taught by a professor, nor does it need a public course, and it is in such a way ordered so as to need no one. A course of natural history is nothing else but a methodic explanation of the various products that are confusedly dispersed here and there upon the earth. [...] In the present, to learn natural history whoever you are in the Royal Museum, all you need is to know how to read. (Fontana, *in* Contardi, 2002, p. 294)

Visitors to the Florentine museum, however ecstatic with admiration for the exhibitions (Olmi, 2006), would have perceived fossils as links that connect the inanimate world of minerals with the living world of animals and plants, with zoophytes such as corals and sponges as intermediates between animals and plants. It was the popular incarnation of the great chain of being (Lovejoy, 1936): in the inextricable labyrinth of the products of nature, an ascending path connected chemistry and mineralogy (ground floor of the museum), with human and zoological sciences (first and second floor), with astronomy (observatory on top of the building), and formed a means to explain the multifaceted expression of God's creation. In this vision, the watchword was to fill in the gaps of the continuum of the exposition (Contardi, 2002).

Lost Species

By the 1790s, Giovanni Fabbroni (1753–1822), who was firmly established as the true administrator of the Florence museum following Fontana, had important contacts with some of the world's greatest naturalists—people such as Thomas Jefferson (1743–1826) in Virginia, Joseph Banks (1743–1820) in London, Richard Kirwan (1733–1812) in Dublin, and Déodat Gratet de Dolomieu in Paris (1750–1801), who were bringing new insights in the vast field of geotheory. An Inventory of the Royal Cabinet of Physics and Natural History was issued in 1790, allowing for an insight in the state of taxonomies during the years immediately following the French Revolution, when Paris became the cradle for the early development of modern geology, chemistry, and mineralogy. Specimens of the Fossil Kingdom were classified in Florence by the botanist and zoologist curator, Attilio Zuccagni (1754–1807), who rightly overcame the classificatory system of Wallerius introducing the concepts of Richard Kirwan and René Just Haüy (1743–1822) and the use of quantitative chemistry and crystallography in the study of minerals (Bardi, 1808; Cipriani et al., 2011). The emergence of modern mineralogy coincidently followed that of a distinct field of enquiry of Earth's history based on the study of rocks and fossils, one that the Swiss savant, Jean-André Deluc (1727–1817), successfully named "geology" in 1778 (*geologie* shifted quite rapidly from denoting a system, or geotheory, to denoting a science: Rudwick, 2005; Heilbron and Sigrist, 2011). When Déodat de Dolomieu sent a small lot of minerals to Florence, exact localities were for the first time reported side by side each specimen (Cipriani et al., 2011), a practice that set them apart from their sole aesthetic value, or relative position in the atemporal great chain of being, to become an instrument for reconstructing Earth's history. As it was clear to Dolomieu, a geotheorist at the center of the revolution, the exact geographic (and stratigraphic) position of a specimen meant in fact its relative age and was of primary importance in shedding light on pre-human history (Rudwick, 2005). Unfortunately, no one in Florence was prepared to explain the potential of this new approach to natural history. All paleontological entries in the 1790s inventory testify to this lack of knowledge, such as "a molar of fossil mammoth from the river Ohio, in America," the American mastodon found at Big Bone Lick, in Kentucky, studied by Thomas Jefferson himself (Cioppi, 2017), or "a natural cast of a turreted turbine of extraordinary size [...] from the Bolca Mountain," belonging to a family of gastropods that Jean Guillaume Bruguière (1750–1798) had described in the *Encyclopédie Méthodique* in 1789–1792 (Bruguière, 1792, p. 472–473); an exact provenance and an informed taxonomy for these entries were missing in the Florence catalogue (Fig. 9).

The approach to taxonomy was to rapidly change everywhere in Europe. In this respect, Fabbroni's most important contact at the Museum National d'Histoire Naturelle of Paris was Georges Cuvier (1769–1832). Building on the conceptual framework established by Jean-André Deluc, Cuvier had shown an early aptitude to see rock formation and fossils as actual documents of Earth's history and which monuments and coins were for antiquarians (Rudwick, 1997, 2005). In 1796, he presented to the public of the Parisian museum his first anatomical comparisons of fossil species (*Mammuthus primigenius*, *Megatherium americanum*, adopting Linnaean binomial names) with their modern living analogues (respectively, the Asian elephant and the ground sloth). In a call for international collaboration in 1800, he expressed in public that the practice of comparative anatomy could open the door to uncover the history of the Earth.

> Like men, the sciences have their stages of life. Given up in youth to brilliant imaginative illusions, they become cooler and more reasoned in maturity. [...] The first [geniuses] guessed at nature rather studying it; the others, while thinking they are only verifying the systems they admire, study it truly; and it is thus that the sciences—like peoples—pass from poetry to history. The theory of the earth has thus taken a new direction in the past twenty years. [...] Mountains, veins, and strata have been penetrated in all directions; one has assembled their materials and compared them with one another, and already we possess a mass of genuine knowledge that far surpasses all that could

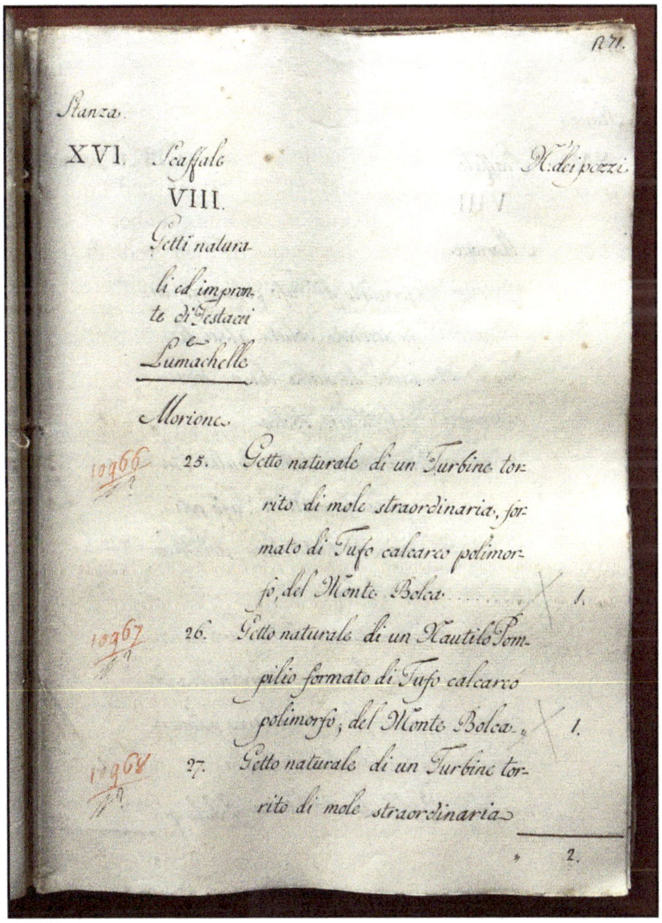

Figure 9. In the inventory of 1793 "organised fossils," exposed in Room 16, were part of the mineral category. This page describes a specimen (*Cerithium giganteum* of authors) belonging to a taxonomic group that had been described in Paris by Bruguière in 1789–1792 (Cerithioidea of modern taxonomy). "Monte Bolca" comprises a series of fossiliferous localities of different age.

have been hoped for when this method began to find favor. There is, however, one part of the animal kingdom whose fossil remains have been less studied, namely the quadrupeds. (Cuvier, 1801, *in* Rudwick, 1997, p. 46–47)

By the study of the bones of large terrestrial animals, without the risk of having missed some of the living analogues, Cuvier could safely provide a proof of the reality of extinction.

We can compare and decide with sufficient certainty whether any fossil bone does or does not resemble the analogous bone in living species. The case is not the same with shells and fish: naturalists are still far from having observed all of them; and every time we find an unknown fossil fish or shell in the earth we can always suppose that the species is still living in distant seas or at inaccessible depths. [...] The number, direction and shape of the bones composing each part of the body determine the movements that that part can make, and consequently the functions it can fulfill. [...] One finds abundantly, under the soil in all countries, bones different from those of the animals that live at the earth's surface today [...] The older the beds in which these bones are found, the more they differ from those of animals that we know today. (Cuvier, 1801, *in* Rudwick, 1997, p. 48–52)

Cuvier's strategy was very effective in forming a global network of learned informants and quickly established him as the genius behind the science that an increasing number of savants were calling "geology," the term having by then gradually spread outside the realm of the French language of its origins (e.g., Fortis, 1778; Olivi, 1792; Correa de Serra, 1799).

The most celebrate foreign naturalists, Messrs. Blumenbach, Camper, Fortis, Fabbroni (...), owners of the most beautiful collections and trustees in public museums in France and abroad, have helped me with their advice and facts that have come to their attention, and have informed me about the specimens that are found at their disposal. Such men should encourage others to follow their example, and I have no doubt that they will find worthy imitators. (Cuvier, 1801, *in* Rudwick, 1997, p. 57–58)

Fabbroni had Cuvier's exhortation translated into Italian by Fabbroni's son, Leopoldo (Fabbroni, 1802) (who also attempted some description of Florentine fossil elephants) and sent detailed drawings to Paris; these drawings were later reproduced in engravings of *Rechérches sur les Ossemèns fossiles* (Fig. 10; see also Rudwick, 2005). Meanwhile, Fontana and Fabbroni were succeeded in the direction of the Florence museum by Girolamo Bardi (1777–1829) during the regency of Maria Luisa of Spain. On 20 February 1807, the queen appointed the teaching of six courses at the museum: astronomy, physics, chemistry, zoology and mineralogy, botanics, and comparative anatomy. In Florence the professor of zoology and mineralogy and curator of fossils, Filippo Nesti (1780–1849), openly encouraged his students to avoid considering the history of fossils as an appendix of mineralogy, praising the innovations introduced by Cuvier and attempting in the first volume of the Annals of the Royal Museum of Natural History the first modern taxonomic study of fossil elephants of Tuscany.

Even though these fossilized remains should be appreciated for the species to which they belonged, and not for the mineral substance that has penetrated them; that might even have been forgiven until the lovely works of Pallas, Camper, Merck and especially Mr. Cuvier closely related the knowledge of fossils, especially mammals, to comparative anatomy, and developed all that could be desired in the history of an order of beings for the best part lost. Therefore, I believed that I had to introduce to each class the history of the remains of those animals belonging to it which are in the state of fossils and to this history I have added a number of observations on the conformational relationships between the solid parts and the soft ones, so that scholars could learn to deduce from certain general forms to which order the animal belonged. (Nesti, 1808a, p. 6)

At the height of his power in the Napoleonic administration, Cuvier traveled to Tuscany in 1809–1810, guided by Filippo Nesti, to meet local collectors and see the upper Arno valley, where most of the celebrated fossil vertebrates came from. No longer needing to convince skeptics of his theory, but only to further prove to himself that anywhere fossil quadrupeds were dug up, they belonged to lost species, Cuvier left the description of the new forms to the young Nesti, in his turn fully aware that he

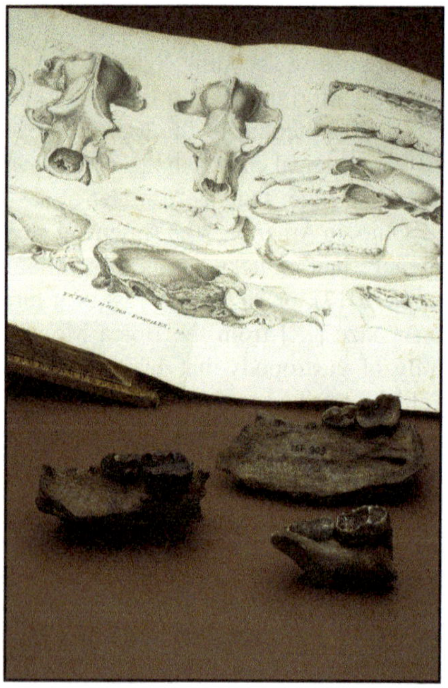

Figure 10. Type material of *Ursus etruscus* from the Florentine collections (Cuvier, 1823, his plate 27, and his figures 8 and 10–11).

was contributing to the unraveling of the eventful and unidirectional history of the Earth.

The earth, in its alluvial strata, contains the monuments of disastrous events; a crowd of organized beings belonging to ancient epochs stems from its bowels, as a gift for the analysis of the naturalist; he sees with surprise that the forms of the animals contained in it, so much differ from analogue species now living as much is old the terrain where they lay. [...] Truth is mingled among many contradictory facts and a thousand doubts, so that the explanation of geological facts is actually an intricate problem that won't be possibly ever untangled. It is however important that the history of fossils is reconstructed in every province where these are buried, and that this history is, as much as possible, described in all the details concerning position, height above sea level, direction of strata, quality and position of surrounding terrains and mountains, so that from the extension of certain particular facts over many points of the earth, one can deduce the way in which the general cause that operated such a great revolution on the globe acted. (Nesti, 1808b, p. 3)

In subsequent works, Nesti fully described rhinoceros bones from Poggio Monte al Pero (Nesti, 1811), mastodon from near the monastery of Montecarlo, and, following criticism by Cuvier, the "southern" mammoth, *Elephas meridionalis* (the first valid description of this species; Nesti, 1825; see Falconer, 1868).

All Is Not Lost

The next notable visitor to the Florentine collections was Giambattista Brocchi (1772–1826). As Cuvier shined a light on vertebrate fossils historically dug up in Tuscany, giving a modernized impulse to the acquisition and study of new collections, so did Brocchi with the invertebrates. Trained in Veneto, following teachings of Fortis and Arduino, appointed curator of the Milan Museum of Natural History in 1808, when Milan was the capital of the Napoleonic Kingdom of Italy, Brocchi was an erudite geologist with wide interests that included historiography, mineralogy, and geology. He visited Tuscany in the summer of 1811 during his geological tour of Italy, collected fossils in the field, and studied specimens found in public museums and in private hands. His largely unpublished diary is a precious source of information, offering an unbiased vision of the state of the Florentine private and public collections.

I stopped particularly to examine the products of Tuscany [...] all have been dug up from the Arno valley. The collection of Mr. Targioni is very instructive, since it offers a series of rocks of Tuscany marked by trivial names such as galestro, bardellone, mattaione, macigno, pietra forte, alberese, names often used by Targioni in his work. The collection of fossil shells is not particularly rich and did not fulfil my expectations; many of them are from unknown localities, only it is known that they are from Tuscany. [...] 16 August. I've examined with more attention the collection of fossils of the Royal Museum. Bones and tusks of elephants are in large number; there are also bones of rhinos, deer horns, teeth and bones of other quadrupeds, all dug up in the Arno Valley. (Brocchi, 1811–1812, entries for 15–16 August 1811, translated by authors)

A description of both marine and terrestrial vertebrates found in Florence at that time is reported on *Conchiologia Fossile Subappennina* (Brocchi, 1814: v. 1, p. 175, 178–184, 189–195, mentioning Targioni Tozzetti's collection, and Steno's specimens on p. 183). More important than this local insight, however, the book played a role in the ongoing geological debates, raising a completely new interest on Tuscan fossil shells and on all marine Tertiary formations. Brocchi's book was soon read by and positively commented upon by leading geologists of Paris, London, Oxford, and Edinburgh (Pancaldi, 1991; Rudwick, 2005; Dominici, 2010). Brocchi's new-found fame was based on three essays of international resonance contained in the *Conchiologia*. Opening the volume was a historiographic essay entitled, "Discourse over the Progress of the Study of fossil Conchology in Italy," the first of his genre, which was almost carbon-copied by Charles Lyell (1797–1865) in the first volume of his *Principles of Geology* (Lyell, 1830; see McCartney, 1976). In tracing this history from the late Middle Ages to the present, consolidating the fame of Tuscan natural philosophers and naturalists of all ages, Brocchi briefly reappraised the pioneer role of Nicolas Steno in proving the organic origin of fossils (overlooking all his other inputs to knowledge and method—understandably, given the topic of the essay). Praising Giovanni Targioni Tozzetti for shedding light on the physical constitution of the Tuscan relief, he proclaimed at the same time the failure of all Italian eighteenth-century taxonomic practice, Targioni Tozzetti's included.

For want of better means, at that time Italian naturalists did not determine fossil species, but with the aid of pictures in books, namely Gualtieri's, Bonanni's and Argenville's. So did Allioni in his Specimen oryctographiae pedemontanae [1757], the first treatise of fossil conchology published in Italy [...] Both Allioni and Targioni confined themselves to point out only the species that corresponded to the figures that they had at hand, and as a consequence many [species] they neglected, or never represented. [...] Should we believe that he [Allioni] never stepped in more than one misunderstanding by simply sticking to a comparison with figures? (Brocchi, 1814, p. XLIII, 144)

The second essay was entitled, "Geological Observations on the Apennines and the Adjoining Soil," a review of all the geological formations known at that time in the peninsular part of Italy, and a geological account of the terrains that he personally visited. In Edinburgh, the book was reviewed by Leonard Horner (1785–1864) of the Geological Society of London (established in 1807). Born and raised in Edinburgh, brother of the whig politician, Francis Horner, Leonard Horner reviewed the geological essay in 1816 for the journal *Edinburgh Review* and personally wrote to Brocchi that the book was "one of the most interesting that has appeared since Geology begun to be studied as a branch of true science" (Pancaldi, 1991; Rudwick, 2005). Sticking to the inductivist philosophy that was by then driving the activities of the Geological Society, and wishing to keep away from all the speculations that had spoiled the debate on the history of the Earth, Horner reviewed the book only for its geological data. On the contrary, he helped promote Brocchi's speculations on the long-term history of species, the subject of the pages that ended the first volume (Eldredge, 2015). In the famous essay entitled, "Reflections on the Losing of Species," Brocchi proposed that species, like individuals, die for natural, inner causes, and that, as individuals of different species have average lives of different duration, so some species last for longer stretches of geological time, others for a more brief interval (Dominici, 2010; Dominici and Eldredge, 2010). The facts collected in the second volume, dedicated to Italian Tertiary marine shelly invertebrates and enhanced by wonderful copper engravings (often portraying fossils collected in Tuscany: Fig. 11), were based on Linnaeus' taxonomy, perfected by Jean-Guillaume Bruguière and Jean-Baptiste Lamarck (1744–1829). With this monograph, Brocchi had entered a debate that was opposing Cuvier and Lamarck. The two French naturalists had been actively publishing, by then for nearly 20 years, on the large collections of fossils from the "younger Secondary" formations of the Paris Basin. Molluscs were the field of expertise assigned to Lamarck, in the footsteps of Bruguière since the opening of the Paris museum in 1794. Focusing on "Tertiary" marine molluscs, and stepping into Lamarck's field, Brocchi came up with an explanation of the fossil record alternative to that of Bruguière, Lamarck, and Cuvier. To Bruguière, species might not be lost, those known as fossils needing only to be discovered in modern seas (Bruguière, 1792). To Cuvier, all fossil species are distinct from modern species, morphologically stable in time, and become extinct because of some environmental catastrophe of a global nature (Cuvier, 1812). To a mature Lamarck (Lamarck, 1809), species are entities that exist only in the mind of the naturalist, not in the real world, and extinction is only apparent, since "species" gradually transform one into the other, under that active force of environmental factors that act during immense stretches of time (Burkhardt, 1977; Rudwick, 2005). In Brocchi's analogical thinking, species, like individuals, have a birth and a death, and a life in between, thus espousing Cuvier's evidence that they are real and stable entities, and that extinction is a real phenomenon. But unlike Cuvier, he proposed that their life is governed in deep time by inner factors that control their destiny. Brocchi was, like Lamarck, a highly competent taxonomist, and his evidence was based on the extended survey of extant molluscs that lived at that time in the Adriatic and the Mediterranean carried out by two other competent taxonomists from Veneto, Giuseppe Olivi (1769–1795; Olivi, 1792) and Stefano Andrea Renier (1759–1830). This comparison proved that ~50% of Italian Tertiary species were still perpetuating their form and, at the same time, that the other half were really extinct—contrary to what Lamarck claimed. Although Brocchi never mentioned it, it was clear to early geologists that none less than transmutation of species was at stake, all evidence for and against being taken from the study of fossils. Effects of his input to early transmutational studies would eventually impinge on the thinking of a young Charles Darwin (1809–1882), a couple of decades later (Dominici and Eldredge, 2010; Eldredge, 2015).

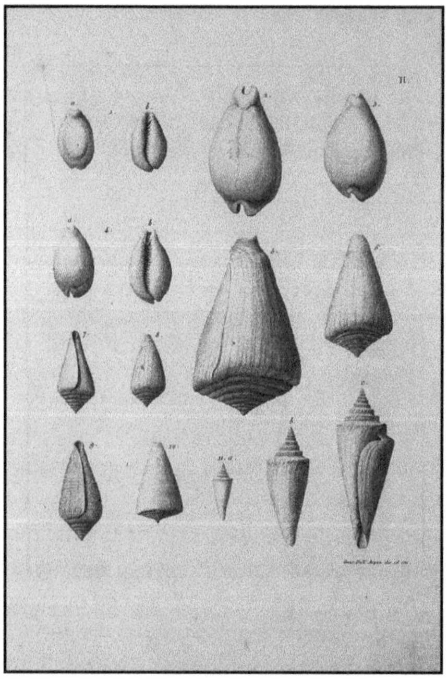

Figure 11. Plate 2 of *Conchiologia Fossile Subappennina,* with the type specimen of *Conus mercati* in the forefront, a species dedicated by Brocchi to Michele Mercati.

Brocchi's fame never reached in his lifetime that of his two French peers, possibly because his book was written in a language (Italian) that ensured oblivion, or because he never engaged in a further public confrontation, or because of his untimely death in Egypt, in 1826. But *Conchiologia Fossile Subapennina* triggered a renewed interest in Tertiary marine invertebrates so plentifully found in Tuscany and their interpretation as tools to measure relative age of rock formations. Geologists visiting Italy following Brocchi's lead were many and important, coming from Germany, England, and France (Rudwick, 2005). In 1828, Charles Lyell, Horner's son-in-law, came to collect fresh data on the stratigraphic distribution of Tertiary molluscs, depending much on Brocchi's monograph (McCartney, 1976) and on the works of other continental geologists (Rudwick, 2008; Secord, 2015). As the second historiographer of fossil conchology, Lyell reviewed in more depth the role of Steno, with some interpretation on Steno's intentions.

> The most remarkable work of [the seventeenth century] was published by Steno, a Dane, once professor of anatomy at Padua [*sic*], and who afterwards resided many years at the court of the Grand Duke of Tuscany. The treatise bears the quaint title of *"De Solido intra Solidum contento naturaliter"* (1669) by which the author intended to express

"On Gems, Crystals, and organic Petrifactions inclosed within solid Rocks." This work attests to the priority of the Italian school in geological research; exemplifying at the same time the powerful obstacles opposed, in that age, to the general reception of enlarged views in the science. Steno had compared the fossil shells with their recent analogues, and traced the various gradations from the state of mere calcination, when their natural gluten only was lost, to the perfect substitution of stony matter. He demonstrated that many fossil teeth found in Tuscany belonged to a species of shark; and he dissected, for the purpose of comparison, one of these fish recently taken from the Mediterranean. [...] Tuscany, according to him, had successively past through six different states; and to explain these mighty changes, he called in the agency of inundations, earthquakes, and subterranean fires. His generalizations were for the most part comprehensive and just; but such was his awe of popular prejudice, that he only ventured to throw them out as mere conjectures, and the timid reserve of his expressions must have raised doubts as to his own confidence in his opinions, and deprived them of some of the authority due to them. (Lyell, 1830, p. 28)

The interest in fossil shells from the hills of Tuscany did not diminish for decades after Lyell's work. In the Florence museum, this is documented by the Tertiary faunas collected and described by Vittorio Pecchioli (1790–1870; Fig. 12), Cesare D'Ancona (1832–1908), Angelo Manzoni (1842–1895), Dante Pantanelli (1844–1913), Carlo De Stefani (1851–1924), and Alberto Fucini (1864–1941), often from outcrops visited by Brocchi himself, such as near San Miniato and Empoli (two localities whose fossils had been mentioned by Leonardo da Vinci and Michele Mercati), Siena and Volterra, or from newly discovered shell beds near Orciano Pisano and Vallebiaia. Thanks to Lyell's extensive work on Tertiary European formations and the introduction of finer stratigraphic subdivisions (epochs), most of these strata are now attributed to the "Pliocene" ("Pleistocene" came in 1839), and their relative age is distinguished with respect to that of Paris Basin formations, belonging to the older "Eocene" epoch (Rudwick, 2008).

Figure 12. Plate of a monograph of Vittorio Pecchioli on Tuscan Pliocene shells, with some of the figured specimens.

Brocchi, the first Italian geologist who described this newer group [Pliocene formations] in detail, gave it the name of the Subapennines, and he classed all the tertiary strata of Italy, from Piedmont to Calabria, as parts of the same system.[...] In a catalogue, published by Lamarck, of 500 species of fossil-shells of the Paris basin, a small number only were enumerated as identical with those of Italy, and only 20 as agreeing with living species. This result, said Brocchi, is wonderful, and very different from that derived from a comparison of the fossil-shells of Italy, more than half of which agree with species now living in the Mediterranean, or in other seas, chiefly of hotter climates [...] We have already satisfactory evidence that the Subapennine beds of Brocchi belonged, at least, to three periods. To the Miocene we can refer a portion of the strata of Piedmont, those of the hill of the Superga, for example; to the older Pliocene belong the greater part of the strata of northern Italy and of Tuscany, and perhaps those of Rome; to the newer Pliocene [Pleistocene formations], the tufaceous formations of Naples, the calcareous strata of Otranto, and probably the greater part of the tertiary beds of Calabria. [...] Cuvier also mentions the remains of a species of lophiodon as occurring among the bones in the Upper Val d'Arno. The elephant of this locality has been called by Nesti *meridionalis,* and is considered by him as distinct from the Siberian fossil species *E. primigenius,* with which, however, some eminent comparative anatomists regard it as identical. The skeletons of the hippopotamus are exceedingly abundant; no less than forty had been procured when I visited Florence in 1828. Remains of the elephant, stag, ox, and horse, are also extremely numerous. (Lyell, 1833, p. 155–157, 221)

The Social Embodiment of Geology

Paleontology started as a field of study distinct from geology in 1822, when the word was coined by Henri de Blainville (1777–1850), successor of Lamarck in teaching at the Paris museum. The discipline had become by then autonomous, not just a tool for stratigraphy. A decade later, with Lyell's three volumes of the *Principles*, geology acquired a wide readership in Great Britain (Rudwick, 2008; Secord, 2015) and, although not ceasing to be controversial, it became "socially embodied" (Rudwick, 1985). The recognition of its role in society, culture, and economy, however, took a longer course in the rest of Europe. At the end of the Napoleonic era, Grand Duke Ferdinand III (1769–1824) had ordered in Tuscany the elimination of the chairs associated with the Royal Museum of Florence, while retaining the more traditional activities of conservation and research, with Nesti in the role of curator. As the old inventory had grown to acquire "a monstrous form," the compilation of a new inventory with "a more exact nomenclature that alone can value objects" was urged by Bardi in 1815, and a new catalogue was started in 1820 (Cipriani et al., 2011). The teaching activities resumed in 1833 when the chair of mineralogy was separated from that of zoology and comparative anatomy, the former entrusted to Nesti and the latter to the Sienese Gaspero Mazzi (1787–1867; since 1838 director of the Royal Museum). The mineralogical-geological and the paleontological collections were then separated. The specimens that had belonged to Micheli and Targioni, passed on to Ottaviano Targioni at the death of his father Giovanni, were acquired by the museum (Fig. 13). In 1839, activities of the First Congress of Italian Scientists held in Pisa were subdivided into six sections,

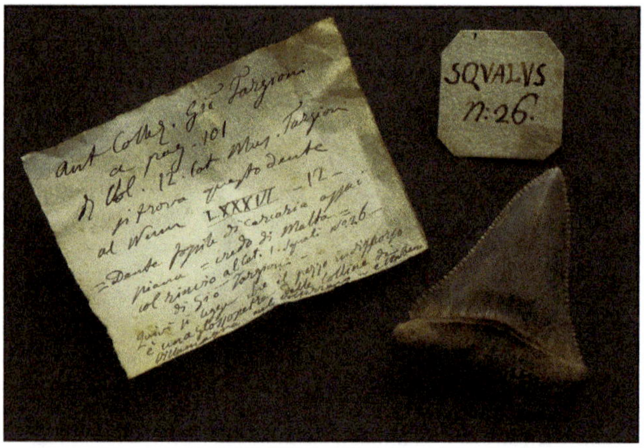

Figure 13. Tooth of the great white shark, *Carcharodon carcharias,* collected in the Pliocene around Volterra and once belonging to the Targioni collections (labels associated with the specimen: "Squalus n. 26," once in cabinet 86-12). Sold by his son Ottaviano Targioni Tozzetti, in 1839, the specimen became part of the Royal Museum: a handwritten note records its past history.

including one called "Geology, Mineralogy, and Geography." Geology had gained, among Italian *scientists,* a status comparable to that attained in England and France, if 20 years later. The term "scientist" was a recently adopted neologism, coined in 1832 by the Cambridge polymath, William Whewell (1794–1866), president of the Geological Society of London in 1837 and author of *History of the Inductive Sciences*: geology was the science of the moment (Rudwick, 2008). This first gathering of Italian geologists did not accomplish much, however, apart from celebrating Steno, Targioni, Arduino, and Brocchi in the role of pioneers of Italian geology, and charging Nesti, Mazzi, and Paolo Savi (1798–1871) from Pisa, to establish an "Italian geological-mineralogical nomenclature." The Third Congress, held at Florence in 1841, in planning the Geological Map of Italy (still far ahead), accepted the proposal to establish an Italian Central Collection of Mineralogy and Geology at the Royal Museum (Cocchi, 1871). At the museum, Mazzi supervised in 1842 the compilation of the catalogue of invertebrates and vertebrates being compiled by Nesti from 1845 until his death in 1849 (a total of 356 entries: Cioppi and Dominici, 2011).

In 1860, the geologist Igino Cocchi (1827–1913) was appointed director of the newly established Royal Institute of Advanced Studies, from which came the University of Florence. Educated at the University of Pisa, Igino Cocchi spent a long period of specialization at Paris and London in the late 1850s (Corsi, 2008), and he held the chair of mineralogy, geology, and paleontology in Florence from 1860 to 1873. The Italian Geological Survey was struggling to accomplish the Geological Map of Italy, paralyzed by lack of funds and by the clash between the party of university geologists, to which Cocchi belonged, and the Royal Mining Corps, led by Quintino Sella (1827–1884; Corsi, 2003). Notwithstanding the substantial defeat of the geologists, at the first meeting of the Italian Geological Committee held in Florence in December 1861, Cocchi obtained approval for the realization of a central paleontological collection and a central library, as prospected in 1841 (Corsi, 2007). During the following decade, many Italian leading geologists contributed many paleontological collections, representing the geology of different parts of Italy, from Lombardy to Sicily (Cioppi and Dominici, 2011). Cocchi himself revised Mazzi's catalogue of fossil shells and set out to expand it with numerous acquisitions from various localities (see Corsi, 2008). The centrality of Florence for the new Italian paleontological collections ended with the establishment of the Royal Geological Bureau in 1873, when all fossils collected during surveys for the Geological Map of Italy started being sent to Rome (Rossi, 2015). As a testimony to the golden age of invertebrate paleontology in Florence, a series of papers dedicated to the taxonomy of Pliocene and Pleistocene invertebrates collected in Tuscany was published in the following decades (e.g., Pecchioli, 1864: Fig. 12; Manzoni, 1868; D'Ancona, 1871; Ristori, 1886, 1890; Fucini, 1891).

The Search for Man's Origin

The second wave of interest for fossil vertebrates was in part connected with the thorny issue of human antiquity (Rudwick, 2008). The arrival of man in the scene of history had traditionally qualified the youngest formation called "diluvium," a name shaped on the hypothesis that a global catastrophe had struck the Earth, and that man populated the Earth only afterwards (Buckland, 1820). During the 1830s, the finding of fossil human bones and human artifacts in caves proposed evidence that man had lived with antediluvian extinct animals. Fossil primates in France, India, and Brazil filled part of the gap between all other vertebrates and man, prompting new arguments in favor of transmutation, while the publication of papers dealing with "fossil man" raised fears in conservative circles of France and England (Secord, 2000; Rudwick, 2008). Paleontologists working on Tertiary and Quaternary formations (the latter concept introduced by Jules Desnoyers [1800–1887] to identify marine terrains that looked younger than the Tertiary: Desnoyers, 1829) attempted to define boundaries of fossil vertebrate associations. By the 1850s, the evidence that man had lived side by side with extinct animal species became a shared knowledge throughout Europe (Rudwick, 2008; Tarantini, 2012). Several protagonists of a new pulse of fossil vertebrate research came to Florence to study the rich paleontological heritage and to excavate Tuscan sites, helping to enlarge collections. The first to cross the Italian border, looking for fossil elephants, was Hugh Falconer (1808–1865), a Scottish physician and expatriate in India, author of a paper on a fossil monkey (Cautley and Falconer, 1840). In Florence since 1856, Falconer praised the quantity and quality of the remains, better and more abundant than any other he had seen in Europe, but he criticized the bad quality of conservation and the poor quality of the studies carried out until that time. He had a low opinion of the expertise of Nesti ("not a professed anatomist": Falconer, 1868, p. 105) and, in presenting the details of his case, concluded that Florence did not have scientists worthy of the quality of its vertebrate fossil collection.

Pliocene deposits of the Val d'Arno [are] unrivalled in Europe both for their abundance and for the perfect condition in which they are preserved. Elsewhere paleontologists are compelled to grope their way by the faint light of mutilated specimens; there the fossil remains of the same forms are presented entire. A good monograph, liberally illustrated, upon the fossil Mammalia of the Val d'Arno would reflect as bright a lustre on the Italian diadem, as do the chefs-d'oeuvre of the Tribune or the Galleries of the Palazzo Pitti. The patronage of the court has been for centuries bestowed upon the wax models of the Museum, but withheld from the magnificent fossil remains that are laid out under the same roof. Except a few and inadequate memoirs by Nesti, nothing worthy of the subject has been brought out in Italy upon these Tuscan collections during the last half century; and it is not overstating the fact to say that the progress of research on the extinct faunas of the upper Tertiary formation in Europe has been retarded a quarter of a century in consequence. Had these collections been yielded either by Siberia or by the northern part of the valley of the Po, the general results would have been familiar knowledge long ago. At present, a journey to Florence is the only means of becoming acquainted with them. (Falconer, 1868, p. 121)

Charles Darwin published "On the Origin of Species by Natural Selection" in 1859, containing his theory of descent with modification. Among the many arguments, he cited paleontological studies on mastodons and elephants and other pieces of evidence amassed by Falconer. In 1863, the topic of man's origin was further popularized by Charles Lyell's, "Geological Evidences of the Antiquity of Man" (Lyell, 1863) and "Evidences as to Man's Place in Nature" by Thomas Henry Huxley (1825–1895; Huxley, 1863).

Cocchi in Florence made original contributions to the debate on man's origins. A professional geologist with insight into the meaning of Tuscan fossils, particularly regarding the stratigraphic position of human bones and artifacts (Cocchi, 1872b, 1867), Cocchi brought together clues from stratigraphy, vertebrate paleontology (including human), and the new discipline of paleo-ethnology, or the study of stone tools (Tarantini, 2012). Cocchi had international connections, including those with Hugh Falconer, and the leading prehistoric archaeologists of his time. Paul Gervais (1816–1879), one of the experts of Tertiary and Quaternary terrestrial vertebrates, recognized then four epochs with remains (either body parts or artifacts) that could be clearly attributed to man, the oldest of which coincided with strata bearing bones of *Elephas meridionalis,* so abundant at the Florence museum. Associated vertebrates, particularly extinct species, were means to define epochs in human evolution.

These four epochs can be paleontologically recognized by means of animal species that had been contemporary [to man], and these come to be added to those that history already recognizes. They share the common character of the presence of flint instruments. They are: 1) The epoch of *Elephas meridionalis,* [...] hard to separate from the following. 2) The epoch of *Elephas primigenius,* large bears, hyenas, large cave felids etc. 3) The epoch of deers [...] 4) The epoch of palafittes. (i.e., lake dwellings built on pilings: Gervais, 1867, p. 34, the three intervals corresponding to lower Pleistocene, upper Pleistocene, Mesolithic and Neolithic, respectively)

The publication of Darwin's book *The Descent of Man,* translated in Italian within months (Darwin, 1871), had further popularized the topic of the origin of humans in Italy. In the same year, the Fifth International Congress of Anthropology and Prehistoric Archaeology was held in Bologna, largely centered on the Italian record (Sommer, 2009) followed by the publication of Cocchi's first volume of the catalogue of the museum, dedicated to prehistoric fossils and artifacts (Cocchi, 1872a).

But Tuscan tertiaries offered also fossils that could shed light on events farther back in the history of human evolution. Cocchi found a mandible of a fossil hominoid in Monte Bamboli, in Miocene deposits of southern Tuscany, and lent it immediately to Gervais, who described the new species *Oreopithecus bambolii* (Gervais, 1872; Cocchi, 1872b; Ristori, 1890). Charles Immanuel Forsyth Major (1843–1923), a physician with a secondary interest in vertebrate paleontology, had been on the hunt for fossil primates when he corresponded with Charles Darwin in 1872 on the fauna associated with *Oreopithecus* at Monte Bamboli (Forsyth Major, 1872; see Burkhardt and Smith, 1985). Forsyth Major had just moved to Florence, following an interest in fossil primates and fossil vertebrate associations soon leading to important results (Forsyth Major, 1875, 1877). Before the publication of the second edition of *The Descent of Man,* he contributed to Darwin information on a female skull of *Leptobos etruscus* from Tuscany (Forsyth Major, 1872).

Dr. Forsyth Major also informs me that a fossil skull, believed to be that of the female *Bos etruscus,* has been found in Val d'Arno, which is wholly without horns. [...] From these various facts we may infer as probable that horns of all kinds, even when they are equally developed in the two sexes, were primarily acquired by the male in order to conquer other males, and have been transferred more or less completely to the female. (Darwin, 1874, p. 327)

In 1880, on behalf of the Geology Laboratory of Florence, Forsyth Major conducted paleontological excavations at Montopoli, not far from San Miniato, in the lower Arno valley, discovering again a large number of new and time-diagnostic species. In 1889, he carried out a second important excavation campaign at Olivola in the Magra valley, in strata younger than those at Montopoli (Forsyth Major, 1890). Forsyth Major remained in Florence until 1892, and his figure remains one of great importance in establishing a chronology of Tuscan mammal faunas, the seed of the modern school of vertebrate paleontology centered on the study of late Miocene–Pleistocene mammals (e.g., Azzaroli, 1977; Azzaroli et al., 1988; Rook et al., 1999). A new

Figure 14. The skull of *Oreopithecus bambolii,* from the skeleton recovered at Baccinello in 1958, by Johannes Hürzeler.

and almost complete skeleton of *Oreopithecus,* a spectacular specimen today on display in Florence, was recovered in 1958 by Johannes Hürzeler (Fig. 14), the Swiss paleontologist author of the paper that started the debate on the phylogenetic position of *Oreopithecus,* not settled yet (Hürzeler, 1954; Tuttle, 2014). In the 1990s (Fig. 15), the first digital catalogue appeared (Cioppi et al., 1996), fit to the mutable nature of taxonomy.

FINALE: BACK TO THE DEEP BLUE WATER

Many university departments, once proudly concerned about "geology," today claim to embrace the wider field of "earth sciences," a category by default, with no historical tradition apart from that of geology itself. This choice forgets that geology has always been about the interaction between biosphere, atmosphere, hydrosphere, and lithosphere, whatever their name in past centuries. Similarly, geology always considered the place of our planet in the solar system, the physics of its interiors, its present and past climates, present and past ecology, and all the many disciplines today embraced by "earth sciences." It would not be unfair to say that geology *is* the earth sciences. However, if wisely interpreted, this change of emphasis underscores the immense quantity of possible data that can be drawn from the study of fossils and their stratigraphic contexts, including taphonomy and paleoecology. It is also a call for an interaction between disciplines that during the past century have grown up as separate fields of knowledge, explored by specialists who no longer read each other's papers.

In 2007, a large skeleton of a Pliocene mysticete was dug up from a field at Orciano Pisano, one of the classical localities made famous by nineteenth-century Florentine and Pisan paleontologists. Together with the fossil whale, an associated fossil fauna has been recovered and described, in part belonging to a so-called "whale-fall community" (WFC, or the community of organisms that feed on whale carcasses as they float or after they sink on the sea floor). Modern WFCs are a recent discovery of marine biology attracting worldwide interest for their many ecological implications in the study of the biota inhabiting the deep sea, one of the frontiers of earth science (Smith et al., 2015), and for highlighting the global character of interactions that take place in the open sea (Roman et al., 2014). The singularity of WFC is epitomized by the strange bone-eating worms of genus *Osedax* (Rouse et al., 2004) and by other weird-looking organisms that live on the rotting tissues of whale carcasses. In this scenario, the interdisciplinary study of the Tuscan find (Dominici et al., 2009; Danise et al., 2010), unique in its genre, has attracted national and international specialists involved in fields such as geology, marine biology, and ecology, who today boast separate interests. In the process, fossil bones from Florentine

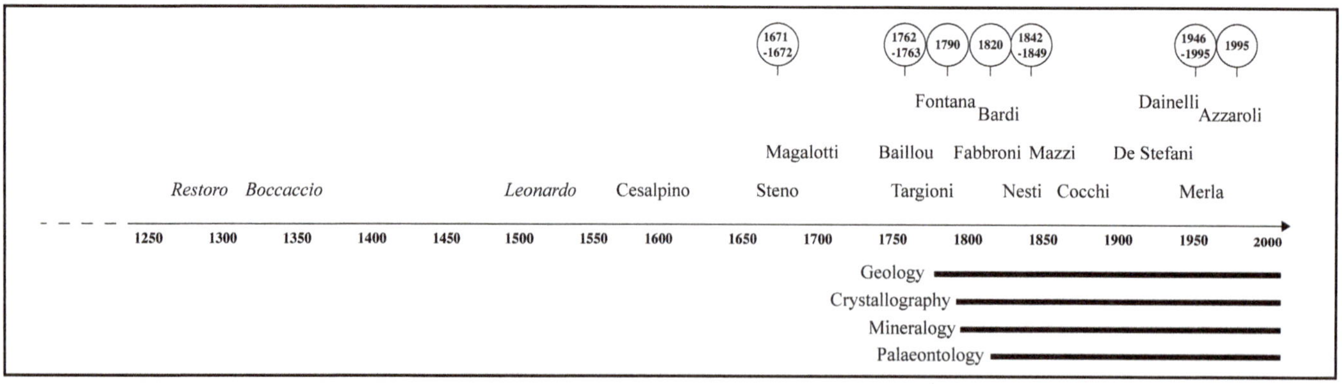

Figure 15. The history of the Florence museum on a timeline, showing authors who first drew attention to the paleontological heritage of Tuscany as a means to reconstruct Earth's history (italics), main curators and directors of museum collections (regular text), museum catalogues (capital text within circles above timeline), and formal names given to distinct fields of knowledge (bars below timeline).

Figure 16. Fossilized skeleton of a Pliocene whale, at the center of the permanent exhibition "Tales of a Whale" opened to the public in May 2016.

historic collections from Orciano Pisano have been pulled out from drawers and studied with modern techniques, with exciting results when traces attributable to *Osedax* were found (Higgs et al., 2012). Today all these exciting aspects of our recent paleontological research, and the serendipitous aspect of science, form the basis for a permanent immersive exhibition inaugurated in the Museum of Natural History, University of Florence, in 2016 (Fig. 16). As visitors enter, they move into the blue, as though they are at the bottom of the Pliocene sea, with a dead carcass at their feet, eaten by a whale-fall fauna including large sharks and bone-eating worms, and with living mysticetes swimming above their head, a scene created by projecting images on the ceiling (Fig. 14). The multimedia exhibition gradually makes visitors aware that each of them plays a part in the modern marine ecosystem that unrolls in front of the Tuscan coastline, reaching Corsica and Liguria—a modern sea where whales live and die the way they did in the Pliocene sea. Visitors are thus led to second thoughts about the many ways they interact with the marine ecosystem (for example, the consequences of their use of plastic and other disposable items of everyday life that eventually make their way down to the sea). And thus they come to appreciate how the geologic history of Italy plays a positive role in ensuring a sustainable future in a world where resources are finite. This is geology today.

ACKNOWLEDGMENTS

We thank Luciana Fantoni, Museum of Natural History, University of Florence, for helping us to browse through old inventories and catalogues and for sharing her findings on some of Steno's specimens. We thank Gary D. Rosenberg for inviting us to make a contribution for the present volume. We have benefited from the constructive critique by Alan Cutler and Troels Kardel. Photographs of fossil specimens are by Saulo Bambi, from the Museum of Natural History, University of Florence.

REFERENCES CITED

Accordi, B., 1980, Michele Mercati (1541–1593) e la "Metallotheca": Geologica Romana, v. 19, p. 1–50.

Arnold, K., 2006, Cabinets for the Curious: Looking Back at Early English Museums: London, Routledge, 310 p.

Arrigoni T., 1987, Uno scienziato nella Toscana del Settecento—Giovanni Targioni Tozzetti: Documenti inediti di cultura toscana, v. 9, p. 1–170.

Azzaroli, A., 1977, The Villafranchian stage in Italy and the Plio-Pleistocene boundary: Giornale di Geologia, v. 41, p. 61–79.

Azzaroli, A., De Giuli, C., Ficcarelli, G., and Torre, D., 1988, Late Pliocene to early mid-Pleistocene mammals in Eurasia: Faunal succession and dispersal events: Palaeogeography Palaeoecology Palaeoclimatology, v. 66, p. 77–100, https://doi.org/10.1016/0031-0182(88)90082-X.

Bardi, G, 1808, Prospetto sugli avanzamenti delle Scienze fisiche in Toscana: Annali dell'Imperial Museo di Firenze, v. 1, p. 3-26.

Boccone, S., 1697, Museo di Fisica e di Esperienza Variato, e Decorato di Osservazioni Naturali, Note Medicinali e Ragionamenti Secondo i Principi de' Moderni: Venezia, Zuccato, 319 p.

Breislak, S., 1811, Introduzione alla geologia: Milano, Stamperia Reale, 490 p.

Brocchi, G., unpublished, Giornale del viaggio mineralogico attraverso vari paesi italiani, Anni 1811–1812: Bassano del Grappa, Biblioteca Civica, folder 31-A-20-1, manuscript M502.

Brocchi, G., 1814, Conchiologia fossile subappennina con osservazioni geologiche sugli Apennini e sul suolo adiacente: Milano, Stamperia Reale, v. 1–2, 712 p., https://doi.org/10.5962/bhl.title.11569.

Bruguière, J.-G., 1792, Histoire naturelle des vers, Encyclopédie méthodique, v. 1: Paris, Panckoucke, 757 p.

Buckland, W., 1820, Vindiciae Geologicae, or the Connexion of Geology with Religion Explained: Oxford, UK, Oxford University Press, 44 p.

Burkhardt, F., and Smith, S., 1985, eds., A Calendar of the Correspondence of Charles Darwin, 1821–1882: New York, Garland, 690 p.

Burkhardt, R.W., Jr., 1977, The Spirit of System: Lamarck and Evolutionary Biology: Cambridge, Massachusetts, Harvard University Press, 285 p.

Cautley, P.T., and Falconer, H., 1840, Notice on the remains of a fossil monkey from the Tertiary strata of the Sewalik Hills in the North of Hindoostan: Transactions of the Geological Society of London, series 2, v. 5, p. 499–504.

Cesalpino, A., 1596, De metallicis libri tres: Roma, Luigi Zanetti, 222 p.

Cioppi, E., 2011, Le tracce di Leonardo, *in* Monechi, S., and Rook, L., eds., The Museum of Natural History of the University of Florence, Volume 3, The Geological and Paleontological Collections: Firenze, Firenze University Press, p. 76–83.

Cioppi, E., 2017, Testimonianze fossili dal Kentucky al Granducato di Toscana: Museologia Scientifica Memorie, v. 17, p. 11–15.

Cioppi, E., and Dominici, S., 2011, Origin and development of the geological and paleontological collections, *in* Monechi, S., and Rook, L., eds., The Museum of Natural History of the University of Florence, Volume 3, The Geological and Paleontological Collections: Firenze, Firenze University Press, p. 18–55.

Cioppi, E., Dorbolò, D., and Berdondini, E., 1996, GEF: Un sistema di catalogazione automatizzata delle collezioni paleontologiche: Museologia Scientifica, v. 13, p. 9–21.

Cipriani, C., and Scarpellini, A., 2007, Un contributo alla mineralogia settecentesca. La Collezione di Giovanni Targioni Tozzetti: Firenze, Leo Olschki, 200 p.

Cipriani, C., Fantoni, L., Poggi, L., and Scarpellini, A., 2011, Le collezioni mineralogiche del Museo di Storia Naturale dell'Università di Firenze dalle origini ad oggi: Studi dell'Accademia Toscana di Scienze e Lettere: La Colombaria, v. 247, p. 1–236.

Cocchi, I., 1867, L'uomo fossile nell'Italia Centrale. Studi paleoetnologici: Memorie della Società Italiana di Scienze Naturali, v. 2, p. 3–80.

Cocchi, I., 1871, Brevi cenni sui principali Istituti e Comitati Geologici e sul Regio Comitato Geologico d'Italia per servire da introduzione al primo volume delle Memorie: Memorie del Regio Comitato Geologico d'Italia, v. 1, p. 3–33.

Cocchi, I., 1872a, Raccolta degli oggetti de' così detti Tempi Preistorici: Cataloghi della collezione centrale italiana di paleontologia, v. 1, p. 104.

Cocchi, I., 1872b, Su di due Scimmie fossili italiane: Bollettino del Regio Comitato Geologico d'Italia, v. 3, p. 59–71.

Contardi, S., 2002, La casa di Salomone a Firenze: L'Imperiale e Reale Museo di Fisica e Storia Naturale (1755–1801): Biblioteca di Nuncius, v. 43, 324 p.

Correa de Serra, J., 1799, On a submarine forest, on the East Coast of England: Philosophical Transactions of the Royal Society of London, v. 89, p. 145–156, https://doi.org/10.1098/rstl.1799.0010.

Corsi, P., 2003, La Carta Geologica d'Italia: Agli inizi di un lungo contenzioso, *in* Vai, G.B., and Cavazza, W., eds., Four Centuries of the Word Geology: Ulisse Aldrovandi 1603 in Bologna: Bologna, Minerva, p. 271–300.

Corsi, P., 2007, Much ado about nothing: The Italian Geological Survey, 1861–2006: Earth Sciences History, v. 26, p. 97–125, https://doi.org/10.17704/eshi.26.1.kq2w1707l43w151x.

Corsi, P., 2008, Fossils and Reputations. A Scientific Correspondence: Pisa, Paris, London 1853–1857: Pisa, Pisa University Press, 411 p.

Cutler, A., 2003, The Seashell on the Mountaintop: A Story of Science, Sainthood, and the Humble Genius Who Discovered a New History of the Earth: New York, Dutton, 240 p.

Cuvier, G., 1801, Extrait d'un ouvrage sur les espèces de quadrupèdes. Dont on a trouvé les ossemens dans l'intérieur de la terre, adresse aux savans et aux amateurs des sciences: Journal de physique, de chimie, d'histoire naturelle et des arts, v. 52, p. 253–254.

Cuvier, G., 1812, Recherches sur les ossemens fossiles de quadrupèdes, où l'on rétablit les caractères de plusieurs espèces d'animaux que les révolutions du globe paraissent avoir détruites, v. 1: Paris, Déterville, 278 p.

Cuvier, G., 1823, Recherches sur les ossemens fossiles, ou l'on rétablit les caractères de plusieurs animaux dont les révolutions du globe ont détruit les espèces, v. 4: Paris, Dufour et d'Ocagne, 514 p.

D'Ancona, C., 1871, Malacologia Pliocenica Italiana: Memorie del Regio Comitato Geologico d'Italia, v. 1, p. 308–358.

Danise, S., Dominici, S., and Betocchi, U., 2010, Mollusk species at a Pliocene shelf whale fall (Orciano Pisano, Tuscany): Palaios, v. 25, p. 449–456, https://doi.org/10.2110/palo.2009.p09-139r.

Darwin, C.R., 1871, L'origine dell'uomo e la scelta in rapporto col sesso, Translation by Michele Lessona: Torino, UTET, 672 p.

Darwin, C.R., 1874, The Descent of Man, and Selection in Relation to Sex, Revised Edition: New York, Merrill and Baker, 688 p., https://doi.org/10.5962/bhl.title.54341.

De Rosa, S., 1986, Niccolò Stenone nella Firenze e nell'Europa del suo tempo: Mostra di documenti, manoscritti, opere nel terzo centenario della morte: Firenze, Leo Olschki, 80 p.

Desmarest, N., 1795, Encyclopédie méthodique, Géographie-physique, v. 1: Paris, Agasse, 857 p.

Desnoyers, J., 1829, Observations sur un ensemble de dépôts marins plus récens que les terrains tertiaires du bassin de la Seine, et pouvant constituer une formation géologique distincte: Annales des sciences naturelles, v. 16, p. 171–214, 402–491.

Di Orio, F., Ferrini, G., and Macchiarelli, G., 2016, Niccolo Stenone dalla medicina alla geologia: Il percorso di un uomo di scienza e di fede nel Seicento: Roma, InItinere, 189 p.

Dominici, S., 2009, Steno, Targioni, and the two forerunners: Journal of Mediterranean Earth Science, v. 1, p. 101–110.

Dominici, S., 2010, Brocchi's *Subapennine Fossil Conchology*: Evolution: Education and Outreach, v. 3, p. 585–594.

Dominici, S., 2011, History of Tuscany, history of the Earth, *in* Monechi, S., and Rook, L., eds., The Museum of Natural History of the University of Florence, Volume 3, The Geological and Paleontological Collections: Firenze, Firenze University Press, p. 2–17.

Dominici, S., 2017, Ricomposizione del mondo in un paesaggio di Leonardo: Giornale di bordo di storia, letteratura ed arte, v. 42–43, p. 1–18.

Dominici, S., and Eldredge, N., 2010, Brocchi, Darwin, and transmutation: Phylogenetics and paleontology at the dawn of evolutionary biology: Evolution: Education and Outreach, v. 3, p. 576–584.

Dominici, S., Cioppi, E., Danise, S., Betocchi, U., Gallai, G., Tangocci, F., Valleri, G., and Monechi, S., 2009, Mediterranean fossil whale falls and the adaptation of mollusks to extreme habitats: Geology, v. 37, p. 815–818, https://doi.org/10.1130/G30073A.1.

Eldredge, N., 2015, Eternal Ephemera: Adaptation and the Origin of Species from the Nineteenth Century through Punctuated Equilibria and Beyond: New York, Columbia University Press, 416 p.

Ell, T., 2011, Two letters of Signor Giovanni Arduino, concerning his natural observations: First full English translation. Part 1: Earth Sciences History, v. 30, p. 267–286, https://doi.org/10.17704/eshi.30.2.22r632j0x453gj45.

Fabbroni, L., 1802, Estratto di un'opera sopra le specie di quadrupedi, le cui ossa si sono trovate nelle viscere della terra: Memorie per i Curiosi di Agricultura e di Economia Rurale, v. 6, p. 3–45.

Falconer, H., 1868, Palaeontological Memoirs and Notes of the Late Hugh Falconer, with a Biographical Sketch of the Author: London, Spottiswode, v. 2, 675 p.

Findlen, P., 1994, Possessing Nature: Museums, Collecting, and Scientific Culture in Early Modern Italy: Berkeley, University of California Press, 449 p.

Findlen, P., 1998, Possessing the past: The material world of the Italian Renaissance: The American Historical Review, v. 103, p. 83–114, https://doi.org/10.2307/2650776.

Fontana, F., 1775, Saggio del real gabinetto di fisica e di storia naturale di Firenze: Roma, Stamperia di Giovanni Zempel, 38 p.

Forsyth Major, C.I., 1872, Note sur des singes fossiles trouvés en Italie, précédé d'un aperçu sur les quadrumanes fossiles en général: Atti della Società Toscana di Scienze Naturali, v. 15, p. 1–17.

Forsyth Major, C.I., 1875, Considerazioni sulla fauna dei Mammiferi pliocenici e postpliocenici della Toscana: Memorie della Società Toscana di Scienze Naturali, v. 1, p. 7–40.

Forsyth Major, C.I., 1877, Considerazioni sulla fauna dei Mammiferi pliocenici e postpliocenici della Toscana: Memorie della Società Toscana di Scienze Naturali, v. 3, p. 202–227.

Forsyth Major, C.I., 1890, L'ossario di Olivola in Val di Magra (Prov. di Massa e Carrara): Processi verbali della Società Toscana di Scienze Naturali, v. 7, p. 57–78.

Fortis, A., 1778, Della valle vulcanico-marina di Roncà nel territorio veronese Memoria orittografica del sig. abate Fortis: Venezia, Carlo Palese, p. 70.

Fucini, A., 1891, Il Pliocene dei dintorni di Cerreto Guidi e di Limite e i suoi molluschi fossili: Bollettino della Società Geologica Italiana, v. 10, p. 49–87.

Gervais, P., 1867, Recherches sur l'ancienneté de l'homme et la période quaternaire: Paris, Bertrand, 132 p.

Gervais, P., 1872, Sur un singe fossile, d'espèce non encore décrite, qui a été découvert au Monte Bamboli: Paris, Comptes Rendu hebdomadaires de Séances de l'Académie de Sciences, v. 74, p. 1217–1223.

Gould, S.J., 1981, The titular bishop of Titiopolis: Natural History, v. 90, p. 20–24.

Gould, S.J., 1998, The upwardly mobile fossils of Leonardo's living earth, *in* Gould, S.J., Leonardo's Mountain of Clams and the Diet of Worms: Essays on Natural History: New York, Harmony, 416 p., https://doi.org/10.4159/harvard.9780674063365.c1.

Heilbron, J.L., and Sigrist, R., eds., 2011, Jean-Andre Deluc: Historian of Earth and Man: Genève, Slatkine Érudition, 368 p.

Higgs, N.D., Little, C.T.S., Glover, A.G., Dahlgren, T.G., Smith, C.R., and Dominici, S., 2012, Evidence of *Osedax* worm borings in Pliocene (~3 Ma) whale bone from the Mediterranean: Historical Biology, v. 24, p. 269–277.

Hürzeler, J., 1954, Zur systematischen Stellung von *Oeropithecus*: Verhandlunden der Naturforschenden Gesellschaft in Basel, v. 65, p. 88–95.

Huxley, T.H., 1863, Evidence as to Man's Place in Nature: London, Williams & Norgate, 159 p.

Kardel, T., 2009, Prompters of Steno's geological principles: Generation of stones in living beings, glossopetrae and molding, *in* Rosenberg, G.D., ed., The Revolution in Geology from Renaissance to the Enlightenment: Geological Society of America Memoir 203, p. 127–134, https://doi.org/10.1130/2009.1203(08).

Kardel, T., and Maquet, P., eds., 2013, Nicolaus Steno: Biography and Original Papers of a 17th Century Scientist: Berlin, Heidelberg, Springer, 738 p., https://doi.org/10.1007/978-3-642-25079-8.

Kemp, M., 2006, Leonardo da Vinci: The Marvellous Works of Nature and Man, 2nd edition: Oxford, UK, Oxford University Press, 432 p.

Laboulais-Lesage, I., 2004, Voir, combiner et décrire: La géographie physique selon Nicolas Desmarest: Revue d'Histoire Moderne et Contemporaine, v. 51–52, p. 38–57, https://doi.org/10.3917/rhmc.512.0038.

Lamarck, J.-B., 1809, Philosophie Zoologique: Paris, Dentu, 428 p.

Lovejoy, A.O., 1936, The Great Chain of Being: A Study of the History of an Idea: Cambridge, Massachusetts, Harvard University Press, 382 p.

Lyell, C., 1830, Principles of Geology: First Edition: London, Murray, v. 1, 438 p.

Lyell, C., 1833, Principles of Geology: First Edition: London, Murray, v. 3, 398 p.

Lyell, C., 1863, The Geological Evidences of the Antiquity of Man: London, Murray, 577 p.

Manganelli, G., and Benocci, A., 2011, Niccolò Gualtieri (1688–1744): Biographical sketch of a pioneer of conchology: Archives of Natural History, v. 38, p. 174–177, https://doi.org/10.3366/anh.2011.0019.

Manni, D.M., 1775, Vita del letteratissimo monsig. Niccolo Stenone di Danimarca vescovo di Titopoli e vicario apostolico: Firenze, Stamperia di Giuseppe Vanni, 335 p.

Manzoni, A., 1868, Saggio di conchiologia fossile subappennina: Fauna delle sabbie gialle: Imola, Galeati, 120 p.

McCartney, P.J., 1976, Charles Lyell and G.B. Brocchi: A study in comparative historiography: British Journal for the History of Science, v. 9, p. 175–189, https://doi.org/10.1017/S000708740001476X.

Mercati, M., 1717, Metallotheca. Opus posthumum, auctoritate et munificentia Clementis undecimi pontificis maximi e tenebris in lucem eductum; Opera autem, et studio Ioannis Mariæ Lancisii illustratum: Romae, Ex officina Jo: Mariae Salvioni, lxiv + 378 p.

Narducci, N., ed., 1859, La composizione del mondo di Restoro d'Arezzo, testo italiano del 1282: Roma, Tipografia delle Scienze Matematiche e Fisiche, 343 p.

Nesti, F., 1808a, Rapporto del professore di zoologia e mineralogia Filippo Nesti: Annali del Museo Imperiale di Fisica e Storia Naturale di Firenze, v. 1, 7 p.

Nesti, F., 1808b, Di alcune ossa fossili di mammiferi che s'incontrano nel Valdarno: Annali del Museo Imperiale di Fisica e Storia Naturale di Firenze, v. 1, 12 p.

Nesti, F., 1811, Sopra alcune ossa fossili di rinoceronte: Lettera del prof. Filippo Nesti al signor dott. Gaetano Savi professore di botanica nello studio di Pisa: Firenze, Tipografia Piatti, 24 p.

Nesti, F., 1825, Lettere sopra alcune ossa fossili del Valdarno non paranco descritte: Nuovo Giornale dei Letterati, v. 11, p. 195–216.

Nova, A., 2015, "Addj 5 daghossto 1473": l'oggetto e le sue interpretazioni, *in* Frosini, F., and Nova, A., eds., Leonardo da Vinci on Nature: Knowledge and Representation: Venezia, Studi e Ricerche, v. 11, p. 285–301.

Olden-Jørgensen, S., 2009, Nicolas Steno and René Descartes: A Cartesian perspective on Steno's scientific development, *in* Rosenberg, G.D., ed., The Revolution in Geology from the Renaissance to the Enlightenment: Geological Society of America Memoir 203, p. 149–157, https://doi.org/10.1130/978-0-8137-1203-1-203.0.149.

Olivi, G., 1792, Zoologia Adriatica ossia catalogo ragionato degli animali del Golfo e delle Lagune di Venezia, preceduto da una dissertazione sulla storia fisica e naturale del Golfo e accompagnato da memorie, ed osservazioni di fisica, storia naturale ed economia: G. Remondini, Bassano, 334 p.

Olmi, G., 2006, "A Wonderful Collection Indeed!": The Royal Museum of Florence in the testimony of two travellers: Nuncius, v. 21, p. 349–368, https://doi.org/10.1163/221058706X00685.

Olmi, G., 2012, Una collezione sul mercato nella Firenze di metà Settecento: A proposito del gabinetto di Jean de Baillou, *in* Pegazzano, D., ed., Scritti di museologia e di storia del collezionismo in onore di Cristina De Benedictis: Firenze, EDIFIR, p. 127–142.

Pancaldi, G., 1991, Darwin in Italy: Science across Cultural Frontiers, Updated and Expanded Edition: Bloomington, Indiana University Press, 222 p. [First edition: Darwin in Italia. Il Mulino, Bologna, 1983.]

Pecchioli, V., 1864, Descrizione di alcuni nuovi fossili delle argille subappennine toscane: Atti della Società Toscana di Scienze Naturali, v. 6, p. 498–550

Pedrini, G., 1791, Gabinetto mineralogico del collegio Nazareno, descritto secondo i caratteri esterni, e distribuito a norma de' principj costitutivi, v. 1: Roma, Lazzarini, 384 p.

Rappaport, R., 1997, When Geologists Were Historians: Ithaca, New York, Cornell University Press, 308 p.

Ray, J., 1673, Observations Topographical, Moral, and Physiological Made in a Journey through Part of the Low-Countries, Germany, Italy, and France: London, Martyn, 499 p.

Ristori, G., 1886, I crostacei brachiuri e anomuri del Pliocene italiano: Bollettino della Società Geologica Italiana, v. 5, p. 93–128.

Ristori, G., 1890, Le scimmie fossili italiane: Bollettino del Regio Comitato Geologico d'Italia, v. 21, p. 178–196.

Roman, J., Estes, J.A., Morissette, L., Smith, C.R., Costa, D., McCarthy, J., Nation, J.B., Nicol, S., Pershing, A., and Smetacek, V., 2014, Whales as marine ecosystem engineers: Frontiers in Ecology and the Environment, v. 12, p. 377–385, https://doi.org/10.1890/130220.

Rook, L., Bondioli, L., Köhler, M., Moyá Solá, S., and Macchiarelli, R., 1999, *Oreopithecus* was a bipedal ape after all: Evidence from the iliac cancellous architecture: Proceedings of the National Academy of Sciences of the United States of America, v. 96, p. 8795–8799, https://doi.org/10.1073/pnas.96.15.8795.

Rosenberg, G.D., 2001, An artistic perspective on the continuity of space and the origin of modern geologic thought: Earth Sciences History, v. 20, p. 127–155, https://doi.org/10.17704/eshi.20.2.n845766556q63672.

Rosenberg, G.D., 2006, Nicholas Steno's *Chaos* and the shaping of evolutionary thought in the scientific revolution: Geology, v. 34, p. 793–796, https://doi.org/10.1130/G22655.1.

Rossi, P., 2001, The Birth of Modern Science: Oxford, Wiley-Blackwell, 288 p.

Rossi, R., 2015, The Type Fossils Preserved in the Paleontological Collections: Roma, Istituto Superiore per la Protezione e la Ricerca Ambientale, 178 p.

Rotschuh, K.E., 1968, Descartes, Stensen und der "Discours sur l'anatomie du cerveau," *in* Scherz, G., ed., Steno and Brain Research in the Seventeenth Century: Analecta Medico-Historica, v. 3: Oxford, UK, Pergamon Press, 302 p.

Rouse, G.W., Goffredi, S.K., and Vrijenhoek, R.C., 2004, *Osedax*: Bone-eating marine worms with dwarf males: Science, v. 305, p. 668, https://doi.org/10.1126/science.1098650.

Rudwick, M.J.S., 1972, The Meaning of Fossils. Episodes in the History of Paleontology: Chicago, University of Chicago Press, 287 p.

Rudwick, M.J.S., 1985, The Great Devonian Controversy: The Shaping of Scientific Knowledge among Gentlemanly Specialists: Chicago, University of Chicago Press, 528 p., https://doi.org/10.7208/chicago/9780226731001.001.0001.

Rudwick, M.J.S., 1997, Georges Cuvier, Fossil Bones, and Geological Catastrophes: Chicago, University of Chicago Press, 320 p., https://doi.org/10.7208/chicago/9780226731087.001.0001.

Rudwick, M.J.S., 2005, Bursting the Limits of Time: The Reconstruction of Geohistory in the Age of Revolution: Chicago, University of Chicago Press, 708 p.

Rudwick, M.J.S., 2008, Worlds before Adam: The Reconstruction of Geohistory in the Age of Reform: Chicago, University of Chicago Press, 614 p.

Savi, G., 1827, Continuazione delle notizie spettanti il Giardino Botanico e il Museo di Storia naturale dell'Università di Pisa: Nuovo Giornale de' letterati, v. 15, p. 30–48.

Scherz, G., 1956, Vom Wege Niels Stensens: Beiträge zu seiner naturwissenschaftlichen Entwicklung, Dissertation, Acta Historica Naturalium et Medicinalium, v. 14, 248 p.

Secord, J.A., 2000, Victorian Sensation: The Extraordinary Publication, Reception and Secret Authorship of Vestiges of the Natural History of Creation: Chicago, University of Chicago Press, p. 624.

Secord, J.A., 2015, Visions of Science: Books and Readers at the Dawn of the Victorian Age: Chicago, University of Chicago Press, 256 p., https://doi.org/10.7208/chicago/9780226203317.001.0001.

Smith, C., Glover, A.G., Treude, T., Higgs, N.D., and Amon, D.J., 2015, Whale-fall ecosystems: Recent insights into ecology, paleoecology, and evolution: Annual Review of Marine Science, v. 7, p. 10.1–10.26.

Sommer, U., 2009, The International Congress of Prehistoric Anthropology and Archaeology and German Archaeology, *in* Kaeser, M.-.A., and Babeş, M., eds., Archaeologists without Boundaries: Towards a History of International Archaeological Congresses (1866–2006): Oxford, UK, Archaeopress, p. 17–31.

Stensen, N., 1667, Nicolai Stenonis Elementorum myologiae specimen, seu, Musculi descriptio geometrica: Cui accedunt Canis carchariae dissectum caput, et Dissectus piscis ex canum genere: Firenze, Sub Signo Stellae, 123 p.

Stensen, N., 1669, Nicolai Stenonis de solido intra solidum naturaliter contento dissertationis prodromus: Firenze, Sub Signo Stellae, 79 p.

Tarantini, M., 2012, La nascita della paletnologia in Italia (1860–1877): Borgo San Lorenzo, All'Insegna del Giglio, 126 p.

Targioni Tozzetti, G., 1754, Prodromo della corografia e della topografia fisica della Toscana: Firenze, Stamperia Granducale, 204 p.

Targioni Tozzetti, G., 1755, Lettre de M. le Docteur Targioni-Tozzetti, célèbre médecin de Toscane, à M. de B*** [*sic*]: Journal étranger, v. 1, p. 228–235.

Targioni Tozzetti, G., 1763a, Notizie appartenenti al catalogo delle cose naturali della Real Galleria di Firenze, o sia tomo 4 si esso catalogo: Firenze, Biblioteca del Museo Galileo, Antico 2378, v. 4, 175 folios: https://bibdig.museogalileo.it/Teca/Viewer?an=302341_4&vis=D#page/80/mode/2up (accessed May 2018).

Targioni Tozzetti, G., 1763b, Catalogo delle produzioni naturali che si conservano nella Galleria Imperiale di Firenze disteso nell'anno 1763: Firenze, Biblioteca del Museo Galileo, Antico 2378, v. 2, 452 p.: https://bibdig.museogalileo.it/Teca/Viewer?an=302341_2&vis=D#page/1/mode/2up (accessed May 2018).

Targioni Tozzetti, G., 1763c, Catalogo delle produzioni naturali che si conservano nella Galleria Imperiale di Firenze disteso nell'anno 1763: Firenze, Biblioteca del Museo Galileo, Antico 2378, v. 1, 500 p., https://bibdig.museogalileo.it/Teca/Viewer?an=302341_1&vis=D#page/1/mode/2up (accessed May 2018).

Targioni Tozzetti, G., 1763d, Catalogo delle produzioni naturali che si conservano nella Galleria Imperiale di Firenze, Parte III che comprende i fossili: Firenze, Biblioteca del Museo Galileo, Antico 2378, v. 5, 164 folios: https://bibdig.museogalileo.it/Teca/Viewer?an=302341_5&vis=D#page/1/mode/2up (accessed May 2018).

Tuttle, R.H., 2014, Apes and Human Evolution: Cambridge, Massachusetts, Harvard University Press, 1072 p.

Vaccari, E., 2006, The "classification" of mountains in eighteenth century Italy and the lithostratigraphic theory of Giovanni Arduino (1714–1795), *in* Vai, G.B., and Caldwell, W.G.E., eds., The Origins of Geology in Italy: Geological Society of America Special Paper 411, p. 157–177, https://doi.org/10.1130/2006.2411(10).

Vai, G.B., 2009, The scientific revolution and Nicolas Steno's twofold conversion, *in* Rosenberg, G.D., ed., The Revolution in Geology from the Renaissance to the Enlightenment: Geological Society of America Memoir 203, p. 175–196, https://doi.org/10.1130/2009.1203(14).

Manuscript Accepted by the Society 1 December 2017
Manuscript Published Online 25 May 2018

Printed in the USA

The Geological Society of America
Special Paper 535

Museum Wormianum: Collecting and learning in seventeenth-century Denmark

Lisbet Tarp*
Postdoc, Art History, Aarhus University, Norde Ringgade 1, 8000 Aahrus, Denmark

ABSTRACT

During the first half of the seventeenth century, the Danish polyhistor, Ole Worm (1588–1654), established a collection in Copenhagen, the Museum Wormianum, consisting of minerals, plants, animals, and man-made objects. The collection attracted visitors and was renowned throughout Europe; however, Worm also used it as a site for teaching his university students. Even though Worm did not contribute significantly to the history of science with new discoveries, this article argues that he played a role in shaping an intellectual environment founded on international exchange in which discussions took place, methods were enhanced, and talents were supported. In this context, his museum had an important function as a site of attraction and exchange, anticipating social interaction and learning, even when Worm himself could not participate.

INTRODUCTION

The past two decades have witnessed an increasing interest in the Danish scholar, physician, and antiquarian, Ole Worm (1588–1654), and his collection, Museum Wormianum. The number of research projects and publications on aspects of Worm's life and pursuits has grown significantly. In 2006, the American artist, Rosamond Purcell, created the installation, "Room One," an interpretation of Ole Worm's museum. It was exhibited at Steno Museum in Aarhus and later at the Geological Museum in Copenhagen (Skydsgaard and Teglhus, 2006). As her point of departure, Purcell used the frontispiece of Worm's collection catalogue, adding to the popularity of this engraving (Fig. 1). The engraving shows a room filled with different specimens from floor to ceiling. Objects such as stones, parts of animals and plants, ethnographic artifacts, small-scale sculptures, and labeled boxes surround a table in the center of the engraving. The frontispiece may depict the actual setting of Worm's museum; however, more importantly in this context, it depicts a site of investigation. The table—metaphorically or in reality—designated Worm's primary activity as a collector: It was the place to study and to develop and share his knowledge of the things he collected.

Worm's project is relevant in relation to the topic of this issue—natural scientific museums—because his pronounced intention was to tie the process of making a collection together with the examination of specimens *and* the act of teaching. He wanted his museum to function as a place for sharing knowledge and learning by doing, not unlike the intentions expressed in the museums today. This article will discuss some of the evidence to Worm's approach to learning in the museum.

OLE WORM (1588–1654)

Ole Worm was born in Aarhus, Denmark, as the son of the town's mayor, Willum Worm (1563–1629), and Inger Olsdatter (1568–1619). Even though Worm was not of noble descent,

*tarp@cc.au.dk

Tarp, L., 2018, Museum Wormianum: Collecting and learning in seventeenth-century Denmark, *in* Rosenberg, G.D., and Clary, R.M., eds., Museums at the Forefront of the History and Philosophy of Geology: History Made, History in the Making: Geological Society of America Special Paper 535, p. 81–89, https://doi.org/10.1130/2018.2535(06).

Figure 1. Frontispiece *Museum Wormianum*, G. Wingendorp. Copyright: Public domain.

he received an education equivalent to a nobleman's son. In his youth, Worm was sent abroad to acquire proper training and to earn his university degrees in medicine, physics, and chemistry (Hovesen, 1987, p. 64; Mordhorst, 2009, p. 33–38).[1] Despite the wars and outbreaks of plagues in Europe, he traveled in Germany, Italy, and France. During the years abroad, he was taught by Caspar Bauhin (1560–1624) and Felix Platter (1536–1614), who gave him access to their collections. Among other places, he visited Ferrante Imperato's (1550–1625) collection in Naples, Ulisse Aldrovandi's (1522–1605) museum and garden in Bologna, and the art collections of Duke Maurice of Hesse-Kassel (1592–1627) as well as the cabinet of natural curiosities of Francesco Calzolari (ca. 1522–1609) and Bernhard Paludanus (1550–1633). In 1613, Worm returned to Denmark in order to take up a position in pedagogy at the University of Copenhagen, where he was later appointed the chair of medicine as well as dean of the faculty several times (Hovesen, 1987, p. 127–144). Related to his training as a doctor, Worm was engaged in testing remedies, prescribing medicaments, and investigating human and animal bodily functions.[2] In the 1620s, inspired by the collections he had seen and the instructions he had experienced, Worm began establishing his own collection, Museum Wormianum, in his home in Copenhagen.[3]

A considerable number of written sources have survived, providing evidence of the wide range of activities that Ole Worm pursued during his lifetime. Besides being a renowned collector and a university professor in medicine for more than 30 years, he mapped remains of Nordic antiquity in his important six-volume publication, *Monumenta Danica*, from 1643. He was busy as a physician, kept a botanical garden, and maintained an extended European network through his correspondence. His three marriages to daughters of the most influential Danish families at the university, Fincke and Bartholin, also enhanced his standing in the academic society (Kragh et al., 2008, p. 105). As a professor, he was engaged in teaching and writing and as a mentor, he supported young family members and sons of friends. Several of them, most notably Thomas Bartholin (1616–1680), became important suppliers of information and objects to his museum, since Worm seldom left Denmark in the last decades of his life.

MUSEUM WORMIANUM

The preserved body of written sources by Worm consists of the catalogue, *Museum Wormianum*, *Monumenta Danica*, his extensive correspondences translated by H.D. Schepelern, his dissertations, as well as a number of other publications.[4] The sources indicate that Worm began collecting systematically in the early 1620s (Mordhorst, 2009, p. 39; Schepelern, 1971, p. 142–143). By the 1630s, the collection had reached a considerable size, and Worm had a short and concise catalogue printed (Worm, 1642). Just before he died in 1654, he managed to finish a major catalogue of his collection printed the year after his death, *Museum Wormianum* (Worm, 1655).[5] Worm listed 1663 items in his catalogue, of which only 40 items are known to have survived (Mordhorst, 2009, p. 31, 43, and 79).[6]

The catalogue consists of the aforementioned frontispiece, title page, full-page portrait of Ole Worm, dedication to King Frederik III (1609–1670), preface, list of references, table of contents, and the inventory of the items in four parts ("books") followed by an index of topics. The preface is an interesting source to Worm's motivation for creating a museum and will be discussed later. The four books describe the kingdoms of minerals, plants, animals, and man-made objects. The man-made objects in book four were furthermore listed according to the materials from which they were made, e.g., amber, stone, iron, and so forth (Liber IV, chaps. 1–12).[7] The four books consist of descriptions of the singular objects in Worm's own collection, but some paragraphs also expand into longer discussions and introductory accounts. Some of the paragraphs can also be recognized in his correspondence or written lectures, of which Worm wrote 18 as part of his duties as a university professor (Hovesen, 1987, p. 157).

[1]Worm wrote an autobiography in 1654. Worm and his life are well described in the research literature (Schepelern, 1971; Hovesen, 1987; Teglhus, 2006; Skydsgaard and Grell, 2007; Kragh et al., 2008).

[2]This aspect of Worm's professional life has been dealt with by, among others, Ejnar Hovesen, Ole Peter Grell, as well as Hanne Teglhus and Morten Skydsgaard, who have also contributed with a general introduction to Worm's life (Hovesen, 1987; Skydsgaard and Teglhus, 2006; Grell, 2007).

[3]For the location of the collection, see Mordhorst (2009, p. 44).

[4]For a full list of Worm's published writings, see H.D. Schepelern (1971, p. 398–399). Schepelern has also listed Worm's unpublished writings (Schepelern, 1971, p. 399–401). See also Mordhorst (2009, p. 290–291).

[5]To this day, only minor parts of *Museum Wormianum* have been translated into Danish or English.

[6]Mordhorst refers to H.D. Schepelern's counting (Schepelern, 1990). For presentations and illustrations of the remaining items from the collection, see Mordhorst (2009).

[7]The man-made items include what might be termed ethnographical items and decorative arts.

Figure 2. Horse jaw in an oak trunk, *Museum Wormianum*, p. 342. The length of the trunk is "a little less than two Roman feet [ca. 58 cm]," and its width is like "a man's forearm" (Worm, 1655, p. 342). Photo credit and copyright: Jens Astrup, Natural History Museum of Denmark.

Figure 3. "Globe" of Florentine marble. Diameter: 7.4 cm, weight: 496.25 g, volume: ~180 cm^3, weight distribution: ~2.76 g/cm^3. Photo credit and copyright: Lisbet Tarp.

Within the text, selected items are illustrated with engravings of varying quality by different artists. Some, such as the illustration of the belemnites, seem to be copied from Conrad Gesner's (1516–1565) catalogue (Gesner, 1565, p. 91; Worm, 1655, p. 70).[8] Other illustrations are of a superior quality and seem to have been made while observing Worm's specimens, such as the bird of Paradise (p. 294), the horse jaw embedded in an oak trunk (p. 342) (Fig. 2), and the Norwegian lemming (p. 325).[9] Some of the items of book four are described twice in the catalogue. For instance, a "marble globe" is listed in the book on human artifacts as pieces of craftsmanship; however, it is also represented with another description in the first book on minerals according to the material, Florentine marble (Mordhorst, 2009, p. 169–171; Tarp, 2013) (Fig. 3). This dual perspective on the marble ball illustrates the fluid borders between what was then considered to be a natural and artificial object. Similarly, Worm treasured the horse jaw in an oak trunk mentioned above because it exemplified how nature could merge plant and animal fragment in a creative process. By meticulously describing every detail of the item, Worm tried to document and thus provide a basis for explaining the phenomenon (Worm, 1655, p. 342; Mordhorst, 2009, p. 162–165). Worm also collected objects with strange stories of origin, for instance, the egg born by a woman, or objects with exceptional features such as stones with the scent of flowers. As the modern taxonomies were established, the interest in these kinds of hybrid phenomena in nature faded, and consequently, objects such as the embedded horse jaw lost relevance as a potential key to understanding nature and its processes.

[8]For other sources for the illustrations, see Schepelern (1971, p. 260 and 267). Thanks to Sachiko Kusukawa for turning my attention to Gesner as a possible source for some of the illustrations.

[9]Besides the frontispiece and the portrait of Ole Worm made by G. Wingendorp, the origins of the catalogue illustrations have not been clarified. Most likely, Worm did not make the illustrations.

Going through the catalogue, the overall impression of Worm's driving force as a collector is his profound interest in materials, their properties and origin. With varying degrees of detail, singular items are presented with information on features such as size, color, texture, shape, fragrance, pharmaceutical use, hardness, and flavor. The descriptions sometimes include information on the acquisition, acknowledgments to donors, or stories connected to the objects, in addition to the scholarly references to known literature on the subject. In this way, Worm combined two fundamentally different sources of information about the items in his catalogue: the input he received from secondary sources such as writings on the subject or his correspondences and firsthand observations he derived from his own study of samples and specimens and the experiments that he conducted. Working on correcting and clarifying knowledge on specimens and materials, Worm did not pursue a major rethinking of how to understand or conceptualize nature in general. His focus was on exchanging objects and information in order to understand and get firsthand experiences with samples. In this process, his museum served an essential function as a site for his own examinations as well as for sharing and helping other people gain knowledge of nature through their own practical endeavors.

THE MUSEUM AS A SITE OF LEARNING BY DOING

Some of the visitors to Worm's museum are mentioned in a book by publisher and bookseller, Jens Lauritsøn Wolf (ca. 1584–ca. 1660) printed in 1654. Wolf's book contains a paragraph on Worm's collection of "strange and curios rarities and diverse things" that attracted "noble people" and "diplomatic representatives."[10] It is no surprise that the nobility and other persons of wealth, social, and political capital paid attention to the museum, but Worm was not a prince, and his purpose of making a museum was not to display political power. Instead, it becomes evident that the museum pieces were regarded as practical devices in the study of nature in order to improve his own understanding as well as that of his guests.

Being a professor at the university, Worm taught students in various subjects. He lectured on traditional subjects such as Aristotle or Galen, but two announcements in the course catalogues indicate that Worm supplemented his teaching in the lecture room with weekly demonstrations at his museum (Panum, 1879, p. 90; Spärck, 1962, p. 14; Hovesen, 1987, p. 149–152; Mordhorst, 2009, p. 90–92).[11] In this way, his students gained access to the collection and its content, which seemingly altered the learning situation. This change of place and teaching style was not just an addition to Worm's lecture plan, but the approach was fundamental to Worm's attitude toward teaching and dissemination of knowledge. This is evident in an early account, where Worm ascribed value to learning by handling and examining things themselves. In the letter from 1639, Worm writes:[12]

> As to the display of curiosities in my museum, I have not yet completed it. I have collected different things on my journeys abroad, and from India and other very remote places, I have brought various things: soil samples, varieties of stones, metals, plants, fish, birds, and terrestrial animals. I carefully store them with the intention that, together with a short presentation of the different background of the objects, I can invite the audience to touch with their own hands and observe with their own eyes, so they will be able to judge to what extent the information matches the objects and thus gain a more exact knowledge of them all.[13]

In this way, Worm made his intentions clear that the audience should not only read or listen to presentations of specimens, but should activate their senses in order to gain a better understanding of the objects presented. Many years later, Worm unfolded this position on learning in the museum even further in the earlier mentioned preface to the catalogue *Museum Wormianum* (Worm, 1655, *præfatio*).[14]

The three-page-long preface is a strong attack on "sophistry" and "superficial and scholastic philosophers," who do not study nature at hand but rather only discuss questions on nature

[10]"At jeg oc icke forbigaar, jo ocsaa med berømmelse skriffver om D. Olai Vvorm: Medicine Konst-Kammer, hvor udi findes, oc med stor Forundring er at see, meget selsomme oc underlige Rariteter, oc Tingester af atskilligt, som mand med Forundring en Deel tilforen icke haffver seet, hvilcket mange Førstelige Personer, saa vel som oc Gesanter der til Kiøbenhaffn ankommer, det begærer at see, for detz store berømmelse Skyld oc Affsagen paa fremmede Stæder, hvilcke sig der ofver høyelig forundrer, oc det icke nocksom loffve oc berømme kand. Hvilcket de til it Vidnesbyrds Tegen, af de det seet haffver, med deres egen Haand hos hannem det udi en Bog testerer oc Vidner." Online transcribed version: http://www.kobenhavnshistorie.dk/bog/encomion/encomion.html section 356 (accessed 3 September 2017) (Wolf, 1654). Mordhorst also refers to this source (Mordhorst, 2009, p. 36).

Translation by Troels Kardel: "With praise I also write about the medicine art cabinet of Ole Worm, in which there are to be found strange and wondrous rarities and several devices not previously seen. Many noblemen as well as diplomatic representatives who arrive in Copenhagen desire to see this cabinet for the sake of its fame and reputation in foreign places, where people are highly astonished and cannot praise and commend it enough. All this is testified and witnessed in a firsthand report of what they have seen, given in their own hand at his place in a book as testimony of witness."

[11]Announcement from 1632–1633: "D. Olaus Worm (Rector) absoluta Therapeutica h. 3. Rariorum et mirandorum quae in Musaeo tenet, tam naturalium quam artificialium succinctam exhibebit historiam *autopsia* singula confirmaturus." Announcement from 1642–1643: "D. Olaus Worm M. Pr. (Rector) Quadrupedum rariorum et exoticorum, quorum vel integra corpora vel partes in Musaeo tenet, Historiam tractabit, quaeque restant artificiose elaborata absolvet, Nomina descriptiones et vires compendiose proponendo *autopsia* singula confirmaturus h. III" (Panum, 1879, p. 90). Not all of the course catalogues concerning Worm's lectures have been preserved.

[12]Valdimar Hafstein also cites this paragraph (Hafstein, 2003, p. 9). The quotations from Worm's letters in this article are my translations of H.D. Schepelern's translations from Latin into Danish (Schepelern, 1965–1968, v. 1–3). This is not an optimal procedure, but unfortunately, I have not been able to view the Latin sources myself. Valdimar Hafstein also cites this paragraph (Hafstein, 2003, p. 9).

[13]Letter to Arngrim Jonsson på Melstaðir, 20 June 1639 (Schepelern, 1967, v. 2, p. 132, letter no. 787).

[14]The *Museum Wormianum* preface is translated into Danish (Schepelern, 1971, p. 213–217). The quotations in the following are supported by Schepelern's Danish translation.

in theory. These sophistic discussions do not "… support our knowledge on nature, its distinctive features and properties on the most humble stone at our feet."[15] According to Worm, this mistaken approach was pursued at schools and academies, thus "… preventing the youth from examining the nature of things."[16] In tune with the general tendency in seventeenth-century Europe, he emphasizes the act of observing and recording natural specimens. In his defense, Worm cites a paragraph of the French philosopher, Pierre Gassendi (1592–1655), elaborating on the subject of the misled students: "When they come into contact with real nature, they are confused and numb, likewise, if they are informed by people, who do not seek answers through contemplation but instead are guided by all things' teacher: experience and observation."[17] In other words, Worm believed that in order to be adequately educated, the students had to have access to the study of specimens—an activity Worm considered the true way of gaining knowledge on nature. He concludes that the "… academics should replace persuasion with proof, discussion with examination, belief with knowledge."[18] In the following, he takes it one step further: "… no one should be forced to swear allegiance to his teacher, rather the objects themselves should consolidate understanding, and you should not trust your teacher more than he can prove with facts obtained by himself."[19] An effect of the emphasis on the empirical approach to knowledge production suggests a change in the relationship between teacher and student. Taken to the extreme, it can be said that both teacher and student are in a process of learning, while studying nature firsthand. After addressing the urgency of improving teaching methods, Worm explains that from the moment he began teaching on nature in Copenhagen, he used "… his means and energy to establish a treasury of natural specimens …."[20] In this statement, he directly connects the making of the museum to his ambition of providing object lessons at the university.

In a letter from 1640, Worm also comments on presenting objects to students in connection with going through the first book on minerals. Subsequently, he proclaims, they would proceed to the books on plants and animals (Schepelern, 1967, v. 2, p. 166).[21] The book he refers to is most likely an earlier version of his catalogue from 1655, and it seems that he used the books to support his object lessons. Unfortunately, we do not have any further substantial evidence of how Worm actually interacted with his students in the museum. A more vivid impression of how the procedures in the museum might have taken place can be found in a firsthand account from a group's visit. In 1634, the French ambassador's secretary, Charles Ogier (1595–1654), visited Worm's museum. In his account, Ogier writes that Worm first lectured at the university and later presented his museum to a selected group of people. Worm showed them a large piece of rock crystal in which a cross and relics had been inserted.[22] Ogier describes how the relics wrapped in silk were taken out of the rock crystal and studied by the group (Schepelern, 1971, p. 158).[23] The description gives an impression of how the group gathered around the presumed table and were allowed to open up and touch the different parts of the item. Supposedly, they scrutinized the components, turned them in their hands, discussing what they saw, rejecting or agreeing on interpretations. Worm probably participated in this social event by providing information on the subject matter. For instance, Ogier adds that Worm tells the story of the acquisition of the rock crystal. Worm received it as payment for curing a Swedish captain, who had acquired it in Germany as war booty. Maybe not wanting to share this information to a wider audience, Worm omitted reporting how he came by the crystal reliquary in the *Museum Wormianum* catalogue. Furthermore, the piece does not appear in the book of artificial things according to its status as reliquary (Schepelern, 1971, p. 159). Instead, Worm focused on its physical qualities as a six-pound piece of rock crystal, placing it in chapter 16 on "chalcedony, carnelian, phengite, amethyst, crystal, flours, and aventurine" (Worm, 1655, p. 100; Mordhorst, 2009, p. 157 and 286–289). The difference Worm made between the oral and the written presentation of the object underscores the possibility for a more open and intimate dialogue on site.

As mentioned earlier, in his educational years abroad, Worm was exposed to famous collections and collectors, who hosted social events such as the one described above, where curiosities were displayed, engaged with, and discussed. From an early age, Worm was influenced by the international habits of exchange and the increasing emphasis on empirical studies in the scholarly environment, which eventually motivated him to establish his own museum (Hovesen, 1987 p. 137–143). In this way, he contributed to nourishing and settling the tradition of object lessons in Denmark (Spärck, 1962, p. 14–20; Schepelern, 1971, p. 134; Hovesen, 1987, p. 149; Findlen, 1994; Wolff, 1999).

---

[15]"Quibus instructus, ne vilissimi quidem lapidis, qui præ pedibus est, naturam, proprietates & vires sciet" (Worm, 1655, p. vii).

[16]"Ita juventus a rerum naturis inquirendis avocatur, …" (Worm, 1655, p. vii).

[17]"Hinc ipsimet, ubi ad hanc veram naturam ventum est, caligant prorsus & obstupescunt, moniti scilicet ab iis, qui non fallaci ratiocinio, sed magistra rerum experientiâ, observationeque omnia prosequuntur" (Worm, 1655, p. viii). Worm quotes Gassendi (1624).

[18]"… ut Acadmiæ definant persuadere, incipient demonstrare, definant disputare, incipient speculari, definant credere, incipient scire" (Worm, 1655, p. viii).

[19]"Nemo jurare cogatur in verba Magistri, sed res ipsæ constringant intellectum, nec plus fidei habeatur præceptori, quam quantum esse habendum re demonstraverit" (Worm, 1655, p. viii).

[20]"Ex eo igitur, quo primum Naturæ Historiam in hac Academiâ Regiâ docere cæpi, nullis parcens sumptibus aus laboribus, rerum naturalium thesaurum haud contemnendum colligere cæpi" (Worm, 1655, p. ix).

[21]Worm mentions his teaching in the museum shortly in letter number 823 and 1050 (Schepelern, 1967, v. 2).

[22]Today, the rock crystal is in the collection of the National Museum of Denmark, Copenhagen.

[23]"Er zeigte uns auch ein viereckigt Stück Kristall […] In demselben waren viele in Seide engewickelte Reliquien verwahrt, welche wir alle herausnahmen und betrachteten" (Schepelern, 1971, p. 158). Schepelern has the quotation from Schlegel (1776).

STUDYING SPECIMENS

Evident in his writings, Worm was engaged in many different discussions on natural phenomena that were related to objects from his museum. But what could he have presented to his students? One plausible topic of an object lesson was fossilized shark teeth, also called tongue stones or *glossopetrae*.[24] Like many other people at the time, Worm was curious about these mysterious stones to which antidote powers were ascribed, and he was the owner of several samples (Worm, 1655, p. 67; Hoch, 2013). As already mentioned, Worm goes through the topic of fossilized shark teeth in a chapter in *Museum Wormianum,* which could be an excerpt from a lecture. Furthermore, he also corresponded with his friend Henrik Fuiren (1614–1659) on the subject (Worm, 1655, p. 67; Schepelern, 1968, p. 36–38, no. 1227). The fact that the fossils of sea creatures appeared at places far from the ocean was puzzling due to the belief in the concept of a static Earth crust. Furthermore, several of the animals that appeared in the stones were extinct species, for instance, belemnites, making the recognition of their origin even more difficult (Etter, 2015).

In his considerations of the concept of fossils, Worm was primarily dependent on other people's descriptions and the specimens sent to him. This left him with the reports he received, theoretical writings on the topic, and the observations he could make of the specimens he had at home. Worm's theoretical approach to fossils was primarily based on the writings of Caspar Bartholin (1585–1629), who was a relative of Worm. Bartholin explicated the ideas of *succus lapidescens*—a lapidifying juice referring to the tradition of Georgius Agricola (1494–1555)—and its underlying principle, *spiritus lapidificans,* in his popular book, *Systema physicum* from 1628, which was widely distributed in Denmark (Garboe, 1959, p. 24; Kragh, 2006, p. 17). Following Bartholin, Worm used the term *succus lapidescens,* and he noted the effect of the supposed petrifying juice in connection with various specimens, for instance, the crystals, stalactites, and stones that looked crafted (Worm, 1655, p. 51, 81, and 99).[25] In the letter to Henrik Fuiren, Worm sums up his knowledge on petrified shark teeth (Schepelern, 1968, p. 36–38, no. 1227).[26] Firstly, he acknowledges that his knowledge on the subject was supported by the reports from his mentee, Thomas Bartholin (1616–1680), son of Caspar Bartholin, as well as what he had heard of ideas by Fabio Colonna (1567–1640), and what he had seen in the museum of Francesco Calzolari (ca. 1522–ca. 1609) (Mordhorst, 2009, p. 34). In the letter, Worm supports Colonna's claim that parts of fish can turn into stone, and then continues that in some places there is a petrifying juice or evaporation capable of entering the pores of the bones and thus changing their nature. Nevertheless, Worm doubts that "all stones sold as tongue stone are teeth of *canis carcharius* changed into stones" (Schepelern, 1968, p. 36–37). He writes that his doubt is founded on the habitual discovery of teeth far from the ocean, and then he gives the examples of findings in Lunenburg in Germany and the Dutch town Deventer, both far from the sea. To this typical objection at the time, Worm adds his own reflections comparing the *glossopetrae* with a shark head and its teeth in his collection.

In his careful inspection of the stones, Worm notes that the *glossopetrae* and the shark teeth differed in size and shape. The *glossopetrae* were much bigger, and some were thicker. Secondly, they differed in color. The shark teeth were white, while the others had colors such as black, purple, and ash gray. Worm proposes that the colors might correspond to the nature and color of the exhalation that caused the petrifaction. The third objection concerns the number of teeth in the mouth. The massive size of the petrified teeth would require an equally large mouth. He writes: "It must have been a tremendous mouth that could hold two hundred of the kind. For this is the number of teeth in the head of the *canis carcharius* that I have in my museum" (Schepelern, 1968, p. 37). Neither material of the stone nor the shape revealed to Worm why the stones appeared as enlarged shark teeth, and the missing living giant shark as a counterpart to the fossils did not make it easier. At the end of the paragraph, Worm concludes that there are two different kinds of teeth made of stone: one kind that was created by mire containing petrifying powers that changed the fish teeth into stone and another kind of tooth-shaped stone, which was created by the "wit of nature." In this way, Worm chose to interpret the seemingly causeless fossils as the result of nature's ability to play and invent images, *lusus naturae* (Findlen, 1990; Adamowsky and Felfe, 2011; Felfe, 2015, p. 113–120).[27]

FACILITATING LEARNING

The case of the shark teeth exemplifies how Worm analyzed and compared the specimens, trying to establish a reliable argument. While reading the description in his letters, it is easy to imagine him performing an object lesson with his students on the topic of shark teeth. Despite Worm's systematic approach to studying specimens, his limited access to dissections, excavation sites, and his loyalty to ideas of a static Earth crust restricted the conclusions he could make. Worm's considerations on *glossopetrae* were neither original nor satisfactory. It was Niels Stensen (Steno) (1638–1686) who would finally argue for the connection between the shark teeth and the *glossopetrae* in his famous

[24]In a letter from 1640, Worm writes that he had taught on the subjects of the first book of minerals (Schepelern, 1967, no. 823). In another letter, he writes that he had presented a whale head to the audience (Schepelern, 1967, no. 1050).

[25]For a general introduction to the question on fossils and *succus lapidescens,* see Rudwick (1972, especially p. 23–26).

[26]There is a likeness between the paragraph on *glossopetrae* in *Museum Wormianum* and the letter (Worm, 1655, p. 67).

[27]One of the most famous exponents of interpreting naturally made images found in stone was Athanasius Kircher (1602–1680) with whose ideas and writing Ole Worm was familiar. Today, Kircher is often quoted for his admiration of the "cities, mountains, clouds" that appeared as if painted in the patterns and colors of marble. "Unde videas & in altero quodam *lapide* marmoreo depictas urbes, montes, nubes, cujusmodi est in meo Museo …" (Kircher, 1678, p. 32).

text on *Canis Carchariae* (1669).[28] Stensen did not go into the ideas on the petrifying juice, which Worm had been twisting and turning in relation to many different kinds of natural specimens. Instead, Stensen's dissection of a huge shark guided his conclusions. Worm's limited empirical study was against him, and his reference to the idea of a petrifying juice also shows that he was attached to an older mindset than Stensen.

In an article from 1999, Jole Shackelford describes some of Worm's investigations leading to correct or incorrect conclusions in modern terms (Shackelford, 1999). He points out several examples, where the most well known is Worm's thesis on unicorn horn that he argued was narwhale tusk.[29] Due to his empirical study, Worm recognized the narwhale tusk; however, Shackelford emphasizes that Worm at the same time maintained his belief in the concept of spontaneous generation (Rudwick, 1972, p. 45; Shackelford, 1999, p. 67). In other words, Shackelford depicts Worm as a researcher in a transitional phase of time with a foot in both camps. Shackelford sums it up: "Worm collected somewhat more selectively [than Aldrovandi] and sought to distinguish fact from fable, not always with success" (Shackelford, 1999, p. 66). Worm never made any important scientific discoveries, and from the perspective of this article, the relevance of highlighting Worm does not lie within his ability to distinguish between what we today consider fable or fact. Instead, Worm is relevant to this issue because he is an early example of an articulated wish to establish a connection between learning and the museum. In Worm's writings, we get the outline of how he included students, friends, colleagues, and guests in discussing, studying, and testing the things he gathered in his museum. In this way, he stimulated the development of a learning environment in Denmark by promoting and providing a site for exchange and hands-on, empirical studies of nature.[30]

THE FATE OF MUSEUM WORMIANUM

When Ole Worm died in 1654, the main parts of his collection were transferred to the Danish King Frederik III, who merged it with the royal collection. This meant that the university lost its study collection, which was not replaced until Worm's mentee and successor, Thomas Bartholin, financially supported the establishment of a new one (Spärck, 1962, p. 20–24; Wolff, 1999). Embedded in the royal collection, many of Worm's items survived for more than two centuries. Additionally, a large number of Worm's object descriptions from the *Museum Wormianum* catalogue were transferred and reformulated in the later inventories. Among them were two major inventories, both called *Museum Regium,* made by Holger Jacobæus (1650–1701) and Johannes Laverentzen (ca. 1648–1729) (Jacobæus, 1696; Laverentzen, 1710).[31] Some of the altered descriptions in the *Museum Regium* catalogues give evidence to changed interpretations. For instance, the *Museum Regium* paragraph on *glossopetrae* does not refer to nature's assumed capability of "drawing" the figures in stone, and Worm's references to further literature as well as his speculations on the topic were omitted (Worm, 1655, p. 67; Jacobæus, 1696, p. 36; Laverentzen, 1710, no. 67).[32] Instead, Niels Stensen was added as a reference for the origin of the fossilized shark teeth.

The royal *Kunstkammer* was established relatively late compared to other European collections, but it remained intact for an unusually long period of time (Mordhorst, 2009, p. 48). Whereas many other collections such as that of the Habsburg family were split up and dispersed in inheritance cases, the Danish royal *Kunstkammer* was handed over from king to king in the order of succession in the royal family. This feature makes it possible to follow the items and the reception of them over a long span of time. Camilla Mordhorst has investigated and analyzed the reception of the 40 known surviving items from their appearance in Worm's collection until the nineteenth century, when the absolute monarchy in Denmark ended and the museum institutions were founded (Mordhorst, 2009). Descriptions of now lost items and later additions are included in her account, widening the scope of her study. She demonstrates how the reception of individual items changed, and she provides examples of how parts of the collection were tied to a wide range of ideas, old and new. One of Mordhorst's conclusions is that *Museum Wormianum* and its afterlife in the royal collection were characterized by continuity of tradition, gradual change, and minor adjustments rather than innovative thinking and radical changes.[33]

In the late eighteenth century, a wish to reorganize the royal collection emerged. Besides Worm's collection, several private collections had been added to the royal chambers during the years, demanding better storage and administration. Eventually, it was decided to establish six commissions responsible for (1) paintings, (2) Nordic antiquities, (3) classical antiquities,

[28]There is a rich literature on Niels Stensen. For literature concerning Stensen's discovery of the origin of *glossopetrae* (Gould, 1983; Hsu, 2009; Kardel, 2009; Kardel and Maquet, 2013, p. 166–172; Bek-Thomsen, 2013) and for Stensen's encounters with collections, see Thomsen (2009).

[29]Worm wrote a dissertation on the unicorn horn using empirical evidence to support his assumption that it was a narwhale tusk (Hovesen, 1987, p. 254–256).

[30]The high cost of sending the young men abroad to be educated motivated an establishment of schools in Denmark, resulting in Sorø Academy 1623. Furthermore, in seventeenth-century Denmark, there was a general interest in improving the reading abilities of laymen (Appel, 2001).

[31]Handwritten inventories, partly in Danish, were made in 1674, 1689, 1690, and 1737. Minor enclosures with recent acquisitions were made in 1775, 1807, and 1827 (Mordhorst, 2009, p. 30). Bente Gundestrup has edited a two-volume catalogue reconstructing the 1737 catalogue of the royal collection (Gundestrup, 1991). These catalogues include English translations. See also Jørgen Hein's work on the Rosenborg collection (Hein, 2009).

[32]"… à Naturâ hac figurâ dotatus" (Worm, 1655, p. 67).

[33]Michel Foucault's theory of developments of episteme has been applied to the history of museums, most influentially by Eilean Hooper-Greenhill (1992). Mordhorst argues that the empirical approach allows for a much more nuanced rendering of Museum Wormianum than a theoretical application of Michel Foucault's episteme can do alone (Mordhorst, 2009, p. 21–26). Mordhorst acknowledges Foucault's theoretical framing as such, but she tones down its usefulness to describe the history of museums, in this case, the development from Museum Wormianum to modern museum institutions in Denmark.

(4) contemporary genuine polished stones and precious objects, (5) ethnographica, and (6) carved and turned pieces (Garboe, 1959, p. 183; Mordhorst, 2009, p. 127 and 270). As an example, the collection of minerals was placed at Rosenborg Castle in 1804, where it was merged with other mineral collections, for instance, Theodor Holm's (1731–1793) and Peter Christian Abildgaard's (1740–1801) collections (Andrup, 1933, p. 25; Callisen 1945; Garboe, 1959, p. 184–188; Mordhorst, 2009, p. 204).[34] In the middle of the nineteenth century, the mineral collection was relocated at the Natural History Museum (Callisen, 1945, p. 529–530). In this way, the commissions dispersed and relocated the items of the collection step by step. Most of Worm's items disappeared in this process, and the royal *Kunstkammer* was officially closed in 1825. By the end of the nineteenth century, the pieces of the royal collection had either been sold, lost in fire, or transferred to the newly constituted museums (Mordhorst, 2009, p. 67–71 and 279–282).

CONCLUSIONS

Ole Worm's museum was not radically different from many other cabinets of curiosities of the sixteenth and early seventeenth centuries. His collection reflected older concepts of nature such as the above-mentioned ideas on petrifying juice, nature's ability to play, and the supposed hybrids between the different kingdoms of nature. The key point of this article is to highlight Worm's contribution as taking a reflected position on learning and the creation of a learning environment. Worm did not only tie together learning and the making of his museum. He also had a need to express why he had an ambition to do so. As argued in this article, Worm enunciated his approach to learning about nature in his preface to *Museum Wormianum* and in his letters. These sources give evidence to Worm as a collector who explicitly aimed at establishing what he considered to be optimal conditions for other people's learning on nature by facilitating object lessons and hands-on activities. In this way, as a teacher and collector—albeit not as a scientist in modern terms, Worm supported the emerging scientific communities, the international exchange of knowledge and students as well as the increasing emphasis on documentation and empirical evidence in the seventeenth century.

REFERENCES CITED

Adamowsky, N., and Felfe, R., eds., 2011, Ludi Naturae: Spiele der Natur in Kunst und Wissenschaft: München, Wilhelm Fink.

Andrup, O., 1933, Den Kongelige Samling paa Rosenborg gennem hundrede Aar: Copenhagen, Gyldendalske Boghandel, Nordisk Forlag.

Appel, C., 2001, Læsning og bogmarked i 1600-tallets Danmark: Copenhagen, Museum Tusculanums Forlag, v. 1–2.

Bek-Thomsen, J., 2013, From flesh to fossils—Nicolaus Steno's anatomy of the Earth, *in* Duffin, C.J., Moody, R.T.J., and Gardner-Thorpe, C., eds., A History of Geology and Medicine: Geological Society, London, Special Publication 375, p. 289–305.

Callisen, K., 1945, Til Mineralogisk Museums Historie. Kong Christian VIII og hans particulaire mineralogiske Cabinet: Meddelelser fra Dansk Geologisk Forening, v. 10, p. 507–531.

Etter, W., 2015, Early ideas about fossil cephalopods: Swiss Journal of Palaeotology, v. 134, p. 177–186.

Felfe, R., 2015, Naturform und bildnerische Prozesse. Elemente einer Wissensgeschichte in der Kunst des 16. und 17. Jahrhunderts: Berlin/Boston, Actus et Imago, 13, De Gruyter.

Findlen, P., 1990, Jokes of nature and jokes of knowledge: The playfulness of scientific discourse in early Modern Europe: Renaissance Quarterly, v. 43, no. 2, p. 292–331, https://doi.org/10.2307/2862366.

Findlen, P., 1994, Possessing Nature: Museums, Collecting, and Scientific Culture in Early Modern Italy: Berkeley, University of California Press, 449 p.

Garboe, A., 1959, Geologiens historie i Danmark: Fra myte til videnskab: Copenhagen, C.A. Reitzels Forlag, 283 p.

Gassendi, P., 1624, Excercitationum paradoxicarum adversus Aristoteleos libri septem: Grenoble, Ex typographia Petri Verderii, 219 p.

Gesner, C., 1565, De omni rerum fossilium genere, gemmis, lapidibus, metallis, et huiusmodi, libri aliqvot, pleriqve nvnc primvm editi: Tiguri, Excudebat Iacobus Gesnerus, 1059 p.

Gould, S.J., 1983, Hen's Teeth and Horse's Toes: Further Reflections in Natural History: London, Norton, 416 p.

Grell, O.P., 2007, In search of True Knowledge: Ole Worm (1588–1654) and the New Philosophy, *in* Smith, P.M., and Schmidt, B., eds., Making Knowledge in Early Modern Europe: Practices, Objects, and Texts, 1400–1800: Chicago, University of Chicago Press, p. 214–232.

Gundestrup, B., 1991, Det Kongelige danske Kunstkammer 1737: Copenhagen, Nationalmuseet, Nyt Nordisk Forlag, v. 1–2, 965 p.

Hafstein, V., 2003, Bodies of knowledge: Ole Worm & collecting in late Renaissance Scandinavia: Ethnologia Europaea, v. 33, p. 5–19.

Hein, J., 2009, The Treasure Collection at Rosenborg Castle: The Inventories of 1696 and 1718: Copenhagen, Selskabet til Udgivelse af Danske Mindesmærker, Museum Tusculanum, v. 1–3, 1720 p.

Hoch, E., 2013, Diagnosing fossilization in the Nordic Renaissance: An investigation into the correspondence of Ole Worm (1588–1654), *in* Duffin, C.J., Moody, R.T.J., and Gardner-Thorpe, C., eds., A History of Geology and Medicine: Geological Society, London, Special Publication 375, p. 307–327.

Hooper-Greenhill, E., 1992, Museum and the Shaping of Knowledge: London, Routledge, 232 p.

Hovesen, E., 1987, Lægen Ole Worm 1588–1654. En medicinhistorisk undersøgelse og vurdering: Aarhus, Aarhus Universitetsforlag, 367 p.

Hsu, K.-T., 2009, The Path to Steno's synthesis on the animal origin of *glossopetrae*, *in* Rosenberg, G.D., ed., The Revolution in Geology from the Renaissance to the Enlightenment: Geological Society of America Memoir 203, p. 93–106, https://doi.org/10.1130/2009.1203(06).

Jacobæus, H., 1696, Museum regium, seu, Catalogus rerum tam naturalium, quam artificialium, quae in basilica bibliothecae augustissimi Daniae Norvegiaeq[ue] monarchae Christiani Qvinti, Hafniae asservantur: Copenhagen, 303 p.

Kardel, T., 2009, Prompters of Steno's geological principles: Generation of stones in living beings, glossopetrae and molding, *in* Rosenberg, G.D., ed., The Revolution in Geology from the Renaissance to the Enlightenment: Geological Society of America Memoir 203, p. 127–134, https://doi.org/10.1130/2009.1203(08).

Kardel, T., and Maquet, P., eds., 2013, Nicolaus Steno: Biography and Original Papers of a 17th Century Scientist: Heidelberg, Springer, https://doi.org/10.1007/978-3-642-25079-8, 762 p.

Kircher, A., 1678 [1665]: Mundus Subterraneus: Janssonio Waesbergiana, Amsterdam, v. 2, 32 p.

Kragh, H., 2006, Inventive nature: Fossils, stones and marvels in seventeenth century Danish natural philosophy, *in* Federhofer, M., ed., Naturspiele: Beiträge zu einem naturhistorischen Konzept der Frühen Neuzeit: Cardanus, Palatina Verlag, p. 15–24.

Kragh, H., Kjærgaard, P.C., Nielsen, H., and Nielsen, K.H., eds., 2008, Science in Denmark: A Thousand-Year History: Aarhus, Aarhus University Press, 607 p.

Laverentzen, J., 1710, Museum regium seu catalogus rerum tam naturalium, quam artificialium, qvæ in basilica bibliothecæ Augustissimi Dania

[34]Gundestrup has visualized the establishment of the institutions, see Gundestrup (1991, v. 1, p. 27).

Norvegiæqve monarchæ, Friderici Qvarti Havniæ asservantur gloriosissimæ memoriæ rege Christiano Qvinto: Copenhagen, 635 p.
Mordhorst, C., 2009, Genstandsfortællinger: Fra Museum Wormianum til de moderne museer: Copenhagen, Museum Tusculanums Forlag, 311 p.
Panum, P.L., 1879, Vort medicinske fakultets oprindelse og barndom: Copenhagen, Gyldendalske Boghandel, 116 p.
Rudwick, M.J.S., 1972, The Meaning of Fossils: Episodes in the History of Palaeontology: London, Macdonald, 287 p.
Schepelern, H.D., 1965–1968, Breve til og fra Ole Worm: Munksgaard, Copenhagen, v. 1–3, 1534 p.
Schepelern, H.D., 1971, Museum Wormianum: Dets Forudsætninger og Tilblivelse: Copenhagen, Wormianum, Odense, 433 p.
Schepelern, H.D., 1990, The Museum Wormianum reconstructed: A note on the Illustration of 1655: Journal of the History of Collections, v. 2, no. 1, p. 81-85, https://doi.org/10.1093/jhc/2.1.81.
Schlegel, J.H., 1776, Sammlung zur Dänishen Geschichte, Münzkentniss, Ökonomie und Sprache: Copenhagen, Sander und Schröder, v. 2, p. 146–48.
Shackelford, J., 1999, Documenting the factual and the artifactual: Ole Worm and public knowledge: Endeavour, v. 23, no. 2, p. 65–71, https://doi.org/10.1016/S0160-9327(99)01177-1.
Skydsgaard, M., and Teglhus, H., 2006, Ole Worm—liv og videnskab: Aarhus, Steno Museet, 56 p.
Spärck, R., 1962, Undervisning i zoologi ved Københavns Universitet. Et tilbageblik over 300 år: Copenhagen, Bianco Lunos Bogtrykkeri, 301 p.
Stensen, N., 1669, Canis Carchariae dissectum caput: Amsterdam, Nicolai Stenonis elementorum myologiæ specimen, p. 90–137.
Tarp, L., 2013, Marble and marvel: Ole Worm's globe and the reception of nature's art in seventeenth century Denmark: Kritische Berichte, v. 3, p. 8–23.
Thomsen, E., 2009, Niels Stensen—Steno, in the world of collections and museums, *in* Rosenberg, G.D., ed., The Revolution in Geology from the Renaissance to the Enlightenment: Geological Society of America Memoir 203, p. 75–91, https://doi.org/10.1130/2009.1203(05).
Wolf, J.L., 1654, Encomion regni Daniæ: Copenhagen. Online version: http://www.kobenhavnshistorie.dk/bog/encomion/encomion.html section 356 (accessed 3 September 2017).
Wolff, T., 1999, The history of the Zoological Museum: Copenhagen, Steenstrupia, v. 24, no. 2, p. 157–176.
Worm, O., 1642, Musæi wormiani catalogus, 49 p.
Worm, O., 1655, Museum Wormianum, seu historia rerum rariorum, tam naturalium, quam artificialium, tam domesticarum, quam exoticarum, quæ Hafniæ Danorum in ædibus authoris servantur, variis et accuratis iconibus illustrata: Amsterdam, Elsevier, 38 p.

Manuscript Accepted by the Society 21 November 2017
Manuscript Published Online 25 May 2018

Printed in the USA

The Geological Society of America
Special Paper 535

Fossil collections and mapping the Silurian: An example from Scandinavia

John A. Diemer
Department of Geography and Earth Sciences, University of North Carolina at Charlotte, 9201 University City Boulevard, Charlotte, North Carolina 28223, USA

ABSTRACT

From 1840 to 1845, Roderick Murchison (1792–1871), an accomplished geologist who was instrumental in developing the Paleozoic portion of the geologic time scale, traveled in Europe, Scandinavia, and Russia, where he documented the occurrence and extent of the Silurian, Devonian, and Permian Systems. A key component of his fieldwork methodology was to examine fossils in both private collections and public museums to assess the types of fossils that occur in a region, in order to guide and expedite his mapping efforts. While working in Scandinavia, he examined fossil collections in Copenhagen, Oslo, Stockholm, Upsala, and Visby, and those collections clearly influenced his choice of travel routes and, ultimately, the geologic map (plate VI) in *The Geology of Russia* (Murchison et al., 1845). Furthermore, he was able to compare the Scandinavian fossils to British collections, thereby confirming the age of the strata he was mapping. The results of Murchison's work appeared in numerous papers as well as two major books, *The Geology of Russia* (Murchison et al., 1845) and *Siluria* (Murchison, 1854), thereby advancing knowledge of Scandinavian stratigraphy. Not only were the fossil collections in the museums of his day essential to Murchison's scientific research, but collections of documents in modern archives are invaluable in reconstructing his movements, observations, and discoveries more than 170 years ago, thereby demonstrating the utility of collections and museums to the construction of knowledge.

INTRODUCTION

While the origins of the science of geology can be traced back several centuries, the discipline developed rapidly in the late eighteenth and early nineteenth centuries (Geikie, 1905; Hallam, 1989; Lyell, 1834; Mather and Mason, 1939; Oldroyd, 1990; Rudwick, 1985b, 1992, 2005; Schneer, 1969; Secord, 1986; Stafford, 1989). One of the important advances of that interval was made by William Smith (1769–1839), who developed a biostratigraphic approach to geologic mapping where distinctive fossil assemblages, rather than similarities in lithology, could be used to correlate and identify the age of strata (Geikie, 1905; Sharpe, 2016; Sharpe and Torrens, 2015). This advance led to the construction of the first geologic map of much of Britain as well as the geologic time scale, a conceptual framework that remains integral today to reconstruction of Earth's history (Diemer, 2016b). From the time of its construction, the geologic time scale has been applied successfully around the world.

Diemer, J.A., 2018, Fossil collections and mapping the Silurian: An example from Scandinavia, *in* Rosenberg, G.D., and Clary, R.M., eds., Museums at the Forefront of the History and Philosophy of Geology: History Made, History in the Making: Geological Society of America Special Paper 535, p. 91–116, https://doi .org/10.1130/2018.2535(07).

One of the geologists who helped construct the geologic time scale using William Smith's biostratigraphic approach was Roderick Murchison (1792–1871). His work led to the recognition of the Silurian, Devonian, and Permian Systems of the Paleozoic Era (Geikie, 1875; Stafford, 2004). His early fieldwork was undertaken in Britain, but he then began to work internationally in an effort to trace rocks of Paleozoic age farther afield (Collie and Diemer, 2004; Rudwick, 1985b). The 1840s were a period of peak activity, during which he undertook fieldwork in Europe, Russia, and Scandinavia (Collie and Diemer, 2004; Diemer, 2008, 2016a; Rudwick, 1985b). We are fortunate to have an extensive record of his work during that interval. In addition to his publications, his scientific correspondence, field notebooks, and autobiographical journal, preserved in several key archives, provide evidence documenting his activities (Collie and Diemer, 1995; Collie and Diemer, 2004; Diemer, 1996, 2008, 2016a).

The importance of archives that preserve primary documents cannot be overemphasized, because those primary documents provide essential evidence for historical scholarship. In particular, the documents held by the Geological Society of London, the British Geological Survey, and the University of Edinburgh were used to reconstruct Murchison's movements and discoveries during his 1840s campaigns in Europe, Russia, and Scandinavia. The documents record the museums that Murchison visited throughout Scandinavia and the nature of the specimens that he examined. Thus, the building and maintenance of museum collections were essential to Murchison's scientific discoveries, and the existence of archive collections is necessary for reconstructing his research program.

Two places where Murchison applied modern stratigraphic "system" terminology for the first time were in Russia and Scandinavia. In 1840 and 1841, Murchison and his colleagues, Edouard de Verneuil (1805–1873) and Alexander von Keyserling (1815–1891), boldly engaged in the Russian field research that allowed them to write the two-volume work entitled *The Geology of Russia* (Murchison et al., 1845). That work documented the widespread nature of Silurian and Devonian systems in Russia and defined a new geologic system, the Permian (Benton et al., 2010; Collie and Diemer, 2004; Thackray, 1978; Torrens, 1990). In order to complete the book, including the map that covers much of Europe in addition to Russia (plate VI of *The Geology of Russia*), Murchison engaged in fieldwork in adjacent territories, first in Poland, Hungary, and Germany in 1843, and then in Scandinavia in 1844 and 1845 (Diemer, 2008, 2016a; Hestmark, 1992, 2008; Narkiewicz et al., 2012).

One significant finding of Murchison's Scandinavian fieldwork was that there is a continuous section of the Silurian System in southern Norway, previously assigned to the "Transition series." Murchison made that discovery in the summer of 1844 when he was invited as a keynote speaker to the Scandinavian scientific association meeting held in Christiania, now Oslo (Boeck, 1847). There he made the acquaintance of many of the leading Scandinavian geologists, friendships that he maintained for the rest of his life. Murchison recognized that the Norwegian section comprised crystalline Precambrian basement rocks unconformably overlain by Silurian rocks that grade up into the Devonian Old Red Sandstone (see Diemer, 2008). He was able to make this important contribution to Norwegian geology because he applied the newly devised Paleozoic stratigraphic systems developed in Britain to the rocks of Norway. He immediately announced the discovery during his keynote address at the Oslo meeting and later confirmed it in a report to the British Association (Murchison, 1844) and in two of his books (Murchison et al., 1845; Murchison, 1854).

After the meeting in Christiania was completed, Murchison made his first tour of Sweden, traveling from Göteborg to Stockholm via the lakes of central Sweden accompanied by the Swedish naturalist, Sven Lovén (Diemer, 2016a). (Note that Murchison's spelling of the names of people and places often differ from modern Swedish. Here, Murchison's spellings are retained in the quotations and on his diagrams, whereas modern spellings are used elsewhere.) The following year, Murchison returned to Sweden, this time accompanied by his coadjutor, Edouard de Verneuil. During both trips, his itinerary was influenced by the 1834 geognostic map of Sweden by Wilhelm Hisinger.

Hisinger (1766–1852) was a chemist, geologist, and owner of manufacturing estates who made many contributions to Swedish geology in the form of maps, papers, and books, and his geognostic map was very useful for the planning and execution of both of Murchison's field seasons. A geognostic map is one that depicts the distribution of lithologies in surface extent and also provides indications of the structure of those lithologies below the surface, without attempting to assign them an age or explain how the rocks came to be distributed the way they are (Rudwick, 2005).

In 1845, Murchison traveled to Lake Siljan in Dalecarlia, Gotland, and the southern coast of Sweden (Murchison, 1846a, 1846b, 1847). Murchison recognized both Lower and Upper Silurian rocks at these localities and correlated the much thinner units of Sweden to equivalent, but thicker, rock units in Britain. He was the first geologist to apply the modern system name, "Silurian," to these rock units.

It is important at the outset to establish what portion of the stratigraphic column Murchison was referring to in the mid-1840s when he assigned rocks to the Silurian System. At that time, it was Murchison's contention that the Silurian System encompassed all fossiliferous strata up to the base of the Devonian System. He divided the Silurian System into a Lower and Upper Silurian, wherein the Lower Silurian included what we would now assign to the Cambrian and Ordovician Systems, and the Upper Silurian would correlate with what we now recognize as the Silurian System.

COLLECTIONS VISITED

In addition to advice received from Swedish geologists and access to Hisinger's seminal geognostic map, Murchison also benefited from inspection of the fossil collections held at various museums as well as in private hands. At the time Murchison worked in Scandinavia, museums across Europe were a relatively recent innovation and in a state of constant flux (Knell et al., 2007). Before the advent of museums, most fossil collections, or cabinets, were in private hands (Rudwick, 1985a). In Scandinavia, early museums generally were housed at universities or, in some cases, attached to scientific societies. Only later did it become common for public museums to be established that were funded by local or national governments. While working in Scandinavia, Murchison examined fossil collections in Copenhagen, Christiania, Stockholm, Upsala, and Visby, and those collections clearly contributed to his rapid compilation of research findings in 1844 and 1845. Those visits will be described chronologically. Furthermore, he was able to compare Scandinavian fossils to British collections, thereby confirming his interpretations concerning the age of the strata he was mapping.

NORWAY 1844

In 1844, Murchison traveled via Copenhagen to Christiania to participate in a meeting of Scandinavian savants (see Boeck, 1847; Diemer, 2008; Hestmark, 1992, 2008; Fig. 1). In preparation of that trip, Murchison wrote to George Forchhammer (1794–1865), Professor of Geology, Mineralogy, and Chemistry at Copenhagen University, to arrange a visit en route to Christiania. This visit was a characteristic example of Murchison's fieldwork methodology (Diemer, 2008) in that he was able to learn in advance as much as possible about the geology of those districts of Norway and Sweden he planned to visit after the meeting. Of particular utility was the opportunity to examine rock and fossil specimens and geologic maps of those new districts that Forchhammer had in his collection. Evidence of this practice is contained in a letter from Murchison to Forchhammer:

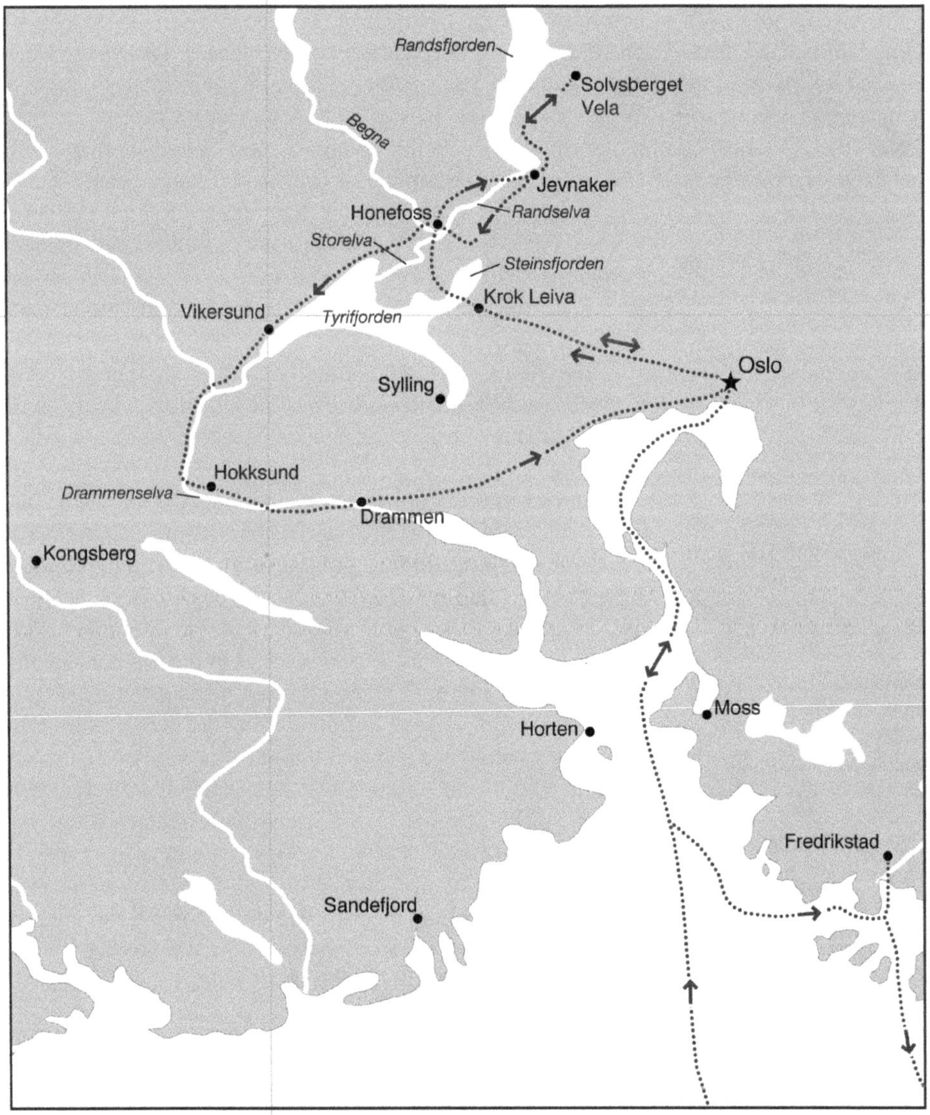

Figure 1. Murchison's travel routes in Norway in 1844 (from Diemer, 2008; used with permission of the History of Earth Sciences Society).

London
6 June 1844

My dear Professor Forchhammer,

I have resolved to visit Christiania at the meeting of the Scandinavian philosophers under Hansteen and his conjoint Presidents. I was for some time undecided about it, as I wished to get my great work on Russia finished, but finding that this is impossible before the early part of the winter on account of various delays which must always occur in extensive scientific publications, I have resolved to take flight for Hamburg and Copenhagen in the first days of July, and hope to find you still at home, that we may go on to Christiania together.

My intention is further to traverse the country from Christiania to Stockholm, and to return by the isles of the Baltic. By this hasty visit and *your instructions* I hope to render my map, which embraces a good part of Sweden, somewhat more perfect. At all events I shall see the source of all my old friends, the erratic blocks, and look at some Silurian relations *in situ.* (Geikie, 1875, v. 2, p. 27–28)

During his stop in Copenhagen, he conferred with Forchhammer concerning the possibility of finding Silurian and Devonian rocks in Norway. Part of the conference took place in the Geological Museum in Copenhagen, an institution founded in 1772 as the Universitets Nye Naturaltheater (the New Natural Theatre of the University), which contained the fossil collections.

From: G. Forchhammer, Copenhagen
To: Roderick Murchison, London
Date: 10 June 1844
Source: F. G. Forchhammer Copibug No 1
September 1841 to April 1846
Register I Copibug No 4. S.
Geology Museum of Copenhagen

My dear Sir,

I am happy to see that we shall have the pleasure to have you at our meeting at Christiania, and I hope you will be satisfied if you take all things in due consideration.

Our tour is arranged. We leave Copenhagen Tuesday 9 of July at 3 o'clock in the afternoon. From Kiel to Copenhagen 3 steamers go every week, one on Wednesday morning, another on Friday, a third on Saturday evening. If you go with the steamer of Saturday 6 of July from Kiel you will arrive at Copenhagen Sunday about 1 o'clock and will have about two days for your stay at Copenhagen which will be a rather short time. If you go with the Wednesday steamer you will have about 5 days to stay at Copenhagen. I should not advise you to go with Friday steamer which is rather a small vessel and often detained by small accidents.

I should advise you to bring your geological apparatus with you from London except very bulky things. Should you want anything we will try to procure it at Copenhagen. About such a servant as you want I shall seek information until your arrival and then the thing will be soon arranged.

I shall try to show you all what we have been able to accumulate. I have thought it necessary to distinguish 2 parts of the transition formation. The one with *Asaphus expansus*, *Orthocera vaginatus* is the oldest and most frequent. The other with *Calymene Blumenbeckii* is decidedly Silurian but has perhaps not a single fossil which agrees with that of the older formation.

Pray pay my respects to Mrs. Murchison and believe me

Yours most truly
G Forchhammer

In this letter, Forchhammer revealed that he was familiar with Murchison's designation of an "Upper" and "Lower" Silurian and key fossils used to make that distinction. In his tabular view of Russian deposits that provided a summary of the stratigraphy of Russia (their plate VI, Murchison et al., 1845) and also in his tables 1 and 2 from Murchison (1845b), Murchison assigned *Asaphus expansus* and *Orthoceratites vaginatus* to the Lower Silurian (for illustrations of these fossils, see Plate 1). For Murchison, *Asaphus expansus* was of sufficient significance that he illustrated it in *The Geology of Russia* (Murchison et al., 1845, p. 37, their figure 6**), where he also reproduced Dalman's description of the fossil (Plate 1). *Calymene Blumenbachii* was one of the fossils characteristic of his Upper Silurian (see Plate 2 and Murchison, 1839, p. 704), where he assigned this fossil to the Wenlock Limestone, lower Ludlow, and the Aymestry Limestone of his Upper Silurian in a summary table of organic remains. He also provided a description of *Calymene Blumenbachii* (p. 653) and illustrations from his plate VII, figures 5–7 of Murchison (1839). Thus, being able to inspect these familiar fossils in the museum collection at Copenhagen would have provided support for Murchison's predictions of the stratigraphy he was about to inspect in Norway and Sweden.

Murchison and Forchhammer traveled together from Copenhagen to Christiania on Tuesday, 9 July, to Thursday, 11 July 1844 (Fig. 1). Later that day, the officers of the Geology and Mineralogy section were elected, and the order of talks was agreed upon (Diemer, 2008). During the meeting, Murchison made several excursions in the vicinity of Christiania. On the evening of the first full day of the meeting (Friday, 12 July 1844), Baltazar Mathias Keilhau (1797–1858), Professor of Geology at Christiania University, took the assembled geologists on a field trip to the raised beaches in the northern suburbs of the city. In his field notebook now held in the archives of the Geological Society of London (M/N 100, p. 20)[1], Murchison produced a sketch that shows Silurian rocks exposed at sea level overlain by (Devonian) red sandstones, which in turn were overlain by a thick porphyry that formed the extensive Ringerigge plateau north of the city. The next day (Saturday, 13 July 1844) Murchison gave the first part of his talk in which he presented the results of his Russian work. He did not finish the talk, and he was then scheduled to complete it on Wednesday, 17 July 1844. On Sunday, 14 July 1844, there were no talks scheduled.

[1]Please note that subsequent references to Murchison notebooks in the Geological Society of London archives will be referred to as M/N 100, M/N 101, and so on.

Plate 1. Selected Lower Silurian fossils

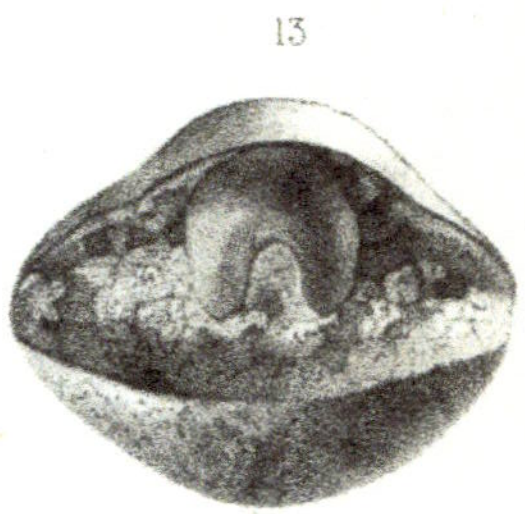

Asaphus expansus (Dalman) from Plate 27, Figure 13 of *The Geology of Russia* (1845).

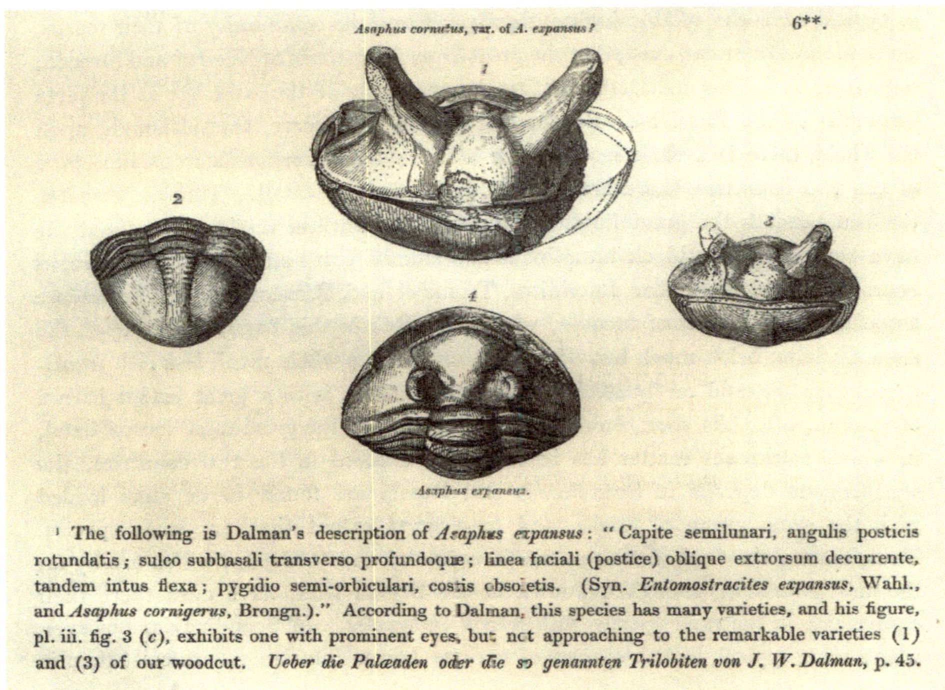

Asaphus cornutus, var. of *A. expansus?* 6**.

Asaphus expansus.

[1] The following is Dalman's description of *Asaphus expansus*: "Capite semilunari, angulis posticis rotundatis; sulco subbasali transverso profundoque; linea faciali (postice) oblique extrorsum decurrente, tandem intus flexa; pygidio semi-orbiculari, costis obsoletis. (Syn. *Entomostracites expansus*, Wahl., and *Asaphus cornigerus*, Brongn.)." According to Dalman, this species has many varieties, and his figure, pl. iii. fig. 3 (*c*), exhibits one with prominent eyes, but not approaching to the remarkable varieties (1) and (3) of our woodcut. *Ueber die Palæaden oder die so genannten Trilobiten von J. W. Dalman*, p. 45.

Asaphus expansus (figures 2 & 4) and *Asaphus cornutus* (figures 1 & 3) from *The Geology of Russia* (1845), p. 37.

Orthoceratites vaginatus from Plate 24 (figures 6a & 6b), *The Geology of Russia* (1845).

Orthoceratites duplex from Plate 24 (figure 13), *The Geology of Russia* (1845).

Plate 1. Selected Lower Silurian fossils that were mentioned in Murchison's publications, journals, field notebooks and letters. In this and subsequent plates, illustrations were derived from *The Geology of Russia* (Murchison et al., 1845) and *Siluria*, third edition (Murchison, 1859), which contains plates of fossils transferred from the original *Silurian System* (Murchison, 1839). Fossil names were updated in *Siluria* (Murchison, 1859); however, the names used in the original work (*Silurian System* [Murchison, 1839]) are retained in brackets.

Plate 2. Selected Upper Silurian trilobites

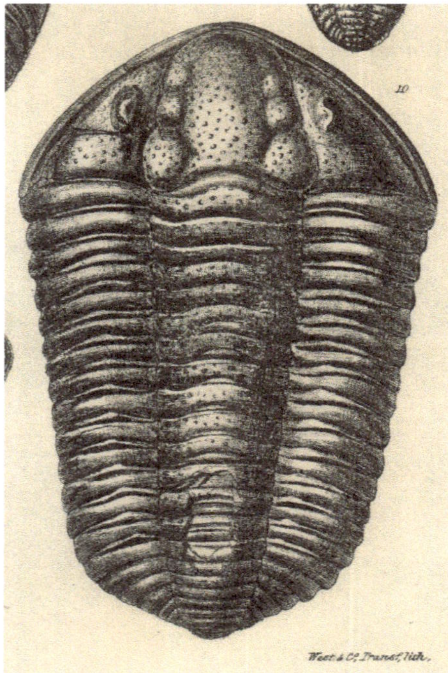

Calymene Blumenbachii, Brong. From the Wenlock Limestone, Dudley. From *Siluria* (1859), plate 18, figure 10.

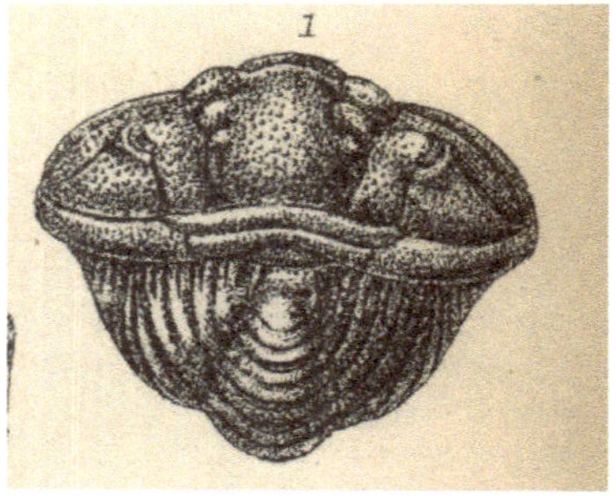

Calymene Blumenbachii, Brong. From the Wenlock Limestone, Dudley. From *Siluria* (1859), plate 17, figure 1.

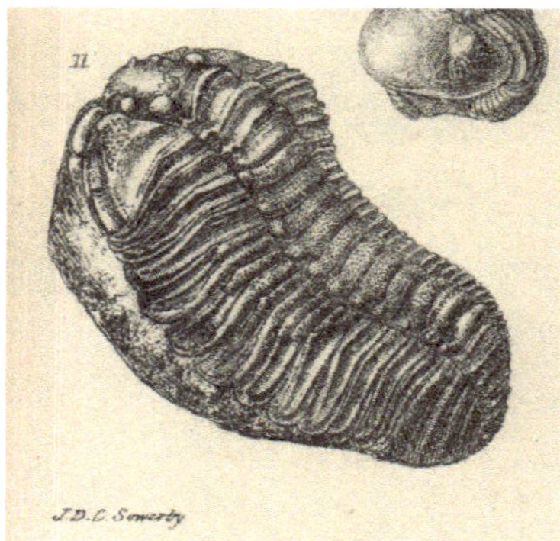

Calymene tuberculosa, Salter (Calymene Blumenbachii), Brong. From the Wenlock Shale, Burrington, Shropshire. From *Siluria* (1859), plate 18, figure 11.

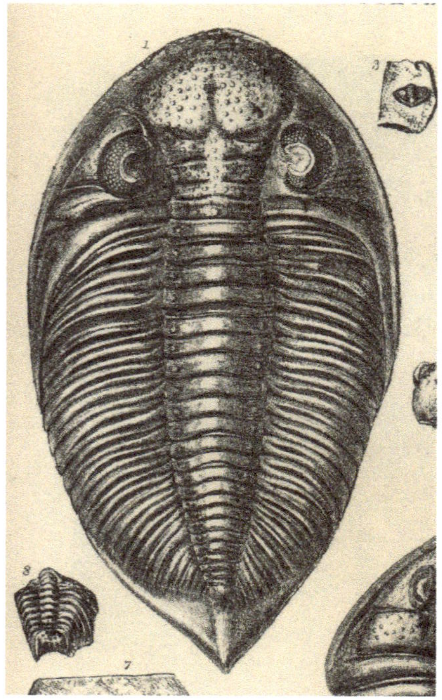

Phacops (Asaphus) caudatus, Brong. From the Wenlock Limestone, Shropshire and Dudley. From *Siluria* (1859), plate 18, figure 1.

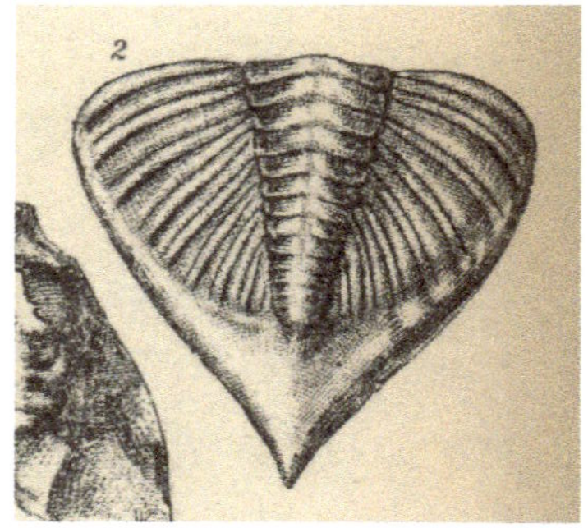

Phacops (Asaphus) caudatus, Brong. From the Wenlock Limestone, Dudley. From *Siluria* (1859), plate 17, figure 2.

Plate 2. Selected Upper Silurian trilobites that were mentioned in Murchison's publications, journals, field notebooks, and letters.

Murchison took the opportunity to hire a boat and, together with Forchhammer and Leopold von Buch (1774–1853), traveled to the islands and eastern shore of Christiania harbor. An objective of this excursion was to examine Silurian fossils along the shore, first noted by von Buch when he resided in Christiania in the early 1800s (von Buch, 1810).

Having established the presence of the Silurian at sea level in Christiania harbor, Murchison then arranged to make a two-day geological traverse to the northwest, accompanied by Keilhau. They had with them a copy of Keilhau's geognostic map for the region (Keilhau, 1844–1849; Fig. 2). On the first day of the excursion, they crossed the Ringerigge plateau and then descended the Krokkleiva escarpment into a valley occupied by a lake known as Steinsfjorden. Keilhau had geognostically mapped the rocks along the western shore of Steinsfjorden as "Thon schiefer, Kalkstein und Kalklager." Murchison quickly realized that these were indeed Silurian strata, similar to those in Christiania harbor, and that the overlying eastward-dipping sandstone on the eastern shore of Steinsfjorden that comprised the steep cliffs of the Krokkleiva escarpment could be assigned to the Old Red Sandstone. Murchison made a detailed section of the Krokkleiva escarpment that documented his observations and interpreted ages for the units where he used for the first time in Norway the stratigraphic terminology that he and others had been developing in Britain (see Diemer, 2008; Fig. 3). It is worth noting that during his stay in Norway, Murchison became aware of the geognostic stratigraphy for Sweden prepared by Wilhelm Hisinger. Murchison reproduced Hisinger's stratigraphy for Sweden on the first page of his field notebook for Norway (Geological Society of London, M/N 100). He likely extracted Hisinger's stratigraphy from Keilhau's 1838 publication entitled, *Gaea Norvegica* (Diemer, 2008).

Murchison and Keilhau then returned to Christiania on the evening of Tuesday, 16 July 1844. The next day, Murchison finished his talk on Russian geology and took the opportunity to announce to the other geologists his discovery of Silurian rocks overlain by the Old Red Sandstone at Krokkleiva. In a letter to his wife, Charlotte, Murchison summarized his discovery:

> I have made a capital section which has thrown a new light on the geological structure of Norway having found "Old Red" *ad libitum* and much Upper as well as Lower Silurian all in succession. (Edinburgh University Library, Special Collections,[2] Gen 2216/70)

Later in the afternoon of Wednesday, 17 July 1844, Murchison inspected Keilhau's fossil collection, including the trilobites, at the museum in the University Botanical Gardens, a collection that most likely was later incorporated into the Natural History Museum at the University of Oslo. He made careful notes of the locations where the fossils had been collected, and that information contributed to his prediction that Silurian strata could be found in the valley occupied by Steinsfjorden and the adjacent lakes of Randsfjorden and Holsfjorden.

In order to test that prediction, Murchison decided to travel with Forchhammer for a week through that valley at the conclusion of the conference.

> I saw so much to interest me in my last day here & so many geological views have unexpectedly opened upon me that I resolved to remain another week in Norway particularly as Forchhammer is here & will travel with me for 6 days when we go together on the steamer stopping myself at Gothenburg whilst he goes on to his home. There I shall meet my friend M Levein a clever young naturalist who will make the journey with me to Stockholm. Thus you see I am well cared for. (Gen 2216/71 * & **)

They traveled to the northwest from Christiania, retracing the route taken earlier by Murchison and Keilhau, and crossed the Ringerigge porphyry plateau (Fig. 1). They then dropped down the Krokkleiva escarpment to the shores of Steinsfjorden where Murchison and Forchhammer confirmed that the carbonates were part of the Silurian System. They then continued northward to Randsfjorden visiting sites such as Solvsberget along the way. They turned south once more and traveled via Tyrifjorden and Drammenselva to Drammen on their way back to Christiania (Fig. 1). During the week-long trip, they cemented their friendship, and Murchison invited Forchhammer to participate in the York meeting of the British Association that was scheduled for early October.

Sweden 1844

Having completed their tour of Norway, Murchison and Forchhammer departed Christiania by steamship on the morning of 27 July 1844. They arrived in Göteborg at 6 a.m. on 28 July 1844, where they parted company. It was at a Mrs. Tod's inn at Göteborg where Murchison was joined by Sven Lovén (1809–1895), who accompanied him on his trip across southern Sweden to Stockholm and Uppsala (Fig. 4). Lovén not only acted as his guide, but he also procured a carriage and a servant for Murchison for the trip (Diemer, 2016a).

The trip took a bit more than a week (29 July to 6 August 1844). During the trip, their route was determined by the several localities that appeared as "Transition" rocks on Hisinger's geognostic map of Sweden (Hisinger, 1834). Most informative of those sites was the hill at Kinnekulle (Fig. 5). Kinnekulle had been visited by Carl Linnaeus in 1746, where he noted the diversity of plants, including orchids, growing in the calcium-rich soils there. It was at Kinnekulle that Murchison believed he could see the full Lower Silurian section resting unconformably on the crystalline basement rocks of the "Azoic." Murchison described

[2]Please note that subsequent references to Murchison's letters held in Edinburgh University Library, Special Collections will be referred to as Gen 2216/70, Gen 2216/71, and so on up to Gen 2216/86.

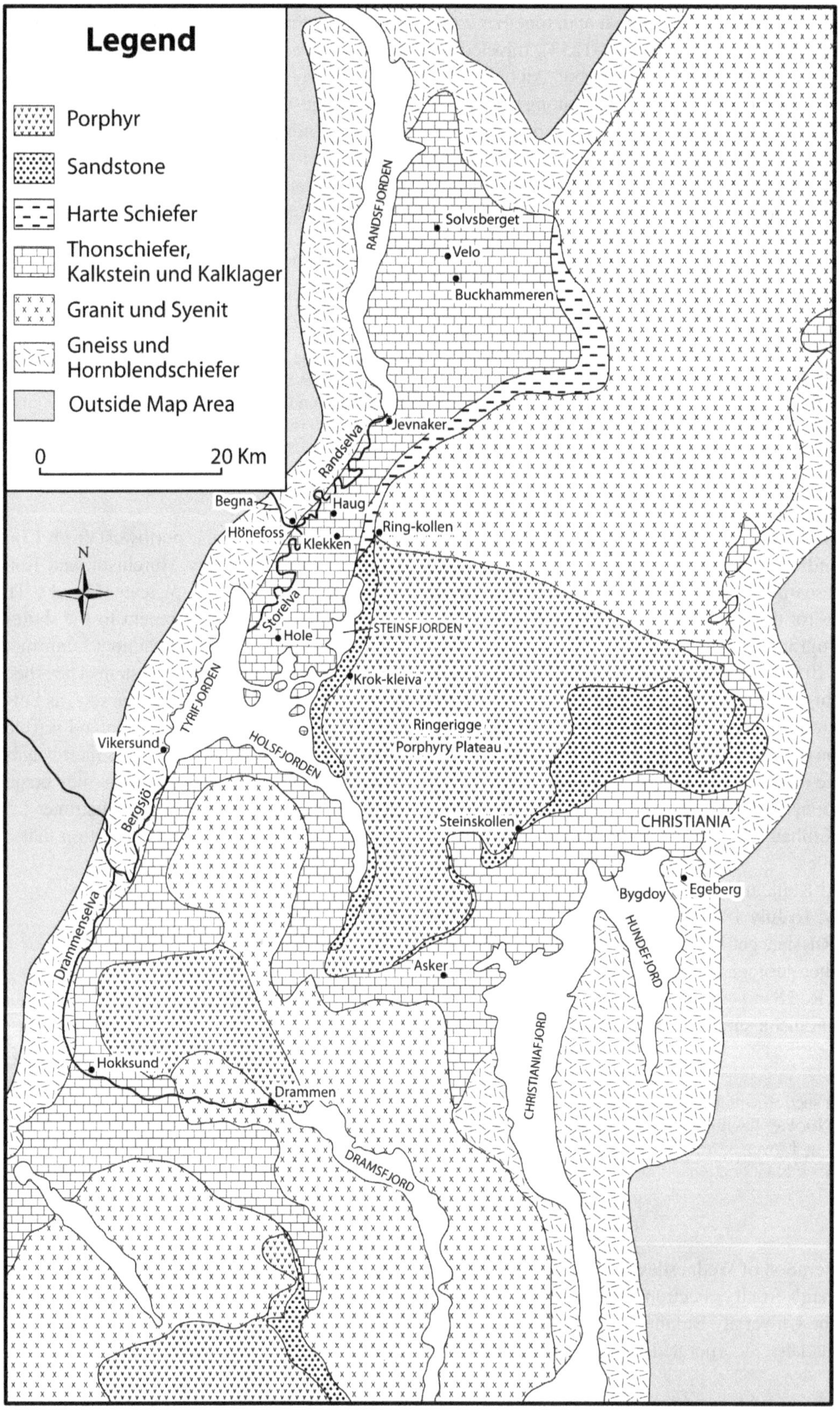

Figure 2. Reproduction of part of Keilhau's 1834 geognostic map of southern Norway used by Murchison during the 1844 field season (from Diemer, 2008; used with permission of the History of Earth Sciences Society).

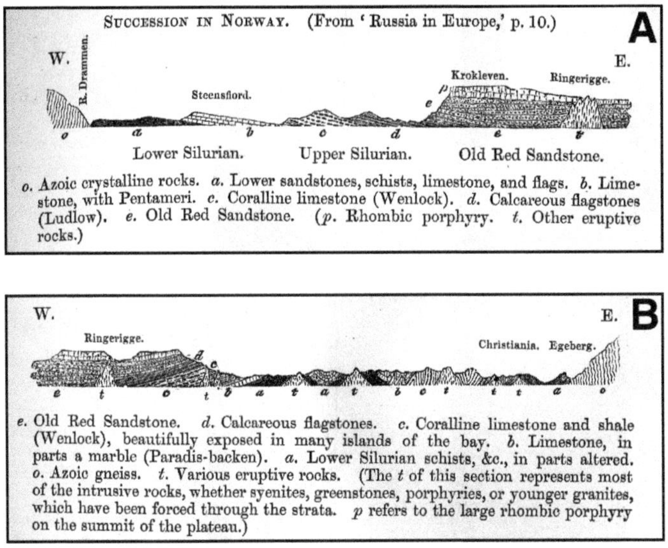

Figure 3. Cross sections of the geology of southern Norway. (A) Cross section through the Steinsfjorden valley from the Drammen River on the west to the Ringerigge porphyry plateau on the east. Note that Murchison mapped the entire Lower and Upper Silurian which graded up into the Old Red Sandstone, unconformably overlain by the porphyry (from Murchison, 1859, p. 367). (B) Eastward continuation of cross-section in Figure 3A, extending from the Ringerigge porphyry plateau on the west to Christiania and Egeberg on the east. Note that much of the section close to sea level in the vicinity of Christiania comprises Silurian strata cut by younger eruptive rocks (from Murchison, 1859, p. 367).

Kinnekulle as follows (Murchison, 1845a, p. 477–478; Diemer, 2016a, p. 358–359):

> Advancing to the next Silurian oasis at Kinnekulle and the hills of Billingen, the same general relations of a low surrounding country of gneiss and granite, to high tabular plateaux of horizontal Silurian strata, usually capped by trap, present themselves to the traveller. In ascending the hills of Kinnekulle, from the low gneissose country of Lidkoping, he is no sooner above the low level of those crystalline rocks, than he meets with a terrace composed of quartzose sandstone, already mentioned as frequently forming the lowest Silurian strata in Scandinavia. This rock, here arranged in beds from a few inches to a foot and a half thick, is light-grey, whitish and fine-grained, in parts freckled with ferruginous stains, and assumes at intervals a quartzose character, with divisions of chloritic shale. Its lowest beds, or those which, as we shall presently show, rest upon the adjacent gneiss of the valleys, are not here visible, owing to a talus of detritus, but in those which are visible, were found branching fucoid-like bodies. This sandstone is, in fact, seen to constitute the prevalent base of all the Silurian strata, and in the hill of Kinnekulle is surmounted, first, by the black alum schists and limestone; next by red Orthoceratite limestone; and lastly, by Graptolite schists with some calcareous courses and Orthoceratites. Though irregularly denuded over a very considerable area, the Orthoceratite limestone occupies a prominent step on the sides of the plateau, and standing out high above the surrounding gneiss, is in its turn covered by black schists, through which basaltic trap has pierced, occupying only a small upper portion of the central part of the tract.

In addition to Kinnekulle, Murchison and Lovén examined several other sites along their route across southern Sweden including Hunneberg and its neighboring basaltic hill, Holleberg, Lugnäs and Bällefors, and Motala and Omberg (Fig. 4). At each site, they confirmed that the stratigraphy was similar to, if not as well exposed and complete as, that seen at Kinnekulle.

Having satisfied himself of the nature of the Lower Silurian stratigraphy in southern Sweden, Murchison and Lovén boarded a steamer bound for Stockholm (Fig. 4). They arrived there on the afternoon of 6 August 1844 and over the next few days Murchison toured the city, visited scientists, and inspected collections. On 7 August, Murchison visited with Jöns Jacob Berzelius, a chemist with an interest in mineralogy, who was professor of medicine and the Perpetual Secretary of the Academy of Sciences. In fact, Murchison dined with Berzelius twice during his stay in Stockholm and also took the opportunity to write up many notes due to the heavy rain that occurred during his stay. During his visits with Berzelius, they pored over maps together and discussed the distribution of Silurian rocks.

On 8 August, Murchison visited several of the sites in Stockholm and described what he saw to Charlotte. Of note is his mention of an informative visit to the natural history museum (presumably this is now the Naturhistoriska riksmuseet, or National Museum of Natural History, founded in 1819 by the Royal Swedish Academy of Sciences), that contained the fossil collections of Hisinger and Dalman (Dalman, 1827; see also Diemer, 2016a, p. 362).

> It is a beautiful northern city & very peculiar & the view of it from the sea front or still better from Moso backe or the Hill of Moses is very striking. Thither I have been this morning — the middle of this day devoted to the Museum in which I have seen many of my old Silurian friends & can have already acquired a capital set of duplicates of Trilobites all well named by Hisinger & Dalman … (Gen 2216/73)

The opportunity to inspect the fossil collections in the museum was evidently a highlight for Murchison because he repeated an account of his visit, or possibly visits, in another letter to Charlotte written five days later:

> I have also done all in the Museums & got a very good collection of duplicates named after the true Swedish Silurian species of Dalman, Hisinger & Dahlman through the intervention of my friend Lovén. (Gen 2216/74)

The collections mentioned in the letters enabled him to accurately predict the distribution of Lower and Upper Silurian strata throughout Sweden. In fact, in the letter to his wife on 9 August 1844, Murchison wrote:

> The sections of all the palaeozoic rocks of the continent of Sweden have convinced me that they are all Lower Silurian — Gottland being Wenlock & Upper Sil. (Gen 2216/73)

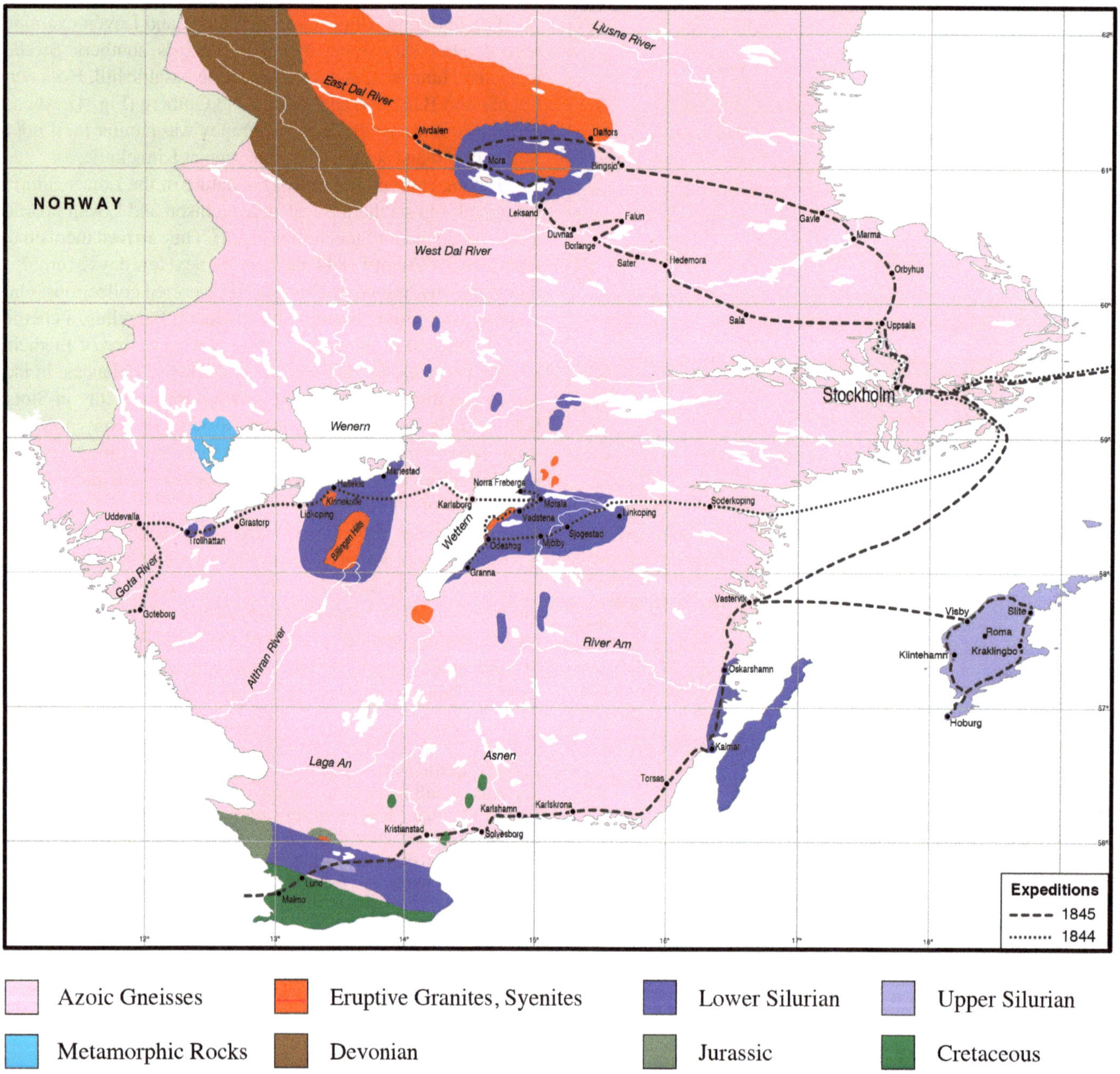

Figure 4. Murchison's travel routes in Sweden in 1844 and 1845. The geology is based on Murchison et al. (1845), *The Geology of Russia*, plate VI.

He later shipped the duplicate specimens, along with other fossils he had collected, to Britain for inspection by experts there, as he wrote in a letter to his wife from St. Petersburg on 3 September 1844:

I have sent home by the Mermaid direct in which this letter goes 3 cases of fossils through Messr Moberly & Co which will be delivered immediately. These & the box from Christiania complete my Summer's harvest which De Verneuil & self are to examine on my arrival. I sail for Stettin on Saturday (Tuesday Morning). (Gen 2216/77)

As reported in *The Geology of Russia* (Murchison et al., 1845) the specialists identified a wide array of Lower and Upper Silurian fossils from the Swedish mainland and Gotland collections, respectively (Diemer, 2016a, p. 362):

In fact, through the kind and liberal arrangements of Baron Berzelius and the assistance of our companion Professor Löven, to whose zeal, intelligence and good arrangements we were singularly indebted, the rich collections of Hisinger and Dalman were laid before us and a selection made from them, for the purpose of comparison with those of

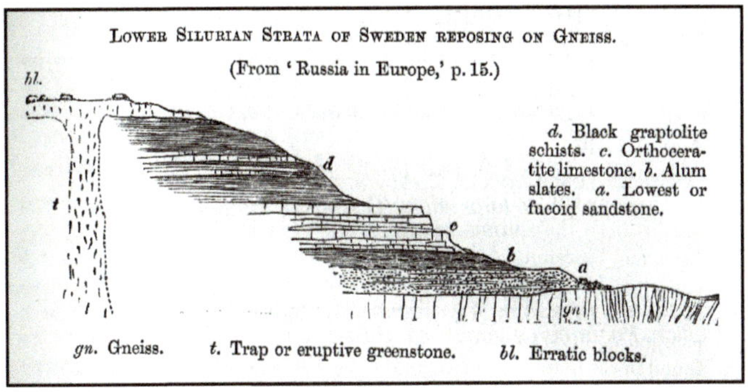

Figure 5. The Lower Silurian strata at Kinnekulle, resting unconformably on Azoic crystalline gneisses, and unconformably overlain by volcanic rocks (from Murchison, 1859, p. 365).

Great Britain, Russia and other countries. A glance at the fossils which were brought to England and examined by other palaeontologists as well as ourselves, is sufficient to convince any one acquainted with the Silurian rocks of the British Isles, that whilst the long island of Öland is essentially composed of the same Lower Silurian group which we had followed over the main land, (its *Orthoceratite* marble limestones having been largely exported to all parts of the Baltic), the island of Gothland is exclusively of Upper Silurian age. (Murchison et al., 1845, p. 18–18*)

It is worth emphasizing that this passage was written in the winter of 1844–1845 when Murchison was extensively revising the chapter on Scandinavia in *The Geology of Russia* (Murchison et al., 1845, chapter 2), prior to Murchison's only visit to Gotland in the summer of 1845. Many of the fossils named by Hisinger and Dalman were incorporated into the list of fossils in the stratigraphic column that accompanied plate VI of *The Geology of Russia* (Murchison et al., 1845) (see Fig. 6 and Table 1). Selected examples of those fossils appear in Plates 1–3. Combined with the fieldwork he had undertaken, the fossil evidence suggested that the Lower Silurian occurred on the mainland of Sweden, whereas the Upper Silurian was confined to the island of Gotland. Thus, one of the most instructive products of the day spent inspecting the fossil collection in the Stockholm museum was that he was able to predict that the strata of Gotland belonged to the Upper Silurian. In order to test that prediction, and that Silurian rocks also occurred near Lake Siljan in the Dalecarlia region, Murchison made plans to return to Sweden the following year.

Winter 1844–1845

Upon returning to Britain, Murchison did not stop thinking about the stratigraphy of Sweden. Instead, he distributed the fossils he collected to various experts for identification, and he continued his reading and correspondence on the subject. A product of that work was the revision, during the winter of 1844–1845, of not only chapters 1, 2, and 3, but also the lower part of the stratigraphic column that accompanied plate VI of *The Geology of Russia* (Murchison et al., 1845) (Fig. 6 and Table 1) as well as emendations to the map itself (Fig. 7).

Although Forchhammer was unable to attend the 1844 British Association meeting held in York, he did prepare a paper on fucoids (which he interpreted as pyritized impressions of algal material, or seaweed, in Lower Silurian shallow marine alum shales); Murchison presented this paper in his stead (Forchhammer, 1844). In his

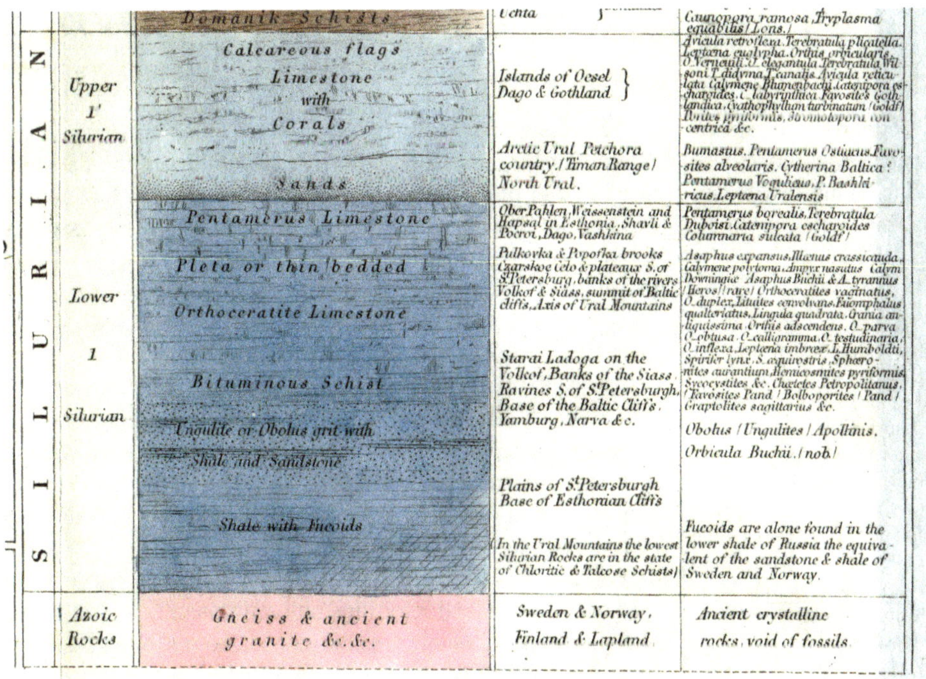

Figure 6. The lowermost portion of the stratigraphic column on plate VI of *The Geology of Russia* (Murchison et al., 1845). The column displays Murchison's thinking in the winter of 1844–1845 concerning the assemblages of fossils that characterized the Lower and Upper Silurian, and where those units occurred in Russia and Scandinavia. See Table 1 for a reproduction of the information in this figure.

TABLE 1. REPRODUCTION OF THE TEXTUAL INFORMATION IN FIGURE 6, FROM THE STRATIGRAPHIC COLUMN ACCOMPANYING PLATE VI OF *THE GEOLOGY OF RUSSIA* (MURCHISON ET AL., 1845)

Formation	Lithologies	Russian localities	Characteristic fossils
Upper Silurian	Calcareous flags limestone with corals	Islands of Oesel, Dago, and Gothland	*Avicula retroflexa, Terebratula plicatella, Leptaena euglypha, Orthis orbicularis, O. Verneuili, O. elegantula, Terebratula Wilsoni, T. didyma, T. canalis, Avicula reticulata, Calymene Blumenbachi, Catenipora escharoides, C. labyrinthica, Favosites Gothlandica, Cyathophyllum turbinatum (Goldfuss), Porites pyriformis, Stromatopora concentrica*, etc.
	Sands	Arctic Ural, Petchora country (Timan Range), North Ural	*Bumastus, Pentamerus Ostiacus, Favosites alveolaris, Cytherina Baltica, Pentamerus Voguliens, P. Bashkiricus, Leptaena Uralensis*
Lower Silurian	Pentamerus limestone	Ober Pahlen, Weissenstein, and Hapsal in Esthonia, Shavli and Pocroi, Dago, Vashkina	*Pentamerus borealis, Terebratula Duboisi, Catenipora escharoides, Columnaria sulcata (Goldfuss)*
	Pleta or thin-bedded orthoceratite limestone	Pulkovka and Popofka brooks, Czarskoe Celo, and plateaus S. of St. Petersburg, banks of the rivers Volkof and Siass, summit of Baltic cliffs, axis of Ural Mountains	*Asaphus expansus, Illaenus crassicauda, Calymene polytoma, Ampyx nasutus, Calym Downingiae, Asaphus Buchii, A. tyrannus, Heros (rare), Orthoceratites vaginatus, O. duplex, Lituites convolvans, Euomphalus qualteriatus, Lingula quadrata, Grania antiquissima, Orthis adscendens, O. parva, O. obtusa, O. calligramma, O. testudinaria, O. inflexa, Leptaena imbraex, L. Humboldti*
	Bituminous schist	Starai Ladoga on the Volkof, banks of the Siass, ravines S. of St. Petersburg, base of Baltic Cliffs, Yamburg, Narva, etc.	*Spirifer lynx, S. aequirostris, Sphaeronites aurantium, Hemicosmites pyriformis, Sycocystites*, etc., *Chaetetes Petropolitanus (Favosites Pand.), Bolborites (Pand.), Graptolites sagittarius*, etc.
	Ungulite or Obolus grit with shale and sandstone	Plains of St. Petersburg, Base of Esthonian Cliffs	*Obolus (Ungulites) Apollinis, Orbicula Buchii (nob.)*
	Shale with fucoids	In the Ural Mountains, the lowest Silurian rocks are in the state of chloritic and talcose schists	Fucoids are alone found in the lower shale of Russia, the equivalent of the sandstone and shale of Sweden and Norway

follow-up report to Forchhammer on 4 October 1844, Murchison gave an account of the reception to the paper. Murchison also reported to Forchhammer his current thinking on the distribution of the Silurian in Scandinavia as reported in his own paper to the British Association (Murchison, 1844):

For my part I gave a general view of the Palaeozoic deposits of Scandinavia as connected with those of Russia on the one hand & those of the British Isles on the other. I showed that whilst there was a [full] order of Lower Silurian, Upper Silurian and Old Red in Norway, the continent of Sweden – i.e. all its central & southern part exposed nothing but Lower Silurian resting upon more ancient crystalline rocks. In like manner I pointed out that all the Silurian of the Baltic provinces of Russia was also Lower; but that the detached masses between those early raised continents & occupying the isles of Gothland, Dago & Orsal were all Upper Silurian. (From a letter from Roderick Murchison, York, to Professor G. Forchhammer, Copenhagen, on 4 October 1844, in the Archive of the Geological Museum, now held at the National Archives in Copenhagen)

His assigning the strata on the island of Gotland to the Upper Silurian could only be based on his inspection during the summer of 1844 of fossil collections in Stockholm. He had not yet traveled to Gotland, a trip that he would make the subsequent summer of 1845. Thus, the existence of fossil collections held in Stockholm was crucial to Murchison recognizing the occurrence of Upper Silurian rocks on Gotland. So confident was he of the stratigraphy of Gotland, that he showed Upper Silurian rocks on that island on plate VI of *The Geology of Russia* (Murchison et al., 1845), a map that was completed in the spring of 1845 prior to his actual visit to Gotland (Fig. 7). It is worth noting that he annotated directly on the map the occurrence of several Upper Silurian fossils on Gotland. That list included: *Calymene Blumenbachii*, *Asaphus caudatus*, *Terebratula Wilsoni*, *Leptaena depressa*, and *Catenipora escharoides* on the central part of Gotland (see Plates 2 and 3 for illustrations of these key fossils). He also listed the following fossil at the southernmost part of Gotland: *Avicula retroflexa* (Plate 3). In addition, he must have been

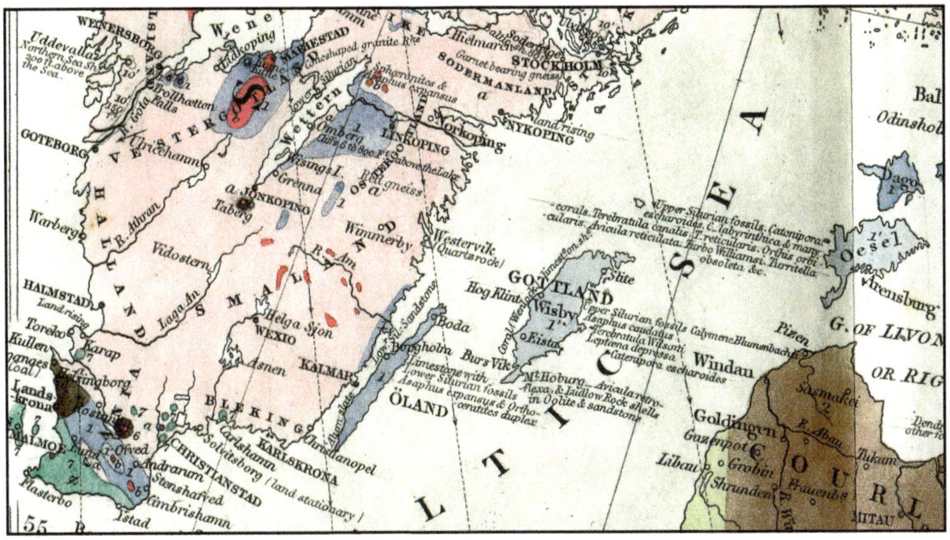

Figure 7. Close-up of the Baltic portion of plate VI of *The Geology of Russia* (Murchison et al., 1845). Note that the text concerning fossils on Gotland, Öland, and Oesel were likely late emendations to the map, added after Murchison's inspection of the fossil collections in Stockholm in the summer of 1844, but prior to publication of the map in the spring of 1845. Note also that Murchison mapped a small patch of Upper Silurian in Skåne near Övedskloster, likely due to the advice received from Forchhammer.

aware of the general southward dip of the strata on Gotland since he labeled the central part of Gotland as "Coral (Wenlock) Limestone on shale" and noted that "Ludlow Rock shells in Oolite & sandstone" occurred at the southern tip of the island (Fig. 7). In contrast, he assigned the island of Öland, another site he had yet to visit, to the Lower Silurian and annotated plate VI of *The Geology of Russia* with the following fossil types: *Asaphus expansus* and *Orthoceratites duplex* (see Fig. 7 and Plate 1).

Upon receiving the 4 October 1844 letter from Murchison, Forchhammer wrote in response:

> To your list of upper Silurian places in Sweden I think I may well certainly add:
>
> Bada on the island of Öland of which the greater part is lower Silurian,
> (the different species of *Sphaeronites*)
> Osmundbying – Dalerna (*Leptaena rugosa & depressa*)
> The uppermost part of the Alleberg in Westgothland
> (*Orthis Pecten Dalm.*, which is a decided Gothland fossil)
> Klinta in Scania (*Avicula retroflexa* and a number of other fossils different from the fossils of the alum slate and the Scandinavian *orthoceratite* Limestone)
>
> In Scania there occurs besides a red micaceous sandstone connected with red argillaceous beds and a red porphyry. I have some misgiving that it may be Old Red. It contains fossils (an *avicula* and a *modiola linae Phil,* no fossils have hitherto been discovered). If I had not fallen ill I should have been able to tell you some thing about the order of superposition of the Scanian Palaeozoic deposits, which I now must defer until next spring. (From G. Forchhammer to Murchison in 1844(?) as recorded in the F.G. Forchhammer Copibug No 1, September 1841 to April 1846, Register I Copibug No 4. S., the archives of the Geological Museum stored in the National Archives in Copenhagen)

In a tardy response to that letter, Murchison responded in February of 1845:

> From: Roderick Murchison, 16 Belgrave Square
> To: Professor G. Forchhammer, Copenhagen
> Date: 10 February 1845
> Source: Archive of the Geological Museum of Copenhagen
>
> My dear Forchhammer,
>
> Though I have often & often spoken of it & thought of you, I have been very remiss in writing to you. In truth I have been buried under the weight of Russia and the Ural Mountains & have neglected my best friends among whom you stand in the front rank. Thank God I now see my way through my "opus magnum" which thanks to my Scandinavian Survey will have two much better ends than if I had not gone amongst you. In the first place, my Introductory chapter is enormously enriched. I feel like a giant who has got his foot firmly placed on a solid rock before he jumps. For small as your vertical Silurian series may be, its base is beautifully and neatly separated from all anterior crystalline rocks – at least in Sweden, while the absolute identity of 19 species out of 20 of the Upper Silurian of England and Gothland strata is marvelous.

Forchhammer's letter clearly had an impact on Murchison's understanding of the distribution of Silurian strata in Sweden. Not only did he include some of the information on plate VI, but he wrote in February 1845 an appendix H entitled, "Palaeozoic Rocks of Scandinavia" to his *Geology of Russia* (Murchison et al., 1845, v. 1, p. 646):

> Some of the fossils from Hysbifiol in Sweden have very much the aspect of being derived from the uppermost beds of the Lower Silurian. In stating our belief, that the chief palaeozoic strata of the continent of Sweden are Lower Silurian, as distinguished from those of Gothland, we allude simply to the zone which fell under our survey. We have grounds to think, that in some places patches of Upper Silurian group will also be found to occur, as in Norway. For example, our friend Prof. Forchhammer has informed us that he believes there are Upper Silurian fossils in the higher part of the Aalleberg in West

Plate 3. Selected Upper Silurian fossils

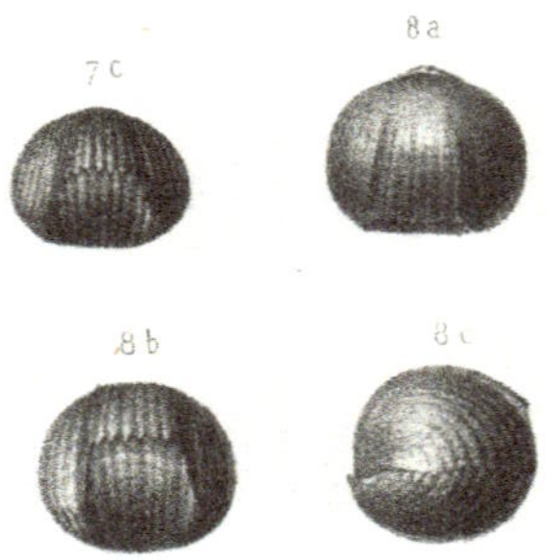

Terebratula Wilsoni, from *The Geology of Russia* (1845), plate 10, figures 8a, 8b and 8c.

Cornulites sepularius, Schloth. From *Siluria* (1859), plate 10, figure 2.

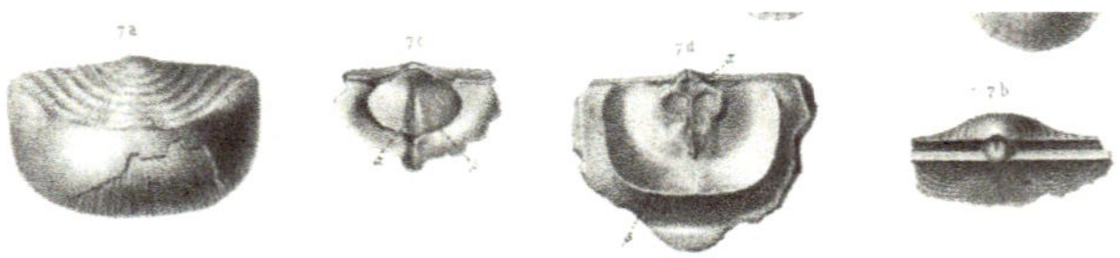

Leptaena depressa, from *The Geology of Russia* (1845), plate 15, figures 7a, 7b, 7c and 7d.

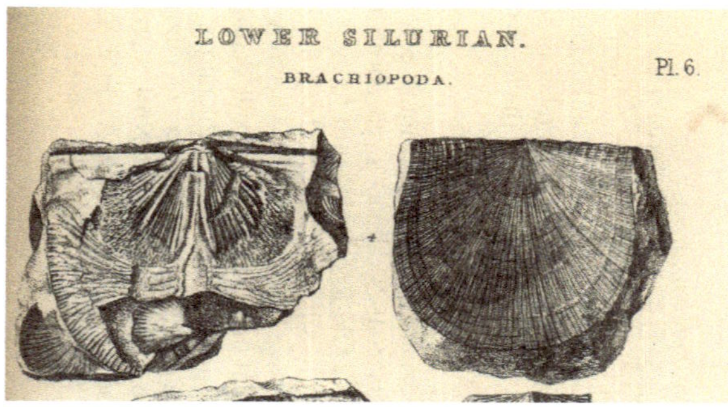

Strophomena expansa, Sowerby (*Orthis expansa* and *Orthis pecten*). From *Siluria* (1859), plate 6, figure 4.

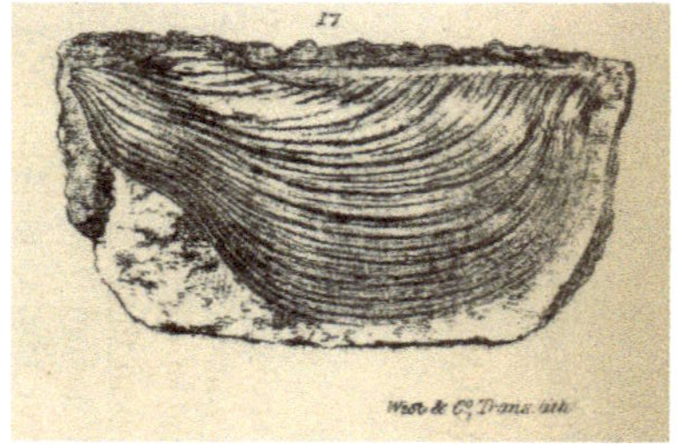

Pterinea (*Avicula*) *retroflexa*. From *Siluria* (1859), plate 23, figure 7.

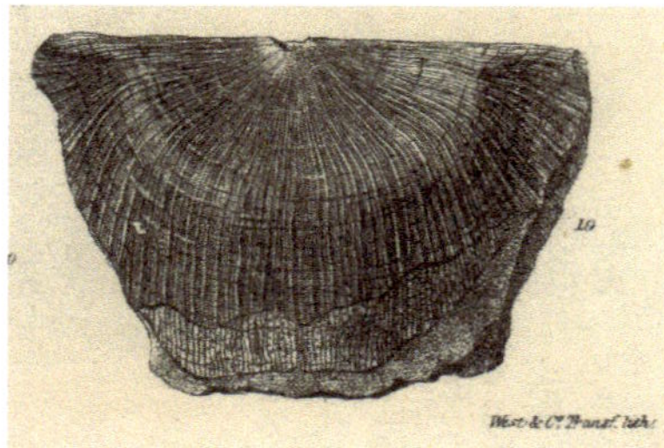

Leptaena euglypha. From *Siluria* (1859), plate 20, figure 19.

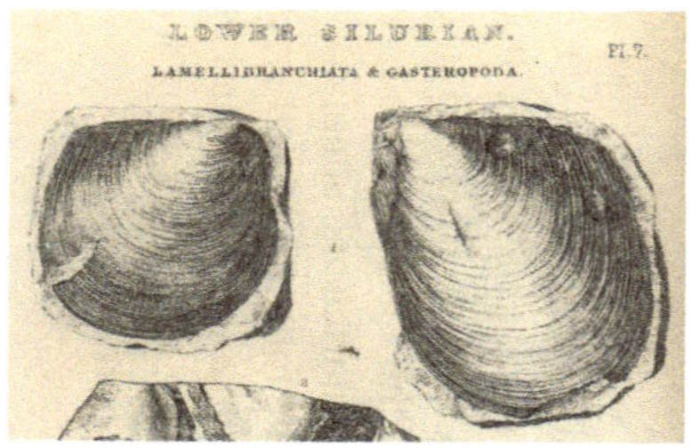

Modiola (*Avicula*) *orbicularis*. From *Siluria* (1859), plate 7, figure 1.

Halysites catenularius, Linn. (*Catenipora escharoides*). From *Siluria* (1859), plate 40, figure 14.

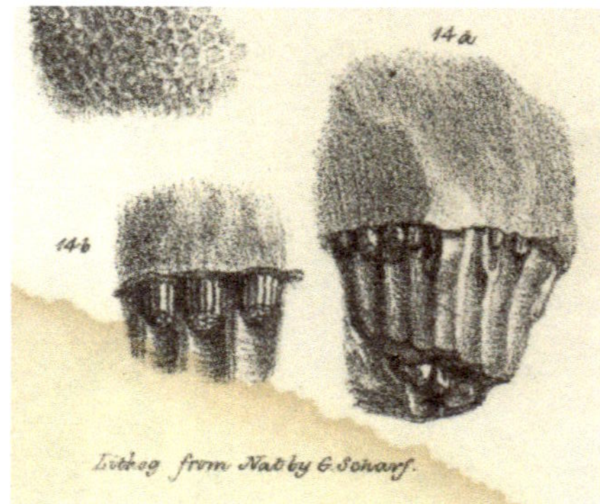

Halysites catenularius, Linn. (*Catenipora escharoides*). From *Siluria* (1859), plate 40, figures 14a and 14b, cast of tubes.

Plate 3. Selected Upper Silurian fossils that were mentioned in Murchison's publications, journals, field notebooks, and letters.

Gothland, and at Klinte in Scania. At the latter place, indeed, the occurrence of the *Avicula retroflexa* and other Upper Silurian fossils quite unknown in the Lower Silurian beds, has already determined the point. In Scania also there is a red micaceous sandstone, connected with argillaceous beds and porphyry (as in Norway), in which forms like *Avicula* and *Modiola* have been discovered, and which Prof. Forchhammer thinks may be of the age of the Old Red Sandstone. Again, from what we saw of the specimens of rocks and fossils from Dalecarlia in the Museum of Stockholm, we have very little doubt that a country which has afforded the *Leptaena depressa*, *L. euglypha*, and a species of the peculiar genus *Brontes* (Goldfuss), must contain Upper as well as Lower Silurian rocks, while a portion of the sandstones which occupy that region may represent the Old Red Sandstone. These points we hope to be able to work out in the course of the ensuing summer (February, 1845).

In this passage, Murchison admits that there may be Upper Silurian outcrops on the mainland of Sweden, but he apparently was reluctant to alter the map to indicate that, without additional supporting evidence. He first declared that he had most faith in the areas that he personally inspected during the 1844 campaign. A manifestation of that is that he did not include Upper Silurian rocks in the area around Kinnekulle even though Forchhammer, in his letter, had written that there may be some occurrences, such as "The uppermost part of the Alleberg in Westgothland" where there were "*Orthis Pecten Dalm.*, which is a decided Gothland fossil." In contrast, Murchison did amend the map by adding a patch of Upper Silurian in Scania at Klinte near Övedskloster (see Fig. 7), presumably because he had word not only from Forchhammer that it contained *Avicula retroflexa,* but also because he had already seen *A. retroflexa* and other Upper Silurian fossils from Klinte in the museum collection in Stockholm. Also of note in this passage is reference to both Upper and Lower Silurian fossils in the museum of Stockholm that came from the Dalecarlia district in the vicinity of Lake Siljan. That locality would become the first region that Murchison visited upon his return trip to Sweden in the summer of 1845.

Sweden 1845

By summer of 1845, *The Geology of Russia,* accompanied by its geologic map of Russia in Europe (plate VI), had been completed. At the top of Murchison's agenda was a trip to St. Petersburg, together with de Verneuil, to deliver copies of their newly completed two-volume book to the Czar and his ministers on 12 August of 1845 (Thackray, 1978). In a letter written to Charlotte from St. Petersburg on that same day, Murchison wrote:

After a week here I may send you a few lines to report progress. We soon got hold of Keyserling & unpacking our 'big box' began to get one or two of the copies of our monster book into circulation. (Gen 2216/77)

After spending most of the month of August in St. Petersburgh, Murchison and de Verneuil departed by steamer and arrived in Stockholm on Friday, 5 September 1845. There they made preparations for an overland trip to Dalecarlia, followed by Gotland and the south of Sweden (see Fig. 4) as recounted in a letter to Charlotte written on Sunday, 7 September 1845:

Having written to you at such length from Petersburgh I have positively nothing to tell you except that a very tedious voyage owing to a head wind & terribly round-about course, we got here two days ago & are off on Monday morning for Dalecarlia. I am perfectly well & have only one sorrow; that you have not written a line to me here. Not a word since you went to France though I specially told you P. Restante Stockholm. However I suppose I shall hear when we get back from Dalecarlia. We then go immediately to Gothland & hope to be in the S of Sweden near Malmoe by the first days of October. Our first journey is in a carriage which we have hired, as well as a most intelligent Swedish servant who talks very good English. Once back to this place all will be by steamers. Forchhammer is to meet me at Malmoe. (Gen 2216/84)

During the few days spent in Stockholm, Murchison and de Verneuil examined the Hisinger fossil collection at the natural history museum associated with the Royal Swedish Academy of Sciences. The first four pages of his 1845 Sweden notebook (M/N 104) are filled with the names and locations of fossils that had been collected and identified by Hisinger. Murchison focused on the fossils from the "Transition" rocks that Hisinger had mapped as occurring not only at Kinnekulle and other localities visited by Murchison in 1844 but also in Dalecarlia, Gotland, Öland, and Skåne. Clearly, the choice of route to be taken by Murchison and de Verneuil in 1845 was influenced by Hisinger's map and fossil collection. By familiarizing themselves with the fossils they could expect to encounter, they could ensure rapid scientific progress during their few weeks in Sweden in September and October of 1845.

Dalecarlia

Murchison and de Verneuil spent about two weeks (7–23 September 1845) traveling to Dalecarlia and returning to Stockholm. The first stop on their trip was in Upsala, where they "collected or rather received fossils from a great collector one Mr. Markham" (Gen 2216/85). While inspecting Mr. Markham's collection, Murchison compiled a "Memoranda" listing fossils from Dalecarlia, Gotland, and Öland (M/N104, p. 11–13); this listing no doubt informed their investigations for the remainder of their time in Sweden.

On 8 September 1845, they departed Upsala. Over the next several days, they traveled via Falun to Lake Siljan:

a great mass of water around which we halted at numerous pet Silurian spots & anchored at Mora pronounced Moorah the very centre of Dalecarlia which was our sleeping place for 3 days and where we made excursions to the Lakes of Orsa & Wenyan revelling in porphyries, orthoceratite Limestones, Isles etc. (Gen 2216/85)

In this letter to Charlotte, Murchison was putting a positive spin on the time spent in Dalecarlia. In contrast, when examining the observations recorded in his field notebook (M/N 104), much of which also appears in his journal account (M841/J17)[3], it is apparent that he found the stratigraphy of that region to be disrupted and obscure. A few of the published cross sections from Dalecarlia capture Murchison's interpretation of the fragmented and disrupted nature of the stratigraphy in Dalecarlia (Fig. 8). At no one site in Dalecarlia did Murchison see the full Lower Silurian sequence clearly exposed. Rather, the strata were commonly tilted at high angles, intruded by igneous rocks, and covered by drift deposits including clays and boulders. In order to map the region as a Lower Silurian sequence, Murchison had to rely heavily on the fossil evidence and comparisons to museum and private collections he had inspected in Stockholm and Upsala. The assistance from de Verneuil in identifying fossils in the field was no doubt a great aid in assigning the section to the Lower Silurian.

In one of his 1846 papers where he described the results of the 1845 field season, Murchison summarized the results of the stratigraphic analysis that he and de Verneuil undertook in Dalecarlia as follows:

On reviewing, then, all the natural evidences which can be collected from this confused and broken district, the geologist who has made himself familiar with the mineral characters, fossils, and order of the strata in other parts of Sweden, necessarily concludes that all the fossiliferous beds which are visible in detached and insulated localities around the dome of erupted rocks above mentioned, or which come out from beneath much crystalline and transported detritus in this part of Dalecarlia, belong to the Lower Silurian group, and are referable in ascending order to the sandstone, bituminous schist, Orthoceratite limestone and Graptolite schist of other parts of the country; the only exceptions being certain sandstones and flagstones, which from evidence, to be given in the sequel, are supposed to belong to the Old red sandstone. (Murchison, 1846b, *Quarterly Journal of the Geological Society,* v. 3, part 1, p. 11)

Part of the frustration that Murchison experienced was likely due to the abundant glacial deposits that covered much of the landscape, making it difficult to trace units over extended regions. In a letter to Charlotte (begun on 17 September and completed on 23 September), Murchison wrote:

No geologist was more opposed to Diluvial action than myself so long as small portions of Europe were known to me but now I am convinced that the detrital phenomena of all the North of Europe are great subaqueous results. One half — nay 4/5ths of the detritus is absolutely the deposit of water, sands, clay with shell etc followed by gravel & rounded rollers which have undergone enormous friction & above these invariably come the matching angular blocks as big as houses which have never been rolled. I am quite satisfied with all I have printed thereon in the work on Russia & am convinced that the great angular superficial blocks were all borne on icefloes in the former sea. There is no vestige of a Mammoth or any thing terrestrial in all Scandinavia. The glacial theory of Aggasiz is perfectly unadmissable for 100 good reasons of which hereafter. The rounded smoothed polished & striated surfaces are invariably to the north & these have all doubtlessly been produced by the torrential rounded blocks, sand & gravel, & the angular blocks & these transports have never produced such appearances (at least very rarely) as suggested by Lyell. (Gen 2216/85 a, b, & c)

Murchison's fixation on the question of the origin of drift, both in Russia and Scandinavia, is evident in the large amount of writing he devoted to the topic, in his field notebooks (e.g., M/N 104), his journal account (e.g., M841/J17), *The Geology of Russia* (Murchison et al., 1845, chapters 20 and 21), and his paper about the superficial detritus of Sweden (Murchison, 1846a). In the case of Dalecarlia, he may have chosen to focus on the drift deposits because the underlying bedrock was disrupted and confusing. It is probable that Murchison was eager to turn his attention to Gotland, where the exposures were much better, especially along the wave-cut coastline, and all indications from Hisinger's map were that the stratigraphy was not disrupted and could therefore be traced over large distances. The indications from Hisinger's map were likely reinforced by the conversations that Murchison had with Helmersen during his time in St. Petersburg. It happened that Helmersen had also visited Gotland earlier in the summer of 1845, and he had shared his observations with Murchison during the latter's visit to St. Petersburg in August of 1845.

GOTLAND

Murchison and de Verneuil departed Stockholm on Wednesday, 24 September, to travel to Gotland via Västervik with the intention of spending about a week there. In the end, they spent nearly two weeks traveling to and around Gotland and returned to Västervik on 6 October 1845. They departed Västervik on the morning of Friday, 26 September, aboard the 16 horsepower steamer, the "Actif," commanded by a Captain Tregår. It was a calm day, and they made landfall at the northern tip of the island and then steamed southward along the west coast of Gotland. They disembarked at the walled city of Visby, which was once part of the Hanseatic League and was constructed on Upper Silurian (Wenlock) limestone (now mapped as the Visby, Högklint, Tofta, and Slite Beds [Fogdestam, 1981]). There they stayed as guests of a Mr. Hagelin in the only decent inn in Visby. They also met the English consul, a Mr. Enquist, who not only introduced them to the Governor, an adjutant general by the name of

[3]Please note that Murchison's 26-volume manuscript autobiographical journal is held in the archives of the Geological Society of London and has the control numbers of M841/J1 through M841/J26.

Figure 8. Six cross sections from Murchison (1846b, plate 1) recording observations made in Dalecarlia when traveling there with de Verneuil in September 1845. Note that all six cross sections have components of the Lower Silurian stratigraphy seen by Murchison elsewhere in Sweden during his 1844 field season. However, in no one locality is the full Lower Silurian section clearly exposed. The captions from Murchison (1846b) are reproduced verbatim.

Figure 8-1. "Section 1. Section near the eastern end of the Lake Siljan in Dalecarlia, between Leksand and Rättvik, showing how the Lower Silurian limestone (*c*) is thrown off by the 'granitello' (*q*) at Alsarby. The slope below the limestone is covered by detritus" (Murchison, 1846b, p. 47).

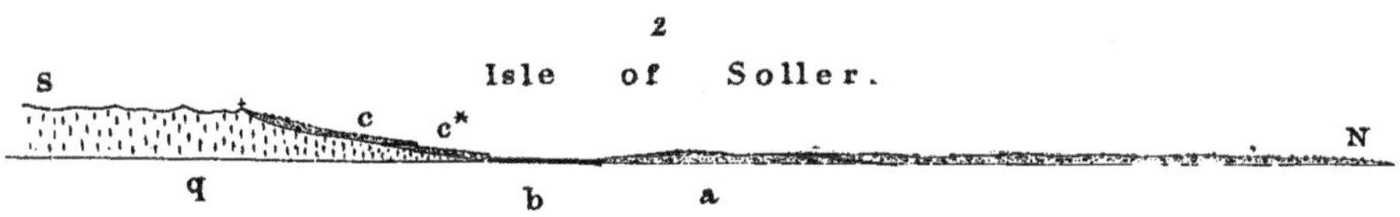

Figure 8-2. "Section 2. Relations in the Isle of Soller of the Lower Silurian rocks (*a*) inferior sandstone, (*b*) schist obscured by morass, and (*c* & *c**) beds of limestone with Orthoceratites and Cystidea in irregular contact with the granitello (*q*). The limestones are broken by transverse faults, and in one part (not represented in this drawing), mantle in an altered condition around bosses of granite" (Murchison, 1846b, p. 47).

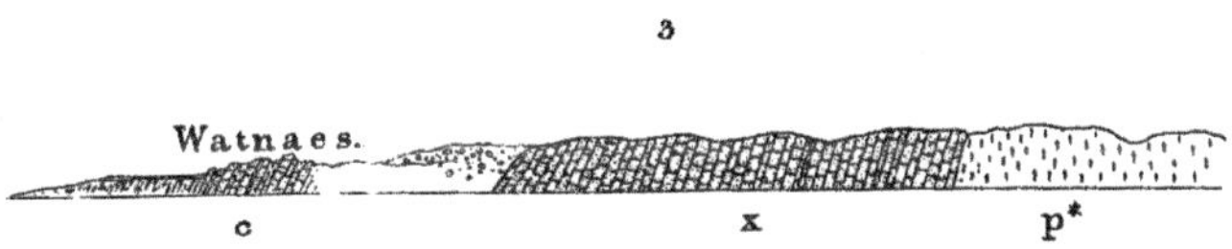

Figure 8-3. "Section 3. Section at Watnaes near Möra. The same Lower Silurian limestone (c) as that of the preceding diagrams, appears at intervals in highly-inclined strata, but is here separated from the eruptive rocks (p*), which form the western part of a great porphyritic dome, by quartzose rocks, siliceous sandstone and fine conglomerate (x), which seem to have been altered by igneous agency" (Murchison, 1846b, p. 47).

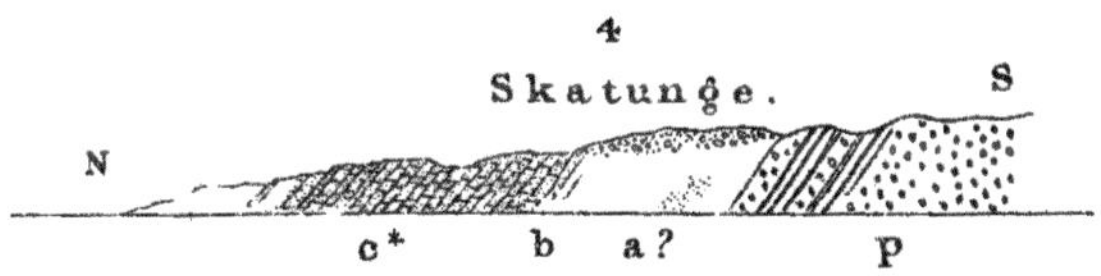

Figure 8-4. "Section 4. Section at Skatunge on the north side of the great dome of porphyry and granitello, exhibiting regularly stratified and jointed, earthy, red porphyry (*p*). A slope obscured and the Lower Silurian limestone and shale (*c**) with Cystidea in highly-inclined positions" (Murchison, 1846b, p. 47).

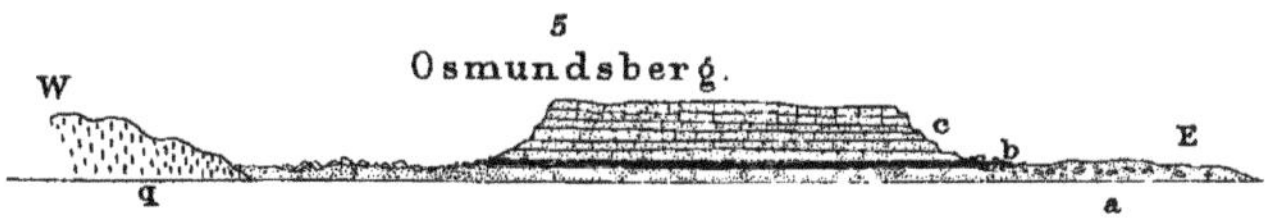

Figure 8-5. "Section 5. Section across the insulated hill called Osmundsberg, which is composed of Lower Silurian limestone (*c*), the inferior schist (lerskiffer) (*b*), and sandstone (*a*), being for the most part covered by detritus" (Murchison, 1846b, p. 47).

Figure 8-6. "Section 6. Relations at the waterfall of Styg-fors near Böda on the eastern flank of the eruptive dome. The subjacent or eruptive rock is here a granitello (*q*), and the sandstone (*a**) in contact with it is a highly-indurated quartz rock, which is surmounted by peculiar nodular shale (*b**)" (Murchison, 1846b, p. 47).

Höhenhausen, and other members of society, but also provided them with a carriage and travel permit during their stay.

Based on observations that he made on board the steamer while making the passage down the west coast of Gotland, and on his knowledge of the Gotland fossil assemblages from the collections in Stockholm, Murchison recorded the following on his first full day (Saturday, 27 September) in Visby:

From a profile of the western coast seen from the steamer … I surmised that an ascending succession from north to south would be found to occur. In this I may prove wrong, as Helmersen, after twelve days here, has arrived at another conclusion, and Hisinger is also against me; but this must be tested. A very slight southward inclination, 1° to 3°, will bring the whole into this order in the space of seventy English miles, or eleven and a half Swedish, even allowing that a large portion is horizontal absolutely. I hang to this opinion, because I already know 1st that the north is Wenlockian in character; 2nd that the mass of the island contains the fossils of that formation and of Aymestry; and 3rd that in Hoburg, the southernmost headland, the *Avicula retroflexa* and Ludlow fossils occur, to the exclusion of the others. We shall see whether English succession will stand in detail through this Upper Silurian mass. (M841/J17, p. 228–229)

His plans to inspect the geology of Gotland were delayed, however, because Murchison was housebound on his first day in Visby due to a bad cold. While Murchison was incapacitated, de Verneuil made an excursion to Lummeland, at the northern end of Gotland, where he found a nodular (Wenlock) shale overlain by Wenlock limestone. This observation was consistent with Murchison's prediction concerning the southward-oriented dip of the Upper Silurian of Gotland.

Meanwhile, Murchison managed "to cure myself by taking colycinth and blue pill; and by sweating" (M841/J17, p. 225). The next day (Sunday, 28 September), while still recovering, he toured the city guided by the local magistrate, during which he made notes on the architecture of the many ruined churches that gave an indication of the earlier grandeur of Visby. During the tour, Murchison visited the museum (possibly a precursor to the current Gotland Museum which was founded in 1875) of which the magistrate was director, and made the following comment:

Museum at Wysby. The good folks here have done exactly what ought to be done in every country town of large size; they have made a good local collection of all the fossils, birds, and quadrupeds of the island, the director of which is my Magister. The fossils are, one and all, Upper Silurian, and nine tenths of them are Wenlock and Dudley and British species. The corals are in finer preservation than any we possess. (M841/J17, p. 235–236)

While visiting the museum, either on that Sunday, or a week later on his return to Visby, Murchison made detailed notes of the fossils that were in the collection (see Table 2). Those notes are in the relevant field notebook (M/N 105) and are organized by localities including: "Fossils at the Paper Mill, N. Gothland" (p. 1), "Fossils of Hoburg—Bursvik" (p. 2–3), "Fossils Hog Klint" (p. 6), and "List of fossils in the Museum of the Town" (p. 12–15). It is worth noting, that many of the fossils are cross referenced to plates and figures in the *Silurian System* (1839), which must have been on hand while identifying the fossils. Whether he was assigning names on-the-spot to the fossils in the museum collection, or relying on identifications made by others, possibly including his companion de Verneuil, is not clear from his records of the day(s) spent in the museum. It is noteworthy that many of the fossils were first described by Hisinger, and those fossils do not appear in the *Silurian System* (1839). Thus it is possible that Hisinger had inspected the collection in the museum sometime after 1839 and identified the range of fossils using his own earlier publications, as well as the *Silurian System* (1839) to do so. If that were the case, Murchison would have benefited tremendously from the efforts of that eminent Swedish geologist. Even if the question of who actually identified the fossils cannot be resolved, the opportunity for Murchison and de Verneuil to further familiarize themselves with the fossils in the collection was good preparation for the fieldwork on Gotland that they were about to begin.

The next day (Monday, 29 September), Murchison and de Verneuil began their excursion to the southernmost tip of Gotland. The Governor assigned a lieutenant of artillery, one Baron Fock, to accompany them. The Baron saw to it that they "were admirably lodged at the house of a rich Pastor one night—with a Judge of the southern district another and so on" (Gen 2216/86). Their route took them first to Hög Klint, a prominent cliff along the west coast of Gotland to the south of Visby, where the Wenlock shale was overlain by the Wenlock limestone (Figs. 9 and 10). They then continued south to Klinteberg, where the next unit up section in the Upper Silurian, the Lower Ludlow rock, was in turn overlain by the Aymestry Limestone (see Fig. 11). They then

… travelled eastward and inland to regain the high road from Wysby to Burswick and Hoburg, and, passing over various diluvial coloured [*sic*] tracts, through which the limestone appears at intervals, we reached the comfortable little inn at Eksta, where I now write, whilst my very damp sheets are toasting and smoking before the wood fire in the usal porcelain stove. (M841/J17, p. 241)

On Tuesday, 30 September, they departed the inn and passed through Bjärgos on their way south to Grötlingbo. There they found

… quite another order of things—walls built of the Downton rock, psammites, blue and yellowish. These are covered in the hollow quarries by beds of fine-grained grey oolite, sometimes white, with thin hillocks of shale dividing, and those above by the surface rock, which is a pisolite, full of shells—*Avicula*, *Nucula* (His.), *Orthis*, *Leptaena*, *Cal.* (M841/J17, p. 247)

TABLE 2. LISTS OF FOSSIL COLLECTIONS IN M/N 105–1845 OCTOBER, SWEDEN AND DENMARK TO FRANCE

Fossil Name	*Silurian System**	
	Plate	Figure
Fossils at Paper Mill, north Gotland		
Leptaena sericea var.	19	2
Spirifer pisum	13	9
Atrypa affinis M. C. and V. Buch	6	1
Leptaena depressa M. C., Dalm., and His.	12	2
Terebratula [decemplicata]	21	17
Terebratula Wilsoni	6	7
Orthoceratites	3	9
Calymene punctata (Wahl)	23	8
Fossils of Hoburg and Burvik		
Orthis Pecten	21	9
Leptaena lata V. Buch	5	13
Terebratula [*pulchra*]	5	21
Terebratula [*furcata*]	21	16
Orthis elegantula Dalman	5	5
Delthyris crispa (*Spirifer crispus*) Dalman	12	8
Orthis rustica	12	7
Turritella		
Murchisonia		
Avicula reticulata	6	3
Avicula var	20	3
Nucula [ ] Hisinger		
Favosites spongites Goldfuss	15 bis	8, 9
Favosites Gothlandica	15 bis	3, 4
Porites pyriformis	16	2
Stromatopora concentrica Goldfuss	15	31
Crinoidea	4	56
Leptaena [*Fletcheri*] Dav.	20	15
Terebratula		
Terebratula affinis		
Pleurotoma articulata	5	25
Murchisonia		
Cypricardia retusa	5	5
Terebratula pulchra	5	21
Turbo corallii	5	27
Homalonotus [*knightii*]	7	1, 2
Gray calcareous grit of Bursvik		
Orthis elegantula Dalman		
Orthis [*Luciena*] Hisinger		
Terebratula Wilsoni M. C.	6	7
Ordinary Wenlock fossils		
Fossils of Hög Klint		
Spirifer (*Delthyris*) *cardiospermiforma*	13	10
Terebratula [*plicatella*]		
Terebratula [*prisca*]		
Terebratula [ ]		
Orthis elegantula Dalman	5	5
Leptaena sericea	19	2
Leptaena transversalis Dalman	13	2
Leptaena depressa M. C., Dalman, Hisinger	12	2
Euomphalus sculptus or funatus	12	7
Euomphalus alatus	13	28
Orthoceras, siphon		
Trilobite, head of		
Stromatopora concentrica Goldfuss	15	31
Catenipora labyrinthica		
Cyathophyllum [habiathiodes]		
Cyathophyllum		
Favosites Gothlandica Lam	15 bis	3, 4
Porites pyriformis	16	2
[*Cyclolites*]		
Crinoidea		

Fossil Name	*Silurian System**	
	Plate	Figure
Fossil List from Museum		
Lituites		
Inachus sulcatus, Hisinger	6	10
Orthoceratites		
Orthoceras imbricatum, Wahl.	9	2
Orthoceras annulatis, Hisinger	5	30
Orthoceratites ibex	5	30
Euomphalus alatus	13	28
Euomphalus [cateadotes] Hisinger		
Euomphalus funatus, M. C.	12	20
Euomphalus [cornariatis]		
Euomphalus sulcata		
Euomphalus [equalateris] Hisinger		
Trilobites		
Calymene Blumenbachii	7	5, 6, 7
Phacops		
Orthis [*basalis*], Hisinger or Dalman		
Pentamerus [*gypridia*]		
Delthyris crispa (*Spirifer*) Dalman	12	8
Delthyris (*Spirifer*) *cardiospermiforma*	13	10
Spirifer [*lynx*]		
Atrypa tenuistriata	12	3
Terabratula Wilsoni	6	7
Terebratula [*plicatata*]		
Terebratula cuneata, Dalman	12	13
Terebratula [*diodanta*], Dalman		
Terebratula bidentata, His.	12	13a
Pentamerus [*galeatus*], Grogarn		
Spirifer [*cardium*]		
Spirifer ptychodes, Dalman	3	13
Avicula retroflexa, Hisinger	5	9
Avicula reticulata	6	3
Bivalves undescribed		
[*Tellina prisca*], Hisinger		
[*Lucina*], Hisinger, de Verneuil		
Orthis Pecten		
Orthis [arachnidea]		
Leptaena lata	3	10b
[*Chonetes sarcinalata*]		
Leptaena [*Fischeri*]		
Crinoidea		
Actinocrinites simplex	18	4
Actinocrinites [*punctates*]		
Hypanthocrinites decorus	17	3
Encrinites		
Corals		
Aulopora serpens, Goldfuss	15	5
Syringopora reticulata, Goldfuss	15 bis	10
Catenipora escharoides, Lam.	15 bis	14
Catenipora [*labyrinthica*]		
[*Sarcenota*]		
Favosites [*basaltica*], Goldfuss		
Favosites Gothlandica	15 bis	3,4
Favosites polymorpha, Goldfuss	15	2
Porites pyriformis	16	2
[*Astrea ananas*]		
Caryophyllia [*truncata*], Hisinger		
Caryophyllia flexuosa, Lam.	16	7
Cyclolites numismalis, Hisinger	15	4
Turbinolopsis [*turbinata*]		
Turbinolopsis [*mitrata*], Hisinger		
Turbinolopsis [*pyramidas*], Hisinger		
[*Siphonica promossa*], Goldfuss		
Millepore repens, Hisinger	15	30
Cyathophyllum [*antedaten*], Hisinger		
Stromatopora concentrica, Goldfuss	15	31

*Plate and figure numbers of fossils as illustrated in Murchison's (1839) *Silurian System.*

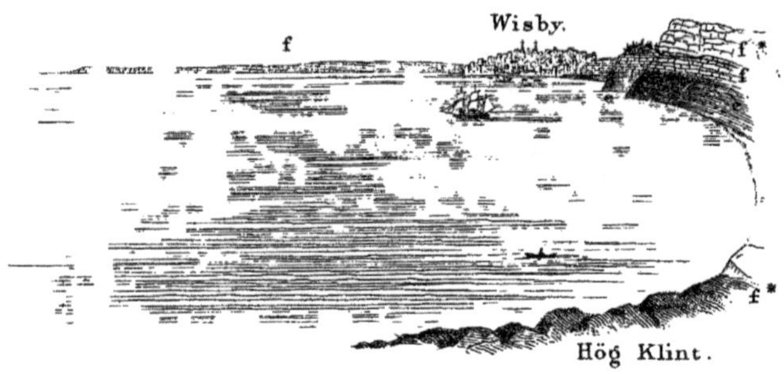

Figure 9. This is section 8 from Murchison (1846b, p. 47) with the following caption: "Geological view, looking from Hög Klint, the highest point of the cliffs of Gothland, and representing Upper Silurian rocks, consisting of the nodular Wenlock shale (*e*) surmounted by the Wenlock limestone (*f*), both to the S. and N. of the city of Wisby" (Murchison 1846b, p. 47).

On the general section for Gotland (Fig. 11), Murchison identified the outcrops near Grötlingbo as Upper Ludlow rocks (*i* and *j*).

The southernmost tip of Gotland is known as Hoburgen, and Murchison illustrated it with a section (Fig. 12). In that section, Murchison noted a thin-bedded micaceous sandstone overlain by a concretionary limestone grading up into a coral limestone, which, in turn, was capped by an encrinite limestone. In the general section (Fig. 11), he interpreted the section at Mount Hoburg as the uppermost Silurian (*k*) which may record a passage into strata of Devonian age.

Figure 10. This is section 9 from Murchison (1846b) with the following caption: "View of 'Hög Klint', or High Cliff, as seen from its northern base, and showing the detailed order of the Wenlock limestone (*f* & *f**), in relation to the subjacent nodular shale (*e*). Two or three erratic blocks (*bl*) are seen on the summit."

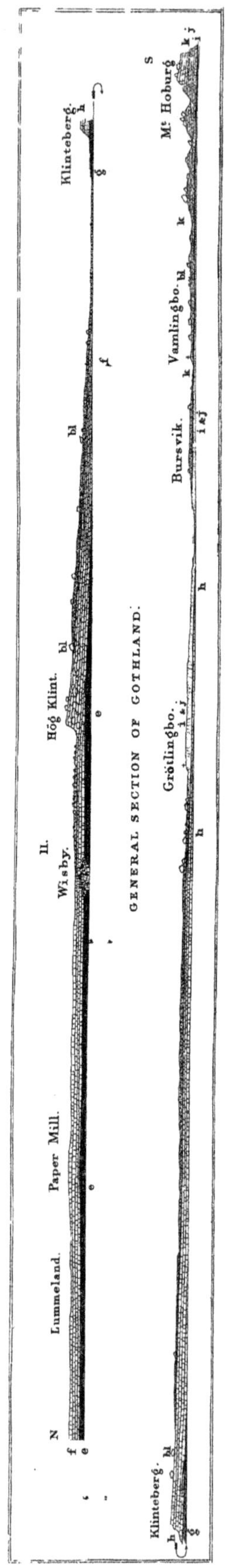

Figure 11. General section of the Upper Silurian, in two parts from north to south along the west coast of Gotland (from Murchison, 1846b, section 11). The southward-dipping stratigraphy ascends from nodular Wenlock shale (*e*) at sea level at the north end of the island, through Wenlock Limestone (*f*), Lower Ludlow rock (*g*), Aymestry Limestone (*h*), Upper Ludlow rocks (composed of oolite, calc grit and sandstone) (*i* and *j*), to the uppermost coralline limestone at Mount Hoburg (*k*).

12

Figure 12. This is section 12 from Murchison (1846b, p. 47–48) with the following caption: "Detailed order of the strata at Mount Hoburg, the southernmost promontory of Gothland, where the preceding section [section 11] terminates; showing how the same sandstone (*i*) which occupies the surface of the plateau at Grötlingbo on the N., is here brought down to the sea level; thus proving a general southerly inclination. The overlying calcareous strata (*j*, *k* and *k**) are described in the memoir."

10.

Rocks of Länna.

Figure 13. This is section 10 from Murchison (1846b, p. 47) with the following caption: "Grotesque forms of the Upper Silurian coralline limestone (Wenlock) at Länna, in the fine bay and anchoring-ground of Slite on the east coast of Gothland."

After reaching Mount Hoburg on 30 September, the account of their route becomes less clear. They apparently visited a site known as Lausberg on the east coast, which may be what is now known as Lausvik. From there they could have headed farther north along the east coast to the protected harbor of Slite. Alternatively, they may have returned to Visby after inspecting Mount Hoburg. From there, they could have made a second excursion to Slite on the east coast. In that harbor they inspected the sea stacks at Länna (Fig. 13), and Murchison assigned them to the coralline Wenlock limestone. From Slite, the geologists returned to Visby by the high road to Othem, where Murchison noted the presence of many erratics, and to Tingstäde, where he praised the architecture of the churches for their lofty towers and spires, "and three tiers of Saxon-like windows" (M841/J17, p. 254). They then traveled on to Visby through forests of oak and hazel.

By Friday, 3 October, the geologists were back in Visby, where the Consul Enquist threw them a going-away party. Their departure from Visby was delayed by bad weather, however, as noted in a letter to Charlotte:

> The weather has been very stormy & bad & our little steamer, the action having only 16 horsepower, could not face a headwind, so we were wind bound in Wisby a day or two beyond our wishes eating a great supper nightly either at the Governor's or at the Consul's. An Evening party in Sweden is as follows: 1st Bows innumerable & a place on the sofa. 2. Tea & cakes (served in 2 relays). 3rd After an hour Cold Punch, [Negas] etc., whilst in one corner of the room—Swedish National glees in chorus in the middle by 4 or more Men of whom the Majestri or Curates & the adjutants of the Governor were the leading folks. These glees are very animated. I most admired the [lyrics] of one which is supposed to be addressed by the romantic Charles 12 to his soldiers when inciting them to the attack of the Russians at Narva where you know he defeated 80,000 with 9,000 "[Avant ye muskovilen] Come on my blue-coat boys." (Gen 2216/86)

In that same letter, Murchison summed up the results of his field and museum work on Gotland:

> I found that what I had printed in the 2nd Chapter of our Russia might stand without the alteration of a word. The whole Island is a calcareous mass 2/3rd of which is made up of Corals & Encrinites—other fossils being scarce to the wayfarer though abundant in Museums. I tried to divide this limestone into 2 members, as separated by a sandstone, & my belief is that Mt. Holberg is the youngest portion of the Isle, since it contains some true Ludlow Rock fossils as well as those of Wenlock. (Gen 2216/86)

Thus, the utility of fossil collections held by museums, both on Gotland and throughout Scandinavia, was confirmed by their inspection of the geology of Gotland.

Once the weather moderated, they made the crossing from Visby to Västervik on Monday, 6 October. There they acquired a used carriage from a local pharmacist for about £5 to take them along the coastline of southern Sweden to Karlskrona, Kristianstad, Lund, and on to Malmö where they planned to take a steamer to Copenhagen (Fig. 4).

SMALAND AND ÖLAND

Murchison and de Verneuil departed Vastervik at 5:45 a.m. on Tuesday, 7 October 1845, on the road to Norby passing through quartzites and granites where Murchison noted the sculpted landscape rich in *roche moutonnées*. They arrived that evening in Norby with enough time to walk along a ridge cored by granite and flanked by valleys filled with sands and eskers. The next day (Wednesday, 8 October), they departed Norby on their route southward along the coast but were detained for a couple of hours in Mönsterås due to the lack of fresh horses for their carriage.

During their trip southward along the coast, they came to a location where they could see the northwestern end of Öland, which they had hoped to visit. However, a strong and persistent wind from the east made that crossing impractical. Moreover, given the fossil collections he had been able to inspect, it was clear that Öland was underlain by Lower Silurian rocks, as shown on his geologic map (Fig. 7) and on the cross section that included Öland (Fig. 14). Because of his prediction that only the Upper Silurian would be found on Gotland (based on the various fossil collections he had inspected prior to his visit), he had no reason to doubt that only Lower Silurian rocks would be found on Öland. Hence, Murchison decided against attempting the crossing, and they continued their journey southward instead, convinced that Öland was composed of the same Lower Silurian strata that he had examined the previous year at Kinnekulle (Figs. 5 and 14). Once they found fresh horses, they continued on to Kalmar, directly opposite Öland. There they spent the night and noted that there were fragments of Lower Silurian sandstone exposed at the surface but no in-place outcrops.

The next morning (Thursday, 9 October), they departed Kalmar on the road to Karlskrona, home to the Swedish navy. At first they passed through a featureless plain with more Lower Silurian sandstone fragments mixed together with northern drift deposits. Upon entering the province of Blekinge, they entered a region underlain by granite and left the Lower Silurian sandstones behind. They continued to travel across granitic rocks all the way to Karlskrona, where they toured the town and spent the night.

The next morning (Friday, 10 October) they continued southward, through granitic terrain at first, passing through Skillinge on their way to Hästeryd, where they spent the night. On Saturday, 11 October, they traveled on to Sölvesborg, which had a certain notoriety for being stable—that is, it was neither rising nor subsiding. The country to the north was known to be rising, while the region to the south, Skåne (or Scania), was known to be subsiding. Also of note were the Cretaceous chalk outcrops just to the west of Sölvesborg, which lay unconformably on the crystalline basement rock of the region (Bergström et al., 1988).

SKÅNE

From Sölvesborg, Murchison and de Verneuil continued on through the rain and across poorly drained and muddy ground to the fortress town of Kristianstad. There they met a Magister Ahlberg, who in turn introduced them to

> a zealous young Swedish fossilist, M. Maln, who has collected about three hundred species of chalk fossils, — at least two hundred more than Nilsson has described. He has supplied Forchhammer in Copenhagen, as well as Professor Haguenau in Pomerania, with many species. (M841/J17, p. 295–296)

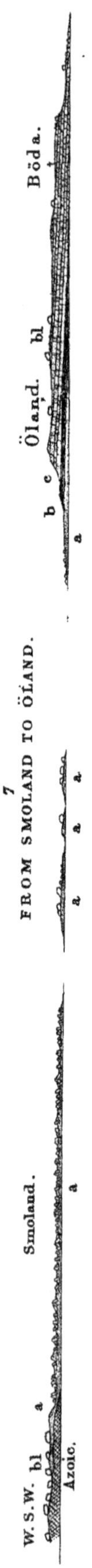

Figure 14. This is section 7 from Murchison (1846b, p. 47) with the following caption: "A general transverse section showing how the Lower Silurian sandstone (*a*) reposes (though chiefly in dismembered lumps) on the azoic or antecedent crystalline rocks of the province of Smoland (*az*); and also, how the same sandstone (fucoid grit of my former memoir) is seen on the western shore of Oland to be surmounted first by the alum schist (*b*), and finally by the Lower Silurian limestone (*c*), which occupies the great surface of the island of Oland in very slightly-inclined strata. The surface of the mainland is obscured by erratic blocks (*bl*)."

The stormy weather prevented them from continuing that day, so they decided to spend the night in Kristianstad.

Early the next morning (Sunday, 12 October), they set off again and soon arrived at Degeberga, the first post station south of Kristianstad. There they chanced upon their new acquaintance, Magister Ahlberg, who was traveling with the chief magistrate of the district to a shipwreck at Juleboda. The wrecked ship was the schooner *Fidelity* from Banff, carrying a load of grain from Danzig. "Finding that my presence might be useful, I willingly gave up half a day to go out of my road, to endeavor to serve my countrymen" (M841/J17, p. 299). Murchison's presence was indeed useful since the captain of the ship, David Deveral, an Aberdeen Scot, had difficulty communicating with the Swedish officials. In the end, Murchison convinced him that he needed to travel to Kristianstad to make his report, leaving his ship under guard during his absence. He also offered to write to the owners of the ship, Nesbitt and Company of Banff, to inform them of the plight of their captain.

From Juleboda, the site of the shipwreck, the two geologists then traveled to Andrarum, where they were to meet Herr Inspector Norblatt of the alum works. They were comfortably housed there, and they took the opportunity to visit the Lower Silurian alum slates and schists in the neighborhood. They had been advised to visit Andrarum by Professor Forchhammer, who thought that both Lower and Upper Silurian rocks occurred in the vicinity. At 8 a.m. on Monday 13 October, Murchison and de Vernuil "proceeded by horrible cross-roads, mud, blocks, and so forth, to make the celebrated transverse section of the Silurian and imagined Devonian strata of Scania, indicated by Forchhammer" (M841/J17, p. 309). In the end, Murchison mapped both Upper and Lower Silurian rocks in a sequence that was dipping gently to the west but did not show any Old Red Sandstone (see Fig. 15). Instead, Murchison thought that a red sandstone in the vicinity of Övedskloster could be assigned to his Lower Silurian Caradoc sandstone. The building and terraces of Övedskloster were constructed of this red sandstone, and, in fact, Murchison and de Verneuil had their cold lunch of tongue and sherry while sitting on the wall at the gate of Övedskloster. From Övedskloster, they traveled via Veberöd and Dalby to the university town of Lund, where they spent the night. While in Lund, Murchison spent time with Professor Nilsson talking about natural history, craniology, runes, and archeology.

On 14 October 1845, Murchison and de Verneuil set off for Malmö, arriving there at sunset. When they arrived, they sold

> our carriage, which cost us 67 rix at Westervick, for 35 rix; so it has cost us 32 rix only, or £1.12 for 350 English miles! Where can one travel so cheaply as in Sweden? Ahlbohm is packing up our alls, previous to taking leave of us. He is an excellent servant, and a great character. I could fill a volume with sketches of him. — Italians are now singing in the café below, and all is gaiety. (M841/J17, p. 323)

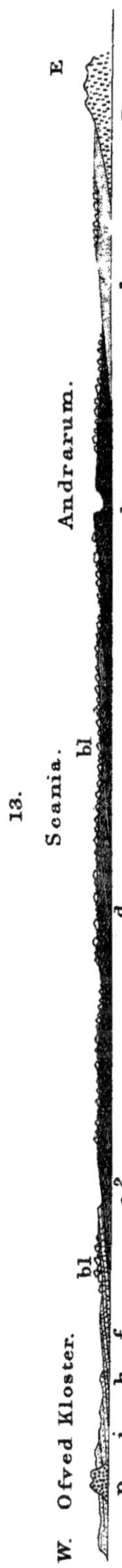

Figure 15. This is section 13 from Murchison (1846b, p. 48) with the following caption: "General section from E. to W. across Scania. This is merely given to indicate, as far as possible, the successive outcrops of Lower and Upper Silurian rocks in a low obscure tract much covered with mud, rolled blocks and gravel (*bl*). The Lower Silurian sandstone (*a*) is partially seen, the alum schist (*b*) with its fine trilobites is copiously exposed, but the prevailing Swedish and Russian Orthoceratite limestone with Cystidea is no where seen, its place being taken by an occasional thin course of black limestone subordinate to graptolite schists (*d*). The Upper Silurian rocks are recognizable in grey limestones and shales (*f*, etc.) which are surmounted by purple sandstones containing casts of Cypricardiae and other fossils apparently belonging to the uppermost zone of the Ludlow rocks. The porphyry of Ofved Kloster is marked (*p*)."

They traveled the next day (Wednesday, 15 October) to Copenhagen, where they planned to spend two nights. During their short stay:

> We had a first inspection of Forchhammer's cabinet, very rich in Gothlandic species, including three or four undescribed species. Saw the gigantic Trilobite, Dimbo (West Gothland), from the alum slate, two spans long, mentioned by Linnaeus. (M841/J17, p. 324)

SUMMARY AND CONCLUSIONS

Roderick Murchison, an early proponent of William Smith's biostratigraphic approach, traveled extensively in Europe and Russia, in part to document the occurrence and extent of the Silurian and Devonian systems (Collie and Diemer, 2004). Murchison also traveled in the Baltic region to correlate the stratigraphy between Europe, Russia, and Scandinavia.

Murchison's first visit to Scandinavia took place in the summer of 1844, when he attended a meeting of savants in Christiania, now Oslo, Norway. While there, he took the opportunity to inspect "Transition" outcrops in the environs of Christiania that reportedly contained Paleozoic fossils. He hoped to document the presence and extent of Paleozoic systems that would help complete the geologic map of Russia in Europe that he was working on as part of the two-volume work entitled, *The Geology of Russia* (Murchison et al., 1845). With the assistance of Professor Keilhau of Christiania University, Murchison made a couple of trips that convinced him that the fossiliferous "Transition" rocks were Silurian in age. The rapidity with which he reached this conclusion was aided by the fossil collection maintained by Professor Keilhau in the museum at the University of Christiania. That collection contained many examples of Silurian fossils familiar to Murchison. By combining the locations from where the fossils had been collected with the information contained on Keilhau's geognostic map of Norway, Murchison was able to predict where Silurian strata could be found in much of southern Norway. Immediately following the meeting, Murchison, traveling with Professor Forchhammer of the University of Copenhagen, was able to confirm his predictions about the distribution of Silurian rocks in southern Norway.

Murchison then undertook a rapid traverse of Sweden, from Göteborg to Stockholm, in the company of Sven Lovén, a Swedish naturalist. Their route was guided by the geognostic map prepared some years earlier by Wilhelm Hisinger. Murchison quickly realized that the carbonate "Transition" rocks that Hisinger had mapped in several parts of southern Sweden could be assigned to the Lower Silurian, largely on the basis of their fossil content. These discoveries were transferred directly to the geologic map that he was preparing. Upon arriving in Stockholm, Murchison inspected the fossil collections of Hisinger and Dalman held in the natural history museum founded in 1819 by the Royal Swedish Academy of Sciences; these collections confirmed his initial assignment of the carbonate rocks on the mainland of Sweden to the Lower Silurian. He also saw in the museum collection fossil assemblages from Gotland that belonged to the Upper Silurian. Again, that information could be directly transferred to the geologic map. Based on the fossil collections at the museum in Stockholm, Murchison made plans to return the following year to inspect the geology of the Lake Siljan region and Gotland, areas that he predicted would exhibit Silurian rocks (see Fig. 7).

Having delivered the freshly published copies of *The Geology of Russia* to Czar Nicolas and his ministers in St. Petersburg on 12 August of 1845, Murchison returned to Sweden in September, accompanied by his coadjutor, Edouard de Verneuil. Their visit began in Stockholm and Upsala, where they inspected fossil collections in both cities. By once more combining fossil localities with the geognostic map of Hisinger, they were able to arrange their travel route to have the most chance of successfully mapping the Paleozoic stratigraphy. They first traveled to Dalecarlia, where they determined that the "Transition" carbonate rocks of that region belonged to the Lower Silurian, despite their structural deformation. They then traveled to Gotland, and as predicted, they recognized a southward-dipping sequence composed of the Upper Silurian. Time spent in the town museum at Visby was very useful because they were able to recognize many of the key fossils that were characteristic of the Upper Silurian, and indeed they were able to cross reference many of them to the fossil figures in Murchison's *Silurian System* (1839).

The final piece of fieldwork they engaged in during the 1845 field season was a traverse through Skåne that verified that most of the Paleozoic rocks exposed in that region were Lower Silurian, but that there was one small inlier of Upper Silurian in the vicinity of Övedkloster. They had visited that locality at the urging of Forchhammer because the fossil collection in the Geology Museum in Copenhagen contained both Lower and Upper Silurian fossils from Skåne.

In summary, a major objective of the month-long 1845 fieldwork campaign was to map Silurian rocks in three parts of Sweden: Lake Siljan in the Dalecarlia region, Gotland, and Skåne. Murchison and de Verneuil were able to rapidly and efficiently accomplish this objective, in part by using fossil collections in museums in Stockholm and Visby and a private collection in Upsala. The fossil collections, both public and private, were among the first resources the geologists would visit when arriving in a new region, often prior to undertaking fieldwork. In those collections, they found familiar Silurian fossils that were geolocated with reference to existing geognostic maps. This made it possible to predict the distribution of Silurian rocks and to rapidly compile up-to-date geologic maps using the latest stratigraphic system nomenclature. By this means, Murchison was able to apply recently developed stratigraphic nomenclature to new territories, thereby advancing the science of geology.

Finally, it is worth emphasizing that reconstructing Murchison's scientific discoveries is possible because the primary documents (letters, notebooks, maps, and journal) are preserved at several key archives. Those primary documents contain invaluable information about the museum collections that Murchison visited and the nature of the specimens in those collections. Thus, reconstructing Murchison's scientific work is possible, not only because mineral and fossil specimens were preserved in contemporaneous museum collections but also because primary documents describing those fossil collections are preserved in modern archives.

ACKNOWLEDGMENTS

The Library of the Geological Society of London is gratefully acknowledged for providing access to the Murchison journals, notebooks, and correspondence in its collections, with special thanks to Wendy Cawthorne, Caroline Lam, and Sheila Meredith. Thanks also go to the staff of the Department of Special Collections at the University of Edinburgh Library and to Patrick Jones (UNC Charlotte), who prepared diagrams. The two reviewers are thanked for their constructive comments on an earlier version of the manuscript. The project was made possible with support from the American Philosophical Society and the UNC Charlotte Faculty Research Fund.

REFERENCES CITED

Benton, M.J., Sennikov, A.G., and Newell, A.J., 2010, Murchison's first sighting of the Permian, at Vyazniki in 1841: Proceedings of the Geologists' Association, v. 121, p. 313–318.

Bergström, J., Kornfält, K.-A., Sivhed, U., and Wikman, H., 1988, Bedrock Geology of Skåne, Geological Survey of Sweden Ser. Ba NR 43/SER. Ah NR 15, scale 1:250 000, 1 sheet.

Boeck, C., ed., 1847, Forhandlinger ved de skandinaviske Naturforskeres fjerde Möde, i Christiania den 11–18 Juli 1844: Christiania, Christian Gröndahl.

von Buch, L., 1810, Reise durch Norwegen und Lappland: Berlin. (Translated by John Black with notes by Robert Jameson, 1813, Travels through Norway and Lapland, during the years 1806, 1807 and 1808: London, Henry Colburn.)

Collie, M., and Diemer, J., 1995, Murchison in Moray: A Geologist on Home Ground: Philadelphia, American Philosophical Society Transactions, v. 85, Part 3, 263 p.

Collie, M., and Diemer, J., 2004, Murchison's Wanderings in Russia: Keyworth, British Geological Survey, 474 p.

Dalman, J.W., 1827, Om Palæaderna eller de så kallade Trilobiterna: Kungliga Svenska Vetenskaps-Akademiens Handlingar, v. 1, p. 226–294.

Diemer, J.A., 1996, Old or New Red Sandstone?: Evolution of nineteenth century stratigraphic debate, northern Scotland: Earth Sciences History, v. 15, p. 151–166, https://doi.org/10.17704/eshi.15.2.uq461282n11qkg61.

Diemer, J.A., 2008, Murchison's research method: An example from southern Norway: Earth Sciences History, v. 27, p. 31–58, https://doi.org/10.17704/eshi.27.1.e876055751500655.

Diemer, J.A., 2016a, Murchison in Sweden: Consolidating Lower Silurian stratigraphy in the summer of 1844, *in* Mayer, W., Clary, R.M., Azuela, L. F., Mota, T.S., and Wolkowicz, S., eds., History of Geoscience: Celebrating 50 Years of INHIGEO: London, Geological Society, London, Special Publication 442, p. 353–366.

Diemer, J.A., ed., 2016b, Special Issue from the William Smith Map Bicentenary Meeting Sponsored by the History of Geology Group and Held at the Geological Society London, 23–24 April 2015: Earth Sciences History, v. 35, part 1, p. 1–236.

Fogdestam, B., 1981, Map of the bedrock of Gotland: Sveriges Geologiska Undersökning Ser. Ab nr 3, Specialkarta 1, scale 1:250 000, 1 sheet.

Forchhammer, G., 1844, On the influence of fucoidal plants upon the formations of the earth, on metamorphism in general, and particularly the metamorphosis of the Scandinavian alum slate: British Association for the Advancement of Science Reports, v. 14, p. 155–169.

Geikie, A., 1875, Life of Sir Roderick I. Murchison Bart., K.C.B., F.R.S., Sometime Director of the Geological Survey of the United Kingdom. Based on His Journals and Letters with Notices of His Scientific Contemporaries and a Sketch of the Rise and Growth of Palaeozoic Geology in Britain. 2 volumes (vol. 1: 387 p.; vol. 2: 375 p.), London, John Murray.

Hallam, A., 1989, Great Geological Controversies (2nd ed.): Oxford, UK, Oxford University Press, 244 p.

Hestmark, G., 1992, Roderick I. Murchison's reisedagbok fra Norge i året 1844: Norsk Vitenskapshistorisk Selskap: Oslo, Årbok 1989–1991, p. 124–139.

Hestmark, G., 2008, "A primitive country of rocks and people"—R.I. Murchison's Silurian campaign in Norway, 1844: Norwegian Journal of Geology, v. 88, p. 117–141.

Hisinger, W., 1834, Geognostiska karta öfver Medlersta och Södra Delarne af Swerige, with explanation, Upplysningar till geognostiska kartan öfver Medlersta och Södra Delarne af Swerige: Stockholm, P.A. Norstedt and Söner, scale 1:800,000.

Keilhau, B.M., 1838–1850, Gaea Norvegica, 3 volumes: Christiania, Johann Dahl.

Keilhau, B.M., 1844–1849, Erster versuch einer geognostischen karte von Norwegen: Christiania, Johann Dahl, scale 1:2,700,000.

Knell, S.J., MacLeod, S., and Watson, S.E.R., eds., 2007, Museum Revolutions: How Museums Change and Are Changed: London, Routledge, 416 p.

Lyell, C., 1830–1833, Principles of Geology. London: John Murray. Volume 1 (1830) 511 p.; Volume 2 (1832) 330 p.; Volume 3 (1833) 398 p. Facsimile edition (1990) with introduction and bibliography by M. Rudwick. Chicago, University of Chicago Press.

Mather, K.F., and Mason, S.L., 1939, A Source Book in Geology: New York, McGraw-Hill Book Company, Inc., 702 p.

Murchison, R.I., 1839, The Silurian System, founded on geological researches in the counties of Salop, Hereford, Radnor, Montgomery, Caermarthen, Brecon, Pembroke, Monmouth, Gloucester, Worcester and Stafford; with descriptions of the coal-fields and overlying formations, 2 volumes: London, John Murray, 768 p., https://doi.org/10.5962/bhl.title.88029.

Murchison, R.I., 1844, On the Palaeozoic rocks of Scandinavia and Russia, particularly as to the Lower Silurian rocks which form their true base: British Association for the Advancement of Science Notices, v. 14, p. 53–55.

Murchison, R.I., 1845a, On the Palaeozoic deposits of Scandinavia and the Baltic provinces of Russia, and their relations to Azoic or more ancient crystalline rocks; with an account of some great features of dislocation and metamorphism along their northern frontiers: Proceedings of the Geological Society of London, v. 4, p. 601–628.

Murchison, R.I., 1845b, On the Palaeozoic deposits of Scandinavia and the Baltic provinces of Russia, and their relations to Azoic or more ancient crystalline rocks; with an account of some features of dislocation and metamorphism along their northern frontiers: Quarterly Journal of the Geological Society of London, v. 1, p. 467–494, https://doi.org/10.1144/GSL.JGS.1845.001.01.96.

Murchison, R.I., 1846a, On the superficial detritus of Sweden, and on the prob able causes which have affected the surface of the rocks in the central and southern portions of that kingdom: Quarterly Journal of the Geological Society of London, v. 2, no. 1–2, p. 349–381, https://doi.org/10.1144/GSL.JGS.1846.002.01-02.51.

Murchison, R.I., 1846b, On the Silurian and associated rocks in Dalecarlia, and on the succession from Lower to Upper Silurian in Smoland, Öland, and Gothland, and in Scania: Quarterly Journal of the Geological Society of London, v. 3, no. 1, p. 1–48.

Murchison, R.I., 1847, Report given to the geology and mineralogy section of the Scandinavian savants, *in* Boeck, C., ed., Forhandlinger ved de skandinaviske Naturforskeres fjerde Möde, I Christiania den 11–18 Juli 1844: Christiania, Christian Gröndahl, p. 287–295.

Murchison, R.I., 1854, Siluria. The History of the Oldest Known Rocks Containing Organic Remains, with a Brief Sketch of the Distribution of Gold over the Earth: London, John Murray, 523 p.

Murchison, R.I., 1859, Siluria. The History of the Oldest Fossiliferous Rocks and Their Foundations, with a Brief Sketch of the Distribution of Gold over the Earth (3rd ed.): London, John Murray, 592 p.

Murchison, R.I., 1867, Siluria. A History of the Oldest Rocks in the British Isles and Other Countries; With Sketches of the Origin and Distribution of Native Gold, the General Succession of Geological Formations, and Changes of the Earth's Surface (4th ed.): London, John Murray, 566 p.

Murchison, R.I., de Verneuil, E., and von Keyserling, A., 1845, The Geology of Russia in Europe and the Ural Mountains, Volume 1 (662 p.), Geology: London, John Murray. Volume 2 (512 p.), Paléontologie: Paris, P. Bertrand.
Narkiewicz, M., Krzywiec, P., and Diemer, J., 2012, Geologia Gór Świętokrzyskich oczami Rodericka Murchisona – wizyta w czerwcu 1843 roku i jej echa (Geology of the Holy Cross Mountains in the Eyes of Roderick I. Murchison—A Visit in June 1843 and Its Echoes): Przegląd Geologiczny, v. 60, p. 220–225.
Oldroyd, D.R., 1990, The Highlands Controversy: Constructing Geological Knowledge through Fieldwork in Nineteenth-Century Britain: Chicago, Chicago University Press, 438 p.
Rudwick, M.J.S., 1985a, The Meaning of Fossils: Episodes in the History of Palaeontology, Second Edition: Chicago and London: University of Chicago Press, 287 p.
Rudwick, M.J.S., 1985b, The Great Devonian Controversy: The Shaping of Scientific Knowledge among Gentlemanly Specialists: Chicago, University of Chicago Press, https://doi.org/10.7208/chicago/9780226731001.001.0001, 494 p.
Rudwick, M.J.S., 1992, Scenes from Deep Time: Early Pictorial Representations of the Prehistoric World: Chicago, University of Chicago Press, 280 p.
Rudwick, M.J.S., 2005, Bursting the Limits of Time: The Reconstruction of Geohistory in the Age of Revolution: Chicago, University of Chicago Press, https://doi.org/10.7208/chicago/9780226731148.001.0001, 708 p.
Schneer, C.J., ed., 1969, Toward a History of Geology: Proceedings of the New Hampshire Inter-Disciplinary Conference on the History of Geology: Cambridge, MIT Press, 469 p.
Secord, J.A., 1986, Controversy in Victorian Geology: The Cambrian–Silurian Dispute: Princeton, New Jersey, Princeton University Press, 386 p.
Sharpe, T., 2016, William Smith's 1815 map, a delineation of the strata of England and Wales: Its production, distribution, variants and survival: Earth Sciences History, v. 35, no. 1, p. 47–61, https://doi.org/10.17704/1944-6187-35.1.47.
Sharpe, T., and Torrens, H.S., 2015, Introduction, *in* Lewis, C., ed., A Memoir to the Map and Delineation of the Strata of England and Wales, with Part of Scotland by William Smith, Engineer and Mineral Surveyor: London, History of Geology Group of the Geological Society of London, p. 1–26.
Stafford, R., 1989, Scientist of Empire: Sir Roderick Murchison, Scientific Exploration and Victorian Imperialism: Cambridge, UK, Cambridge University Press, 293 p.
Stafford, R., 2004, Roderick I. Murchison, Dictionary of National Biography: Oxford, Oxford University Press; https://doi.org/10.1093/ref:odnb/19555.
Thackray, J.C., 1978, R. I. Murchison's Geology of Russia (1845): Journal of the Society of Bibliography of Natural History, v. 8, p. 421–433, https://doi.org/10.3366/jsbnh.1978.8.4.421.
Torrens, H.S., 1990, The scientific ancestry and historiography of the Silurian System: Journal of the Geological Society, v. 147, p. 657–662, https://doi.org/10.1144/gsjgs.147.4.0657.

Manuscript Accepted by the Society 21 November 2017
Manuscript Published Online 6 July 2018

Printed in the USA

The Geological Society of America
Special Paper 535

Exhibiting life history at the Paris Muséum d'Histoire Naturelle (nineteenth–twenty-first centuries)

Claudine Cohen*
PSL École des Hautes Études en Sciences Sociales,
CRAL 96 Boulevard Raspail, Paris, France, and *EPHE, chaire "Biologie et Société,"*
Laboratoire Biogéosciences, UMR 6282 UB/CNRS, Université Bourgogne Franche-Comté, Dijon, France

ABSTRACT

The notion of a history of life has been the subject of extensive debates in France, from the turn of the nineteenth century to the dawn of the twentieth century and beyond. In this paper, I analyze how the organization of buildings and collections within the Paris National Museum of Natural History (MNHN), and their successive transformations up to the present, have reflected concepts and controversies over this history. After a brief overview of the constitution of the Museum, this chapter studies the organization of the Museum's paleontological, zoological, and comparative anatomy collections, and the devices through which scientific ideas and debates about evolution have been displayed and presented to the public.

This paper will focus more particularly on the settings of three major collections of zoological specimens: (1) Georges Cuvier's "Museum of Comparative Anatomy," which was opened to the public in 1806; (2) the Palaeontology and Comparative Anatomy Gallery, which was founded by Albert Gaudry in 1898; and (3) the Great Gallery of Evolution, which was inaugurated in 1994. I will argue that organizing these exhibits involved theoretical choices, taxonomic methods, scientific practices and rhetoric of displays, as well as ideological choices that connected natural history to its intellectual and sociopolitical background.

Through the examination of these different settings over a period of more than two centuries and their present use, I will question today's choices of the Paris Museum regarding the meaning and roles of these galleries, involving patrimonial conservation, public entertainment, and diffusion of scientific knowledge in the public.

*Cohen@ehess.fr

Cohen, C., 2018, Exhibiting life history at the Paris Muséum d'Histoire Naturelle (nineteenth–twenty-first centuries), *in* Rosenberg, G.D., and Clary, R.M., eds., Museums at the Forefront of the History and Philosophy of Geology: History Made, History in the Making: Geological Society of America Special Paper 535, p. 117–129, https://doi.org/10.1130/2018.2535(08).

INTRODUCTION

The Paris National Museum of Natural History (MNHN) carries to the present day the legacy of a long history, beginning in the seventeenth century. Numerous works have studied its development (Deleuze, 1823; Dorst, 1980; Laissus, 1995), following in the "longue durée" its institutional and scientific choices (Limoges, 1980), upon which state policy and colonialism had an important impact (Schnitter, 1996; Bonneuil, 1999), and in which intellectual organization and the stakes of power played, by turns or together, crucial roles (Outram, 1984; Sloan, 1997). Other works have studied the biographies of its major scientific actors (Appel, 1987; Laurent, 1997; Taquet, 2006) and the constitution of its collections (Spary, 2000) and underscored the Museum's importance in the shaping of French biological and geological disciplines (Laurent, 1987; Blanckaert et al., 1997; Loison, 2012).

Originally, a botanical garden, "the Jardin royal des plantes médicinales," was created by an edict of King Louis the XIII in May 1635 at the request of one of his personal physicians, Guy de la Brosse. The site was devoted to the cultivation of pharmaceutical plants. By the end of the seventeenth century, the Jardin du Roi expanded to become a place for the presentation and study of the whole natural world, and due to its growing prestige, specimens and curiosities were sent and accumulated there. Stuffed animals, skeletons, shells, minerals, and fossils were gathered both from France and from all over the world through voyages of exploration (Spary, 2000). These specimens were stored and classified in a building named "le Cabinet du Roi" (the term being also used for the collection itself). Georges-Louis Leclerc, Comte de Buffon, who was appointed as an "Intendant du Jardin du Roi" in 1739 and remained in charge until his death in 1788, gave the institution a glorious international reputation (Deleuze, 1823; Roger, 1989), relying on his own scientific prestige, the accumulation and ordering of collections, and the many enlargements and improvements he organized in the gardens and in the King's Cabinet (Fig. 1).

Soon after the French Revolution, the National Convention[1] abolished the Jardin du Roi and the Cabinet du Roi and created by a decree of the Assemblée Nationale on 10 June 1793 a new institution, which was named the "Muséum National d'histoire naturelle" (Hamy, 1893). Some of the old structures of the Jardin du Roi, its buildings, gardens, and collections, remained as a part of the new institution. The Museum, which was created in a time of Jacobin supremacy, was ruled by the assembly of its professors (initially 12), all equal in rank, which constituted what has been called a "République des Savants" responsible for the election of their director. The territory of the Museum was then expanded far beyond Buffon's additions. The new buildings would be used as residences for the professors and provided more space for scientific use. By then, the Museum's scientific attributions covered the whole spectrum of non-mathematical sciences: astronomy, chemistry, geology, zoology, and mineralogy. Scientific teaching and research, along with collecting and classifying specimens, were the major aims assigned to the Museum by the Convention (Dhombres and Dhombres, 1989, p. 86). Displaying the natural order then appeared both as a source of prosperity for the nation and of virtue for the individual (Spary, 2000).

Today, the original status of the Museum has been changed to a regular state institution, the administration of which is shared by three different ministries (education, research, and environment). The different parts of the Jardin des Plantes occupy the space of a vast quadrilateral limited on its eastern side by the Seine River and the Quai Saint-Bernard and on the three other sides by streets named after Buffon, Georges Cuvier, and Geoffroy Saint-Hilaire. Through the richness of its gardens bordered with avenues lined with linden and plane[2] trees; botanical collections and herbaria; vast buildings exhibiting comparative anatomy and zoology, paleontology, geology, and mineralogy collections; greenhouses and live animal menagerie; and libraries and amphitheaters decorated by nineteenth-century naturalist painters; the Paris National Muséum d'histoire naturelle offers to the public a vision of nature's diversity and history along with its own history (Figs. 1 and 2). The statue of Buffon sitting in a majestic armchair and overlooking the gardens, Cuvier's bust anonymously standing in the middle of the Comparative Anatomy Gallery, the statue in homage to Jean-Baptiste Lamarck, "the founder of the evolution doctrine"[3] (erected in 1909) (Richard, 1997), the bust of Albert Gaudry enthroned at the entrance of the Paleontology Gallery, the medallions representing Étienne Geoffroy Saint-Hilaire, Haüy, and Joseph Louis Gay-Lussac on the pediment of the Great Gallery of Zoology—all recall illustrious figures and great actors of French natural sciences and evoke memories of major episodes in their history and important debates that occurred during the successive centuries of its existence.

Among the debates that animated the Museum's research and teaching and resulted in structuring of its buildings and collection displays, the concept of a history of the living world was a central question and complex issue. Indeed, the acceptance of the notion of a history of life and arguments over the mechanism of its changes were subject to heated controversies from the turn of the nineteenth century to the dawn of the twentieth century and even beyond (Conry, 1974; Laurent, 1987; Cohen, 2017). The specifics of the French situation regarding these questions and especially the particular understanding of the processes of life's history that involved the choices of Cuvier's fixism, Lamarck's or Neo-Lamarckian transformism, and the difficult and belated acceptance of Charles Darwin's evolutionary concepts have often been discussed (Conry, 1974; Bowler, 1983; Cohen, 2017). Relying on

[1]The National Convention (21 September 1792–26 October 1795) was created after the French Revolution. It was a revolutionary government that established the Constitution of the first French Republic. This assembly had a major institutional role; it was founded on the novel basis of the system of public education and created a number of important public institutions.

[2]Plane trees belong to the family Platanaceae and look like the American Sycamore.

[3]All quotes from French have been translated by the author.

Figure 1. Buffon's statue overlooking the garden of the Paris National Museum of Natural History (photo © Xénia Cohen; used with permission).

Figure 2. A view of the gardens and the Great Gallery of Evolution, previously Gallery of Zoology, built in 1889 (photo © Xénia Cohen; used with permission).

Henri Daudin's studies of nineteenth-century zoological taxonomies (Daudin, 1926) and on Michel Foucault's reflections on the epistemological meaning of classifications (Foucault, 1966), I will analyze how the classification of zoological and paleontological collections in the Paris Museum, and their successive reorganizations up to the present, have embodied debates that were essential to the development of the natural sciences in France.

This paper will focus on the settings of three major zoological collections exhibits: (1) Cuvier's "Museum of Comparative Anatomy," which opened to the public in 1806; (2) the Palaeontology and Comparative Anatomy Gallery founded by Albert Gaudry in 1898; and (3) the Great Gallery of Evolution, inaugurated in 1994. I will argue that organizing these exhibits involved theoretical choices, taxonomic methods, scientific practices, and rhetoric of displays as well as ideological choices that connected natural history to its intellectual and social background. Through the examination of these different exhibits over a period of more than two centuries, and of their present use, I will question today's choices of the Paris Museum regarding the meaning and roles of these galleries, involving scientific research, patrimonial conservation, public entertainment, and diffusion of scientific knowledge in the public.

CUVIER'S "MUSEUM OF COMPARATIVE ANATOMY" AND CONTROVERSIES OVER LIFE'S HISTORY (1796–1830)

The early nineteenth century was a time of intense scientific elaborations and debates at the Paris MNHN, with accumulation of collections and emergence of the notion of a history of nature and discussions over its processes. The new institution housed scientific collections that were unequalled in the whole world. In addition to extensive botanical collections, it held more than 400 specimens of stuffed mammals; more than 2000 birds; innumerable insects; and riches in mineralogy, geology, and fossils. In 1795, Napoleon's wars of conquest resulted in the transfer to the Museum of all the zoological collections belonging to the Stadtholder of Holland, including a number of fossil specimens and two living elephants. The Museum's collections were also enriched through scientific expeditions, such as the voyages of Bougainville, Laperouse, and Levaillant, and the expedition to Egypt led by Napoleon (Laissus, 1998). The menagerie of living animals, created in 1794, soon became a public attraction and contributed to making the Jardin des Plantes a favorite place for Parisian walks and leisure time (Burkhardt, 1997; Efrat, 1998). In May of 1827, the arrival of the giraffe from Africa to the menagerie was a public event that aroused immense popular interest and curiosity (Burkhardt, 2013).

By the end of the eighteenth century, new disciplines had emerged and transformed the traditional and descriptive framework of traditional natural history into a vision of the history of nature (Whewell, 1837). Geology was created and named as a discipline (Gohau, 1990; Rudwick, 2005), different from speculative "theories of the Earth" that flourished during the previous centuries, and now relying on field research and observation. France was the cradle of vertebrate paleontology (Rudwick, 1997; Cohen, 2002) and the place of the invention of its name (Ducrotay de Blainville, 1822). Stratigraphic paleontology was developed by French scientists, in particular through Cuvier's expeditions in the Paris Basin with Alexandre Brongniart from 1808, which resulted in a volume entitled, *Essai sur la Géographie minéralogique des Environs de Paris* (Cuvier et Brongniart, 1811), illustrated with maps and cross sections combining the description of successive strata and the observation of the fossils they contained. The study eventually served as an introduction to the four volumes of Cuvier's paleontological studies (Cuvier, 1812) which were to expand during the following decades. By mid-century, invertebrate paleontologist Alcide d'Orbigny (1802–1857) established a succession of 22 stratigraphic layers, each one being characterized with distinct "fossiles directeurs." Through these observations, fossil fauna and flora were inserted into a historical framework provided by geological chronology.

During the first third of the century, often named as "the golden age" of the French Museum (Limoges, 1980), debates over the history of the living world involving Jean-Baptiste Lamarck (1754–1829), Georges Cuvier (1769–1832), and Étienne Geoffroy Saint-Hilaire (1772–1844), shaped the structure of the French scientific community (Appel, 1987; Laurent, 1987; Corsi, 1988) and were widely echoed in French and European general culture (Cohen, 1999). These debates, which dealt with the question of the fixity or the transformations of living organisms and the question of the unity of the animal realm, played an important role in the organization of the Museum, in the distribution of its different buildings, and of their contents.

It was Cuvier who took prominence in the natural sciences of the time—as well as in the space of the Jardin des Plantes. In his early works (Cuvier, 1800–1805, 1816), Cuvier introduced the ideas and methods of German naturalists, particularly oriented toward comparative anatomy, geology, and the study of fossils, which had become familiar to him during his studies at the Caroline Academy of Stuttgart (Taquet, 2006). He defined the practical methods and theoretical principles of comparative anatomy,[4] relying on a view of the organism as a closed and finalized whole (Cohen, 2011). Every organism is ruled by the laws of "correlations of parts" and "subordination of functions" within a teleological concept of adaptation to its life's conditions (Russell, 1916). Cuvier used Linnean taxonomy to describe species as fixed types, and he divided the animal realm into four

[4]In this, he followed the Convention's guidelines, which formally defined the responsibilities of the holder of the chair of animal anatomy to include the exposition of a "general idea of the internal organization of the various classes of animals; He will choose in each of them what he thinks fit to do an anatomical demonstration and seize the opportunities of putting before the eyes of the students those whose organization is less common. And will especially attend on the comparative anatomy of either animals between them, or animals with man. A private room destined for his preparations will likewise be assigned to him near the place of the lessons" (cf. Dhombres, 1989, p. 85–88).

distinct *embranchements*[5] with no connections or affinities between them.

His works in comparative anatomy also enabled him to recognize that fossil remains were vestiges of beings that were *extinct,* a demonstration he had performed as early as 1796, as he spectacularly demonstrated that Russian mammoths belonged to an extinct species, different from extant Indian and African elephants (Cohen, 2002). In so doing, Cuvier founded vertebrate paleontology as a historical discipline. The structural principles of his comparative anatomy enabled him to propose a method for the identification and reconstruction of extinct species, which he claimed made one able to reconstruct any extinct and unknown animal from a single claw, tooth, or bone and to determine its relative antiquity relying on its stratigraphic position (Cuvier, 1812, 1825).

Cuvier strongly rejected "evolutionary" views that were then advocated by Lamarck and Geoffroy Saint-Hilaire, and he explained discontinuities in the fossil record by immense catastrophes that periodically marked the history of life and destroyed successive faunas and floras and were followed by new appearance or creation of species. According to Cuvier, the paleontologist's aim was to recognize and reconstruct extinct beings and to establish their relative antiquity relying on their stratigraphic succession. Although his world vision was historical by the irreversible succession of events he described (appearance and disappearance of species and cataclysms periodically destroying faunas and floras), Cuvier imposed a fixist and catastrophist framework on French zoological and paleontological research (Laurent, 1987).

The ordering of Cuvier's collections would reflect his scientific principles. As he settled in the apartment of his predecessor, anatomist Jean-Claude Mertrud (1728–1802), he acquired a large building where he established his personal "museum," as an extension to his private apartments. This building was composed of two floors—a large, open, ground floor and on top a series of eight smaller rooms, lit by windows along the roof line. By piercing the wall of one of his apartment's rooms, Cuvier could manage direct access to the building.

Cuvier's Museum of Comparative Anatomy was open to the public from 1806. Its geographical location, 300 m away from the main zoological galleries, gave a material expression to the controversies that were to develop between Cuvier on the one side, Lamarck (1809) and Geoffroy Saint-Hilaire (1831) on the other, over the transformation of beings and the "unity of composition" of the living world (Deleuze, 1823; Appel, 1987; Sloan, 1997). Thus Cuvier's museum appeared to be quite different from the other galleries organized on the basis of classifications by Daubenton, Lacépède, and Lamarck. In the former Cabinet du Roi, Louis-Jean-Marie Daubenton (1716–1800), who long played a major role in the organization of these collections, had classified natural objects of the three reigns of nature according to a "methodical order," in classes, genera, and species. After the creation of the MNHN in 1793, the main natural history galleries were arranged by Gotthelf Fischer, who classified mammals, birds, reptiles and fishes, according to Bernard Germain de Lacépède's (1756–1825) principles, based on organs of motion and feeding habits. Invertebrate collections were organized by Lamarck, whose classification of fossil and extant invertebrates led to observing *analogies* and concluding possible filiations between extinct and present-day forms. A follower of Buffon, he criticized the reductive character of Linnaeus' nomenclatures and rejected the notion of species, in favor of the idea of a "series" that could account for possible changes and a gradation in complexity of living organisms. Therefore, Lamarck classified fossil invertebrates according to his "series of complexity" (Lamarck, 1801), descending from most complex mollusks to simplest polyps and infusoria.

Cuvier's own Museum of Comparative Anatomy proposed yet another *arrangement,* following the order adopted in his *Leçons d'anatomie comparée* of 1800–1805. It was essentially based on inner anatomy, showing structural relationships between physiology, skeletal mechanics, and conditions of existence. When introduced into his collections, "visitors ... were carefully confronted with a theoretical arrangement of organisms. A theory of the relationships between form and function developed as they were guided through the series of rooms in a pre-established order. The visit usually started on the ground floor with the whales, large quadrupeds, and entire human skeletons, then with a series of rooms on the upper floor, which confronted the visitor with whole bodies and presentations of superficial muscles. Only then was he introduced into the functional organization, with presentations of parts of disarticulated skeletons, and single organs preserved in alcohol" (Sloan, 1997, p. 615).[6] Animal exhibits were organized functionally, in respect to locomotion, feeding, respiration, sensation, and digestion. According to his concept of the general classification of living beings, Cuvier never drew into comparative display animals from different *embranchements.*

As a founder of vertebrate paleontology, Cuvier also introduced the novel use of fossil collections, which until then had mainly served as stratigraphic markers. His methods to identify and reconstruct extinct beings made him able to conduct extensive studies relying on the fossil vertebrates kept in the galleries, on the Tertiary fossils from Montmartre's gypsum quarries near Paris, and on a number of specimens that were sent to him by collectors, scientists, curious people, and amateurs who passed on to him their discoveries (Rudwick, 1997). However, his own Museum did not display paleontological specimens; fossils remained either in his private cabinet or in the main Gallery,

[5]Namely, vertebrates, mollusks, articulated animals (arthropods and annelids), and zoophytes (cnidaria and other phyla).

[6]Cuvier's collections made a deep impression on a number of scientists, anatomists, and paleontologists around the world, among them, William Clift, Curator of the Hunterian Museum; English paleontologist Richard Owen who eventually founded the London Natural History Museum (Sloan, 1997); and young Louis Agassiz, who eventually set up his own Museum of Comparative Zoology at Harvard University (Agassiz, 1869; Winsor, 1990).

making it impossible for the visitor to acknowledge relationships between extinct and living forms. Eventually in 1841, fossil vertebrates were displaced to another building located along Buffon Street, and they were partly stored with collections of mineralogy, while the rest of the collections were dispersed in different locations. Visitors had to wait until the end of the century to see fossil vertebrates exhibited in a new setting that associated paleontology with comparative zoology and reused Cuvier's specimens, along with those that were accumulated later, to embody the notion of an evolution of the living world.

EXHIBITING EVOLUTION AT THE PALEONTOLOGY GALLERY

"Paleontology teaches us that we must no longer speak of the fixity of species. All beings, however powerful they may be, have been ephemeral: the law of change is the great law that dominates the world," proclaimed Jean Albert Gaudry (1827–1908) in his inaugural lecture as a chair professor of paleontology at the Paris Museum in 1873. In a time when Cuvier's disciples had become academic authorities and still argued in favor of their master's ideas (Laurent, 1987)[7], Gaudry was an evolutionist and an admirer of Darwin, whose work he read as early as 1862[8]. First appointed as an "aide-Naturaliste" at the Museum in 1853, he was barred for a long time at the Museum because of his Darwinian positions, and he had to wait until 1872 before he was elected to professorship.

Gaudry believed that the living world evolves and animal species have been subject to modifications through geological time. His studies aimed to shine a light on "intermediary" species between known fossil genera. For this demonstration, he used specimens that had been accumulated by his predecessors as well as fossils he had discovered in the Miocene levels of Pikermi (Greece), where he had conducted extensive digs since 1855, resulting in his *animaux fossiles de l'Attique* (fossil animals from Attica) (Gaudry, 1867) and later in Tertiary deposits of the Lubéron (Vaucluse, southeastern France), yielding a similar fauna (Gaudry, 1878). As a result of his finds, Gaudry was able to draw phylogenetic trees representing several families of terrestrial vertebrates, hyenae, proboscides, horses, rhinoceros, and suiforms. These genealogical trees, taking Darwin's abstract evolutionary scheme (Darwin, 1859, chapter 4) and applying it to concrete fossil evidence, were among the first of their kind (Cohen, 2002, 2017; Tassy, 2006).

Gaudry started a project to build a new gallery to exhibit the Museum's paleontological collections in a setting that would express his own evolutionary concepts. "Today, Paleontology has taken on immense proportions and has become the history of the development of organic nature. To embrace this history, we will have to build a long gallery where we will follow the march of life, starting with its first manifestations on the globe and ending with the shining of human intelligence," he wrote in the volume celebrating the centenary of the foundation of the Paris Museum of Natural History (Gaudry, 1893, p. 327; Museum, 1893).

Under Gaudry's incentive, with support of comparative anatomist Georges Pouchet (1800–1872) and anthropologist T.H. Hamy (1842–1908), the creation of this gallery was part of the will to reassert the original mission of the Museum—collecting and displaying specimens—upon novel, "evolutionary" grounds. The "Galerie de paléontologie et d'anatomie comparée" was built during the final decade of the century, from plans of architect Ferdinand Dutert (1845–1906), who had enjoyed great success with the "galerie des machines" he had built for the Paris Universal Exhibition of 1889. The new gallery, which was located on the eastern end of the Garden, by Place Valhubert near the Austerlitz railway station, was a three-story *art nouveau* building constructed of bricks, metal, and glass, erected to gather in one place the collections of comparative anatomy, vertebrate, invertebrate, and plant paleontology.

Although the building only partly fulfilled Gaudry's and Dutert's initial grand project—the exhibition halls were 80 m long, instead of the 320 m and the double nave initially planned; therefore, the architect refused to participate in the opening—the Palaeontology Gallery, which was inaugurated on 21 July 1898, immediately became an immense public success: 11,000 visitors crowded into the gallery on the first Sunday, and 10,000 visitors attended on the following Sunday. The paleontology exhibit offered contemplation of the developments of the fossil world to the curiosity of the Parisians. Ascending through the great staircase with elegant wrought-iron flowers adorning its banister, visitors discovered in the immense rectangular room the relics of former worlds, a majestic procession of fossil animals with brown bones and strange shapes and names—mammoths and mastodons, *Palaeotheriums* and *Anoplotherium,* and *Megatherium* and *Edentates* (Fig. 3). The organization of specimens through the space of the gallery was meant to illustrate "the history of the developments of organic Nature" through the Primary, Secondary, Tertiary, and Quaternary eras, according to Gaudry's evolutionary concepts, as exposed in the three volumes of his *Enchaînements du monde animal* (1878–1890). "It contains on the ground floor Comparative anatomy, on the first floor Palaeontology, and on the second Anthropology, arranged in galleries which circle around the palaeontology collections, leaving the center empty so that the gigantic fossils may have place and light. Beside the gallery of palaeontology are the laboratories, somewhat cramped, but furnished with all modern instruments; They

[7]Cuvier's intellectual influence extended well beyond his death in 1832 and imposed until late in the century his theoretical framework. His worldview of a fixist and catastrophic framework was long influential in French paleontological research. For example, Alcide d'Orbigny, who was the first holder of the chair of paleontology, created in 1855, settled his whole classification of paleontological stratigraphy relying on his classification of fossil invertebrates within a Cuvierian framework (d'Orbigny, 1849). Well into the 1860s, in France, Cuvier's disciples had become academic authorities and continued to argue in favor of their master's ideas, for example against the existence of "fossil Man" (Cohen and Hublin, 2017).

[8]Gaudry read Darwin's *On the Origin of Species* (Darwin [1859], 1872, chapter IV) in Clémence Royer's considerably biased translation. His recently discovered annotated copy reveals his enthusiastic reading (Tassy, 2006).

Figure 3. A view of the Palaeontology Gallery, built in 1898: according to Gaudry, Tertiary elephants were "the most beautiful, the most active, the most intelligent" of all quadrupeds (photo © Momoko Seto; used with permission).

have the invaluable advantage of being placed near the collections" (Gaudry, 1898, p. 800). Thus Gaudry described the Palaeontology Gallery in 1898.

However, Gaudry's evolutionary positions were in fact closer to American Neo-Lamarckism (illustrated, for example, by Cope's *Theology of Evolution* [1897]) than to Darwin's principles explained his book *On the Origin of Species by Means of Natural Selection*. Gaudry's "evolutionary" views and quest for "intermediary forms" (Gaudry, 1878–1890) involved a metaphysical framework (Gaudry, 1896). For him, all living beings were organized and connected in perfect harmony, and evolution was ruled by transcendent laws and determined by a divine "plan," as a necessary progress culminating in mankind. Each geological era witnesses a certain type of perfection. In the Secondary era, by the end of the Jurassic and in the early Cretaceous, dinosaurs, "the most gigantic of all continental quadrupeds," represented "the reign of brutal force." Then, "the real apogee of the animal world [...] appears by the end of Tertiary Times, during the Miocene and Pliocene, that is immediately prior to the advent of Man." Tertiary elephants were "the most beautiful, the most active, the most intelligent" of all quadrupeds (Gaudry, 1896, p. 206). And finally, man appears—the crown of creation.

The arrangement of fossils in the paleontology gallery reflected these ideas. Their order, from the oldest to the most recent, gave dinosaurs, elephants, and men successive places of honor in the exhibit, while the famous "Elephant of Durfort" (*Mammuthus meridionalis*) standing on its gigantic legs, dominated by its enormity the troop of fleshless monsters. The next epoch culminated in the appearance of the Hominine family; molds of human fossil skulls were exhibited at the end of the immense hall, the back wall of which would later be decorated with a reproduction of Lascaux Cave's painted frescoes. However, the largest part of the anthropological collections was later removed from these galleries to occupy a museum of its own.[9]

Today, the paleontology gallery allows us to contemplate much of the history of the living world as well as the history of paleontology in the first century of its existence. Still presented in the same order as Gaudry organized it, the gallery resonates with the great names of eighteenth- and nineteenth-century paleontology and exhibits famous specimens that marked its major episodes. A whole window is devoted to Cuvier, exhibiting specimens he published in his *Recherches,* such as the "*Mosasaurus*"[10] studied in 1808 and identified as a giant lizard, and "Pachyderms from the gypsum of Montmartre," including the full skeleton of a *Paleotherium,* an *Anoplotherium,* and the *Oppossum* he had spectacularly dissected and identified in public (Cuvier, 1804). Tertiary fossils from Sansan (Gers), such as the gigantic *Macrotherium* excavated by Filhol in 1887, are also exhibited, as well as Gaudry's own finds from Pikermi (Greece) and Mount Lubéron (Vaucluse, France), illustrated by vestiges of *Hipparion gracilis*, *Machairodus giganteus*, *Hyaenidis graeca*, and a large fragment of ossiferous "breccia from the Mount Luberon" with remains of *hipparion, gazella, antilope,* and *rhinoceros*. The top floor exhibits fossil invertebrates and plants displayed according to traditional classifications in families, genera, and species. A hexagonal window is devoted, like a shrine, to Lamarck's collections of shells of fossil and extant

[9] Le Musée de l'Homme is now located in the Palais de Chaillot on the Place du Trocadero. It was founded in 1938 by Paul Rivet and was recently entirely renovated; the new Musée de l'Homme was inaugurated in 2015.

[10] This original specimen, which came from the Stadtholder's collections, was recently given back to Holland. Its mold is today exhibited at the same place in the gallery.

Figure 4. A view of the Comparative Anatomy Gallery (photo © Xenia Cohen; used with permission).

mollusks ordered in "series." Another, larger window exhibits d'Orbigny's collections of invertebrates, including many types of specimens that he defined as "fossiles directeurs" in his biostratigratigraphic column (d'Orbigny, 1849).

The paleontology gallery also witnesses international collaborations and exchanges. A mastodon's femur from Big Bone Lick site, in Kentucky, was sent from America to France by Baron de Longueil in 1739 and later studied by Daubenton (Daubenton, 1764; Cohen, 2002); the mold of a huge *Iguanodon* standing on its hind legs, discovered in the mine of Bernissart in 1878, studied by Louis Dollo (Dollo, 1882), was offered to the Museum by the Royal Academy of Belgium one year after the inauguration of the gallery in 1899; several molds of American dinosaurs, among which the skull of a *Tyrannosaurus rex* were gifts of Henry Fairfield Osborn from the American Museum of Natural History (AMNH) in New York, along with a miniature reconstruction of the animal by AMNH artist Charles Knight. The life-size *Diplodocus* that occupies the center of the room was offered by American steel magnate Andrew Carnegie in 1910. It is a mold of the original skeleton displayed at the Carnegie Museum of Natural History in Pittsburgh (Nieuweland, 2010). Several South American fossil specimens, such as *Megatherium, Glyptodon,* and Edentates reveal the important scientific interactions with these countries since Cuvier's time; and a number of fossils from West Africa and Madagascar recall the French colonial empire and its impact upon the Museum's collections. The remains of a wooly mammoth from the Liakhov islands (Siberia), excavated in 1910 by Russian paleontologist and geologist K.A. Vollosovitch (Vollosovitch, 1914), were subsequently donated by Count Aleksander Stenbock-Fermor to the Museum of Paris in 1912 (Cohen, 2004). Stored for a long time in the gallery's drawers, the mammoth's skeleton, now reconstructed, stands in the room, while its soft parts are exhibited in an adjacent window. Some tissues have been used for experiments on ancient DNA (Debruyne et al., 2003), demonstrating that old paleontological collections can be used again for novel studies through high-technology devices.

On the lower floor, the comparative anatomy gallery provides yet another spectacular sight (Fig. 4). Under the guidance of a flayed man[11] standing with his left hand pointed to the sky, marches another troop of vertebrates skeletons, those of extant animals, many of which came from Cuvier's Museum of Comparative Anatomy, with additions of specimens later collected by his successors from exploration voyages or from the menagerie's deceased animals. In the central part of the room, a procession of full mounted skeletons illustrates the adaptive diversity of vertebrates: carnivores, equids, and rhinoceroses; artiodactyles, proboscids, and sirenians; the largest aquatic

[11]This "écorché," originally intended as an anatomical model for art Aix en Provence students, was the work of sculptor Jean-Pancrace Chastel (1758).

mammals occupying the whole rear part of the exhibit room, while a sequence of several windows in its center features marsupials and monotremes, birds, reptiles, and primates. The lateral windows all around the hall offer a comparative view of the anatomical parts of various animals. Finely dissected organs preserved in alcohol, or their wax replicas, illustrate physiological functions—livers and stomachs, arteries and hearts, brains, successively representing the digestive, circulatory, and nervous systems—while displays of osteological parts recall their mechanical functions, such as adaptations of the limb bones to running, swimming, or flight.

In different parts of the room, several open "boxes" show skeletal dissections of different species of fishes (monkfish, sturgeon, etc.); these boxes and their extraordinarily fine dissections were made up and used by Cuvier himself for his teachings in comparative anatomy; his own handwriting is still readable on the labels. In its overall arrangement, the exhibit essentially presents vertebrate comparative anatomy without departing much from the principles that were defined by Cuvier, whose marble bust stands in the center of the room, while those of Étienne Serres, Paul Gervais, and Geoffroy Saint-Hilaire are more discreetly posted in the gallery's corners.

The exhibit is restricted to vertebrate anatomy and physiology, thus remaining within the limits of Cuvier's own practice and his strict definition of four separate branches (*embranchements*). No reference is made in this exhibit to comparative anatomy concepts such as homology or connections of parts, which were advocated by Étienne Geoffroy Saint-Hilaire and later morphologists. If embryology and teratology are present in the exhibit, it is in reference to the works of Serres and Isidore Geoffroy Saint-Hilaire; however, there is no mention of contemporary biology, i.e., molecular biology, or developmental genetics and their impact on the understanding of unity of the living world.

As it stands today, the whole building and its exhibits remain for the most part faithful to what it originally was. The renovation, which was planned for its centennial in 1998, resulted in minimal changes. The premises were refurbished for public safety, specimens were dusted off, a few obsolete labels were changed, but many "historical" labels remain intact, with their faded ink and citations of names that have now fallen into oblivion. In the Vertebrate Paleontology Gallery, some evolutionary relationships are suggested ("From Dinosaurs to Birds," "Evolution of Equids," "Origin and Evolution of Elephants," "From the Origin of Primates to Human Evolution"); but for the most part, the gallery is a testimony to nineteenth-century scientific practices, research, and debates: a museum of museum.

The presentation of fossils and living animals in the galleries of paleontology and of comparative anatomy, with their accumulation of specimens, their homage to the founders of the discipline, their strange and austere beauty, and the superposition of the comparative zoology hall with the fossils galleries, provides a descriptive and analytical view rather than a synthetic and explanatory presentation of evolution. In the three-floor succession, fossil diversity and comparative anatomy are spatially displayed, without an attempt to explicate evolutionary concepts and debates, perhaps as a result of the intricate controversies over evolution and its mechanisms, which long animated French natural history and lingered and remained unsolved until the mid-twentieth century (Cohen, 2017).

In order to see evolution presented and its principles explained, one must walk to the other end of the Jardin des Plantes and access yet another gallery.

EXHIBITING LIFE WONDERS: THE GREAT GALLERY OF EVOLUTION (1994)

Alongside Geoffroy Saint-Hilaire Street, the "Grande Galerie de l'Évolution" today occupies a building, the previous name of which, "Galerie de Zoologie," remains inscribed in large letters on its exterior wall (Fig. 2). This imposing building was erected under the supervision of Jules André, a rationalist architect and disciple of Gustave Eiffel. The interior architecture offers, with its glass ceiling that provides zenithal light to the immense central room and its balconies running around its three floors, a profusion of metal pillars and vaults that recall the triumph of cast-iron architecture at the turn of the nineteenth century. Like the Eiffel Tower, it was built to celebrate the anniversary of the French Revolution in 1889. It once housed one of the richest zoological collections in the world. Thousands of specimens of stuffed animals, which had been accumulated ever since the early eighteenth century, were exhibited there, as an object of wonder and entertainment for the Parisian public (Efrat, 2002). In 1965, the building was crumbling and had to be evacuated for security. A large reserve (called "zoothèque") was built underground in front of the gallery to house its endangered collections. Eventually in 1986, a project was planned to rehabilitate the building and organize a new gallery. Following a vote of the MNHN scientists, the theme of evolution was chosen as a leading thread for the general presentation.

The conceptual organization of the new gallery was directed by geneticist André Adoutte and museologist Michel Van Praet (Van Praet, 2005), and the building's renovation was carried out by architect Paul Chemetov. The restoration included creating a new entrance, rehabilitating the roof and the colonnaded galleries, installing transparent elevators, digging foundations, and adding a ten-meter hollowed area to the eleven-meter height of the main room. In parallel with the building's renovation, restoration of the stuffed specimens started in 1991. In three years, a thousand animals were restored by taxidermists—350 mammals, 500 birds, and more than a hundred reptiles, fish, and amphibians.

The organization of specimens was entrusted to theater stage designer René Allio, who organized the exhibit to fit with the scientific ideas carried out by the Grande Galerie de l'Évolution's project, implementing a scenography with "animated walls" rather than dioramas, and skillfully organizing contrasts of light and darkness and emptiness and fullness. The scenography took advantage of the building's architecture to display the specimens on three floors,[12] in order to represent life's evolution both through its effects and its mechanisms. The gallery presents, in three "acts," life's diversity and unity. At the entrance, the visitor is faced with an extraordinary profusion of animal life, the presentation of specimens combining the reconstructions of fauna and ecosystems (seafloor, pelagic environments, African savannah, and Antarctic ice sheet) with taxonomic arrangement (different classes of invertebrates, fish, and mammals) (Fig. 5). As on a platonic mode, the top floor provides theoretical keys enabling visitors to understand the unity and logic of the profusion of animals of the bottom floors. After a quick presentation of the great figures of the history of transformism (e.g., Lamarck and his shells, Darwin and his pigeons, Mendel and his peas), the principles of evolutionary theory are recalled along an itinerary leading from DNA and molecular transmission to the mechanisms of natural selection and adaptive diversification. Going around the gallery, one follows in a linear mode the main stages of the history of the animal realm—the appearance of animals in the Vendian; the adaptive diversification of primitive invertebrates in the Cambrian; the emergence of terrestrial vertebrates; the origin of the amniotic egg; and the emergence of birds and mammals. At the end of this itinerary are presented comparatively different adaptations to quadruped locomotion, from coelacanth to Galapagos turtles, from the bat to the beaver, from the giraffe to the snake-python, thus echoing the more dry-looking presentation of vertebrate locomotion in the Comparative Anatomy Gallery at the other end of the Garden.

Man enters stage in-between, on the second floor, not to embody the triumphant goal of evolution, but as an actor who owns responsibility for the transformations of nature (Blandin, 2000). For ten thousand years, man's activities of predation (hunting and gathering) and of the domestication of animals and plants have transformed landscapes and ecosystems, created new animal varieties (the pig, the ox, and the hairy donkey of Poitou) or vegetable varieties (the varieties of maize). But human activities are also sources of pollution and destruction. In the ancient gallery of birds, there is now a "sanctuary" of rare and endangered species. The thylacine (the "marsupial wolf") from Australia, the black emu of Madagascar, and the tiger of China, in their glass cages, give the feeling of the beings of nature's fragile beauty.

With its wings of birds and butterflies, iridescent fishes (Fig. 6), giraffes and hippopotamuses, and herd of animals, all of whom seem to be leaving Noah's ark, the Great Gallery's exhibit is of great plastic beauty. Its magic is reinforced by the architectural splendor of the place and the subtlety of sounds and lights. But, due perhaps to the fragility of the specimens, most of the exhibit is plunged in semi-darkness, making it difficult to read labels and follow a rather tough intellectual itinerary—the quasi-absence of fossils making it difficult to understand fully the demonstrative intentions of this staging of the evolution of life.

The Great Gallery is conceived both as an entertaining display of a profusion of animal specimens and as a didactic presentation. Criticism has addressed the fact that theoretical issues are confined to the top end of the building and presented in a general mode, with little access to discussions, modes of reasoning, and issues in today's science (Laissus, 1995). The exhibit's motto is "biodiversity," but this could well be just "a new word on an old reality" (Pinna, 1998), with beautiful displays but a lack of thorough reflection on the meaning and implications of the concept. However, since its opening in 1994, the "Grande Galerie" has enjoyed tremendous public success. It can be said, for the most part, this staging of evolution fulfills its goal and continues a long tradition of science education through aesthetics and theatricality—a tradition that has endured in France since the nineteenth century (Efrat, 2002).

CONCLUSIONS

Representing and explaining the evolution of life is today a central issue for natural history museums, involving not only research and scientific discussions, but also efforts to educate the general public, an important goal in a time when fanaticism and withholding information tend to overtake rationality and scientific understanding. However, exhibiting and explaining the evolution of life never was an easy task, because it presents a number of paradoxes (Pinna, 1997). It requires presenting the concept of deep time through spatial dimensions, juxtaposing discrete objects in order to materialize the continuity of descent, often representing a branching process in the form of a linear succession. It also requires dealing with the complex events of speciation, natural selection, and extinction, without featuring the totality of the causes involved, and evoking molecular-scale processes through the sole aspect of macroevolution.

In order to overcome these difficulties, several museums in the world have chosen to experiment with novel modes of presentation, namely to organize their displays in the shape of cladograms. This was the case for the dinosaur exhibit at the London Museum of Natural History in 1979, and it is today the case for the American Museum of Natural History in New York, which from 1995 organized the different displays of its "Hall of Vertebrates" according to the genealogical structure of the cladogram. Visitors are invited to walk on the branches of the evolutionary tree, and they meet on their way evolutionary "knots"—events from which species successively diverged. This general presentation of evolutionary patterns allows fossil specimens to be presented in relationship with living animals. Undoubtedly such a presentation gains in clarity. However, it is incomplete, because its explanatory power regarding the contexts and processes of evolution is extremely limited. It is also somewhat dry and tends to erase the originality and specific culture of a given museum.

[12]The four following paragraphs are translated by the author from a paper written on the occasion of the double inauguration of the Paris Grande Galerie de l'Évolution and of the recently reorganized Hall of Vertebrate Paleontology at the American Museum of Natural History (AMNH) in New York (C. Cohen, 1994 (Sept.), "Paris-New York l'Évolution mise en scène," *La Recherche*, no. 268, p. 932–933).

Figure 5. A view of the Great Gallery of Evolution: the troop of African animals, exhibited on the lower floor, by the entrance of the gallery (photo © Xénia Cohen; used with permission).

Figure 6. The exhibition of fishes at the Great Gallery of Evolution: a scenography on an "animated wall" (photo © Xénia Cohen; used with permission).

As we have seen, the Paris Museum followed a different path. Its presentation of life's evolution remains spread out between three exhibit halls with quite different structures, meanings, and appeal. The Paris Museum remains strongly attached to its own history and culture, with its buildings reflecting the history of the organization and division of knowledge and the constitution of disciplines. Its specimens tell a story of traveling and collecting, of debates and scientific exchange, of individual ambitions and scholarly networks; and its displays carry the memory of practices involved in ordering objects and of controversies over the classifications and history of the living world. It also carries the memory of changes in its political and administrative status. Originally reserved for the kings' private use, then a democratic institution conceived as a "République des Savants," it has become today, a place for research at an international level, for public entertainment and education, and a place of memory.

The plurality of modes of presentation of evolution, which is a result of the Museum's long history, has its faults but is also its richness. Through its laboratories, gardens, buildings, various exhibits, and artistic works, the Paris Museum as it stands today reflects positively the efforts to present and popularize the natural sciences as part of culture and to fulfill the deep goals of natural history, in the most modern sense of the word—a scientific endeavor, which "makes it possible to break through the limits of time and space, to trace back and understand the history of Earth and Life, to inventory biodiversity, to analyze the complexity of ecosystems, to understand how the living world evolves, and to identify humankind's place on the planet" (David et al., 2017, p. 62).

ACKNOWLEDGMENTS

I owe many thanks to Gary Rosenberg and colleagues from the GSA History and Philosophy of Geology Division for their encouragement to write this paper. I would like to thank in particular Michel Veuille for our many visits to the Paris Museum's galleries and for our discussions over the years; Ken Taylor for his precise reading and thoughtful comments; Tomer Efrat for his inspiring work; Giovanni Pinna for sharing with me his thoughts on natural history museums and their organizations; Philip Sloan for providing his publications and enlightened analyses; my colleagues of the Centre Koyré and of the Institut Systématique, Évolution et Biodiversité Laboratory of the Paris Museum for our exchanges; and the current President of the MNHN, Bruno David, for his invitation to contribute in a stimulating reflection group on the future of natural history at the MNHN during the year 2016–2017.

REFERENCES CITED

Agassiz, L., 1869, De l'Espèce et de la classification en zoologie: Paris, Baillière, 400 p.

Appel, T.A., 1987, The Cuvier-Geoffroy Debate, French Biology in the Decades before Darwin: Oxford, UK, Oxford University Press, 305 p.

Blanckaert, C., Cohen, C., Corsi, P., and Fisher, J.-L., eds., 1997, Le Muséum au premier siècle de son histoire: Paris, Editions du Muséum d'histoire naturelle, 687 p.

Blandin, P., 2000, Des "relations Homme-Nature" à "l'Homme facteur d'évolution": Genèse d'un projet muséal, *in* Eidelman, J., and Van Praet, M., eds., La muséologie des sciences et ses publics: Paris, Presses Universitaires de France, p. 31–52.

Bonneuil, C., 1999, Le Muséum national d'histoire naturelle et l'expansion coloniale de la Troisième République (1870–1914): Revue Française d'Histoire d'Outre-Mer, v. 86, p. 143–169, https://doi.org/10.3406/outre.1999.3720.

Bowler, P.J., [1983] 1986, The Eclipse of Darwinism: Anti-Darwinian Evolution Theories in the Decades around 1900: Baltimore, Maryland, Johns Hopkins University Press, 291 p.

Burkhardt, R., 1997, La Ménagerie et la vie du Muséum, *in* Blanckaert, C., Cohen, C., et al., eds., Le Muséum au premier siècle de son histoire: Paris, Éditions du Muséum d'histoire naturelle, p. 481–508.

Burkhardt, R., 2013, Civilizing specimens and citizens at the Muséum d'histoire naturelle, 1793–1838, *in* Prince, S.A., ed., Of Elephants and Roses, French Natural History 1790–1830: Philadelphia, APS Museum, Memoirs of the American Philosophical Society, v. 267, p. 1–15.

Cohen, C., 1994, Paris-New York l'Évolution mise en scène: La Recherche, v. 268, Sept., v. 5, p. 932–933.

Cohen, C., 1999, L'Homme des origines, savoirs et fictions en préhistoire: Paris, Éditions du Seuil, 311 p.

Cohen, C., 2002, The Fate of the Mammoth: Fossils, Myths and History [Preface by S.J. Gould]: Chicago, University of Chicago Press, 336 p.

Cohen, C., 2004, Le Destin du Mammouth: Paris, Éditions du Points-Seuil, 424 p.

Cohen, C., 2011, La Méthode de Zadig: La trace, le fossile, la preuve: Paris, Éditions du Seuil, 352 p.

Cohen, C., 2017, How nationality influences opinion: Darwinism and palaeontology in France (1859–1914): Studies in History and Philosophy of Science Part C Studies in History and Philosophy of Biological and Biomedical Sciences, v. 66, December, p. 8–17, https://doi.org/10.1016/j.shpsc.2017.10.003.

Cohen, C., and Hublin, J.-J., 2017, Boucher de Perthes, les Origines romantiques de la préhistoire: Paris, Belin, 400 p.

Conry, Y., 1974, L'introduction du Darwinisme en France: Paris, Éditions du Vrin, 480 p.

Cope, E.D., 1897, Theology of Evolution: A Lecture: Philadelphia, Arnold, 54 p.

Corsi, P., 1988, The Age of Lamarck, Evolutionary Theories in France, 1790–1830: Berkeley, University of California Press, 361 p.

Cuvier, G., 1800–1805, Leçons d'anatomie comparée, 5 vols.: Paris, Déterville.

Cuvier, G., 1812, Recherches sur les ossemens fossiles de quadrupèdes, où l'on rétablit les caractères de plusieurs espèces d'animaux que les révolutions du globe paroissent avoir détruites. Discours préliminaire: Paris, Déterville, 4 vols.

Cuvier, G., 1817, Le Règne animal distribué d'après son organization: Paris, Fortin, Masson et cie, libraires, 4 vols.

Cuvier, G., 1825, Discours sur les Révolutions de la Surface du Globe: Paris, Edmond d'Ocagne, 400 p.

Cuvier, G., and Brongniart, A., 1811, Essai sur la Géographie minéralogique des environs de Paris: Paris, Baudouin, 303 p.

Darwin, C., 1859, On the origin of species by means of natural selection, or the preservation of favoured races in the struggle for life: London, John Murray (facsimile of the first edition, with an introduction by Ernst Mayr, 1964, Harvard University Press, sixth printing 1981), 540 p.

Darwin, C., 1872, The origin of species by means of natural selection, or the preservation of favoured races in the struggle for life: London, John Murray (sixth edition, with additions and corrections), 443 p.

Daubenton, L.J.M., 1764, Mémoire sur des os et des dents remarquables par leur grandeur: Mémoires de l'Académie Royale des Sciences de Paris, an. 1762, p. 206–229.

Daudin, H., 1926, Les classes zoologiques et l'idée de série animale: Paris, Éditions Alcan, 798 p.

David, B., Paquet, T., et al., 2017, Muséum Manifesto: What Future without Nature?: Paris, Éditions du Muséum d'histoire naturelle, 80 p.

Debruyne, R., Barriel, V., and Tassy, P., 2003, Mitochondrial cytochrome *b* of the Lyakhov mammoth (Proboscidea, Mammalia): New data and phylogenetic analyses of Elephantidae: Molecular Phylogenetics and Evolution, v. 26, no. 3, p. 421–434, https://doi.org/10.1016/S1055-7903(02)00292-0.

Deleuze, J.P.F., 1823, Histoire et description du Museum d'histoire naturelle: Paris, Royer, https://doi.org/10.5962/bhl.title.122312, 720 p.

Dhombres, N., and Dhombres, J., 1989, Naissance d'un nouveau pouvoir: Sciences et savants en France, 1793–1824: Paris, Payot, 938 p.
Dollo, L., 1882, Première note sur les Iguanodons de Bernissart: Bulletin du Musée Royal d'Histoire Naturelle de Belgique, v. 1, p. 161–180.
Dorst, J., dir., 1980, Le Muséum d'histoire naturelle: Paris, Fernand Nathan, 252 p.
Ducrotay de Blainville, H., 1822, De l'Organisation des animaux, ou Principes d'anatomie comparée: Paris, F.-G. Levrault, 252 p.
Efrat, T., 1998, Le public du Muséum d'histoire naturelle au 19e siècle: Mémoire de maîtrise, EHESS, 652 p.
Efrat, T., 2002, L'objet naturalisé et la nature objectivée: Curiosités urbaines et collections d'histoire naturelle en France (1830–1930) [Thèse de Doctorat]: Paris, École des Hautes Études en Sciences Sociales, 579 p.
Foucault, M., 1966, Les Mots et les choses: Paris, Gallimard; (English translation) 1970, The Order of Things: An Archaeology of the Human Sciences: New York, Pantheon Books, 404 p.
Gaudry, A., 1862–1867, Animaux fossiles et géologie de l'Attique, d'après les recherches faites en 1855–56 et en 1860 sous les auspices de l'Académie des Sciences: Paris, Savy, 96 p.
Gaudry, A., 1878, Animaux fossiles du Mont Lubéron (Vaucluse): Étude sur les Vertébres: Paris, Libraire de la Société Géologique de France, 180 p.
Gaudry A., 1878–1890, Les enchainements du monde animal dans les temps géologiques: Paris, F. Savy, 3 vols. (vol. 1: 293 p.; vol. 2: 317 p.; vol. 3: 336 p.).
Gaudry, A., 1893, L'éléphant de Durfort, in Centenaire de la fondation du Muséum d'histoire naturelle, 10 juin 1793–10 juin 1893, volume commémoratif: Paris, Museum d'histoire Naturelle, p. 328–347.
Gaudry, A., 1896, Essai de paléontologie philosophique: Paris, Savy, 230 p.
Gaudry, A., 1898, le Nouveau musée de paléontologie: Revue des Deux Mondes, 4e période, tome 149, p. 799–824.
Geoffroy Saint-Hilaire, E., 1831, Recherches sur des grands sauriens trouvés à l'état fossile vers les confins maritimes de la Basse-Normandie, attribués d'abord au crocodile, puis déterminés sous les noms de Téléosaurus et de Sténéosaurus: Paris, F. Didot frères, 138 p.
Gohau, G., 1990, A History of Geology: New Brunswick, New Jersey, Rutgers University Press, 259 p.
Hamy, E.T., 1893, Les derniers jours du Jardin du Roi et la fondation du Muséum d'histoire naturelle, *in* Centenaire de la fondation du Muséum d'histoire naturelle, 10 juin 1793–10 juin 1893, volume commémoratif: Paris, Muséum d'histoire Naturelle, p. 1–162.
Laissus, Y., 1995, Le Muséum d'histoire naturelle: Paris, Gallimard, 144 p.
Laissus, Y., 1998, L'Egypte, une aventure savante: Paris, Fayard, 611 p.
Lamarck, J.-B., 1801, Système des animaux sans vertèbres: Paris, Déterville, 467 p.
Lamarck, J.-B., 1809, Philosophie zoologique: Paris, Dentu, 2 vols.
Lamarck, J.-B., 1815–1822, Histoire naturelle des animaux sans vertèbres; reprinted by Forgotten Books, 2018, v. 3, 764 p.
Laurent, G., 1987, Paléontologie et évolution en France, 1800–1860. De Cuvier-Lamarck à Darwin: Paris, Éditions du Comité des travaux historiques et scientifiques, 553 p.
Laurent, G., 1997, Lamarck, 1744–1829: Paris, Éditions du Comité des travaux historiques et scientifiques, 760 p.
Limoges, C., 1980, The Development of the Museum of Natural History of Paris, 1800–1914, *in* Fox, R., and Weisz, G., eds., The Organization of Science and Technology in France 1808–1914: Cambridge, UK, Cambridge University Press, p. 211–240.
Loison, L., 2012, Le projet du Néolamarckisme français (1880–1910): Revue d'Histoire des Sciences, v. 65, no. 1, janvier–juin 2012, p. 61–79.
Museum, 1893, Centenaire de la fondation du Muséum d'histoire naturelle, 10 juin 1793–10 juin 1893, volume commémoratif: Paris, Muséum d'histoire Naturelle, 571 p.
Nieuweland, A., 2010, The Colossal Stranger. Andrew Carnegie and *Diplodocus* intrude European Culture, 1904–1912: Endeavour, June, v. 34, no. 2, p. 61–68.
d'Orbigny, A., 1849, Cours élémentaire de paléontologie et de géologie stratigraphique: Paris, Masson, 310 p.
Outram, D., 1984, Georges Cuvier: Vocation, Science and Authority in Post-Revolutionary France: Manchester, UK, Manchester University Press, 299 p.
Pinna, G., 1997, Fondamenti teorici per un Museo di Storia Naturale: Milan, Jaca Book, 152 p.
Pinna, G., 1998, Scientific Research versus Public Exhibits: A schizofrenic aspect of Natural History Museums: http://giovanni.pinna.info/exhibits.html.
Richard, N., 1997, Des Dîners Lamarck au monument: La construction d'une mémoire, *in* Laurent, G., dir., Lamarck, 1744–1829, Editions du Comité des Travaux Historiques et Scientifiques, p. 631–649.
Roger, J., 1989, Un philosophe au jardin du roi: Paris, Fayard, 645 p.
Rudwick, M., 1997, Georges Cuvier, Fossil Bones and Geological Catastrophes: Chicago, University of Chicago Press, 318 p.
Rudwick, M., 2005, Bursting the Limits of Time: The Reconstruction of Geohistory in the Age of Revolution: Chicago, University of Chicago Press, xxiv + 708 p., https://doi.org/10.7208/chicago/9780226731148.001.0001.
Russell, E.S., 1916, Form and Function. A Contribution to the History of Animal Morphology: London, J. Murray, 396 p.
Schnitter, C., 1996, Le développement du Muséum national d'histoire naturelle de Paris au cours de la seconde moitié du XIXe siècle: se transformer ou périr: Revue d'Histoire des Sciences, v. 49, no. 1, p. 53–98.
Sloan, P., 1997, Le Muséum de Paris vient à Londres, *in* Blanckaert, C., Cohen C., et al., Le Muséum au premier siècle de son histoire: Paris, ed. du Muséum National d'Histoire Naturelle, 687 p.
Spary, E., 2000, Utopia's Garden, French Natural History from Old Regime to Revolution: Chicago, University of Chicago Press, xv + 321 p., https://doi.org/10.7208/chicago/9780226768700.001.0001.
Taquet, P., 2006, Cuvier, Naissance d'un génie: Paris, Éditions Odile Jacob, 539 p.
Tassy, P., 2006, Albert Gaudry et l'émergence de la paléontologie darwinienne au xixe siècle: Annales de Paléontologie, v. 92, no. 1, p. 41–70, https://doi.org/10.1016/j.annpal.2005.09.003.
Van Praet, M., ed., 2005, La Grande galerie de l'Évolution: Paris, Éditions du Muséum National d'Histoire Naturelle, 94 p.
Vollosovitch, N.K., 1914, The Mammoth from Bolchï Likhovskij Island, a Geological Sketch: Zap. Mineral. Obsch. 2nd series 50, p. 305–339.
Whewell, W., 1837, History of the Inductive Sciences: From the Earliest to the Present Times: London, Parker, 3 vols. (vol. 1: xxxii + 394 p.; vol. 2: xii + 488 p.; vol. 3: xiv + 563 p.).
Winsor, M., 1990, Reading the Shape of Nature: Comparative Zoology at the Agassiz Museum: Chicago, University of Chicago Press, 342 p.

Manuscript Accepted by the Society 5 February 2018
Manuscript Published Online 10 July 2018

The Geological Society of America
Special Paper 535

History and importance of the geoscience collections at the Natural History Museum Vienna

Christian Koeberl*
Natural History Museum Vienna, Burgring 7, 1010 Vienna, Austria, and
Department of Lithospheric Research, University of Vienna, Althanstrasse 14, 1090 Vienna, Austria

Franz Brandstätter*
Department of Mineralogy and Petrography, Natural History Museum Vienna, Burgring 7, 1010 Vienna, Austria

Mathias Harzhauser*
Department of Geology and Paleontology, Natural History Museum Vienna, Burgring 7, 1010 Vienna, Austria

Christa Riedl-Dorn*
Department Archive for the History of Science, Natural History Museum Vienna, Burgring 7, 1010 Vienna, Austria

ABSTRACT

The Natural History Museum Vienna is one of the most important museums of natural history in the world. Its collections date back to the year 1750, when the Emperor Franz Stephan of Lorraine (Franz I. Stephan) purchased (from Italy) what was then the largest and most famous collection of natural history specimens. The meteorite collection of the Natural History Museum in Vienna, Austria, has the longest history of all comparable collections in the world. In the second half of the eighteenth century, soon after the foundation of the Imperial Natural History Cabinet in 1750, the Viennese curators began to collect meteorites. Although the first curators neither believed in the extraterrestrial origin nor accepted—in several cases—the written and witnessed histories of these allegedly "heavenly" stone and iron masses, they preserved them in the Natural History collection. Among the first acquisitions were the historical important meteorites Hraschina (Agram), Tabor, Krasnojarsk (Pallas Iron), and Eichstädt. These and other well-documented specimens from the Vienna collection were, for example, used by E.F.F. Chladni for his seminal treatises of 1794 and 1819, respectively. The central figure in the early history of the collection is Carl von Schreibers (1775–1852). After the fall of the Stannern meteorite in 1808, he availed himself of every opportunity to acquire meteorite specimens. His continued interest in meteorites laid the foundation for the Vienna collection to be of the historical and scientific importance it is today. Due to the efforts of Schreibers, who also is regarded as founder of meteoritic science in Vienna, and his successors,

*E-mails: christian.koeberl@univie.ac.at; franz.brandstaetter@nhm-wien.ac.at; mathias.harzhauser@nhm-wien.ac.at; christa.riedl-dorn@nhm-wien.ac.at.

Koeberl, C., Brandstätter, F., Harzhauser, M., and Riedl-Dorn, C., 2018, History and importance of the geoscience collections at the Natural History Museum Vienna, *in* Rosenberg, G.D., and Clary, R.M., eds., Museums at the Forefront of the History and Philosophy of Geology: History Made, History in the Making: Geological Society of America Special Paper 535, p. 131–161, https://doi.org/10.1130/2018.2535(09).

the Vienna collection became the largest and most extensive in the course of the nineteenth century.

In terms of the geological and paleontological collections, early expeditions and collecting campaigns were mainly targeting exotic animals and plants, while paleontological objects were welcome but subordinate. It was only in the early nineteenth century that the paleontological collections were—literally and figuratively speaking—systematically enlarged. Internationalization and diversification became the focus of the collection strategy. The paleontology collections at the Vienna museum also became important in the Darwinian view of evolution.

INTRODUCTION: THE NATURAL HISTORY MUSEUM VIENNA

The Natural History Museum Vienna (NHMW) is one of the most important museums of natural history in the world. Its collections date back to the year 1750, when the Emperor Franz Stephan of Lorraine (Franz I. Stephan) (1708–1765) purchased what was then the largest and most famous collection of natural history specimens in the world from the Florentine scholar Jean de Baillou (1684 or 1686–1758). Following the death of Franz Stephan in 1765, Maria Theresia (1717–1780) decided to give this collection to the state. The objects were displayed in a wing of the imperial palace and opened to the public twice a week. The collection grew steadily through further activities, above all from many overseas expeditions. This led to a shortage of space in the imperial palace. In the course of the revolution in 1848, parts of the very crammed collections were burned.

After the demolition of the old city walls in 1858, initiated by Emperor Franz Josef I (1830–1916), the court museums were conceived as one part of the development of the Ringstrasse (the road encircling Vienna's most central district). For the first time, all collections could be accommodated in one large building in the 1st district of Vienna. The building offered enough room not only for the safekeeping and study of the specimens but also for their public display. At its opening in 1889, the Natural History Museum of Vienna thus became one of the first museum buildings of its kind in the world. Despite severe setbacks caused by the two world wars, it has been possible to preserve both the building itself and the valuable collections, which today consist of ~30 million collection numbers and many more individual items.

The historical building (Fig. 1) and its fixtures and fittings of course present a variety of problems today, not only in preparing new exhibitions and educational events, but also, for example, in meeting modern requirements for working conditions, fire protection, and access for the disabled. Naturally, the leaders, curators, and also the Austrian government as owner can and must address these challenges today. However, it is precisely these historical considerations that make the Natural History Museum Vienna so attractive, in contrast to modern science centers that have no historical furnishings or extensive collections of this kind. The Natural History Museum Vienna is today an important center of knowledge for questions related to natural science and one of the largest non-university research institutions in Austria. It houses some modern research laboratories, such as state-of-the-art electron microprobe, electron microscopy, and DNA facilities. The museum has a total staff of ~330 people, with its research departments being home to ~60 scientists (plus project staff) carrying out basic research in a wide variety of fields of earth science, biology, and human sciences. The museum scientists publish (or are involved in) over 200 peer-reviewed scientific publications every year, participate in dozens of Austrian and international third-party–funded research projects, host hundreds of scientific visitors from all over the world who use the collections and facilities, organize and participate in scientific meetings, and are involved in teaching classes at or hosting students from most of the Viennese and Austrian universities (even though in terms of administration, the museum belongs to the portfolio of the culture ministry in Austria, whereas the universities are affiliated with the science ministry).

Research departments and collections at the museum include anthropology, botany, geology, meteorites, mineralogy, petrology, paleontology, prehistory, and zoology, as well as an archive and extensive libraries. Several technical departments (such as exhibits, education, outreach, ecology, library, event management, communications, and media), scientific laboratories, and taxidermy facilities complete the portfolio of the museum. The museum has three external branches: one is in Hallstatt (Upper Austria), where NHMW staff has performed archeological excavations for more than 50 years. The ecological station in Petronell (Lower Austria), near the River Danube, allows hundreds of groups and school classes to participate in a variety of ecological courses and excursions to the Danube wetlands. The pathological-anatomical collection (a collection of objects—mostly human anatomical and pathological specimens, wax models, and antique medical equipment) in the so-called "Narrenturm" building in the 9th district of Vienna is part of the anthropological department of the NHMW. In 2016, despite continuing interior renovations, more than 36,000 visitors were counted at the "Narrenturm." The collections there are the largest of their kind in the world, and updated exhibits are planned for when the renovations of the building are completed.

In 2017, the Natural History Museum Vienna counted more than 750,000 visitors, making it one of the most visited museums in Austria. Over the past several years, the permanent exhibits on dinosaurs, meteorites, and prehistory were thoroughly renovated (in many cases, for the first time in decades) and reopened to the public in 2011, 2012, and 2015, respectively. In addition, a new permanent exhibit on anthropology (focusing on the origin and evolution of humans) was established in 2013. Additional smaller adaptations to the permanent exhibits were made in the mineralogical and zoological halls. For example, in mineralogy, the collection of building stones and decorative stones (the largest at least in Europe) was presented in a modern new exhibit space, and just recently, in April 2017, some showcases that present introductory mineralogy were refurnished and modernized and now show the concept of "mineral evolution," i.e., how the abundance and composition of minerals on Earth changed with time as a result of changing environmental conditions (cf. Hazen et al., 2008).

In all cases of modernizing permanent exhibits, several guiding concepts were used, besides renovating the historic showcases (where available and necessary) or adding fitting new showcases and installing state-of-the art lighting and alarm systems. Brief but specific information (often lacking from old displays) explaining the general importance of objects or collections and their scientific value was included. The philosophy of the NHMW is that at the center of attention should be the objects themselves. In some cases, important selections were made as opposed to earlier attempts to show as many objects (often without explanation) as possible. All these newly designed exhibits turned out to be major attractions for the public (and for experts and students as well). In late 2014, as part of the 125th anniversary celebrations of the museum building, a digital planetarium was installed in Hall 16. This new facility features full-dome projection technology and supplements the outreach facilities of the museum. In summary, the Natural History Museum Vienna has not only a long history and tradition of its very valuable collections, but also is a major and modern research institution and a very attractive place for the public to learn about and be entertained by the importance of the natural sciences.

The following sections describe the general (early) history of the establishment of the collections, followed by the history and importance of the meteorite collection (the oldest in the world, the largest meteorite display in the world, and one of the top three or so in terms of importance and numbers), and the role of the paleontological collections for the concept of evolution.

Figure 1. The Natural History Museum Vienna today with the monument of Maria Theresia.

EARLY HISTORY OF THE "NATURALIENSAMMLUNG"

The Early Years of Franz I. Stephan's Collections Told through a Painting

Before describing how the early natural sciences collections in Vienna were established, we start with an illuminating story. The famous *Kaiserbild* (a painting showing the emperor), located in the grand staircase of the Natural History Museum, depicts the Emperor Franz I. Stephan (seated) with the directors of the imperial collections. The collections are not only represented by their directors but also by objects that are either held or are standing next to the collections directors. At the center of the painting is the director of the "Cabinet des Medailles et des Monnaies" (Cabinet of Medals and Coins), Valentin Jamery Duval (1695–1775), who is holding a drawer with gold coins and medals; standing to the left of Duval is the director of the "Cabinet des Curiosités Naturelles" (Cabinet of Natural Curiosities), Jean de Baillou, wearing the dress uniform of an artillery officer in the rank of lieutenant-colonel and holding an ammonite; the director of the "Cabinet des Machines" (Physical and Astronomy Cabinet), Abbé Jean François de Marcy (1707–1791), is positioned in the lower right corner, standing next to a globe and a pair of compasses, and the court physician and prefect of the imperial library, Gérard van Swieten (1700–1772), is standing in the lower left side of the painting and holds a book (see Fig. 2).

According to the inscription beneath the painting, it was made eight years after the emperor's death and was considered by contemporaries to be one of the best portraits of Franz I. Stephan. It even served as reference for the glass painting created for the "Lothringersaal" of Laxenburg Castle in 1830 (Wandruszka, 1962). The painting was commissioned by Maria Theresia, the widow of Franz I. Stephan, was executed by the court painters, Franz Messmer (1728–1773) and Jakob Kohl (1734–1788), and was completed in 1773 (Lhotzky, 1941–1945). Based on an entry in the *Geheimen Kammerzahlamtsbücher* (privy chamber payments office books), Messmer had received 300 ducats in 1767 for the production of a painting depicting the emperor with the court physician van Swieten, Abbé Marcy, Duval, and Baillou in the "Naturalien-Cabinet" (Fleischer, 1932). According to Fitzinger (Fitzinger, 1856), Messmer, one of the premier portraitists of his time, was responsible for painting the heads, while the remaining parts such as clothes, background, etc., were done by Jakob Kohl (Lhotsky, 1941–1945). Among the portrayed persons, only Duval, who returned to the Netherlands in 1765, would have been able to see the finished work, because the other directors and the court physician died several years before the completion in 1773; incidentally, 1773 also is the year of Messmer's death. The furniture shown in the painting partly sprang from the artist's imagination, and the Naturalien-Cabinet wasn't moved to the Augustinergang of the imperial palace until after the emperor's death (Stütz, 1807; Fitzinger, 1868b; Lhotsky, 1941–1945). Some objects and specimens that served the painters as models such as the emerald held by Franz I. Stephan and the snails and shells on the table and in the cabinets were part of Jean de Baillou's original collection, which was bought by the emperor. Some other objects, such as the quartz crystal lying at the feet of Abbé Marcy, were added ca. 1935 nearly one and a half century after the painting was completed. The quartz crystal, which is still part of the museum's mineralogical collections, was most likely added by conservators to cover up damage to the painting. A historical drawing of the painting made by A. Obsieger in 1859, as well as historical photographs and movies from before 1935, show a bare floor where the crystal now lies (Figs. 3 and 4). In the course of the restoration and conservation work done on the painting in 1992, infrared and X-ray analyses were conducted and revealed five men who were painted over (Figs. 5 and 6). One of the persons wears a clerical collar, which leads to the conclusion that the person portrayed had been a member of the Jesuit order; another person was identified as a servant (Riedl-Dorn, 1998). The clergyman was the only fully completed figure; the other four people were in different stages of completion, mostly sketches with some paint. Using the technique of image matching, Christa Riedl-Dorn and Nora Pärr were able to identify the Jesuit as Maximilian Hell (1720–1792), the first director of the University observatory in Vienna (Fig. 7) (Pärr, 2013).

Maximilian Hell was invited in 1769 by the king of Denmark to observe the Venus transit. Following the invitation, he traveled, together with his assistant, the linguist János Sajnovics, to Vardø in Norway, and then to Denmark and Norway. The Venus passage enabled Hell to calculate the distance between the sun and the moon. He published his findings in the book, *Observatio Transitus Veneris Ante Discum Solis Die 3. Junii Anno 1769,* which was financed by the Royal Danish Academy of Sciences and Letters. Meanwhile the mood turned against the Jesuits, culminating in a papal decree in 1773 effectively dissolving the order. Hell was posthumously accused by the later dean of the University of Vienna, Carl Ludwig von Littrow (1811–1877), of having falsified the data; however, later research has absolved Hell of that accusation. A strong argument can be made that the dissolution of the order and the emperor distancing himself from its members might be the reason that Hell was painted over (Riedl-Dorn, 1998). A fact that gives some credence to this theory is that the final dissolution of the order coincides with the completion of the painting. The emperor was also very much in favor of the dissolution because the property and possessions of the order went, among others, to the state and the emperor's coffer. The vast scientific collections of the Austrian Jesuits were divided among different institutions and museums.

The anachronism the painting inadvertently brings to light also characterizes the historical accounts of the foundation of the natural history collection. Similar to the painting, things have been painted over, added, and/or distorted and could only be revealed and corrected by intensive research.

Figure 2. Painting by Franz Messmer and Jakob Kohl of Franz I. Stephan with the directors of his collections and the court physician; the painting is called the *Kaiserbild* and was created in 1773. Standing at the center of the painting is the director of the "Cabinet des Medailles et des Monnaies" (Cabinet of Medals and Coins), Valentin Jamery Duval; to the left of Duval is the director of the "Cabinet des Curiosités Naturelles" (Cabinet of Natural Curiosities), Jean de Baillou. The director of the "Cabinet des Machines" (Physical and Astronomy Cabinet), Abbé Jean François de Marcy (1707–1791), is positioned in the lower right corner; and the court physician and prefect of the imperial library, Gérard van Swieten (1700–1772), is standing at the lower left side behind the Emperor, who is sitting in a chair.

Precursors of the Natural History Collections

The Habsburg rulers of the early modern age were already interested in natural sciences and had their own natural history collections. These early collections had more in common with curiosity cabinets than natural sciences collections. The cabinets of Ferdinand I (1503–1564) and Rudolph II (1552–1612) are still known today. A common characteristic of the cabinets was to show everything that god and men had created and make it available to interested people for research. Antiquities, artwork, armor, and automatons were exhibited next to astronomy tools, minerals, corals, animals, and strange objects from nature. An

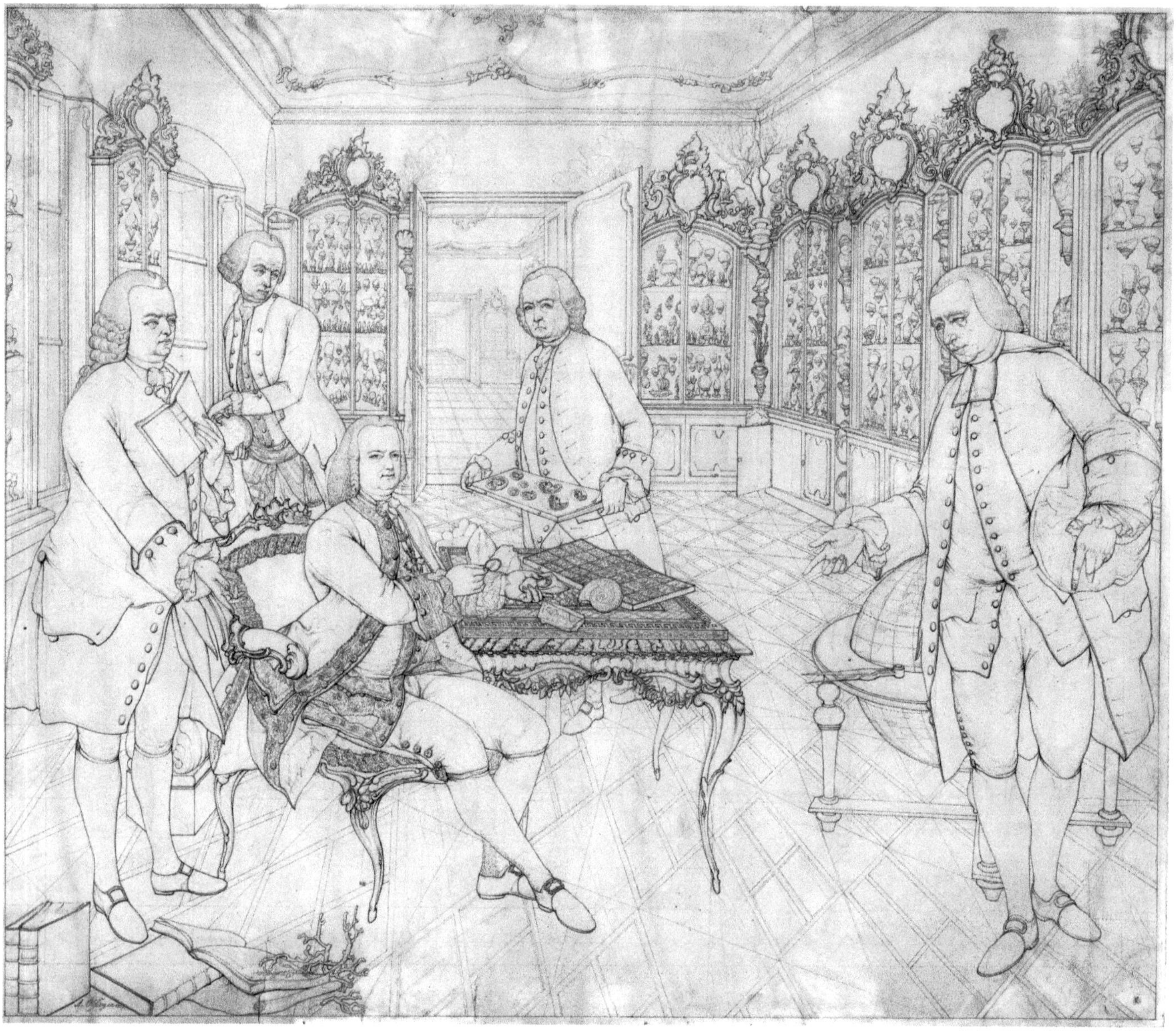

Figure 3. A. Obsieger's copy of the *Kaiserbild,* 42.7 × 50.5 cm, pencil drawing, 1859; the drawing shows the *Kaiserbild* without the quartz crystal at the feet of Abbé Marcy.

example of a "Kunst-und Wunderkammer" (Cabinet of Art and Wonders) that still exists is the collection of Archduke Ferdinand II (1529–1595) of Tyrol located at Schloss Ambras. A large number of objects from the Ambras Cabinet, such as the emerald stage from Colombia or an ammonite referred to as "Schlangenstein" (serpent stone), are kept at the Natural History Museum Vienna (Fig. 8). The first natural history collections mentioned in relation to the "Kaiserliche Kunst-und Schatzkammer" (Imperial Art Cabinet and Treasury) in Vienna were those of Emperor Maximilian II (1527–1576). A report about the "kayserliche Schatz und Kunst Cammer" from 1730 (Küchelbecker, 1730) describes the cabinet as housing all kinds of curiosities, red corals, artwork made from corals, as well as horns from different species of land- and water-based "unicorns." While the aristocrats had their "Kunst-und Wunderkammer," which were geared toward the extraordinary, scholars, pharmacists, and physicians had their own reference collections that they used for education and for scientific studies; these collections were referred to as "Studioli." These collections were often precursors to natural history cabinets. Even after Franz I. Stephan founded an official natural history collection, "Naturalien-Cabinet," extraordinary objects such as "Aeroliths" (meteorites) were kept in the treasury instead of the natural history collection. An example is the iron meteorite of Hraschina, which fell on 26 May 1751 and was presented to the emperor, who had it transferred to the treasury. The meteorites were moved from the treasury to the "Naturalien-Cabinet"

Figure 4. Photograph of the *Kaiserbild,* 1905. This historic photograph of the *Kaiserbild* shows the painting without the quartz crystal at the feet of Abbé Marcy.

in 1778, 13 years after the emperor's death (see section "On the Early History and Importance of the Meteorite Collection at the Natural History Museum Vienna").

Emperor Franz I. Stephan's Collections

Three of the collections represented on the painting by their directors were established by Franz I. Stephan, who was already familiar with scientific collections from his time in Lorraine and Florence where scientific collections had already been established for quite some time. The three collections were (1) the "Physikalisch-Astronomisches Kabinett" (Physical and Astronomy Cabinet), headed by the Jesuit priest and mathematician, Abbé Jean Francois de Marcy, which housed globes, automatons, machines, and astronomical instruments, such as a heliocentric mechanical planetarium; (2) the "Münzkabinett" (Cabinet of Medals and Coins), headed by the well-known numismatist, Valentin Jameray-Duval, which housed coins and medals; and (3) the "Naturaliensammlungen" (natural history collection), headed by Jean de Baillou. According to Fitzinger, all three cabinets were founded in 1748; however, the founding date of the natural history collection does not hold up to close scrutiny. The famous historian, Alfons Lhotzky (1903–1968) had doubts about the founding date of the "Naturaliensammlungen" established by inhouse historian Fitzinger (Lhotzky, 1941–1945), who considered the acquisition of the natural history collection of the Florentine nobleman, Jean de Baillou, by Franz I. Stephan, the starting point or beginning of the "Naturaliensammlung." In an article about the history of the "Hof-Mineraliencabinet" (imperial minerals cabinet), Ferdinand von Hochstetter, the first "Intendant" of the "k.k. Naturhistorisches Hofmuseum" mentioned 1747 (Hochstetter, 1884) as the founding year of the "Naturalien-Kabinett." A diary entry of Prince Johann Joseph Khevenhüller (1706–1776) from 29 January 1752 says: "… I had the honor of visiting the Naturalien-Cabinet with the emperor who had bought the cabinet one year ago from M.[onsieur] Baillou in Florence…." (Lhotzky, 1941–1945; Riedl-Dorn, 1998). The diary entry suggests that the cabinet was acquired in 1751 and not in 1748 as Fitzinger claimed. Also wages for the staff off the "Cabinet des Machines" (Physikalisch Astronomisches Cabinet) and for the staff of the "Cabinet des Curiosités Naturelles" (Naturalien-Cabinet) aren't listed separately in the expenditure list of the imperial household until 1752 (ÖStA/HHStA, Posch-Akten, jüngere Serie, Karton 8).

The Acquisition of Chevalier Jean de Baillou's Collections by the Emperor

Facts about the history of the Natural History Museum: On 11 January 1749, Emmanuel de Navy Count Richecourt (1697–1759) in Florence received a dispatch (ASF Consiglio di Reggenza, Fasz. 6: Depeches au Conseil des Finances de l'anneé 1745, fol 82r.) stating the desire for acquiring the "cabinets de Jean de Baillou." Count Richecourt should make an offer over 40,000 scudi, but if necessary, the offer could be raised. On 28 January 1749, he wrote in a letter to his "sacré Majesté" (ASF, Consiglio die Reggenza 23 Carta 27) that he had made an offer to Jean de Baillou to buy his cabinet but that Baillou wanted first to seek advice before committing to the sale. Duval acted most likely as negotiator for this sale (Tromballa, 1953).

Under the budget item "unsafe and extraordinary expenses" in the *Ristretto dell' Incassato, e Speso dalla Depositeria generale de S.M.J. in Firenze in un anno dal primo gennaio a tutto dicembre 1749,* we find the entry 42,000 scudi for "Cav. Gio. Baillou per conto del Museo" (ASF Consiglio di Reggenza 24, p. 287 "Per causa ordinaria" "Spese incerte e Straordinarie"), and between January and end of July of 1750, The *Ristretto dell'incassato* shows expenses of 21,000 scudi for "Cav. Gio. Baillou per conto del Museo" (ASF Consiglio di Reggenza 24, p. 286, *Ristretto dell'Incassato in mesci setti—cive dal primo gennaio a tutto luglio 1750).* Instructions given on 3 October 1750 state that Baillou is to be paid 42,000 scudi right away (ASF Consiglio di Reggenza, 24 p 295r Dimonstrazionen della differnza che si ved dal Resultato de le Ristrello dell Entrata e Spese della Deposita S. equile nellánno 1749. …). Apart from the purchase price, it was agreed that Chevalier de Baillou would

Figure 5. Detail of an X-ray image of the *Kaiserbild.* The X-ray was made in the course of restoration work done to the *Kaiserbild* in 1992 and shows a painted-over person who was later identified as Maximilian Hell.

receive the post of director of the newly founded "Mineralien-Cabinet" and that the directorship should be hereditary in his family and to be transferred to the eldest son upon death or retirement (Born, 1780; Haidinger, 1782). Franz I. Stephan showed interest in the natural sciences even before the acquisition of the Baillou collections; he was especially interested in experiments for which he had bought a large number of diamonds (Grauer, 1932). He also acquired one of the largest burning mirrors of that time (Stütz, 1807) to conduct his experiments about the combustibility of diamonds, which, according to Abbé Andreas Xaver Stütz (1747–1806), was proven without a doubt, when the emperor burned diamonds worth several thousand guilder in 1751 (Stütz, 1807). Franz I. Stephan and his brother Carl (1712–1780) enclosed diamonds and rubies in a pyramid-shaped earthen jar and left the jar in a strong fire for 24 hours. They observed that the diamonds had vanished, whereas the rubies remained unchanged (Bauer, 1883). Other sources tell of experiments Franz I. Stephan had conducted together with the Jesuit and professor of mathematics and astronomy of the University of Vienna, George Joseph Franz (1704–1776), at the chemistry laboratory of the Jesuits. The goal of the experiments was to find whether it was possible to create a large diamond or emerald by melting several small diamonds or small emeralds using focused sunlight (Fitzinger, 1856). The priest, Georg Joseph Franz, was also responsible for establishing a Jesuit museum and the laboratories connected to the museum. Together with Maximilian Hell, he founded the first public observatory in Vienna in 1755. He also headed the "Orientalische Akademie" for several years. The mineralogical department of the Natural History Museum Vienna possesses some burned diamonds that were thought to be connected to Franz I. Stephan's experiments, but modern research has cast some doubt on this assertion. Johann Georg

Figure 6. Photograph of the *Kaiserbild* with delineated contours of the painted-over persons who were discovered in the course of restoration work done to the painting in 1992.

Megerle von Mühlfeld (1780–1831) mentions 1807 in his supplement to Stütz (Stütz, 1807) that the "Mineralien-Cabinet" has a diamond that was transformed into carbon from the collection of Cosimo III de' Medici (1642–1723) in its holdings. Megerle refers to the diamond as monumental to the beginning of the era of scientific studies. Cosimo III had conducted research on the effects of using the focused beam of a burning mirror on a diamond. The interest in conducting experiments using burning mirrors didn't end with Franz I. Stephan's death. On 8 March 1775, Maximilian Hell, Nicolaus Joseph von Jacquin (1727–1817), Georg Joseph Franz, and Joseph Anton Nagel (1717–1804) (ÖStA/HHStA, Sonderreihe, Karton 371, Mappe 8, Bericht) assessed an expensive burning mirror that the University of Vienna had considered buying.

The First Director of the Natural History Collections, Chevalier Jean de Baillou

How did Jean de Baillou, who was often portrayed as a lieutenant colonel, become director of the "Naturalien-Cabinet"? Jean de Baillou was born either 1684 (Blöchlinger, 1868) or 1686 (Bergmann, 1856; Tromballa, 1953) as son of Marguerite de Gonet and Sebastian de Baillou in France or Italy (Fitzinger,

Figure 7. Johannes Esaias Nilson (1721–1788), Portrait of Maximilian Hell, engraving, ca. 1770.

Figure 8. Ammonite *coroniceras rotiforme* from the Ambrasian Collection.

1856). The family moved from France to Italy. The family can be traced back to the thirteenth century and originally stems from a small village in Flanders (ASF, Deputazione sopra la nobiltá e cittadinanza Fasc. 18). Jean studied mathematics and natural sciences in Paris. Prior to that, he worked for the prince of Lorraine-Vaudemont, who had him instructed in mathematics, riding, and fencing. In 1710, he married the Marchesa Monti della Scrivia, who bore him five daughters and three sons.

Sometime prior to 1718, he moved to the court of Duke Francesco Farnese in Parma (1678–1727). He started out as court architect, and in 1725, he received the positions of general commissioner of the artillery of the duchy of Parma and general engineer, as well as general director of the mining operations. Following Francesco Farnese's death in 1727, Baillou continued to serve Francesco's brother, Antonio. In 1731, he moved to Florence to serve the last Medici Grand Duke Gian Gastone. In 1735, he was made director of the famous Uffizi galleries, and in 1736, he advanced to director of all gardens, fortresses, buildings, and mines in the Grand Duchy of Tuscany. After Gian Gastone's death in 1737, Francis of Lorraine, the later emperor Franz I. Stephan, succeeded him as Grand Duke of Tuscany. Francis confirmed Baillou's positions and dignities and made him a member of the Habsburg-Lorraine court.

During this time, wonders about the knowledge and skills of Baillou are described (Mambriani, 1994), especially a strange little book written by a former "student" of his (Joannon de Saint-Laurent; see below); the book extols the virtues and knowledge of Baillou. According to the book, Baillou was not only a brilliant physicist who could recite Newtonian laws of optics as if he had written them himself, but he was also a brilliant inventor without an equal in Italy. It is not clear if the description of Baillou's achievements in the field of optics is the result of the author confusing Jean de Baillou with his brother François, who worked as an optician in Milan and was honored in 1750 by Empress Maria Theresia with the title of "Regio Cesare Optico" (court optician; Lualdi, 1996). Furthermore, Jean de Baillou's inventions were usually in the field of mechanics. His best known creation by all accounts was the "magic grotto" of Colorno, a series of automatons retelling stories from Greek mythology. Little is known about the author of the book, Joannon de Saint-Laurent (1714–1783); he seems to have been an adventurer, philosopher, archaeologist, physician, and naturalist. He was born in Lyon and later moved to Italy. In 1745, he came in contact with the literary circles of Florence, where he most likely met Baillou (Garms-Cornides, 1999). In 1746, Saint Laurent had offered Baillou the opportunity to produce an inventory of his collections, but Baillou declined. In the same year, Joannon de Saint Laurent wrote the *Déscriptions abregée du fameux cabinet des M. le Chevalier de Baillou pour servir a l'histoire naturelle des pierres précieuses, métaux, minéraux et autres fossiles (A short description of the famous cabinet of M. the Chevalier de Baillou to function as guide to the natural history of precious stones, metals, minerals, and other fossils)* without Baillou's knowledge or authorization. Through the intervention of Count Leopold Anton von Firmian (1679–1744), whom he also met in Florence, he was employed in the Lombardian administration.

Baillou had started his natural history collection while working for Francesco Farnese in Parma. The foundation of the cabinet is probably related to Baillou's position as general director of the mining operations as which he had to conduct scientific research in the mines. Over time, the collection grew to over 30,000 objects covering a wide range of themes. The collection included a grain of sand, artificial gemstones, mineral samples, fossils, snails, seashells, and a large number of corals. The collections became famous all over Europe and were visited by noblemen and scientists alike. The electoral prince of Saxony was an often-seen guest and was especially interested in experiments on magnetism. The prince supposedly told Baillou that Baillou was the true interpreter of nature (Blöchlinger von Bannholz, 1864). Unlike other natural history cabinets, Baillou's collections were not meant to show off the owner's wealth but to educate and fight ignorance and superstition. In the previously mentioned book by Joannon de Saint-Laurent, the author praises the principles of arrangement of the objects, based on an assumed nature-immanent system and the use for mankind. According to Saint-Laurent, the principles were also supposed to serve as a basis for a series of illustrated books about natural history written by Jean de Baillou under the title, *Traité universel des pierres précieuses, métaux, minéraux et autres fossiles, le tout distribué dans ses propres classes. Où l'on trouve leurs diférents noms, leurs origines, leurs analises, leurs principes, leurs qualités, leurs étimologies, leurs choix, et tout ce qui'il y a de particulier et remarquables en les exposant au foier de al lentille astronomique et du mirroir ardent. On y traite aussi des plantes marines pierreuse, crustacés et coquillages de mer, soit dans leur état naturel, soit en diférens degrés des Calcination et de pétrification. On donne a la fin: La maniere de travailler toutes sortes de pierres, et celles de les imiter par les voies le plus courtes et dans leur plus grande perfection* (Saint-Laurent, 1746) (*Universal treatise on precious stones, metals, minerals and other fossils, all distributed in its own classes. Where we find their different names, their origins, their analisms, their principles, their qualities, their etimologies, their choices, and all that is particular and remarkable by exposing them to the foil of the astronomical lens and the burning mirror. It also treats stony marine plants, crustaceans and seashells, either in their natural state, or in different degrees of Calcination and petrification. At the end, the manner of working all sorts of stones, and those of imitating them by the shortest and most perfect ways*). For whatever reason, the books never had been published (Lualdi, 1996).

The cabinet was divided into three thematic groups: *pétrifications* (fossils), *métaux* (metals), and *pierreries* (stones), which comprised 24 collections. The collections were subdivided into classes. The term classes is used by Baillou in a sense different from Linné's system and describes units. The earth pigment ochre, for example, was grouped with the *pierreries* (stones) and could be found under the class *les terres des peintres* (soil used by painters) which was part of the *Collection: les terres* (soils), *les sables* (sands), and *les graviers* (gravel), and another example of fossilized madrepore corals was grouped with the *pétrifications* (fossils) and could be found under the class of *Madrepores petrifiés* (fossil madrepore, coral), which was part of the *Collection les plantes marines petrifiés* (marine plant fossils).

The "Naturalien-Cabinet"

The location of Franz I. Stephan's "Naturalien-Cabinet" is not exactly known. Likewise unknown is how long the natural history collections were stored at the "Kaiserhaus." In 1752, Prince Johann Joseph Khevenhüller notes that he was allowed to visit the cabinet, which was only half finished at the time and was located close to the library (Riedl-Dorn, 1998, p. 15). The room he had in mind was situated in the right wing of the Hofbibliotheks Building and was the former reading room of the Austrian National Library. In a diary entry for 7 April 1755, he states that the emperor invited him to visit the "Naturalien-Cabinet" and that he spent some time there. Upon arrival of the collections from Florence, Jean de Baillou and his son started to organize the "Naturalien-Cabinet" and supplemented the collection with minerals from the treasury. The objects were arranged in large oak cabinets with glass doors. The structure followed the system Baillou had used in Florence. The emperor is said to have visited the cabinet almost daily.

In order to enlarge the collections, Franz I. Stephan sent out scientists to collect new objects. He also gave orders to the directors of his mines and the mining administration to deliver samples of special rocks and minerals to the imperial collections. Franz I. Stephan's interest in natural science was mainly driven by economic reasons, especially to support dwindling mining activities (Riedl-Dorn, 2001). In 1751, the "hofmathematici," Nagel, and Baillou traveled to the High Tatra Mountains (southern Poland) to collect minerals for the Vienna collection (Schönburg-Hartenstein, 1987). In 1754, the young "médecin hollandaise," Nikolaus Joseph Jacquin, was sent to the Caribbean and South America, together with a gardener and two taxidermists. The Dutch-born Jacquin received precise instructions that were drafted by the emperor himself (Zedinger, 2009), describing what he and the other members of the expedition were to collect. This included living birds and quadrupeds—with the exception of carnivores—for the menagerie, flowers, woods, and other plants for the Dutch gardener, Adrian van Stekhoven (1705–1782), and stones, shells, and fossils for the collection of "M[onsieur] Baillou" (ASF, Segreteria Finanze Affari primo del 1788, Anni precedenti, 479: Toussain an Richecourt 4.11.1754, p. 1).

During the expedition, Jacquin not only collected minerals but also bought minerals and metals from all parts of the world. He was the first to bring platinum to Austria. In 1754, Baillou's reputation was still held in such high regard that the cabinet secretary of Franz I. Stephan, Baron Francois-JosephToussaint (1689–1792), referred to the cabinet not as "cabinet de sa Majesté" but as the cabinet of M. Baillou. Also Jacquin sent his report, *Memoire des Production naturelles pour sa Majesté Imperiale,* to Baillou and not to the emperor. Under the item, "production naturelles," Jacquin lists not only natural history objects but also ethnographical objects and coins that found entry into the collections of the "Mineralien-Cabinet." After Jean de Baillou's death in 1758, Jacquin continued to send reports to Baillou's son, Luigi Balthasar (1731–1802), who had succeeded his father in accordance with the purchase contract as director of the Cabinet.

The Transfer of the Collections and Opening Them to the General Public

Luigi Balthasar, who together with his brother Joseph, was ennobled by Emperor Joseph II on 4 April 1766, was unable to properly manage the collections. After the death of her husband, Franz I. Stephan, in 1765, Empress Maria Theresia had the Naturalien-Cabinet transferred to a single exhibition room and an anteroom in the "Augustinergang" of the imperial palace. The same year, she handed over the collections to the state as public property and placed them under the administration of the "Oberstkämmereramt" (office of the High Chamberlain). In 1766, the famous gemstone bouquet and 57 pietra dura (an inlay technique using precisely cut stones to form an image; the stone pieces are usually much larger than those used for mosaics) from the "Kaiserhaus" in the Wallnerstrasse were transferred to the natural history cabinet (Fig. 9). The costs for the move were borne by the imperial household.

According to Fitzinger, in 1765, the "Naturalien-Cabinet" was opened to selected visitors (scientists and noblemen) twice a week; although no sources could be found to prove that claim. From 1769 onward, the cabinet was opened to artists and members of all classes, including women and children. The changes made to the cabinets were related to the Empress's reorganization of the educational system, which made, among other things, the attendance of schools for children obligatory. Maria Theresia was very interested in enlarging the collection as well as the publication of a catalogue of the "Naturalien-Cabinet." In 1773, Luigi Balthasar asked the Empress that the cabinet be closed for visitors so that he could write the inventory catalogue the Empress had requested, but he also stated that the mineral collection was smaller than the collections of other cabinets and that writing the catalogue was too expensive and would take too much time (Hassmann, 2015). Maria Theresia partly answered Baillou's request and closed the cabinet to the general public. Two years later, the president of the "Hofkammer in Berg- und Münzwesen," Count Franz Kolowrat-Novohradsky (1739–1802), recommended that given the bad condition of the collection and many errors in the inventory *(Inventaire raisonné),* Nicolaus Joseph von Jacquin, who was the first professor for mining and metallurgy at the newly founded mining academy at Schemnitz (Banská Štiavnica in the Slovak Republic) and who possessed a large mineral collection himself, be given the assignment of putting the collection in order and writing a catalogue (ÖstA/HHStA, OkäA Akten Serie B, Karton 4, CCC4), but to no avail. Luigi Balthhasar was once again tasked with the project. After a second negative report in 1776, the task was reassigned to the Transylvanian mineralogist and geologist, Ignaz von Born (1742–1791).

Figure 9. Gemstone bouquet of Maria Theresia. The gemstone bouquet was a gift from Maria Theresia to her husband, Franz I. Stephan, and consists of 2102 diamonds and 761 colored stones; the leaves were made of silk. The bouquet is now part of the mineralogical collections of the museum.

The First Printed Catalogue and Reorganization of the Collections

Born, who had received a three-year contract and was to be paid 2000 florin per annum, suggested that the collection inventory should follow the design of Jacquin's "Flora Austriaca" (Hassmann, 2015). It was planned that the catalogue would be published in several parts with each part describing a single collection. In 1778, the first volume was published under the title, *Index rerum naturalium Musaei Caesarei Vindobonensis Testacea. Vol. 1.* The book was a description of the snails and shells collection and used the Linnean system. It was followed by an illustrated luxury edition, *Musaei Caesarei Vindobonensis Testacea, quae jussu Mariae Theresiae disposuit,* two years later. The book's title was the first instance in which the Latin designation "Musaei Caesarei Vindobonensis" or imperial museum was used for the Naturalien-Cabinet. It is also interesting to note that Born used the English, French, Latin, and German name in the objects description. One of the main points of contention between Kolowrat and Luigi Balthasar was that the latter had written the inventory only in French and not in the language of science—Latin—or in German. Born's contract was extended for another three years in 1780. In 1787, Emperor Joseph II asked for a report about the changes and work done in the cabinet since the beginning of his regency in 1765. Ignaz von Born's reply (ÖStA/HHStA, OKäA Akten Serie B, Karton 9, Nr 41 ex 1787) offers some interesting insights into the inner workings and history of the "Naturalien-Cabinet." According to the report, it seems that Born not only took over the responsibility of cataloguing the collections from Luigi Balthasar but was also tasked by the "Oberstkämmerer" (High Chamberlain), Count Franz Xaver Orsini-Rosenberg (1723–1796), with furnishing a new second exhibition room for the cabinet, which he finished in 1780. The report also describes the acquisition of zeoliths and chalcedons from the German jewel trader, Pierre Laporterie, as well as the acquisition of stones and minerals from Joseph von Damm (ca. 1745–1790) for, at the time, the extremely large sum of 10,000 florins.

On 1 March 1780, Carl Haidinger was appointed assistant of Ignaz von Born. The report also shows that there had been a flourishing exchange of objects (mostly duplicates) between the various museums and cabinets since the 1780s. It's also interesting to note that in 1785, a fund was created to increase the number of foreign minerals in the "Naturalien-Cabinet." The cabinet employees' wages are also listed in the report. Born's wages, for instance, were almost as high as those of the cabinet director, Luigi Baillou, who received 3000 florins per year, and higher than those of the vice director, Abbé Stütz, who received 800 florins annually. For comparison, the curator, Megerle, received 350 florins per year. This fact alone shows the importance attached to Born and his work.

According to Carl Haidinger (Haidinger, 1782), the stones and minerals were arranged following the systematic order of Wallerius and Kronstädt, the conch collection and "rare" animal specimens following the systematic order of Linné, and finally, the zoophytes following the systematic order of Pallas. The most popular exhibition objects were stored in glass cabinets, and the less popular objects were kept in drawers beneath the cabinets. The objects in the drawers were arranged by systematic order.

After the physikalisches cabinet was closed in 1791, a third exhibition room was added to the "Naturalien-Cabinet." The room was furnished by Abbé Stütz following plans made by Ignaz von Born before his death in 1791. A fourth room that was already part of the "Naturalien-Cabinet" was refurbished and housed the pietra dura and the gemstone bouquet. The gemstone bouquet (Fig. 9) was a gift from Empress Maria Theresia to her husband Franz I. Stephan and consists of 2102 diamonds and 761 precious stones that were arranged in a way to imitate a real flower bouquet, with the leaves made of silk (Niedermayer, 1984).

Since 1779, the cabinet was opened to the public on Mondays, and in 1793, the opening day was changed from Monday to Tuesday. The mosaic room was accessible only by prior appointment. Three years later, the public was allowed to visit the cabinets Monday through Friday until noon, except for holidays, on which the cabinet was closed (Hassmann, 2015).

"Thiercabinet" (Animal Cabinet)

From 1796 to 1802, two natural history cabinets were located in the building of the imperial library at the Josephsplatz: (1) the "Mineralien-Cabinet" (Mineral Cabinet) was in state ownership, and (2) the "k.k. Physikalisches-Astronomisches, Kunst- und Natur-Thier-Cabinet" was owned by the imperial household. Abbé Simon Eberle (1756–1827), who was appointed director of the Physikalisches Cabinet in 1795, suggested that since zoological specimens were already stored on the premises of the "Physikalisches Cabinet," the collections of astronomy and physical instruments and the animal collections should be united. One year later, the emperor had the antler and horn collection moved from Kaiserebersdorf (today part of Vienna's 11th district) castle to the Hofburg. This mix between the "Physikalisches Cabinet" (Physical Cabinet) and the future "Thiercabinet" (Animal Cabinet) was to be housed in the wing of the imperial library adjacent to the "Augustinergang." To accomplish this move, the apartments of the prefect of the library and of higher positioned civil servants had to be vacated. The rooms in the narrow and relatively tall building received little light and were not really suitable for a museum. The goal of the new museum with the convoluted name, "k.k. Physikalisches-Astronomisches, Kunst- und Natur-Thier-Cabinet" (Imperial Physical-Astronomy, Art, Nature and Animal Cabinet) was to promote education and to allow the public access to the private collections of the imperial household. Eberle had made several very detailed and expensive dioramas showing the animals in their natural habitat. The concept for the exhibition was mostly directed toward success with the public. Some macabre details certainly contributed to this kind of success. Stuffed human bodies, especially Africans, were included with the animals in their appropriate habitat. The humans used for Eberle's exhibit had all died a natural death. The display of stuffed humans, considered as barbarian today, was perfectly in line with the time's taste and by no means a peculiarity of Eberle's exhibition. We find the same in museums and cabinets all over Europe, e.g., Paris, Torino, and other places as well (Riedl-Dorn, 1998). Shortly after his appointment as new director of the "Vereinigte k.k. Naturalien-Cabinete"(United Natural History Cabinets) in 1806, Carl von Schreibers had the stuffed humans removed from the exhibit and moved to the attic. The bodies were destroyed by fire in 1848. Due to inappropriate use of funds, Eberle was forced into early retirement in 1802.

"Vereinigte k.k. Naturalien-Cabinete" (United Imperial and Royal Natural History Cabinet)

In 1797, the assistant of Luigi Balthasar Baillou, Abbé Andreas Stütz, was promoted to second director of the "Mineralien-Cabinet." After Eberle's premature retirement, Stütz was put in charge of the "k.k. Physikalisches-Astronomisches, Kunst- und Natur-Thier-Cabinet." Luigi Balthasar Baillou died on 23 February 1802. His son Josef refused the hereditary directorate and thus allowed the merging of the two natural history collections into a single one. The new cabinet was aptly named, "Vereinigtes Naturalien- und Physikalisches-Astronomisches Cabinet" (United Natural History- and Physical-Astronomical Cabinet). Stütz soon began to reorganize the collections and the exhibition based on scientific criteria; with the exception of crabs and starfish for which he used the system developed by Johan Christian Fabricius (1745–1808) and the system developed by Johann Friedrich Wilhelm Herbst (1743–1807), Stütz used Linnéan taxonomy for the crustaceans, conches, and fossils.

Stütz died before he was able to finish his work. After Stütz's early demise in 1806, the Emperor Franz I (II HRR) ordered a restructuring of the collections, and the cabinets were once again split into a "Physikalisches-Astronomisches Cabinet" (Physical-Astronomical Cabinet) and a "Naturalien-Cabinet"

(Natural History Cabinet). The zoological and botanical collections were now housed at the Josephsplatz, part of the imperial palace. In 1810, the united natural history cabinets were divided according to the three kingdoms of nature into botanical, zoological, and mineralogical cabinets. Emperor Franz I (II HRR) himself had laid the foundation of the botanical cabinet in 1803. Franz's personal herbarium was integrated into the botanical cabinet in 1808. The following year, the emperor handed over the "Tier und Pflanzenkabinett" to the state.

Carl Franz Anton von Schreibers (1775–1852), a doctor of medicine by profession but well trained in other natural sciences, followed Stütz as director of the "Vereinigte k.k. Naturalien-Cabinete" in 1806. The new director was tasked with modernizing the zoological exhibition according to the newest scientific and educational knowledge of the time and to create an organizational and working plan for the rationalization of the scientific administration for each collection. He was also charged with enlarging the collections while adhering to the systematic order of the cabinets. Until his time, mineralogical and geological objects had been dominant in number by far. The modernization of the zoology exhibition included the creation of object labels for the hitherto unlabeled objects and specimens; the labels stated the scientific name, place of origin, and general distribution of the animals. Schreibers interpreted the task he was given as an order to transform the cabinets into a scientific institution that would, among other purposes, serve the education of future scientists.

He assembled a number of eminent scientists as curators, such as the physician and helminthologist, Johann Gottfried Bremser (1767–1827), who collected ~60,000 specimens of intestinal worms; Leopold Trattinick (1764–1849), a botanist and mycologist, who was made curator of the herbarium, which was founded in 1807; Friedrich Mohs (1773– 1839), who was professor of mineralogy at the time and was responsible for the rearrangement of the minerals collection, for which he used his own system; he separated the mineralogical objects from the geological objects. The latter were arranged and cared for by Paul Maria Joseph Partsch (1791–1856), who was the author of the first geological map of Austria. The mineralogical cabinet also included a center for research on meteorites, a favorite subject of study by Schreibers. The Saxonian physicist, Ernst Florens Friedrich Chladni (1756–1827), and the Vienna industrialist and scientist, Alois Beckh von Widmanstätten (1753–1849), were often seen as visitors at the meteorite collection (see below).

Due to the emperor's interest in plants and horticulture, several expeditions were launched to collect new plants and animals as well as minerals. Among these, the most outstanding was the expedition in 1817 to Brazil, accompanying archduchess Maria Leopoldina (1792–1826) following her marriage to Dom Pedro de Alcântara (1798–1834), the later emperor of Brazil. Before the expedition departed Austria, every step was carefully planned, and the scientists received special instructions to ensure the greatest possible success. The scientific part of the instructions was composed by Schreibers, while the Lord Chancellor Klemens Wenzel Lothar von Metternich (1773–1859), who took great interest in the development of natural history in general, had contributed to the planning of equipment, routes, etc. Many famous scientists were associated with the expedition: the botanist and mineralogist, Johann Emmanuel Pohl (1782–1834); the botanist and entomologist, Johann Christian Mikan (1769–1844); the horticulturist, Heinrich Wilhelm Schott (1794–1865); the botanist, Carl Philip von Martius (1794–1868); the zoologist, Johann Baptist von Spix (1781–1826); and above all, the zoologist and taxidermist, Johann Natterer (1787–1843), who remained in Brazil until 1836. The endeavor was not only accompanied by scientists but also by artists, such as the landscape painter, Thomas Ender (1793–1875), and the botanical illustrator, Johann Buchberger (+1821). To exhibit all the specimens and other items that were brought home, a special Brazilian museum, "Brasilianum," with nearly 133,000 single objects distributed in 13 rooms, was created. Von Schreibers' request for a new building to house the large collections was denied, and the objects had to be included in the already crowded "Naturalien-Cabinet."

The revolution of 1848 also affected the natural history collection. Von Schreibers' private quarters and parts of the collections were destroyed when the last canon shot fired during the revolution hit the "Augustinertrakt" of the imperial palace. Von Schreibers never fully recovered from that loss. He retired one year before his death in 1852. In 1851, the "Vereinigte k.k. Naturalien-Cabinete" was divided into three independent cabinets: the "k.k. Botanisches Cabinet" (Imperial and Royal Botanical Cabinet); the "k.k. Zoologisches Cabinet" (Imperial and Royal Zoological Cabinet); and the "k.k. Mineralogisches Hof-Cabinet" (Imperial and Royal Mineralogical Court-Cabinet). The formal separation into five departments happened in 1876 and included a separated geological department.

ON THE EARLY HISTORY AND IMPORTANCE OF THE METEORITE COLLECTION AT THE NATURAL HISTORY MUSEUM VIENNA

Meteorite Collection: Background

The Natural History Museum Vienna (NHMW) possesses one of the largest and most important meteorite collections in the world. In addition, the NHMW contains by far the largest meteorite display in the world. The display in Hall 5 of the museum, which consists of ~1100 specimens comprising 650 different meteorites, cannot be matched anywhere else in the world (Fig. 10). Of special note are the many well-documented historical meteorite falls, most of which are presented to the public and which add to the collection's richness and uniqueness. Moreover, the Vienna collection is the oldest, worldwide, of its kind. The origin of the collection goes back to the second half of the eighteenth century. Meteorites were already collected in Vienna before the scientific community began to accept the existence of extraterrestrial stones. Although the early curators regarded meteorites as earthly phenomena and did not believe in their extraterrestrial origin, they kept these objects in the natural

Figure 10. View of the meteorite collection in Hall 5 of the Natural History Museum, Vienna. Approximately 1100 meteorite specimens from 650 localities are on display. In 2012, the meteorite hall was completely renovated and modernized. The cabinets in the middle of the hall contain the systematic collection, and they are surrounded by thematic stations along the walls.

history collection. The rise of the Vienna collection to its present-day historical and scientific importance is attributed to curators Carl von Schreibers, Paul Maria Partsch, Moritz Hörnes, Gustav Tschermak, Aristides Brezina, and Friedrich Berwerth. They all were successful curators as well as being dedicated scientists in the field of meteoritics. Due to the efforts of Schreibers and his successors, the Vienna collection became the most important one in the course of the nineteenth century (Brandstätter, 2006). The collection also played an important role in the history of meteorite studies and became one of the centers of the newly established science of meteoritics. However, the breakdown of the Austro-Hungarian monarchy and the outbreak of the First World War brought all collection-related activities to an abrupt halt. In the following years until World War II, only modest activities took place. Due to the precaution of the curators of that time, the meteorite collection had not been subject to any significant damage or loss during World War II and the years after. At present, the collection comprises more than 7000 catalogued specimens from more than 2400 different localities from 96 countries and lies in third place behind the collections of the Smithsonian Institution in Washington, D.C. (USA), and the National Institute of Polar Research in Tokyo (Japan).

Here we give an overview on the early history of the collection from the beginning to the middle of the nineteenth century. Above all, this historical account is dominated by renowned personalities. A central figure is Carl von Schreibers, who is regarded as the founder of meteoritic science in Vienna. Furthermore, a few selected examples of meteorite acquisitions that entered the Vienna collection are used to illustrate their importance for the history and/or scientific development of meteorite studies.

The Beginning

The First Meteorites

In 1751, after the appearance of a fire ball, which divided into two parts with detonations, two masses of iron fell from the sky at Hraschina near Zagreb, Croatia. Emperor Franz I ordered a report on the fall from the episcopal syndicate in Zagreb. The fall report, written in Latin, contained the sworn statements of witnesses from different localities who said that they saw "a brilliant ball of fire that split into two balls linked by fiery chains." In the same year, the fall protocol and the iron masses of 39 kg and 9 kg, respectively, were at first sent to Bratislava, where the emperor met with Hungarian authorities. There and earlier in Zagreb, the mass of the smaller iron piece was repeatedly reduced by sampling. Finally, the entire piece was lost, and nothing is known about any remains (Berwerth, 1918). The larger piece was then sent to Vienna together with the fall report and kept in the "k.k. Schatzkammer" (Imperial and Royal Treasury).

The noted earth scientist, Born, was able to engender a steady flow of mineralogical samples to the Vienna collection. Born also organized the transfer of the Hraschina iron and another meteorite previously kept in the Imperial Treasury to the Natural History Cabinet in 1778. In this way, the first two meteorites, Hraschina and Tabor (stone, fall 1753, Bohemia), entered the collection (Fitzinger, 1856). The 39 kg Hraschina iron (Fig. 11) is regarded as the founding piece of the Vienna collection.

After the death of the second director, L.B. Baillou in 1802, Stütz became sole director of the Natural History Cabinet. During his directorship, the number of acquired meteorites increased to a total of seven pieces: Hraschina (Croatia, 39 kg); Krasnojarsk (Russia, 2.5 kg); Tabor (Czech Republic, 2.7 kg); Steinbach (Germany, 1.1 kg), Eichstädt (Germany, 126 g); L'Aigle (France, 1.1 kg); and Mauerkirchen (Austria, 429 g). Stütz also produced the first handwritten catalogue of the natural history collection, the so-called "Catalogus Stützianus" (Fig. 12).

Güssmann's Hypothesis on the Origin of the Pallas Iron and the Hraschina Iron

A decade before Ernst Florens Friedrich Chladni (1756–1827) published his seminal treatise (Chladni, 1794), in which he linked meteorite falls with fireballs and concluded that meteorites are of extraterrestrial origin, Franz Güssmann (1741–1806) published a systematic mineralogy text in Latin entitled, *Lithophylacium Mitisianum* (Güssmann, 1785). In the chapter, "Ferrum Nativum" (p. 127–131), he discussed the origin of the so-called Pallas iron and of the Harschina iron meteorite. Güssmann, a Viennese mathematician and professor of natural sciences, believed that both iron masses had fallen from the sky, but he concluded that they had originated on the Earth from where they were launched into the sky by enormous "electric fires" (Güssmann, 1785). Thus, he hypothesized, just as the Jesuit, Troili, had argued in a publication two decades before on the fall the Albareto stone (Troili, 1766), that the Hraschina meteorite was thrown skyward by a "terrestrial force."

In a later treatise (Güssmann, 1803), the author mentioned in a note that in 1780, he had tried to convince Ignaz von Born that Hraschina indeed fell from the sky, and that he also had tried to explain to him his ideas on the origin of the iron masses as published later in 1785. Apparently, Güssmann's attempts were not successful, and despite being a renowned person in the scientific

Figure 11. The meteorite Hraschina; main piece of the fall of 1751 near Zagreb, Croatia. In 1778, the 39 kg iron mass was transferred from the Imperial Treasury to the Natural History Cabinet and became the founding object of the Vienna meteorite collection. Inventory no. A2.

Figure 12. "Catalogus Stützianus," the seven-volume catalogue of the natural history collection, produced by Abott Andreas Xavier Stütz (1747–1806). In 1806, the handwritten catalogue contained the entries for seven different meteorites.

circles in Vienna, his discussion of native irons seems to have passed unnoticed (Berwerth, 1918; Marvin, 2006).

Stütz's Disbelief of Stones That "Allegedly Fell from the Sky"

In 1789–1790, a collection of papers mainly related to mining topics (editors I. v. Born and F.W.H. v. Trebra) appeared as "Bergbaukunde" in a two-volume publication. Therein, Stütz published an article entitled, "On Some Stones Allegedly Fallen from Heaven," with the purpose of discrediting the idea that stones could fall from the sky. In his discussion, Stütz (1790) referred to the three witnessed meteorite falls—Eichstädt (Bavaria, 1785), Tabor (Bohemia, 1753), and Hraschina (Croatia, 1751). From the Eichstädt stone, Stütz had received a small specimen from his friend, Baron von Hompesch of Eichstädt, together with a report on how the stone was found (Fig. 13). He described the specimen as ash-gray sandstone, containing small grains of native iron with, in places, iron ochre scattered through it. He further remarked that the whole mass showed evidence for an interaction with fire. Stütz then cited his predecessor, Ignaz von Born, who had described in a catalogue note the Tabor stone (Fig. 14) as consisting of refractory iron intermingled with a greenish stone and having a slaggy surface and then had remarked "… from which some credulous people claim that the stone had fallen from heaven during a thunderstorm." The reports from Eichstädt and Tabor reminded Stütz of the Hraschina iron mass in the Natural History Cabinet that also was said to have fallen from Heaven. He then compared Eichstädt and Hraschina and remarked that the effect of fire was even more pronounced on the latter one. On Hraschina, he also recognized the presence of "spherical depressions" and their similarity to the ones observed on the metal of the Pallas iron. After retrieving the fall report that had been sent with the iron mass to Vienna, he translated the original text from Latin into German. In his conclusion, he said that new writings on "electricity and thunder" that recently had come into his hands motivated him not to simply deny witnessed phenomena that one cannot explain. In particular, he referred to the experiments of a certain "Mr. Komus," who had reduced iron oxides to metal by electric discharges. Stütz then reasoned that lightning, that is an electric stroke on a large scale, might have produced the Hraschina metal by a huge discharge from clouds. He further argued that his hypothesis would give an explanation for "allegedly fallen stones" without the need to regard the reports on these witnessed phenomena as fairy tales.

Interestingly, Berwerth (1918, p. 722) stated in a remark that Stütz had changed his opinion on fire balls after the Chladni's treatise of 1794 and that Stütz then had fully accepted Chladni's theory without any restrictions.

The Era of Schreibers

Carl von Schreibers and the Reorganization of the Natural History Cabinet

In 1802, the Natural History Cabinet and the Physical-Astronomical Cabinet and Nature-Animal Cabinet, both situ-

Figure 13. The largest still-existing fragment of the stone that fell in 1785 at Eichstädt, Bavaria. As one of the earliest meteorite acquisitions, the 122 g specimen is listed in the "Catalogus Stützianus" (1806) and depicted in the book by Schreibers (1820; see also Fig. 20). Inventory no. A14.

Figure 14. The Tabor stony meteorite. Largest single stone (weighing 2.8 kg) from the meteorite shower that occurred in 1753 in Bohemia, Czech Republic. In 1778, it was transferred together with the Hraschina iron from the Imperial Treasury to the Natural History Cabinet. It is listed in the "Catalogus Stützianus" (1806) and depicted in the book of Schreibers (1820; see also Fig. 20). Inventory no. A10.

ated at different locations at the imperial castle, were combined to form the "United Natural History, Physical- and Astronomical Cabinet." Immediately after the death of Stütz in 1806, the "United Natural History, Physical- and Astronomical Cabinet" was divided again with the "United Imperial and Royal Natural History Cabinet" as a separate unit of collections, headed by the newly appointed director Carl von Schreibers (1775–1852) (Fig. 15). The main task of Schreibers was, at the request of the emperor, to reorganize the natural history collections, particularly the zoological and botanical specimens, using the Paris Natural History Museum as a model. Schreibers succeeded also in raising the annual purchase budget for the mineral cabinet significantly (Fitzinger, 1868a). The inventory of the earth science collection increased within a few years by several thousand specimens. In particular, a series of interesting meteorites were acquired, and their number in the natural history collection grew steadily. Acting on Schreibers' proposal, a separate display room devoted to meteorites was arranged, and the foundation for today's collection of the natural history museum was thus laid.

The Fall of the Stannern Meteorite

On 22 May 1808, after explosions were heard, a shower of stones fell in the area of Stonarov (Stannern) in today's Czech Republic (Fig. 16). On 25 May, a Viennese citizen who by chance had gotten a fragment of one of the fallen stones into his hands visited the Natural History Cabinet and presented his acquisition to Schreibers. On the following day, Schreibers traveled to Stonarov as a member of an imperial commission (together with Alois Beck von Widmanstätten) to investigate the fall. From there, he brought to Vienna 61 stones comprising complete individual stones (i.e., single stones that are fully covered by fusion crust) and fragments, totaling ~15 kg. In the same year, he published a detailed report about the Stannern meteorite shower (Schreibers, 1808). In his concluding remarks, Schreibers mentioned that the presence of a black nonmagnetic fusion crust and the general appearance of the Stannern stones had convinced him about their extraterrestrial origin, although all other stony meteorites known at that time had a dull and magnetic fusion crust. In addition, he noted that Stannern did not contain any grains of nickel iron. It turned out that Stannern was completely different than all stony meteorites previously known. At the request of Schreibers, the Viennese apothecary, Josef Moser, performed a chemical study of Stannern and published a bulk analysis (Moser, 1808) that yielded significantly higher contents of calcium and aluminum than the other stony meteorites analyzed prior to that. The French chemist, Louis Nicolas Vauquelin (1763–1829), made another chemical analysis of a Stannern sample. Vauquelin (1809) confirmed Moser's results and classified Stannern as a new type of stony meteorites, which Aristides Brezina (1848–1909), a later curator of the Vienna collection, named "achondrites" (Brezina, 1885). Schreibers (1809) published a second detailed paper about the Stannern stones in which he focused on exterior features, such as shape, surface, and fusion crust.

The Discovery of the Widmanstätten Pattern

In 1808, Alois Beck von Widmanstätten (1754–1849), director of the so-called "Kaiserliches Fabriksprodukten Kabinett" (Imperial Industrial Products Cabinet) obtained from his close friend, Schreibers, a slab of the Hraschina iron for "technical"

Figure 15. Carl von Schreibers (1775–1852), director of the United Imperial and Royal Natural History Cabinet from 1806 to 1851. He is regarded as the founder of meteoritic science in Vienna. For the first time, Schreibers also arranged a separate display room devoted to meteorites.

Figure 16. One of the largest stones of the Stannern meteorite shower, weighing ~2 kg. It was collected by Carl von Schreibers a few days after the fall event near Stonarov, Czech Republic, in 1808. The investigations of Schreibers (1808), Moser (1808), and Vauquelin (1809) revealed that Stannern was a new type of stony meteorite; later this meteorite type was named achondrites. Inventory no. A21.

experiments (Berwerth, 1918). In the course of heating experiments that Widmanstätten had performed on that slab to study the evolving tempering colors, he discovered an unusual pattern of crisscrossing lamellae (Fig. 17). Soon he found out that etching with nitric acid revealed the same pattern in a much more pronounced way. Because the depth of the etched figures depended on the duration of the etching process, he had the idea to ink the etched surface and to print it on paper (Schreibers, 1820). Doing so, he was able to display minute details of the pattern at a full scale. In addition to Hraschina, he then printed the pattern of several other irons, such as Toluca (in 1810), Elbogen (in 1812), and Lenarto (in 1815), all of which are included in Schreibers book from 1820 (Fig. 18). Although Widmanstätten never published his prints, he showed them from the very beginning of his discovery to his Viennese colleagues and to visitors of the Natural History Cabinet (e.g., to Chladni in 1812). Schreibers himself also distributed the findings of Widmanstätten within the scientific community by oral communication and written notes (Schreibers, 1820). Thus, many contemporary scientists began to name the pattern "Widmanstätten figures," a term that has dominated the meteorite literature ever since. In 1939, it turned out that a few years before Widmanstätten, the English geologist, G. Thompson (1761–1806), had discovered a similar pattern on an etched slab of the Pallas iron; this finding had been described in an apparently unnoticed publication (Marvin, 2006). Although Thompson had clear priority of publication, his discovery completely failed to attract the attraction of his contemporaries and, thus, did not contribute to the advancement of science. In contrast, Widmannstätten gained recognition within the scientific community because of the early and widespread distribution of his prints (e.g., Paneth, 1960).

Chladni and the Vienna Collection

In his landmark paper from 1794, the famous physicist, Chladni, compiled 18 reports of witnessed falls, spanning a wide range, from ancient times to the eighteenth century (Marvin, 2006). Among these 18 witnessed events are seven observed falls of which a specimen still exists (Hoppe, 1982), including Eichstädt, Tabor, and Hraschina. For the discussion of these three meteorite falls, Chladni referred to the description of Stütz (1790). In particular, he made extensive use of Stütz's German translation of the fall report of the Hraschina iron mass. The continuous activities and efforts of Schreibers on meteorite issues and the availability of meteorites at the Natural History Cabinet prompted Chladni to come to Vienna in 1812. The main purpose of his visit was to study the Vienna meteorite collection and to make preparations for his second major publication on meteorites (Berwerth, 1918). In 1819, he again came to Vienna to finish the treatise, *Über Feuer-Meteore und über die mit denselben herabgefallenen Massen* (Chladni, 1819), with an appendix written by Schreibers (1819) listing 36 different meteorites for the Vienna collection. One year later, an illustrated meteorite book (Schreibers, 1820) was published as a separate appendix to the book by Chladni mentioned above (Figs. 19 and 20).

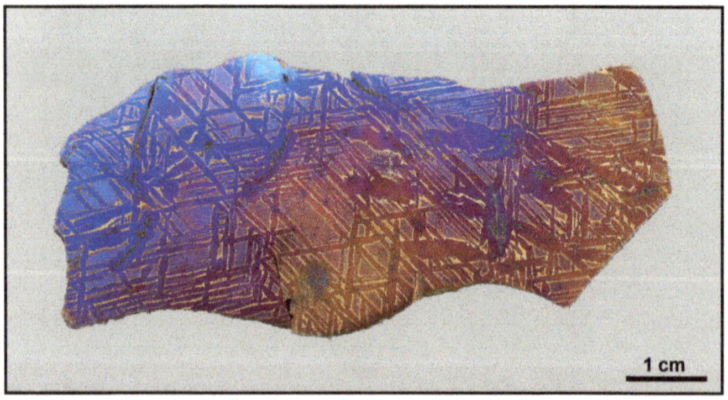

Figure 17. One of the original platelets of the Hraschina iron meteorite. In 1808, Alois Beck von Widmanstätten used these platelets for his flame-heating experiments. In the course of these experiments, he discovered the pattern that is characteristic for most iron meteorites and that ever since bears his name.

Paul Partsch

Paul Maria Joseph Partsch (1791–1856) was educated in philosophy, law, and natural history at the University of Vienna. In 1816, he came into contact with the mineral cabinet via the curator, Abbot Rochus Schüch (1788–1844) (Fitzinger, 1868b), and worked as a voluntary clerk at the imperial natural history collection until 1824. Partsch worked intently on improving his mineralogical training, traveled for more than ten years throughout Europe meeting the foremost European mineralogists, and wrote reports on geological matters for different imperial commissions. Several attempts by Schreibers to secure the post of curator then becoming vacant for Partsch were without success because of the intrigues of the all-powerful State Counsellor Freiherr von Stifft (Fitzinger, 1868b). Finally, in 1835, the state counsellor's intrigues came to an end with the death of Emperor Franz I (II HRR).

The Continuous Growth of the Collection

In 1827, the famous mineral collection of the merchant, Jacob Friedrich van der Nüll (1750–1823), within which were a number of meteorites, was purchased for the mineral cabinet. The collection had been revised and described by Friedrich Mohs (1773–1839), one of the most respected mineralogists in Europe at that time. From 1827 onward, Partsch assisted Mohs with the task of reorganizing the mineral collection in the mineral cabinet and also gave a description of this revised collection. In 1835, Mohs left Vienna to take a position at what is today's Mining University in Leoben. At the same time, Partsch was appointed custodian of the mineral cabinet. Eight years later, he published (Partsch, 1843) a detailed description of the meteorites in the imperial mineral cabinet (Fig. 21). With this book, Partsch appears to have been the first curator to publish a complete list of the meteorites in a museum collection (Burke, 1986). Partsch wrote, in 1843, that the Vienna collection included 94 localities with 258 specimens. The localities consisted of 69 "stones"

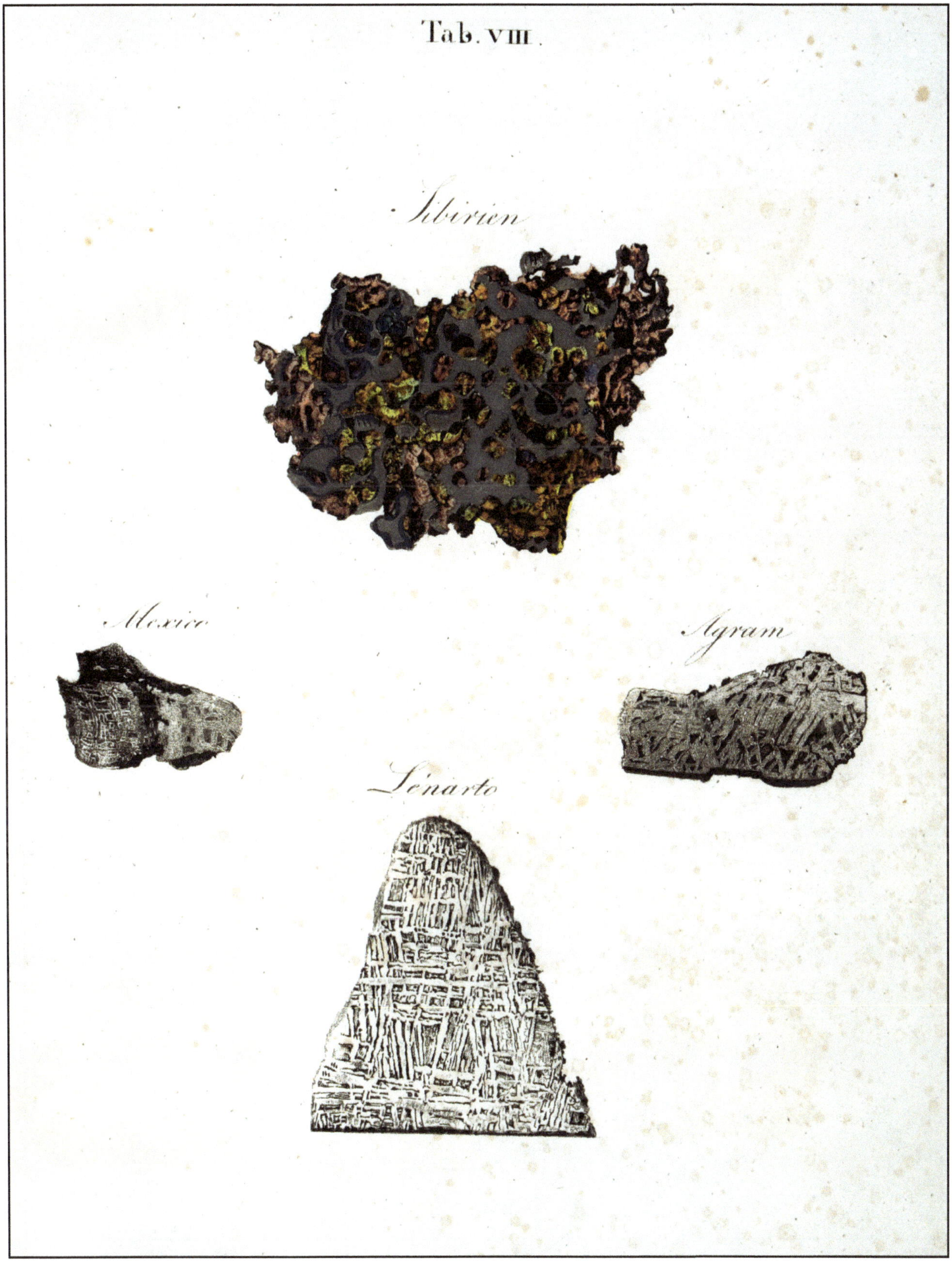

Figure 18. Plate VIII of Schreibers' book from 1820, depicting a color print of the pallasite Krasnojarsk (Sibirien) (top); and early prints of the irons; Toluca (Mexico) (left); Lenarto (Lénarto) (bottom); and Hraschina (Agram) (right).

Beyträge
zur
Geschichte und Kenntniß
meteorischer
Stein- und Metall-Massen,
und der Erscheinungen,
welche deren Niederfallen zu begleiten pflegen.

Von

D. Carl von Schreibers,

der österreichischen Erblande Ritter und Landstande in Nieder-Oesterreich, k. k. Rathe und Director der Hof-Naturalien-Cabinette, Mitgliede der medicinischen Facultät und der k. k. Landwirthschafts-Gesellschaft in Wien; der königl. Akademie der Wissenschaften zu München; der königl. Gesellschaft der Wissenschaften zu Göttingen; der ehemahls kaiserl. Leopoldinisch-Carolinischen Akademie der Naturforscher zu Bonn; der königl. Akademie nützlicher Wissenschaften zu Erfurt; der Societät für National-Industrie und der philomatischen Gesellschaft zu Paris; der Gesellschaft für Künste und Wissenschaften zu Lille; der kaiserl. Gesellschaft der Naturforscher zu Moscow; der Gesellschaft naturforschender Freunde zu Berlin, und der naturforschenden Gesellschaften zu Jena, Leipzig, Hanau, Marburg; der mineralogischen Gesellschaften zu Jena, Petersburg, Dresden; der Werner'schen Societät für Naturkunde zu Edimburg; der physisch-medicinischen zu Erlangen und der pharmaceutischen zu St. Petersburg; der niederrheinischen Gesellschaft für Natur- und Heilkunde zu Bonn; der Societät für Forst- und Jagdkunde zu Dreißigacker, u. s. w. Mitgliede, und der mineralogischen Societät zu Jena ordentlichem Assessor.

Wien.
Im Verlage von J. G. Heubner.
1820.

Figure 19. Title page of Schreibers' book from 1820 entitled, *Contributions to the History and Knowledge of Meteoritic Stone and Iron Masses and of the Phenomena That Usually Accompany Their Fall.* The lower part of the title page depicts three polished and etched objects (one ring and two cubes) made from the Elbogen iron mass.

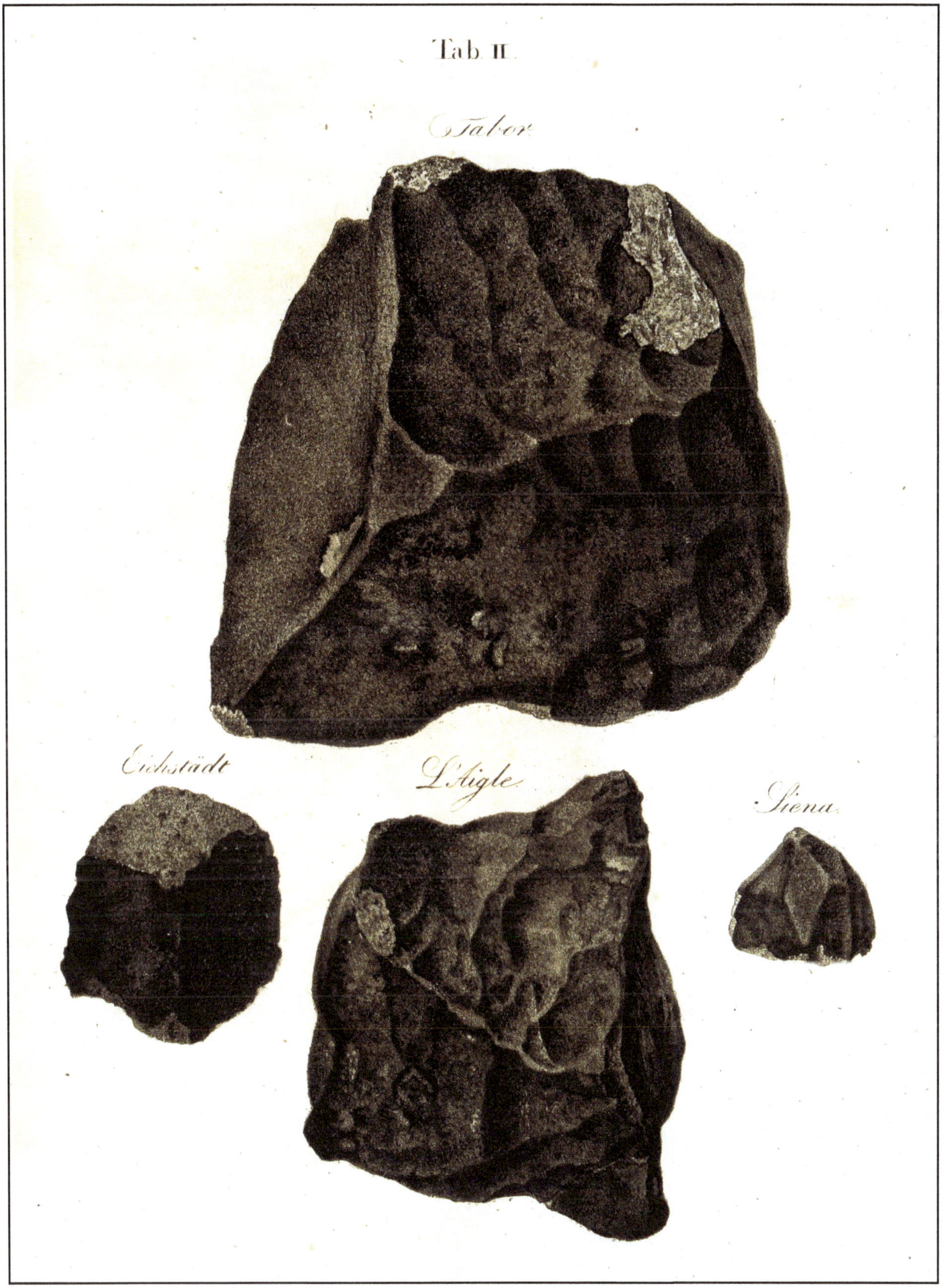

Figure 20. Plate II of Schreibers' book from 1820, depicting specimens of the historical important stony meteorites: Tabor (top), Eichstädt (left), L'Aigle (bottom), and Siena (right).

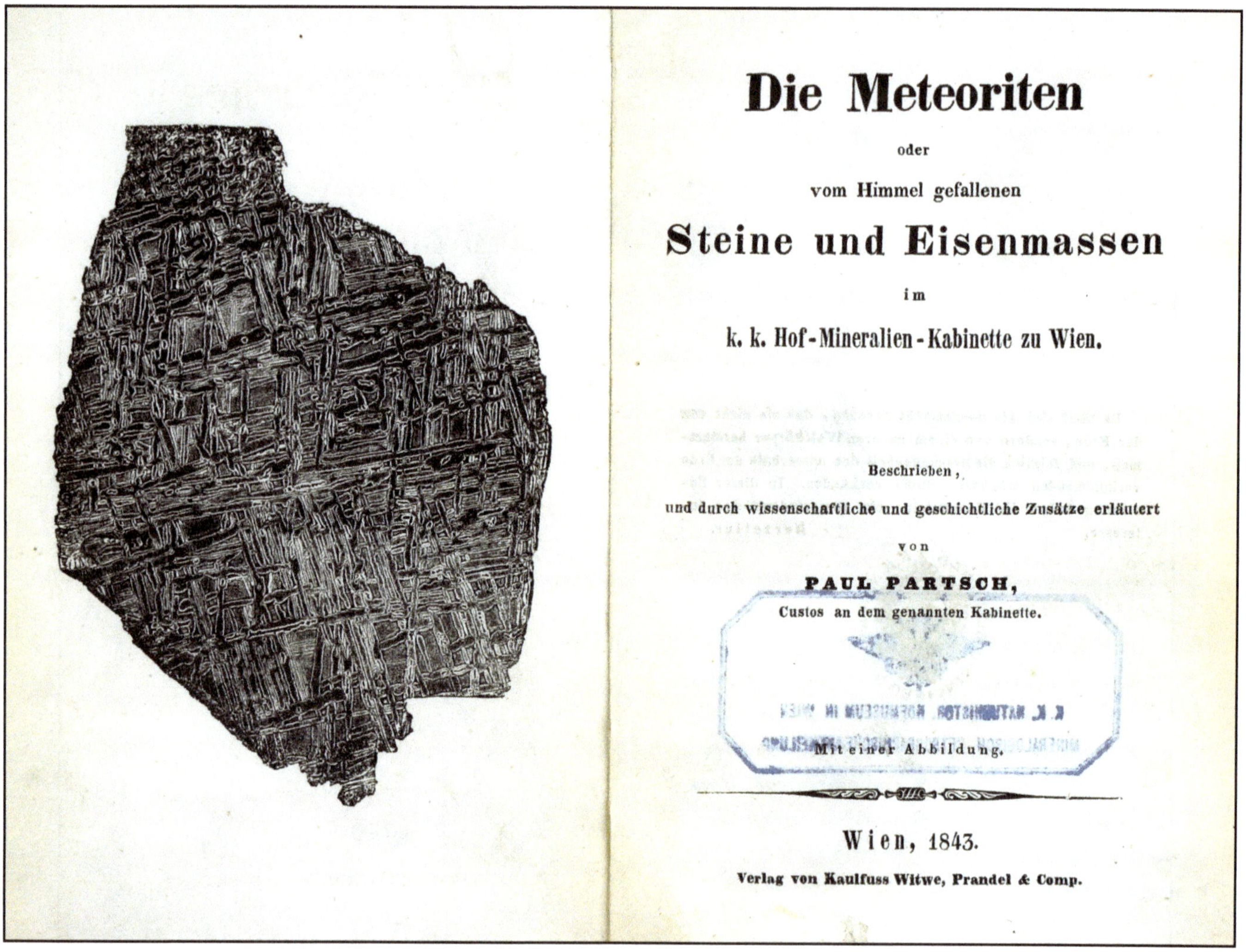

Die Meteoriten

oder

vom Himmel gefallenen

Steine und Eisenmassen

im

k. k. Hof-Mineralien-Kabinette zu Wien.

Beschrieben,

und durch wissenschaftliche und geschichtliche Zusätze erläutert

von

PAUL PARTSCH,

Custos an dem genannten Kabinette.

Mit einer Abbildung.

Wien, 1843.

Verlag von Kaulfuss Witwe, Prandel & Comp.

Figure 21. Title page (right) of Partsch's 1843 publication, which appears to be the first printed meteorite catalogue of a museum collection. On the left, the etched surface of a plate of the Lenarto iron with a pronounced Widmanstätten pattern is depicted.

and 25 "irons," whereas the "iron" specimens comprised iron and stony iron meteorites. Some of the meteorite acquisitions were gifts, whereas the majority were acquired by purchase and exchange. In particular, the practice of Partsch and the later custodians to actively encourage collectors and dealers to propose exchanges was the major reason why the Vienna meteorite collection became the largest and most extensive in the course of the nineteenth century (Burke, 1986).

The Collection during the Revolutionary Year 1848 and After

In October 1848 of the so-called revolutionary year, the natural history cabinets suffered painful losses when portions of the collections and the depots in the attic of the imperial library were set on fire through artillery shelling. In particular, the precious private library of Schreibers and many of his scientific notes were lost in the fire, although the mineral cabinet itself was saved from destruction. A few days before the fire, Partsch had begun to save the most valuable specimens of minerals and meteorites by transferring them to other depots, including his own apartment (Hamann, 1976). Schreibers, who was devastated to see part of his life's work go up in flames, retired at the end of 1851 and died a few months later in May 1852 (Scholler, 1953). Unfortunately, Partsch survived Schreibers by only four years. Immediately after Schreibers' retirement, an administrative separation of the natural history cabinet into three individual "Imperial Royal" court cabinets took place, and Partsch became director of the mineral cabinet and held this position until 1856.

The death of Schreibers marks the end of an era that laid the foundation for two major developments of the meteorite collection in the second half of the nineteenth century. On the one hand, the collection grew very quickly under his successors, thus placing the Vienna collection into the forefront of the great European museum collections. On the other hand, Schreibers initiated the scientific investigation of meteorites at the NHMW. In the decades after Schreibers death, renowned personalities, such as Gustav Tschermak (1836–1927), Aristides Brezina

(1848–1909), and Friedrich Berwerth (1850–1918) made significant contributions to the advancement of the emerging scientific discipline of meteoritics.

PALEONTOLOGICAL COLLECTIONS AND THE DAWN OF EVOLUTIONARY SPECIES CONCEPTS

Of the ~30 million specimens housed by the Natural History Museum Vienna, ~10% of the objects are curated by the scientists of the Geological-Paleontological Department. Only a small part of the huge paleontological collection is displayed in the five public galleries, which combine modern scientific contents with the historical architecture. While the galleries attract visitors with eye-catchers, such as a true-to-life animatronics model of an *Allosaurus,* the "real" treasures are often stored in the scientific collection; these comprise various systematic collections with thousands of type specimens along with large collections devoted to regional geology and stratigraphy.

Like the museum's architecture, the scientific collections were strongly influenced by the history of the Habsburg Monarchy. Previously, in 1750, Emperor Franz Stephan of Lorraine (Franz I. Stephan) (1708–1765), husband of Empress Maria Theresia (1717–1780), laid the cornerstone for the Vienna collections. As mentioned above, he bought the famous "museo" of Jean de Baillou, who was the general manager of the Medici gallery in Florence and of the gardens and mines in Tuscany. The collection was one of the largest of its time, consisting of ~30,000 objects, such as minerals, molluscs, corals, and fossils (Riedl-Dorn, 1998). In Vienna, the collection was displayed in a room of the Augustine Tract of the Hofburg Palace, and Baillou was named the managing director of this new "Naturalien Cabinet" (Natural History Cabinet). During the next decades, the collections were repeatedly reorganized and renamed (see above and Riedl-Dorn, 1998, 2000, for details).

The scientific upheaval of the Age of Enlightenment (ca. 1715–1798) was soon reflected in the structures of the collections as well. The idea of a "natural system" became common knowledge, and consequently, Ignaz von Born (1742–1791) rearranged the imperial collections from 1778 to 1780 according to the newest scientific standards. He established the systems of the Swedish botanist and zoologist, Carl Linnaeus (1707–1778), and the German natural scientist, Peter Simon Pallas (1741–1811), for displaying the invertebrates. The systems of the Swedish mineralogists, Johan Gottschalk Wallerius (1709–1785) and Axel Frederic von Cronstedt (1722–1765), were used to systematically structure the mineralogical collection (Fitzinger, 1856). Espousing such systems, however, raises the quest for completeness, e.g., to acquire a specimen of each kind of minerals, fossils, and living organisms.

Since then, the collections of the museum were greatly enlarged by numerous expeditions and collecting campaigns. The targets of many of these expeditions were exotic animals and plants, while paleontological objects were welcome but subordinate. It was only in the early nineteenth century that the paleontological collections were—literally and figuratively speaking—systematically enlarged. Internationalization and diversification became the focus of the collection strategy. These goals were achieved by intense purchasing from private collectors and especially from market-dominating professional sellers such as the "Heidelberger Mineralien-Comptoir," the "Freyberger Mineralien Comptoir," and the "Mineralien-Geschäft von Krantz." Large collections arrived in Vienna also as donations as well as by exchange with or purchase from other important institutions, such as the Leverian Museum in London (1807) and the Musée d'Histoire Naturelle in Paris (1827). Therefore, the Vienna collection was soon representing all localities and taxonomic groups, which were "en-vogue" at that time.

The next major change in collection strategy, coinciding with the step into modern science, occurred during the mid-nineteenth century. The "Gründerzeit"—a time of industrialization and urbanism between the 1840s and 1870s—began to reshape the cities of the Austrian Empire. Vienna, as its capitol, saw an enormous building activity. This required new outcrops for building stones and clay pits for millions of bricks—all leading to a renaissance of fossil collecting and paleontology. Geology became established as an important science, and geologists were recognized to be of tremendous economic importance in terms of exploration, construction work, and even water supply. The famous Baden Tegel, a Middle Miocene green-blue clay with a rich marine fauna, was intensively exploited (Fig. 22). The slightly younger Late Miocene Inzersdorf Tegel, with abundant endemic lake molluscs, was used in huge quantities as well. Even more prominent in the imperial architecture were Middle Miocene corallinacean limestones, quarried in numerous outcrops in the vicinity of Vienna and near the border to Hungary (Kieslinger, 1972). Geologists of the Natural History Museum began to buy the excellently preserved and enormously diverse fossils from local workers (Hoernes, 1890) and conducted first scientific excavations with early attempts at quantitative analysis (Stur, 1870).

Soon, the scientists became fascinated by the wealth and tropical character of the marine mollusc fauna found throughout the Habsburg Monarchy territory in Middle Miocene strata. The French geologist Louis-Constant Prévost (1787–1856), a pioneer in studying the geology of the Vienna Basin, was the first to recognize the importance of the mollusc faunas for international stratigraphic correlations (Prévost, 1820).

What was the starting point for Austrian paleontologists? The avant-garde in describing Tertiary mollusc faunas was a group of outstanding French and Italian natural scientists, who inspired the Viennese scientists. The enormously species-rich faunas from the Paris Basin and the Aquitaine Basin in France were described during the early nineteenth century by Jean-Baptiste de Lamarck (1744–1829), Jean-Pierre Sylvestre de Grateloup (1782–1861), and Gérard Paul Deshayes (1795–1875). No less pioneering were the descriptions of Eocene, Miocene, and Pliocene mollusc faunas from northern Italy by Giambattista Brocchi (1772–1826), Alexandre Brongniart (1770–1847), Giovanni Michelotti

Figure 22. Collection drawer with hundreds of turrid gastropods (from the Vienna Basin), which were eponymous for the so-called *Pleurotoma* clays.

(1812–1898), and Luigi Bellardi (1818–1889). Their descriptions, and even more so their illustrations, were most influential and epoch making for European earth scientists. The iconic French and Italian monographs were all available in the Viennese library, and exchanges of material from the Vienna Basin with those from other important European localities allowed a direct comparison of putatively related taxa. At the Natural History Museum, in those days called the "Vereinigte K.K. Naturalien-Cabinete," an important group of scientists focused on the Miocene faunas. Paul Maria Partsch (1791–1856) started to arrange the paleontological collection, followed by Moriz Hörnes (1815–1868) and his assistant, Mathias Auinger (1811–1890). First comprehensive lists were published, culminating in the seminal publications on the mollusc fauna of the Vienna Basin by Hörnes (1851–1856, 1859–1867).

Despite these nearly ideal framework conditions, in terms of collections and accessibility of fossil-bearing localities, the identifications in these early monographs were severely flawed due to the still quite ambiguous species concepts. Numerous taxa were erroneously considered to be conspecific with older French species or with the Pliocene Italian taxa (Fig. 23). Obvious morphological differences between the species were often simply considered to be varieties within a broad morphologic species concept. This approach had the clear disadvantage of overemphasizing the stratigraphic ranges of species and of obscuring paleogeographic patterns.

At that time, micropaleontology was at its very beginnings and (bio)stratigraphical considerations were often based on comparisons of mollusc faunas in marine deposits. In his *Principles of Geology* (1830–1833), the British geologist, Charles Lyell (1797–1875), proposed a novel statistical method to reconstruct the relative age of fossil-bearing strata based on the percentage of living molluscs represented within the fossil assemblages. Lyell's method was based on data of Gérard Paul Deshayes, published as a long list of taxa in volume II of the *Principles of Geology*. According to that census, Pliocene faunas yield ~45% of still-living species, Miocene faunas comprise ~17%, and Eocene faunas only 3.4% of extant species. Due to the broad species concept, in which Miocene shells from the Vienna Basin were equated with Pliocene species from Italy, Moriz Hörnes could not reproduce Lyell's results, and thus he rejected this method in the prologue of his monograph in 1851. Because he was unable to differentiate unambiguously between Miocene and Pliocene faunas, he proposed to discard the terms Miocene and Pliocene and instead unite both in his "Neogene." Two years later, Hörnes (1853) formally introduced the term Neogene and again emphasized the impossibility to distinguish between Miocene and Pliocene mollusc faunas. Unfortunately, he felt encouraged in his erroneous opinion by the spectacular faunas from Lăpugiu de Sus in Romania, which had just been discovered. For Hörnes, it was unacceptable that the exceptionally well preserved shells, representing typical Vienna Basin species, could be of Miocene age—which he called an imposition—and assumed that they had lived in a much younger sea. Curiously enough, Hörnes' misconception of the Neogene—based on a pre-Darwinian species concept, neglecting evolution—has survived until today as a formal period in the International Chronostratigraphic Chart (Gradstein et al., 2012; Cohen et al., 2013).

After Hörnes' death, a new generation of paleontologists, including his son, Rudolf Hoernes (1850–1912), started to reevaluate his data. The collections of the museum were strongly enlarged by new findings from other Neogene basins in the territory of the Austrian monarchy. In 1876, the museum was reorganized as the "K.K. Naturhistorisches Hofmuseum" (Imperial Royal Court-Museum). In the late 1870s, Emperor Franz Joseph I decided to build a new museum complex for the natural history collections and the fine arts collections. In the fall of 1871, the excavations for the construction of the imperial buildings at the Burgring started. About ten years later, the new building was completed, and the collections were successively transferred. This entailed rearranging the collection of fossil molluscs. Mathias Auinger (1810–1890), an autodidact who started as a servant

Figure 23. Facsimile of a plate depicting cancellariid gastropods from the comprehensive monograph on the Miocene mollusc fauna from the Vienna Basin by Hörnes (1851–1856) with a picture of the syntype of *Trigonostoma exgeslini* (Sacco, 1894). Half of the species on the plate were erroneously equated with Early Miocene or Pliocene species, thus blurring the stratigraphic significance of the Middle Miocene faunas (see Harzhauser and Landau, 2012, for a revision).

in the paleontological collection of the museum and successively achieved a position equivalent to a collection manager today, sorted the new acquisitions. He soon recognized that many of the former "varieties" represent distinct morphologies and undescribed species. Nonetheless, this morphological species concept lacked an overarching and explanatory theory.

In the meantime, however, a revolutionary idea had stimulated the leading Viennese paleontologists—the Darwinian theory of evolution, referred to as "Descendenztheorie" in contemporary Austrian and German literature. Now, species were "allowed" to change, and morphological varieties of older authors were recognized to represent evolutionary successions. The morphologic species concept became distinctly more elaborate, and the understanding of genera as groups of phylogenetically related species required a complete rethinking. The Miocene mollusc faunas were revised by Rudolf Hoernes and Mathias Auinger in several monographs (Hoernes and Auinger, 1879–1891). The Middle Miocene assemblages turned out to be highly discrete, though often intermediary between Early Miocene and Pliocene faunas. Moreover, the enigmatic endemic mollusc faunas from the various long-lived Miocene and Pliocene lakes on Austrian territory suddenly "made sense in the light of evolution"—to rephrase Dobzhansky (1973). For decades, paleontologists were exasperated with the enormous morphologic disparity observed in certain freshwater gastropod genera such as *Melanopsis* and *Viviparus*. The obvious plasticity was a contradiction to the assumed stability and immutability of species. Even the otherwise ingenious Theodor Fuchs (1872) wrote that "… die scheinbare Veränderlichkeit in Wirklichkeit gar nicht existire, sondern nur durch eine ungewöhnlich reiche Erzeugung von Bastarden hervorgerufen werde" ["in reality, the ostensible variability does not exist but is caused by an unusual production of bastards"], and he referred to the phenomenon as "chaotic polymorphism." In sharp contrast, Melchior Neumayr (1845–1890) (Fig. 24), in the seminal monograph by Neumayr and Paul (1875), provided a new interpretation of the rapid shift in morphologies observed in viviparid gastropods in the lignite mines of Slavonia in Croatia: they reflect extrinsic factors such as changes in environmental parameters. The book became known to Charles Darwin, who immediately recognized that Neumayr's results perfectly fit his theory. On 9 March 1877, he sent a letter to Neumayr stating: "… It seems to me to be an admirable work; and is by far the best case which I have ever met with, showing the direct influence of the conditions of life on the organization. …There can now be no doubt that species may become greatly modified through the direct action of the environment. I have some excuse for not having formerly insisted more strongly on this in my 'Origin of Species,' as most of the best facts have been observed since its publication" (Darwin, 1887, p. 232).

More than 100 years later, a detailed morphometric analysis of hundreds of gastropod shells from the late Miocene of Lake Pannon by Neubauer et al. (2013) fully confirmed Neumayr's concept. Shifts in the geochemical composition of the shells, linked with changes in the environment, clearly coincided with phenotypic changes. The observed expansion of the morphospace was a result of adaptive radiation (Fig. 25).

During the past decades, the stratigraphy of the European Neogene basins was elucidated and constrained by integrated stratigraphy, including micropaleontology, magnetostratigraphy, astrochronology, and absolute dating. Within that advanced framework, the mollusc faunas lost their prominent position as stratigraphic tools. Instead, they turned out to be perfect mirrors of climate change in the past, tracers of paleoenvironmental shifts, and excellent indicators for paleogeographic developments (Harzhauser and Piller, 2007; Zuschin et al., 2013). A recent revision of the cone shells by Harzhauser and Landau (2016) revealed more than 13% of the taxa to be undescribed species. Moreover, their unexpected close relation to modern faunas from Western Africa raises serious doubts about the traditionally assumed Indo-Pacific affinity of Miocene European faunas. This makes the collections of the Natural History Museum invaluable archives that are available to exploration from ever new perspectives.

SUMMARY AND CONCLUSIONS

The Natural History Museum Vienna has had a long and illustrious history in terms of its geoscience collections. The meteorite collection is the oldest in the world and still one of

Figure 24. Melchior Neumayr (1845–1890) was one of the first Darwinists among the Austrian paleontologists. Image courtesy of Archive of the Geological Survey of Austria.

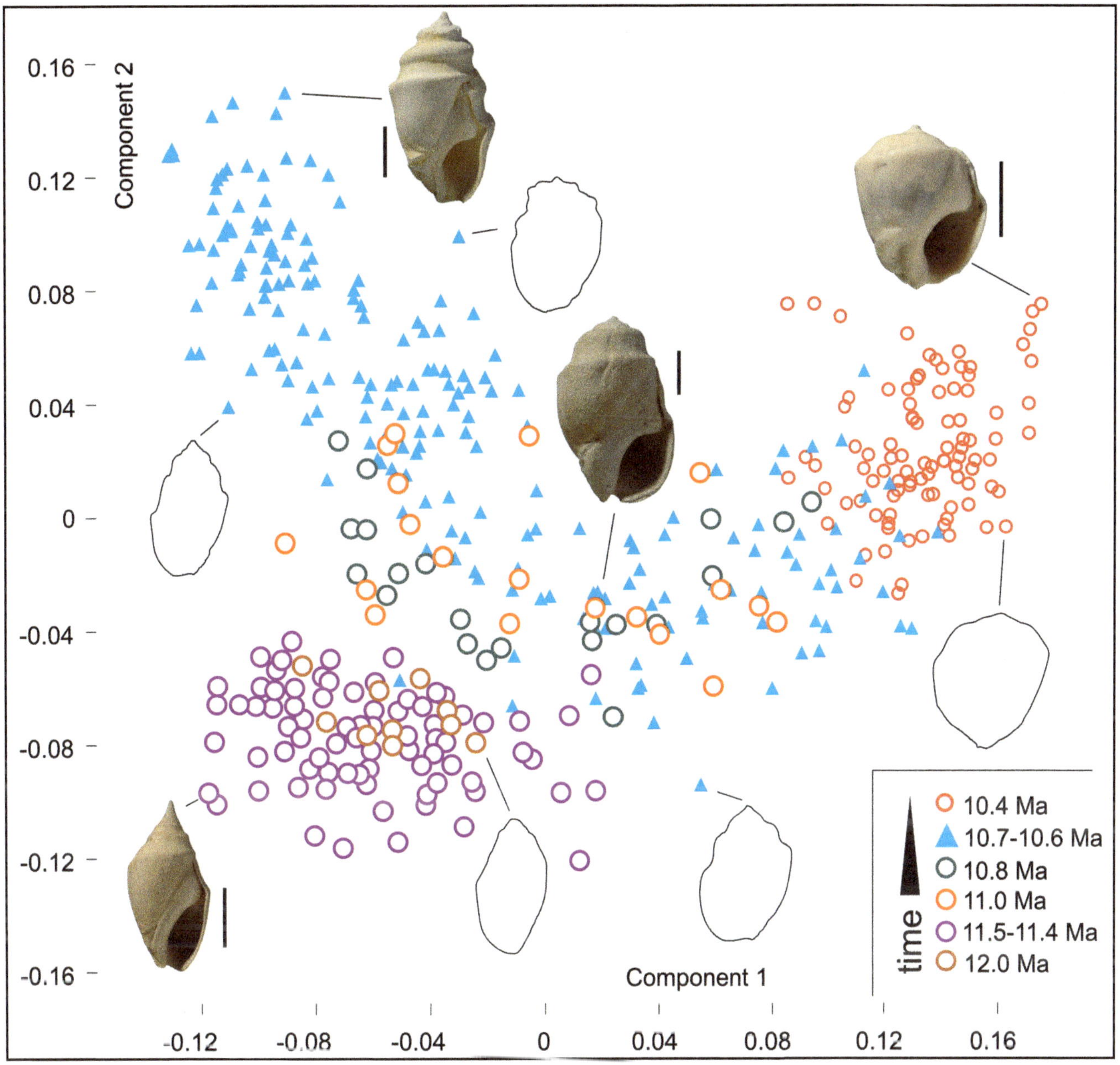

Figure 25. Principal component analysis of Fourier coefficients reflecting morphological differences in shell outlines of the Late Miocene *Melanopsis* species, modified from Neubauer et al. (2012). To visualize shifts in the morphologic spectrum over time, some typical specimens are illustrated (scale bar = 10 mm).

the most important ones that exist, both in terms of the historical objects and the total number. The museum's paleontological collections were of great importance in the development of Darwinian evolution in general and Austrian paleontology in particular.

ACKNOWLEDGMENTS

We thank Ludovic Ferrière for providing high-quality photographs of meteorite specimens.

We are grateful to Thomas Hofmann (Geological Survey, Vienna) for his critical review and comments, to a second anonymous reviewer, and to editor G. Rosenberg for many helpful comments.

Unpublished Sources

AfW NHMW = Archiv für Wissenschaftsgeschichte Naturhistorisches Museum Wien

ASF = Archivio di Stato di Firenze

ÖStA/HHStA = Österreichisches Staatsaarchiv Abteilung Haus-, Hof- und Staatsarchiv Wien/Vienna

ASF, Consiglio die Reggenza, Faz.6: Depeches au Conseil des Finances de l'année, fol 82r

ASF, Consiglio die Reggenza, 23 carta 27

ASF, Consiglio die Reggenza, 24, p. 287, "Per causa ordinaria"
ASF, Consiglio die Reggenza, 24, p. 286 Ristretto dell'Incassato.. in mesci setti—cive dal primo gennaio a tutto luglio 1750, August 1750
ASF Consiglio di Reggenza, 24, p. 295r Dimonstrazionen della differnza che si ved dal Resultato de le Ristrello dell Entrata e Spese della Deposita S. equile nellánno 1749
ASF, Segreteria Finanze Affari primo del 1788, Anni precedenti, 479: Toussain an Richecourt 4.11.1754, p. 1
ASF, Deputazione sopra la nobiltá e cittadinanza Fasc. 18
ASF, Segreteria Finanze Affari primo del 1788, Anni precedenti, 479: Toussain an Richecourt 4.11.1754, p. 1
ÖStA/HHStA, Posch-Akten, jüngere Serie, Karton 8
ÖStA/HHStA, Sonderreihe, Karton 371, Mappe 8, Bericht
ÖstA/HHStA, OKäA Akten Serie B, Karton 4, CCC4
ÖStA/HHStA, OKäA Akten Serie B, Karton 9, Nr 41 ex 1787

REFERENCES CITED

Bauer, A., 1883, Chemie und Alchemy in Österreich bis zum beginnenden XIX: Jahrhundert: Vienna, R. Lechner, 85 p.

Bergmann, J., 1856, Pflege der Numismatik in Oesterreich im XVIII, und XIX: Wien, Jahrhundert mit besonderem Hinblick auf das k. k. Münz- und Medaillencabinet in Wien, 80 p.

Berwerth, F., 1918, Die Meteoritensammlung des Naturhistorischen Hofmuseums als Born der Meteoritenkunde: Sitzungsberichte der kaiserlichen Akademie der Wissenschaften, phil.-hist. Classe, v. 19, p. 31–108.

Blöchinger von Bannholz, C.F., 1864, Der Oberstlieutenant Chevalier de Baillou und die Entstehung des k.k: Hof-Naturalien-Cabinets zu Wien: Vienna, Österreichische Militärzeitschrift, v. 3, p. 273–284.

Blöchlinger von Bannholz, C.F., 1868, Chevalier Jean de Baillou, erster Director des k. k. Hof-Naturalien-Cabinets in Wien und Oberstlieutenant in der Artillerie. Ein Beitrag zur Geschichte der Gelehrten, der Kunst und der Erfindungen: Wien, E. Schlieper, 52 p.

Born, I., 1780, Musaei Caesarei Vindobonensis Testacea, quae jussu Mariae Theresiae disposiuit: Vienna, Gerold, 442 p., doi:10.5962/bhl.title.119974.

Brandstätter, F., 2006, History of the meteorite collection of the Natural History Museum of Vienna, *in* McCall, G.J.H., Bowden, A.J., and Howarth, R.J., eds., The History of Meteoritics and Key Meteorite Collections: Fireballs, Falls and Finds: Geological Society, London, Special Publication 256, p. 123–133.

Brezina, A., 1885, Die Meteoritensammlung des Kaiserlichen königlichen mineralogischen Hofcabinetes in Wien am I. Mai 1885: Wien, Jahrbuch der Kaiserlichen königlichen geologischen Reichsanstalt, v. 35, p. 151–276.

Burke, J.G., 1986, Cosmic Debris: Meteorites in History: Berkeley, University of California Press, 445 p.

Chladni, E.F.F., 1794, Über den Ursprung der von Pallas Gefundenen und anderer ihr ähnlicher Eisenmassen, und Über Einige Damit in Verbindung stehende Naturerscheinungen: Riga, Johann Friedrich Hartknoch, 63 p.

Chladni, E.F.F., 1819, Über Feuer-Meteore und über die mit denselben herabgefallenen Massen: Wien, J.G. Heubner, 434 p.

Cohen, K.M., Finney, S.C., Gibbard, P.L., and Fan, J.-X., 2013, The ICS International Chronostratigraphic Chart: Episodes, v. 36, p. 199–204.

Darwin, F., 1887, Charles Darwin: Life and Letters: London, John Murray, v. 2, 398 p.

Dobzhansky, T., 1973, Nothing in biology makes sense except in the light of evolution: The American Biology Teacher, v. 35, p. 125–129, doi:10.2307/4444260.

Fitzinger, L.J., 1856, Geschichte des kaiserlich-königlichen Hof-Naturalien-Cabinetes zu Wien 1. Abteilung: Älteste Periode bis zum Tode Kaiser Leopold II. 1792: Sitzungsberichte der mathematisch-naturwissenschaftlichen Classe der kaiserl: Akademie der Wissenschaften, v. 21, p. 433–479.

Fitzinger, L.J., 1868a, Geschichte des kais. kön. Hof- Naturalien-Cabinetes zu Wien: Wien, Sitzungsberichte der mathematisch naturwissenschaftlichen Klasse der kaiserlichen Akademie der Wissenschaften, v. LVII, p. 1013–1092.

Fitzinger, L.J., 1868b, Geschichte des kais. kön. Hof- Naturalien-Cabinetes zu Wien: Wien, Sitzungsberichte der mathematisch naturwissenschaftlichen Klasse der kaiserlichen Akademie der Wissenschaften, v. LVIII, p. 35–120.

Fleischer, K. 1932, Das kunstgeschichtliche Material der geheimen Kammerzahlamtsbücher in den staatlichen Archiven Wiens von 1705 bis 1790: Quellenschriften zur barocken Kunst in Österreich und Ungarn: Wien, Krystall-Verlag, v. 1.

Fuchs, T., 1872, Ueber den sogenannten "chaotischen Polymorphismus" und einige fossile *Melanopsis*-Arten: Verhandlungen d. k. k. zoolog.-botanischen Gesellschaft, v. 22, p. 3–6.

Garms-Cornides, E., 1999, Funktionäre und Karrieren in Italien Karls VI, *in* Mazohl-Wallnig B., ed., Italienisches Österreich?: Vienna, Interkulturelle Gemeinsamkeiten und nationale Differenzen vom 18. Jahrhundert bis zum Ende des Ersten Weltkrieges, p. 207–225.

Gradstein, F.M., Ogg, J.G., Schmitz, M.D., and Ogg, G.M., eds., 2012, The Geological Time Scale 2012: Amsterdam, Elsevier, 2 vols., 1144 p.

Grauer, K.J., 1932, Franz I. Stephan von Lothringen-Toskana, römisch-deutscher Kaiser. Sein Leben dargestellt bis zu seiner Kaiserwahl [Ph.D. thesis]:Vienna, University of Vienna, 97 p.

Güssmann, F., 1785, Lithophylacium Mitisianum—Dissertatione Praenia et Observationibus Perpetuis Physico Mineralogicis Explicatum: Viennae, Josephi Nobilis de Kurzbeck, 632 p.

Güssmann, F., 1803, Über die Steinregen an den jungen Grafen Eugen Wrbna: Wien, Johann Thomas Edlen von Trattern, 50 p.

Haidinger, K., 1782, Eintheilung der kaiserl. königl. Naturaliensammlung zu Wien: Vienna, Friedrich Christian Wappler, 104 p.

Hamann, G., 1976, Das Naturhistorische Museum in Wien. Die Geschichte der Wiener naturhistorischen Sammlungen bis zum Ende der Monarchie: Wien, Veröffentlichungen aus dem Naturhistorischen Museum, Neue Folge 13, 98 p.

Harzhauser, M., and Landau, B., 2012, A revision of the Neogene Cancellariid Gastropods of the Paratethys Sea: Zootaxa, v. 3472, p. 1–72.

Harzhauser, M., and Landau, B.M., 2016, A revision of the Neogene Conidae and Conorbidae (Gastropoda) of the Paratethys Sea: Zootaxa, v. 4210, p. 1–178, doi:10.11646/zootaxa.4210.1.1.

Harzhauser, M., and Piller, W.E., 2007, Benchmark data of a changing sea—Palaeogeography, palaeobiogeography and events in the central Paratethys during the Miocene: Palaeogeography, Palaeoclimatology, Palaeoecology, v. 253, p. 8–31, doi:10.1016/j.palaeo.2007.03.031.

Hassmann, E., 2015, Quellen und Regesten zur Schatzkammer, Gemäldegalerie und zu den drei Kabinetten aus dem Archivbestand des k.k. Oberstkämmereramtes 1777 bis 1787 mit einem Nachtrag zu den Jahren 1748 bis 1776: Wien, Jahrbuch des Kunsthistorischen Museums Wien, v. 15–16, 403 p.

Hazen, R.M., Papineau, D., Bleeker, W., Downs, R.T., Ferry, J.M., McCoy, T.J., Sverjensky, D.A., and Yang, H., 2008, Mineral evolution: The American Mineralogist, v. 93, p. 1693–1720, doi:10.2138/am.2008.2955.

Hochstetter, F., 1884, Die Geschichte des Hof-Mineraliencabinets und seiner Sammlungen. Jahrbuch der k.k: Geologischen Reichsanstalt, v. 34, p. 263–298.

Hoernes, R., 1890, Versteinerungen aus dem miocänen Tegel von Walbersdorf: Verhandlungen der k.k: Geologischen Reichsanstalt, v. 1890, p. 129–131.

Hoernes, R., and Auinger, M., 1879–1891, Die Gastropoden der Meeresablagerungen der ersten und zweiten Miozänen Mediterranstufe in der österreichischen-ungarischen Monarchie: Abhandlungen der k.k. geologischen Reichsanstalt, v. 12, p. 1–382.

Hoppe, G., 1982, E.F.F. Chladni—Über den kosmischen Ursprung der Meteorite und Feuerkugeln (1794), *in* Goetz, D., ed., Ostwalds Klassiker der exakten Wissenschaften (zweite Auflage): Leipzig, Akademische Verlagsgesellschaft Geest & Portig K.-G., 104 p.

Hörnes, M., 1851, 1856, Die fossilen Mollusken des Tertiär-Beckens von Wien. I. Band. Univalven: Abhandlungen der Geologischen Reichsanstalt, v. 3, p. 1–736.

Hörnes, M., 1853, Mittheilungen an Professor Bronn gerichtet: Neues Jahrbuch für Mineralogie, Geologie: Geognosie und Petrefakten-Kunde, v. 1853, p. 806–810.

Hörnes, M., 1859, 1867, Die fossilen Mollusken des Tertiär-Beckens von Wien. II. Band. Bivalven: Abhandlungen der Geologischen Reichsanstalt, v. 4, p. 1–430.

Kieslinger, A., 1972, Die Wiener Ringstraße—Bild einer Epoche: Die Erweiterung der Inneren Stadt Wien unter Kaiser Franz Joseph: Band IV: Die Steine der Wiener Ringstraße: Wiesbaden, Steiner, 665 p.

Küchelbecker, J.B., 1730, Johann Basilii Kuchelbeckers allerneueste Nachricht vom Römisch-Käyserlichen Hofe: Nebst einer ausführlichen historischen Beschreibung der kayserlichen Residentz-Stadt Wien, und der umliegenden Oerter, theils aus den Geschichten, theils aus eigener Erfahrung zusammen getragen und mit saubern Kupffern ans Licht gegeben: Hanover, Förster, 884 p.

Lhotsky A., 1941–1945, Festschrift des Kunsthistorischen Museums. Zur Feier des fünfzigjährigen Bestandes 2. Teil: Die Geschichte der Sammlungen, Horn, Austria, Verlag Ferdinand Berger, 412 p.

Lualdi A., 1996, Francois de Baillou, un ottico della Milano Teresiana: Nuncius Annali di Soria della Scienza XI, fasc. 2, Firenze, p. 613–630.

Lyell, C., 1830–1833, Principles of Geology: Being an attempt to Explain the Former Changes of the Earth's Surface, by Reference to Causes Now in Operation: London, John Murray, v. 1 (1830), 511 p., v. 2 (1832), 398 + 109 p., v. 3 (1833), 330 p.

Mambriani, C., 1994, La Versaglia dei duchi di Parma: Prototipi e metamorfosi die giardini di Colorno: I giardini del "Principe" IV Convegno Internazionale. Parchi e giardini storici, parchi letterari, Roma, p. 149–158.

Marvin, U., 2006, Meteorites in history: An overview from the Renaissance to the 20th century, *in* McCall, G.J.H., Bowden, A.J., and Howarth, R.J., eds., The History of Meteoritics and Key Meteorite Collections: Fireballs, Falls and Finds: Geological Society, London, Special Publication 256, p. 15–71.

Moser, J., 1808, Darstellung der physisch-chemischen Eigenschaften der, welche am 22. Mai 1808 bei und in Stannern in Mähren aus der Luft gefallen sind: Annalen der Physik, v. 29, p. 309–327, doi:10.1002/andp.18080290705.

Neubauer, T.A., Harzhauser, M., and Kroh, A., 2013, Phenotypic evolution in a fossil gastropod species lineage: Evidence for adaptive radiation?: Palaeogeography, Palaeoclimatology, Palaeoecology, v. 370, p. 117–126, doi:10.1016/j.palaeo.2012.11.025.

Neumayr, M., and Paul, C.M., 1875, Die Congerien- und Paludienschichten Slavoniens und deren Faunen: Abhandlungen der k.k.: Geologischen Reichsanstalt, v. 7, p. 1–106.

Niedermayer, G., 1984, Der Blumenstrauß aus Stein: Magma, v. 2, p. 82–83.

Paneth, F.A., 1960, The discovery and earliest reproductions of the Widmanstätten figures: Geochimica et Cosmochimica Acta, v. 18, p. 176–180, https://doi.org/10.1016/0016-7037(60)90085-5.

Pärr, N., 2013, Maximilian Hell und sein wissenschaftliches Umfeld in Wien des 18. Jahrhunderts: Religionsgeschichte der Frühen Neuzeit, v. 14, 333 p.

Partsch, P., 1843, Die Meteoriten oder vom Himmel gefallenen Steine und Eisenmassen im k.k. Hof-Mineralienkabinette zu Wien: Wien, Kaulfuss Witwe, Prandel & Company, 162 p.

Prévost, C., 1820, Essai sur la constitution physique et geognostique du bassin a l'ouverture duquel est situee la ville de Vienne en Autriche: Journal de Physique, Chimie, d'Histoire naturelle et sur les Arts et Métiers, v. 92, p. 428–434.

Riedl-Dorn, C., 1998, Das Haus der Wunder. Zur Geschichte des Naturhistorischen Museums in Wien: Vienna, Holzhausen, 308 p.

Riedl-Dorn, C., 2000, Chevalier de Baillou und das Naturalienkabinett, *in* Zedinger, R., ed., Lothringens Erbe, St. Pölten, self-published, p. 111–123.

Riedl-Dorn, C., 2001, Franz Stephan als Förderer naturwissenschaftlicher Aktivitäten, *in* Kowalská, E., and Celková, M., eds., Zlatá a strieborná cesta cisára Frantiska Stefana Lotrinského po stredoslovenských banských mestách: Slovakia, Banská Stiavnica, p. 29–39.

Sacco, F., 1894, I molluschi dei terreni terziarii del Piemonte e della Liguria. Parte 16: Fam. Cancellariidae H. e A. Adams 1853: Bollettino dei Musei di Zoologia ed Anatomia comparata della Reale Universita di Torino, v. 9, no. 171, p. 68–70.

Saint-Laurent, J., 1746, Déscriptions abregée du fameux cabinet des M. le Chevalier de Baillou pour servir à l'histoire naturelle des pierres précieuses, métaux, minéraux et autres fossils: Luques [Lucca, Tuscany], Sauveur & Jean-Dominique Marescandoli, 156 p.

Scholler, H., 1953, Carl Franz Anton Ritter von Schreibers. Zur 100. Wiederkehr seines Todestages: Annalen des Naturhistorischen Museums in Wien, v. 59, p. 23–48.

Schönburg-Hartenstein, J., 1987, Josef Anton Nagel—ein Direktor des physikalischen Kabinettes: Vienna, Verlag der österreichischen Akademie der Wissenschaften, 113 p.

Schreibers, C., 1819, Anhang-Verzeichnis der Sammlung von Meteor=Massen, welche sich im k.k. Hof-Mineralien-Cabinette in Wien befindet, *in* Chladni, E.F.F., Über Feuer-Meteore und über die mit denselben herabgefallenen Massen: Wien, Heubner, p. 425–434.

Schreibers, C., 1820, Beytrage zur Geschichte und Kenntnis meteorischer Stein-und Metall-Massen, und der Erscheinungen, welche deren Niederfallen zu begleiten pflegen: Wien, Heubner, 97 p.

Schreibers, K. v., 1808, Nachrichten von dem Steinregen, der sich am 22sten Mai 1808 in und um Stannern in Mähren ereignet hat: Annalen der Physik, v. 29, p. 225–250, doi:10.1002/andp.18080290702.

Schreibers, K. v., 1809, Beschreibung der mährischen Meteorsteine nach ihrem Aeußern, vorzüglich der Rinde, und nach ihrer Masse, und einige Folgerungen, auf welche diese Beschreibung führt: Annalen der Physik, v. 31, p. 23–77, doi:10.1002/andp.18090310103.

Stur, D., 1870, Beiträge zur Kenntniss der stratigraphischen Verhältnisse der marinen Stufe des Wiener Beckens: Jahrbuch der k.k: Geologischen Reichsanstalt, v. 20, p. 303–342.

Stütz, A., 1790, Ueber einige vorgeblich vom Himmel gefallene Steine, *in* Born, I., and Trebra, F.W.H., eds., Bergbaukunde, zweyter Band: Leipzig, Georg Joachim Goeschen, p. 398–409.

Stütz, A., 1807, Mineralogisches Taschenbuch enthaltend eine Oryctographie von Unterösterreich; zum Gebrauche reisender Mineralogen: Wien and Triest, Geistinger's Buchhandlung, 394 p.

Troili, D., 1766, Della Caduta di un Sasso dall'Aria, Ragionamento: Modena, 120 p.

Tromballa, T., 1953, Franz Stephan und sein Kreis: Beiträge zur Kulturgeschichte des Hauses Habsburg-Lothringen [Ph.D. thesis]: Vienna, University of Vienna, 288 p.

Vauquelin, L.N., 1809, De la aerolite tombée à Stannern en Moravie, le 22 Mai 1808: Annales de Chimie et de Physique, v. 70, p. 321–330.

Wandruszka, A., 1962, Die Habsburg-Lothringer und die Naturwissenschaften, Mitteilungen des Instituts für Österreichische Geschichtsforschung 70: Vienna, Böhlau Verlag, p. 355–364.

Zedinger, R., 2009, "Kaiserliche Wunschliste": Die Instruktion für Nikolaus Joseph Jacquin, *in* Zedinger, R., and Schmale, W., eds., Franz Stephan von Lothringen und sein Kreis/L'empereur François Ier et le réseau lorrain/L'imperatore Francesco I e il circolo lorenese: Das Achtzehnte Jahrhundert und Österreich Band 23, p. 293–300.

Zuschin, M., Harzhauser, M., Hengst, B., Mandic, O., and Roetzel, R., 2013, Long-term ecosystem stability in an Early Miocene estuary: Geology, v. 42, p. 7–10, https://doi.org/10.1130/G34761.1.

Manuscript Accepted by the Society 21 November 2017
Manuscript Published Online 25 May 2018

Printed in the USA

The Geological Society of America
Special Paper 535

Different functions of learning and knowledge—Geology takes form: Museums in the Habsburg Empire, 1815–1848

Marianne Klemun*
Department of History, University of Vienna, Universitätsring 1, 1010 Vienna, Austria

ABSTRACT

Few institutions are as multifaceted and successful as museums. The three main objectives of the museum—collecting, researching, and teaching—always existed next to each other, but they were not necessarily equally weighted. They changed in the course of the history of the museum depending on location and concept. How different was the character of these institutions at the beginning of their existence, when many of them were newly established in the Habsburg Empire. In contrast to modern museums, which focus apart from preservation on exhibitions for a broader public, the concept of the museum of the Vormärz period (1815–1848) was based on different forms of teaching. In 1811, Archduke Johann founded the Joanneum, a universal museum (a museum that included objects from nature and culture as well as crafts) in the province of Styria as a state museum (Landesmuseum) for the region, which became the model for other institutions in the Habsburg provinces. It has been largely overlooked that this outstanding early museum, first and foremost, provided regular instructional lectures in botany and *mineralogy* and, in so doing, made natural history and earth sciences known to interested people. At a time when *mineralogy* was not taught at secondary schools and was rarely part of the university curriculum, scholars such as Friedrich Mohs, Matthias Anker, and Karl Haidinger filled this gap and gave lectures in these subjects at museums. This contribution examines the shift from *mineralogy* as a science of specimens to geognosy (geology) as a science of stratigraphy during the Vormärz period, and I will argue that museums played a pivotal role in this transition process.

INTRODUCTION

Few nineteenth-century institutions were, in general, as multifaceted and successful as museums. It seems to be a consensus among scholars of museum studies that museums emerged from a variety of cultural practices, institutional forms, and concepts. Museology defines research, mediation, and, first and foremost, preservation as the tasks of modern museums (Waidacher, 1999). Cultural studies on museums are in particular interested in the performance of a reality in the making (Davis, 1996; te Heesen, 2012). In general, the authority of museums of a society or a state does not rely on the wealth and the number of objects in their possession but on claims to political, moral, and epistemological features derived from the fostering of knowledge, cultivation,

*marianne.klemun@univie.ac.at

Klemun, M., 2018, Different functions of learning and knowledge—Geology takes form: Museums in the Habsburg Empire, 1815–1848, *in* Rosenberg, G.D., and Clary, R.M., eds., Museums at the Forefront of the History and Philosophy of Geology: History Made, History in the Making: Geological Society of America Special Paper 535, p. 163–175, https://doi.org/10.1130/2018.2535(10).

education, and professionalization. However, in the past 50 years, museums have changed considerably, not just by adopting new techniques of presentation and new forms of display, but also by reflecting changes in attitudes and new cultural approaches. The three main objectives of the museum—collecting, researching, and teaching—always existed next to each other but were not necessarily equally weighted. They changed in the course of the history of the museum depending on location and concept.

How different was the triple claim of museums at the beginning of their existence, when new institutions were established in the Habsburg Empire during the Vormärz period (1815–1848)! This paper addresses the functions of the museum in this period, when it was defined apart from preservation as a vital educational tool. It is fair to consider it as an ideal knowledge space (Livingstone, 2003) focused more than any other institution on different forms of learning and teaching with an emphasis on research. This marks a sharp contrast to the functions of modern museums, the main objective of which is—apart from education, preservation, and research—to attract a broad audience by spectacular events, displays, and exhibitions.

In general, museums are based on research, education, and entertainment; but are all these aspects equal in balance, and how do they fit together? When examining the role nineteenth-century museums played in regard to the general fascination with natural history at that time, it is necessary to include their involvement in teaching. I will focus on the newly established museums in the capitals of the provinces (state museums, *Landesmuseen,* and national museums, *Nationalmuseum*) and how they presented the earth sciences. I will then look at the scientific specialization in earth sciences as represented by Friedrich Mohs at the Court Museum (1826–1835) as well as, briefly, the Mining Museum (*Montanistisches Museum*), both situated in Vienna (1835–1848). The lectures held at these three museums illustrate the strong conceptual connection between teaching and research. Thanks to this connection, all museums were able to make a significant contribution to the earth sciences and the development of geognosy (geology). It would be possible, at this juncture, to introduce a definition of the various parallel terms and their meanings. But such a venture would be problematic since at that time, mineralogy did not yet exist as an independent discipline but was only beginning to develop. There are various branches of earth sciences for which mineralogy serves as an umbrella but for which (in the sense of natural history) it lost its supremacy to nature and was forced to share the field with the newly emerging discipline of geology.

"Geognosie" was a term that had been used in German territories since the time when Abraham Gottlob Werner used it for the description of mountain chains, while in other languages, "geology" or "géologique" gradually defined themselves as a field of study focusing on changes in the Earth's crust over time. Even though in German-speaking countries, the differences between geognosy and geology diminished in the course of time, German geognosts from Werner's school emphasized the difference as follows: "whereas geognosy limits itself to the spatial juxtaposition of mountain ranges, geology investigates the origin of and changes in these ranges in successive periods of time" (Glocker, 1839, p. 5). This orientation had one very useful dimension in that it was directed at the training of miners in the upper levels of the civil service. That is a specific feature of the German and Austrian history of earth sciences, which were socially, politically, and institutionally linked to mining in a way that was unknown in Great Britain. In this respect, museums not only supported the emergence of geology as a science but also fostered the public understanding of this new knowledge field. This paper will stress this connection and explain why regular lectures and a scientific curriculum of museums disappeared at the same time that universities included natural history and geology as subjects of their own rights in their curricula after 1848. Although universities held collections from the eighteenth century onward as mixed accumulations of different objects and *physica naturalia,* they did not develop geology as a scientific subject during the eighteenth century. They also played a minor role in creating public awareness regarding the importance of natural history or geology. Furthermore, only a small group of people, such as students and professors, had access to university collections.

The lectures held at public museums and court-collections, on the other hand, attracted a wide interest in natural history and accentuated the need for earth sciences for a broader public, mostly from the middle classes of the society.

Historians of science have stated a shift from *mineralogy* to geognosy or geology (as the most common terms for what would be called the *earth sciences* today) within the time from the end of the eighteenth to the mid-nineteenth century (Rudwick, 1996). Hence, it is fruitful in this regard to focus on museums in the Habsburg Empire, which were established in this time span and can be seen as spaces where the shift from *mineralogy* as a science of specimens to geognosy as a science of fieldwork, from questions of mineral distribution to a science of rock formation, took place. I will argue that it was the museum space that enabled this process.

THE JOANNEUM IN GRAZ (STYRIA): THE MUSEUM AS A TEACHING INSTITUTION

With the foundation of the Joanneum (1811) in Graz in the province of Styria, Archduke Johann of Austria implemented a universal museum (a museum that includes objects from nature and culture as well as crafts) for the region (state museum or *Landesmuseum*). The specimens originated from his private collections, but the holder of the museum was the estates (*Stände*) of Styria. Culture, arts, crafts, and nature were included in this modern institution. It was supposed to initiate progress and innovation in the province and to support the process of industrialization. The archduke expressed this goal in a written declaration, which museologists today consider the most important conceptual foundation of museums of that time (Fliedl, 1992, p. 14; Jahresbericht, 1811). Instead of being oriented toward the emperor or the archduke, the museum's cultural and natural collections were to represent and define the character of the country. As an instructional tool, the museum was at the core of a new cultural and educational concept that was supposed to close the gap between terminology and perception (*Anschauung*). According to enlightenment pedagogy, realia can create a link between seeing and thinking. The German word "Anschauung" expresses this relationship, but it has a double meaning: looking at something (Anschauen) results in an opinion (Anschauung). Due to its outstanding role and in contrast to the majority of other museums at the time, it became a model for similar institutions in the provinces of the empire. The museums in Cividale (1817, Friuli, today Italy), Prague (1818, Bohemia, today Czech Republic), Ljubljana (1821, today Slovenia), Innsbruck (1823, Tyrol), Linz (1833, Upper Austria), and Klagenfurt (1844, Carinthia) followed this example. All these museums were based on the involvement of the nobility and were supported by an elitist society.

Historiography on these institutions is legion but focuses mainly on their role as social representations or their nation-building impact (Raffler, 2007). Their significance regarding natural history has been largely underestimated as well as the fact that these early museums, first and foremost, provided regular instructional lectures, mainly in sciences. In so doing, they made many fields of knowledge—for instance, natural history and, in particular, *mineralogy*—accessible for civil servants, the middle classes, and young people. Let us focus on natural history and how geology emerged as a new field of knowledge. At a time when neither *mineralogy* as a traditional science nor geology as a new, emerging field were taught at secondary schools or found at universities, museums filled this gap. Museum lectures were well organized, and the students could take exams and get certificates. The shift from *mineralogy* to geognosy (geology) happened precisely at this time.

At the Joanneum, Friedrich Mohs was responsible for the curriculum from an early stage. Mohs (1773–1839), born in Gernrode in the Harz district, studied mathematics at the Prussian university of Halle for a short time and completed a training course at the Freiberg Mining Academy. On his mineralogical travels across Central Europe, he visited Vienna and came into contact with representatives of the upper classes who were interested in mineralogy and earth sciences (Wurzbach, 1868). The Viennese banker, Van der Nüll, commissioned him to rearrange his famous private mineralogical collection (Flügel et al., 2011). Mohs developed a new classification system especially for this purpose (Mohs, 1804). His travels through Prussia, Saxony, the Harz district, and Hessen in the north and through Carinthia in the south of Central Europe made him a much-traveled, experienced field naturalist (Mohs, 1807). Thus, he had experience classifying objects, an indispensable foundational knowledge for rearranging collections. He also possessed empirical knowledge regarding the geographical distributions of *minerals* among different territories as well as personal experiences in mining. It was presumably his versatility that convinced Archduke Johann to appoint Mohs as "professor of mineralogy" and as custodian at the Joanneum in Graz. During the first year in his new position, Mohs developed his first classification concept and started organizing the collection accordingly. He also published the first attempt of a basic textbook on mineralogy (Mohs, 1812). The task of identifying and classifying specimens, the key tools of *mineralogy* at this time, led Mohs to develop a consistent system as the foundation of the collection. The decision on how to arrange the objects was based on distinctions developed by his Freiberg teacher, Abraham Gottlob Werner (1749–1817) at Freiberg.

Mineralogy included four different collections in the Joanneum. The so-called "Kennzeichensammlung" (collection according to identifying features) presented different characteristics of the "fossils" ("Foßilien" = *minerals*, everything dug from earth). The "Systematische Sammlung" (systematic collection) illustrated the classification according to families, genera, and species. Werner referred to the "Vaterländisch-geognostische und oryctognostische Sammlung" (literally: "patriotic-geognostic and oryctognostic collection") (Jahresberichte, 1811–1826) as the collection of rocks from different types of mountains and distinguished it from the geographical collection (Guntau, 1996), whereas Mohs combined both and presented different geological formations manifested at different places. Werner referred to the "technische Sammlung" (technical collection) as "Ökonomische Sammlung" (economic collection) and demonstrated the purpose and use of the mineral kingdom. By naming and defining the geographical "Suitensammlung" (suite collection), which referred to different regions of the Habsburg Empire, the born Saxon proved at the Joanneum to be familiar with the patriotic Habsburg discourse in Austria. The "patriotic discourse" meant being cosmopolitan in knowledge while at the same time maintaining a focus on regional interests.

THE RECEPTION OF WERNER'S CONCEPT IN THE HABSBURG MONARCHY VIA THE JOANNEUM

Werner's ideas on the structure and design of collections did not come directly from Freiberg but traveled via Graz to the Habsburg provinces. In a modified version, this concept became the role model of almost all earth science collections of state museums in the Crown Provinces during the following years and decades, including the museum in Klagenfurt established in 1844–1848 (Klemun, 1998). From 1826, the concept was also applied in the Mineralogical Court Cabinet in Vienna established in the mid-eighteenth century, the predecessor of the Natural History Museum. Even the choice of furniture of the Joanneum was emulated. The Viennese custodian, von Schreibers, reported to his superior that he had asked for a "detailed description and drawing of the boxes and other pieces of furniture"(HHStA, Haus-, Hof- und Staatsarchiv Vienna, OKäA Oberstkämmereramt, B Akten, box 228, Z. 2196/1826) in the Joanneum. It was decided to equip all objects at the Joanneum with labels referring to three different classification systems by A.G. Werner, F. Mohs, and R.J. Haüy (a crystallographer), respectively. This practice was also adopted in the Mineralogical Court Cabinet in Vienna. In 1821, it was stressed that the labels of the objects in the Joanneum served as "instructing auxiliary means for self-research" (Jahresbericht, 1821, p. 6) for visitors and students. From a historiographical perspective, this remark is striking, because previously it had been the custodian who had explained and commented on the objects. For an intensive approach, the visitor had not been left alone in the collection. As long as the custodian came between the objects and the visitor, a focused and self-determined process of learning was not possible.

The systematic and the "patriotic" collections were the first ones established in the Joanneum. Already the first public report stated that teaching was the core competency and the main objective of the museum. "It was not the intention of His Imperial Highness to establish a museum as a monument of royal liberality for the mere glory of the city [...] but for education and teaching that shall spread from the institution to all estates of society, to all fields of civil life: that is the main purpose of the institution" (Jahresbericht, 1811, p. 5). *Mineralogy* was taught for one hour on a daily basis. As early as 1813, an additional lesson in the afternoon was implemented to consolidate what had been learned by reference to the empirical objects (specimens). The lectures drew a large audience. By 1813, 30 people were already enrolled. This figure reached a first peak of 150 students in 1823 (Jahresbericht, 1823, p. 25). Exams at the Joanneum were soon accepted as a supplement of the *Lyzeum*, the remedial course for university students, especially in medical schools. Mohs had put his mark on the museum regarding mineralogy and the structure and design of the collections, but his career led him away from Graz. In 1818, while he was on a geognostic trip through Scotland together with Count Breuner, Mohs was appointed as professor of mineralogy at the Mining Academy of Freiberg by the Saxon King Friedrich August I. Thus, he became the successor of Abraham Gottlob Werner, his world famous teacher, who had arguably been the most renowned expert on earth sciences in the German-speaking area and beyond it in Europe (Mathé, 1989; Rösler, 1989). His stratigraphical classification included different groups of rocks: primitive rocks (Urgebirge), transitional rocks (Übergangsgebirge), secondary rocks (sandstones, limestones, etc.), and the youngest rocks (lose gravel, sand, and limestones). Furthermore, he developed a classification method concentrated on color, glitter, etc., and the shape of minerals (Werner, 1774). At the very beginning of his time as professor in Freiberg, Mohs published his ambitious book on mineralogical classification in Dresden (Mohs, 1820a). He started writing this book while still in Graz. It also appeared in English translation at the same time (Mohs, 1820b). Within his broad range of interests, the Freiberg Mining Academy ranked first throughout his whole life. The fact that Mohs' name (Wagenbreth, 1999) still has its place in the pantheon of science is more likely due to a useful tool of field geology for a simple procedure—the so-called Mohs' hardness scale, developed in Graz between 1812 and 1818—than to his classification method. Mohs was committed to a method of classification, originally developed for botany and zoology based on "externally visible features" ("characteristics") (Mohs, 1812, Vorrede [Preface]), which he transferred to mineralogy. At the same time, he completely refrained from chemical methods of analysis that became more and more popular among mineralogists.

TEACHING AT THE JOANNEUM

In Graz, Mohs found a group of students who followed and continued his approach: Matthias Joseph Anker (1772–1843), who attended Mohs' lectures at the museum and became his successor, and Karl Haidinger, the co-founder of the Mining Museum (1840–1848) and one of the driving forces behind the Geological Survey in Vienna after 1848.

Born in Styria, Matthias Joseph Anker was a trained surgeon and was commissioned by Archduke Johann to rearrange the collection of the *Lyzeum*, originally owned by the Jesuit Leopold Biwald (1731–1805), who taught *physica generalis* there. Familiar with the local conditions, Anker had already conducted excursions commissioned by the archduke before he started his position as custodian. His weekly personal letters sent to Johann between 1816 and 1843 show that he humbly followed his teacher's footsteps. He attracted a group of interested students; among them, for instance, was the very creative paleontologist and evolutionist, Franz Unger, who became custodian at the Joanneum 1836 and professor at the university of Vienna in 1849 (Klemun, 2016). Administrative civil servants, members of the nobility and the middle classes, military officers, and representatives of trade and industry as well as students came together for the daily classes at the museum Joanneum that Anker described in a handwritten letter as follows:

> My second voluntary class on *mineralogy* for the non-obligatory goes well so far. The number of those students is 40; among them physicists, also several students who belong to the field of mining and construction, some educators, legal practitioners, important personalities such as, among others, Hofrath Nittel, Professor Wrbna from the house of cadets, First Lieutenant Huebmann from the Genie Corps. As the start shows I hope to be rewarded for my small effort to give these voluntary lectures insofar that this useful science will spread. (handwritten letter, 26 October 1832, Anker to Archduke Johann—henceforth A.J.—Graz, Styrian Archive, Stmk LA, Archiv (A) Meran (M), box 19, no. 47, fol. 194–199)

The love for *mineralogy* was infectious. Even historians developed an interest in stones and rocks—for example, Georg Göth, who initially worked as Archduke Johann's secretary and about whom Anker wrote to the archduke: "A few days ago I received a letter from Göth in Vordernberg in which he asked my advice on what book on mineralogy he should buy for his self-teaching on this field of science? I had to answer: that he could not master the science without attending the lectures and seeing the characteristics of the minerals with his own eyes" (letter, 2 November 1831, Anker to A.J., Graz, Stmk LA, A. M., box 18, no. 133, 201–204). Anker kept insisting on the significance of studying the object itself in the museum.

Afterwards, special attention was given in the Joanneum to the technical collection, which came closest to fulfilling the archduke's wish of turning the museum into a place of progress. It was the first public collection in Europe that exclusively addressed *mineralogy* as an applied science. It presented rocks in six sections: mining, construction, trade, dye-work industry, agricultural use, and chemical analysis (Jahresbericht, 1823, p. 4).

Thanks to contacts and exchanges with foreign scholars and donations from the local population, the number of objects continuously increased. Archduke Johann sent questionnaires and appeals to the relevant authorities resulting in further official donations to support the objectives of the museum in many places of the province of Styria. All these donated objects from mining facilities and civil servants were mentioned in the published reports of the museum and thus publicly appreciated.

Many institutions and individuals participated, thus proving a significant patriotic effort. As early as 1824, it was explicitly stressed that the "mineral" collection was a "highly important depository of natural history and regional knowledge" (Jahresbericht, 1823, p. 25). Thus, the museum firmly pursued the strategy to "draw attention to our fatherland" (Jahresbericht, 1824, p. 3), which should be achieved by the "overview of the types of mountains in Styria," established and expanded from 1821 onward (Jahresbericht, 1821, p. 1). The students took part in excursions to enlarge this collection. Initially, this special collection was structured according to political and administrative entities (the districts of the province). However, in 1826, this focus was abandoned in favor of Graz as the point of orientation of the political entity. The collection was arranged with Graz as the center of the four points of the compass featuring the "world regions of Styria" (Jahresbericht, 1826, p. 4). This arrangement was also justified by pointing out that visitors from abroad would not know the political structure of districts. Thus, optimizing the perception of the visitors was given priority over the reference to administration.

The work on the collection of rocks according to different mountain types, which gave an "overview of mountain types in Styria," was oriented to the "utility postulate." It contributed to the shift from *mineralogy* as a science of objects to geognosy with a focus on the stratigraphy of the country. In Graz, Matthias Anker played an important role in this respect with more than 50 brief publications on Styria (Flügel, 2004). However, Anker remained a follower of Mohs also in epistemic terms, which is shown by a letter from 1831, in which he expressed his doubts toward the increasing significance of geognosy: "Mohs sent me a letter the day before yesterday and confided to me that he was planning a work on geognosy, in which he might put a stop to the premature conclusions of the so-called traveling geognosts (: from which he does not exclude the English). I cannot wait to see his work" (letter, 2 August 1831, Anker to A.J., Graz, Stmk, LA, A. M., box 18, booklet 3, no. 108, fol. 101–104).

Traveling was associated with geognosy, but there was a narrow connection to geology too, a science which brought a theoretical consciousness into the earth sciences. Mohs' concept of describing rocks was based on the works of Werner, who had coined as well the term geognosy.

GEOGNOSY AS A BRIDGE BETWEEN *MINERALOGY* AND GEOLOGY: FROM A SCIENCE OF SPECIMENS TO A SCIENCE OF STRATIGRAPHY

Werner's school had a sustained impact. He had exhibited a rather skeptical attitude toward general theories of the growing field of geology, the history of the Earth, but proceeded on the assumption of the third dimension of the Earth, today called stratigraphy (Oldroyd, 1996, p. 101). Many scholars of his time adopted Werner's instructional system of describing rocks, because it was based on the visible surface of objects similar to and in line with Linnaeus' botanic description methods, which were highly recognized and even admired (Guntau, 1984). Describing rocks meant to identify them by sight (color, glitter, and shape) according to a learned experience rather than examining them chemically.

In order to disassociate his concept from what was commonly known as *geology* (Dean, 1979), Werner coined the term *geognosy* (Klemun, 2015) for exploring the third dimension of the Earth. He considered geology a new, uncertain, and speculative science in contrast to geognosy, which was based on precise terms and well-established traditional practices. His careful choice of wording also reflects his language skills as he pointed out: "Let us not be careless with our choice of words so that we can do justice to the things" (Rahden, 1992).

Werner's concept of geognosy, which he developed in addition to his description method within the culture of mining, was neither obliged to reason nor to geognosy as a total theory of the Earth (Hofbauer, 2003). Unlike some English scholars, who had separated from traditional *mineralogy* as a science of merely collecting specimens, Werner was no *geo-philus*. His stratigraphy was based exclusively on rock identification and determination and not on fossils. And in contrast to a large number of geologists of this new generation, who were quite fond of traveling (Klemun, 2010), Werner rather stayed at home. While Anker stuck to Mohs' approach in general, he moved from merely identifying and classifying *minerals* to questions of stratigraphy. The museum enabled this transformation. The new tasks Anker was addressing were not least influenced by the archduke's visit to London in 1815–1816. Johann was not only received at court in London but also visited the places of industrialization in Glasgow, Liverpool, and Manchester. He met with the geologist William Buckland at Oxford University as well as with the inventor James Watt and the astronomer William Herschel. He also saw a map compiled by William Smith—the first of its kind—based on the correlation of fossils specific to different strata of the Earth and noted in his diary that this was "a beautiful large-scale geognostic map of England" (Flügel, 2004, p. 63).

Whether Archduke Johann was in fact influenced by this encounter with Smith's map when he commissioned Anker to compile a geological map of Styria is a matter that must remain open. However, it is quite likely. In addition, Leopold Buch (1774–1853), one of the most renowned German geologists and the first German member of the Geological Society of London, visited Anker in Graz in 1819 at the Joanneum. Their joint excursions corroborated Anker's initial findings. For instance, he divided the Paleozoic in Graz into limestone of the transition series (Übergangsgebirge) and the older stratified series (*Flözgebirge*). Buch had already identified a sequence of gneiss, shale, and limestone. Here is not the place to outline the increase in stratigraphic knowledge in the context of Anker's transition into a geognost and at least into a geologist, but suffice it to say that he soon extended his classification system from three to five groups. Anker explicitly and in great detail discussed the difficulty of where the primitive series (*Urgebirge*) ended and the transition series (Übergangsgebirge) started (Flügel, 2004).

In his annual reports until 1830, Anker used the term *mineralogy* to subsume his activities. He mentioned geognosy or geology as equal terms for the first time in 1831 (Jahresbericht, 1831). It was in the same year that Anker rearranged his collection according to stratigraphic aspects. The layout of the collection and the scientific principle of its structure were mutually dependent. If the objects displayed in a museum can be considered an indicator of the curator's ideas, then Mohs' and Anker's activities were characteristic insofar as they represented different approaches. Nevertheless, Anker was still critical toward scientific change. He wrote to the archduke:

> My love and enthusiasm for geology is still a bit impaired; as I realize that all geognosts' identification of rocks is based on shaky ground – fossil deposits in specific formations and divisions based on those are also questioned. In so doing, we lose any fixed formation principle: we cannot determine where one formation begins and another ends. The identification and classification of compound rocks also presents many problems, first and foremost with those rocks whose compounds merge and amalgamate. I compare our current geognosts with bees, which have to make more observations and gather more facts to justify this science. (letter, 21 September 1832, Anker to A. J., Graz, Stmk, LA, A. M., box 19, booklet 1, no. 40, fol. 162–169)

Anker's statement reveals his uncertainty, since he no longer found Werner's method of determining mountain ranges adequate, for it depended exclusively on lithographic criteria that did not permit any differentiated identification of strata, although this had already been achieved by other geologists.

THE MUSEUM AS "CONTACT ZONE" (PRATT, 1991) FOR INTERNATIONAL GEOLOGISTS AND THE CIRCULATION OF GEOLOGICAL CONCEPTS

What caused Anker's change of mind? It is fair to assume that there were several factors, beginning with the huge quantity of findings gained by field geology. In 1828, Friedrich Keferstein (1784–1866) visited Graz as the first German geognost from abroad who undertook excursions together with Anker (Keferstein, 1828). In 1829, the geologists, Roderick Impey Murchison (1792–1871) and Adam Sedgwick (1785–1873), president of the Geological Society of London, came to Graz, as well as the famous paleontologist, Caspar Maria Count Sternberg (1761–1838) from Prague. They were attracted by the reputation of the museum as a key institution for traveling geologists from abroad. The museum's specimens and objects on display represented important aspects of different geological sites in Styria. Anker reported to Archduke Johann about visitors of the museum on a regular basis and about the impact by the visitors. Thus, the museum was not only an educational institution but also a space of communication to share new knowledge on geology. Also, the Graz museum held the most recent publications on geological topics as shown in a letter from Anker to the archduke from 1833, in which Anker expected that even his teacher Mohs would address geognosy in his next publication: "Mohs was probably delighted about the opportunity to talk to Your Imperial Highness; when will his geognosy arrive, while at present almost every month a new book on geognosy appears and every one of them is full of inconsistencies typical for this science. Among the new ones I like the books by: De la Beche, Lyell and Meyer" (letter, 13 January 1833, Anker to A.J., Graz, Stmk LA, A.M., box 19, booklet 2, 48 letters, 55c, fol. 9–12).

These three personalities represented very different scientific approaches. The *Principles of Geology* (1830–1833) by Charles Lyell (1797–1875) was a theoretical foundation of geology as a science by postulating the significance of currently observable occurrences for geological phenomena. Henry Thomas de la Bèche was the president of the first Geological Survey in London, and Hermann Christian Erich Meyer (1801–1869) established vertebrate paleontology in the German-speaking area (Wild 1999). Their scientific positions led Anker to refuse geology as a speculative science. Although Anker was in general skeptical of a type of geology that gave answers concerning the process of change in the Earth's crust and the reasons for this, he examined intensively new concepts of geology that looked at geological processes. It is striking that he slowly adopted the most modern concepts of geology in its concern to history. The foreign visitors of the Joanneum brought these insights from abroad and evoked debates on these new aspects of geology.

The visit of Ami Boué (1794–1881) to the Joanneum underpinned the important role of the museum in enabling contacts between geologists from different countries (Klemun, 2011). The German scholar had studied in Edinburgh and co-founded the first geological society in Paris in 1830. He offered geologists from Austrian provinces the opportunity to publish in his journal and the museum as a space of communication where scientists could meet personally despite the restrictions of Metternich's surveillance. Klemens Wenzel Nepomuk Lothar Metternich was Chancellor of the Habsburg Empire from 1821 until the liberal revolution of 1848 and was famous for his censorship and the repressive treatment of attempts at democratization. Anker reported to Archduke Johann: "Yesterday I received a visit from the most laudable and well-known geognost Boué. The collection of the institution, first and foremost the fossils, attracted his attention. He also mentioned that he would like to receive the Styrian journal in exchange for the French geological journal" (letter, 17 July 1832, Anker to A.J., Graz, Stmk, LA, A. M, box 19, booklet 1, 55 letters, no. 30, fol. 120–123).

The Styrian journal, *Der Aufmerksame* and the *Steiermärkische Zeitschrift,* included mixed topics about culture, crafts, and arts as well as nature, whereas the French *Bulletin de la Societé Geologique de France* mentioned by Boué was one of the first specialized journals in earth sciences.

SCIENTIFIC RETROGRESSION AND ORGANIZATIONAL INNOVATIONS IN VIENNA

In Graz, Anker was gradually leaning toward geology as a result of his stratigraphic studies, although without being able to discuss the issue of age determination. In Vienna, by contrast, Mohs was again given a privileged position within the academic world, and *mineralogy* was prioritized over geology as a science of classification. A handwritten letter from Mohs to Franz Laurenz X. Riepl (1790–1857), professor of commodity economics at the Vienna Polytechnic (predecessor of the Technical University of Vienna), which has been largely ignored by historiography (handwritten letter, German National Museum Nürnberg, Germany, Autographen, 7 October 1825, Mohs to Riepl, Freiberg), shows that Mohs himself pursued his move from Freiberg to Vienna. The court finally granted his wish: "I feel called upon to take the Royal Saxon *Bergrath* (mining inspector) and Professor in Freiberg, Friedrich Mohs, in my service and to allow him an annual salary of 2000 guilders convention money as full professor of mineralogy" the court decree from July 1826 reads, with which Emperor Franz I (II) brought Friedrich Mohs from Freiberg to Vienna (Vienna, HKA, Münz- und Bergwesen, Nr. 12544/1826, Laxenburg 10.7.1826; and AVA, Studienhofkommission, Medizin, Vienna, box. 243, Z. 3459/297, 12.7.1826. Emp. Franz I to Count Saurau). It is no surprise that the emperor met the wish of the efficient mineralogist, who was already well known, even beyond German-speaking areas (Oldroyd, 1996). But that does not explain why the emperor in fact established a chair of mineralogy at the University of Vienna in the first place. And what was Mohs' reason to leave the renowned Freiberg Mining Academy, where he as Abraham Gottlob Werner's successor held the most prestigious chair of this subject in the German-speaking area (and probably in the whole of Europe) and come to Vienna?

Under the pledge of secrecy, Mohs revealed his conditions for a potential move from Freiberg to Vienna to his confidant, Riepl. It is fair to assume that Riepl, who had contacts, pulled the strings to appoint Mohs. Vienna was generally intriguing for Mohs, because here he saw the possibility to establish his specific classification method of minerals at a high level and to spread it from the capital into all parts of the monarchy (Klemun, 2004b), so that the entire teaching staff would have to adopt it unquestioned, as shown in a letter to Riepl: "If the intention is to educate science teachers who will afterwards work with and spread the method to Banská Štiavnica, Pest, Lviv etc., this is a completely different matter and it seems to me that under this condition I could not be of more use in any other place of the world than in Vienna" (handwritten letter, German National Museum Nürnberg, Autogr. 7 October, Mohs to Riepl, Freiberg). Mohs expected that the Viennese academic world would accept his specific method of describing minerals in principle, as his objective was its establishment and distribution within the monarchy. In addition, Mohs demanded access to the Mineralogical Court Cabinet (predecessor of the Natural History Museum in Vienna), the most important collection of the Habsburg Empire. This highly prestigious place and its richness were supposed to be available for knowledge transfer. However, Mohs neither wanted to be custodian nor subordinate to the director of the collections (the "Vereinigtes k.k. Naturalien-Cabinet," "United Imperial and Royal Natural History Collection"). He was not interested in a position within the hierarchy of the court service. Instead, he was keen to use the rich collections and the results of 70 years of work. This intention was a novelty for the imperial collections. For 20 years, collections had been Mohs' scientific starting point for developing a classification method. Now, the emperor's museum in Vienna was supposed to serve the purpose of a space of science policy in connection with the university (Klemun, 2004b).

In what follows I will discuss the paradox that even though the principal (emperor) and agent (Mohs) had conflicting interests, they could be brought down to a common denominator, because both sides expected advantages from this arrangement. Politics and science got fruitful "resources for one another" (Ash, 2002, p. 32). As already mentioned, Mohs pursued a classification method based on "externally visible features" ("characteristics") (Mohs, 1820a) and transferred it to *mineralogy*. His *Grundriss der Mineralogie* (*Foundations of Mineralogy*) (Mohs, 1822 and 1824) had been met with disapproval in Freiberg and Berlin (Mohs, 1829). But Mohs had reason to hope that his ideas would be welcome at the Viennese court, especially with Franz I (II), an expert of Linnaean botany. In general, the elites considered the Linnaean reforms the beginning of a substantial rise of natural history.

Naturalists of Romanticism, on the other hand, had already pursued the divide within the subjects of natural history (von Engelhardt, 1997)—even more so between geology and the aspiring biology—or defended their similarities.

In his memorandum to the Keeper of the Privy Purse, Mohs argued in favor of the well-known bond between the three kingdoms of nature:

> The scientific reason of the humbly proposed institution is the present state of mineralogy that it is seen as part of natural history, and thus this institution is no other than the zoological cabinet which already exists. The same principles that are applied in zoology and botany are to be applied in mineralogy and just as these two parts of natural history are taught, experience will certainly help make the lessons effective

and fruitful; mineralogy has to be taught in the same manner and will also be effective and fruitful, whereof experience has shown not only to the signatory himself (especially in Graz) but also to some of his followers at home (Prof Anker, Riepl and Zippe) and abroad (Jameson, Haidinger in Edinburgh, Naumann in Jena and Leipzig). It has been a tiresome work to put mineralogy in the same scientific state zoology and botany have already reached, which has occupied the signatory for more than twenty years [...] (Vienna, HHStA, ÖKäA, B Akten, box 228, 2005 ex 1826)

The emperor and his advisers expected that Mohs' expertise and competence would bring the Mineralogical Court Cabinet up to date. Already in the mid-eighteenth century, a fortune had been spent to purchase the Florentine cabinet of Jean de Baillou with roughly 30,000 objects for the court collection newly designed by Emperor Franz Stephan, who had established the international reputation of the wealth of the imperial collections. Ignaz Born (1742–1791), who was appointed as custodian in 1776, again improved the collection during the second half of the eighteenth century. The display of the objects and the furniture in the rooms of the Augustinian Corridor within the Castle was completely rearranged, and the collection was opened to the general public once a week. Also, Born convinced the Court Chamber of Coinage and Mining to urge mining towns and areas to add specimens to the museum's collection (Klemun, 2004a). Minerals were arranged according to the systems developed by the Swedish scholars, Cronstedt and Wallerius. However, since that time, nothing much had been changed, apart from cataloguing the objects of the collection by Abbé Andreas Stütz in 1806. As a zoologist, Director Carl Franz Anton von Schreibers (1775–1852) was no expert on earth sciences but fought for the approval of reforms, as he pointed out in a letter from 1826:

Six years ago, when I had the opportunity to empty and rearrange the great hall of the Imperial and Royal Mineralogical Cabinet in order to undertake the planned refurnishing and fitting of the imperial mineralogical collection I took great care that the new boxes which were to be designed and built, 26 in number, would not only serve their mere purpose but also meet the demands of the current fashion. Since in this respect the furnishings of the museum in Gratz [*sic*], which have been fitted recently according to the design and under the direction of Prof Mohs, has been praised by my esteemed and expert principal Count von Wrbna His Excellency of blessed memory and was even met with applause by His Majesty himself, I decided to take those as an example and for that purpose asked the principals of this museum for detailed descriptions and drawings of the boxes and other furniture. After I had received these descriptions, I ordered to furnish our hall accordingly, only I was forced to make some minor modifications so that these new furnishings would not clash—at least at first glance—with the older rooms, which have been in existence for 40 years, since I did not dare to order the amendment of those as well due to the substantial increase in costs, especially at that time. (Vienna, HHStA, OKäA, B Akten, box 228)

It was presumably due to a shortage of money in the times of the Napoleonic Wars that the emperor did not want to change the budget for the Mineralogical Cabinet—that was rather low compared to the other departments—but then he accepted the financial demands of the director, since with Mohs as a new employee, this investment seemed profitable, which the director of the collections did not fail to mention:

Professor Mohs considers the current amendments of those rooms with old furnishings absolutely indispensable in order to meet expectations and to enable his design and his planned curriculum and I cannot forbear to agree with him, all the more as for a long time I have been of the opinion that those amendments will be useful and highly desirable and under the current circumstances I find them even more desirable and necessary. According to the preliminary estimates of the involved workers and professionals, those amendments and the uniform production of the entire premise at the Imperial Mineralogical Cabinet according to the demands and wishes of Professor Mohs will probably amount to roughly 3000 fl Mz for which I ask your approval [...] even more urgently since the construction works have to be completed within a few months in order to not delay the completion of the furnishings and the display which have constantly been worked at in the meantime. (Vienna, HHStA, OKäA, B Akten, box 228, 29 December 1826)

A new classification system necessitated rearranging the exhibition structure and design and changing the furnishings, because showcases and system were interwoven. The display of objects went hand in hand with textual representations (MacGregor, 1994). The formulaic expression of a "purposeful furnishing" represented a quality mark for a well-organized and well-structured collection, which was mirrored by the furnishings. Collections had an epistemic function—they provided order within the disorder of natural historical knowledge. A mineral in a showcase did not gain meaning as an isolated object but in reference to others next to it. It would be naïve to assume that order and structure could be caught at a glimpse—just as Goethe aptly stated. Only the "inner eye" (Hamm, 2000, p. 106) was able to conceive the carefully designed structure. It was thought that what became settled as knowledge functioned as an "inner eye" after a complex intertwining of observing and learning structures. An exhibition catalogue guiding the perception was to grant true insight. Mohs followed this practice. His comprehensive classification system was supposed to have the same function as a catalogue. This system, geared toward totality, had to stand the test as guideline for education based on the versatile collection, if this totality was considered in the concept of the collection:

As the systematic structure of the Imperial and Royal mineralogical collection is a matter on which Not only the lectures depend, with which the signatory has been commissioned, but that also influences the spreading of scientific mineralogy within the provinces of the Imperial and Royal Austrian monarchy, which are immensely rich in minerals, and has an impact on the public opinion, the signatory dares to present his humble view on the arrangement of the Imperial and Royal mineralogical collection to be examined. The arrangement requires **simplicity** and **consistence** [author's emphasis]. This is the basis of the impression of sublimity and greatness, which the design of the exhibition is supposed to create at first glance, without any distracting elements. (Vienna, HHStA, OKäA, B Akten, box 228, 2005)

Thus, distraction should be avoided. But what did this mean? Goethe even went as far as criticizing the collections at the Jena court in principle, because they were equipped with showcases and thus only attracted the "gawking crowd who are led to believe" (Hamm, 2000, p. 106) and who disrupted the process of true studying. Mohs too championed

a close look, [provided the fact] that the sections are clearly and firmly distinguished according to class, order and genus, because this is the true intention of the exhibition layout, be it for the purpose of a mere systematic overview of the products of the mineral kingdom or for the purpose of teaching. Very often teaching mineralogy remains fruitless and this has to have a general reason. Many years of experience in teaching mineralogy under different circumstances have shown the signatory that it is difficult to clearly convey the concept of classes, orders and genera to beginners of this science if one is unable to present these classes, orders etc. in front of the students. Since in that case the beginners have to learn every mineral as an individual object, which is difficult and time consuming and only possible when a collection is at hand [...]. If the teacher is in the position to give his audience a clear idea of the classes [...] etc based on a purposefully equipped and well arranged collection, he will be able to help them by using the means that science currently offers with minimal costs in time and money and it can be expected that his students will gain the knowledge that is required to be useful to the state and to science. (Vienna, HHStA, OKäA, B Akten, box 228, 2005)

In this way, the display of the collection voiced the curators' values and disclosed their mental and political concepts.

MOHS' ACADEMIC TEACHING AT DIFFERENT MUSEUMS IN VIENNA

The connection between collections (= museums), research, and teaching was evident in Mohs' eyes.

For him, fundamental research was the elementary value of *mineralogy* as a science. Although he did not explicitly refer to the educational concept of Wilhelm von Humboldt (1767–1835), the Prussian founder of the Humboldt University of Berlin, Mohs' ideal link between knowledge based on research and the inclusion of the material world was reminiscent of Humboldt's influential reform idea of "education through research" (Bruch, 1999, p. 29).

Mineralogy should not, according to Mohs, exclusively be geared toward mining. Instead, it was to be taught by using the same methods as in botany and zoology, but independent from them. Education and training of higher-grade mining personnel in mining academies that were founded especially for that purpose (Freiberg in Saxony in 1775; Banská Štiavnica in Upper Hungary in 1770; St. Petersburg in 1773; Almaden in 1777; Paris in 1783; and Mexico City in 1792) had taken on a life of their own long before Mohs started his career. He was not interested in a reform of these institutions; his focus was on the universities. In the wake of the higher education reform under Emperor Maria Theresia in the eighteenth century, universities in the Habsburg provinces taught mineralogy only under the umbrella of natural history, while the traditional *historia naturalis* had already separated into individual subjects at many universities elsewhere. Despite a general willingness for reform, Habsburg curricula did not initially undergo differentiation (Egglmaier, 1988). The fruitless debates on these reform plans evolved around the key question: was the university to provide the theoretical foundations of sciences in general, or should students only gain practical knowledge to prepare them for their professional lives? In contrast to the Humboldt University reform in Berlin, educational political decisions in the Habsburg provinces during the entire eighteenth and the first half of the nineteenth centuries were oriented toward "utility" and practical applicability instead of pure science.

This pragmatism hampered, in my view, serious discussions on the significance of mineralogy within the educational system and the establishment of geology as an independent university subject, which only happened in the course of the Leo Count Thun-Hohensteins reforms in 1848 (Lentze, 1962). The reputation of mineralogy improved not before the Vormärz period, and this was thanks to semi-public collections and museums. Since the 1780s, Viennese elites showed great interest in mineralogy (Huber, 1981), starting with the establishment of private collections and intense exchange practice among one another. Rochus Schüch (1788–1844), "secular priest [not belonging to a religious order] and custodian at the local Imperial and Royal Natural History Cabinet," responded to this trend and asked the emperor in 1816 "for the permission to give popular lectures on mineralogy for a fee to a small group of people from the educated classes three times a week during four months. The main purpose of these lectures is the technical and economic use of minerals to improve industriousness, which is taken into consideration by the national legislation in an excellent manner" (Vienna, AVA, Allgemeines Studienhofkom., 4, Philosophie, box 245, 178 ex 1816).

Despite the anti-commercial position of the emperor based on distrust of and opposition to liberalization and democratization currents, isolated measures were taken to support some branches of industry. Polytechnic institutes were founded in Prague and Vienna (predecessors of the technical universities). Schüch's focus on utility went down well, supported by the beginning economic depression (Sandgruber, 1980). Schüch also assured the emperor that he would not "treat mineralogy in a speculative manner—not mere historical tellings together with displaying minerals—and thus reduce it to a mere memory work." The "speculative manner" indirectly hinted at geology, which was associated with revolutionary science or untenable explanations by English and French scientists. "Mere memory work" referred to traditional classification methods for the sake of classification only.

And "historical tellings" pointed at the structure of knowledge within the traditional corpus of natural history—in the sense of history as a listing of events and phenomena. These three aspects Schüch mentioned as arguments against contemporary discourses were innovative only in appearance and matched in fact the conservative-restorative paradigm of the new "patriotic" Habsburg state ideology. "Speculation, mere memory work and historical tellings" were denounced by Mohs as well, which his textbooks proved well enough.

In Freiberg, Mohs had been entrusted with the task to continue the life work of his world-famous predecessor Werner as an epigone. His superiors clearly failed to realize that Mohs had in fact distanced himself from Werner's concept already before his appointment. Mohs insisted that he had his own "method of natural history." Apparently, Freiberg was no ideal place to establish this method.

In order to improve his professional profile, Mohs had to find a new position that was not overshadowed by a leading figure such as Werner. Although Mohs' classification concept differed from Werner's theory in many singular aspects, its fundamental structure was still based on Werner's principles. The core of Mohs' concept was his method of description. Similar to Werner, it was based on characteristics or "external appearances" such as form, hardness, weight, and denseness. Mohs not only appreciated crystallography but also refrained from using chemical methods of identification and the popular blowpipe analysis, because he opposed mixing methods from different disciplines in principle. Mohs made every effort to create a fundamental concept of natural history—which he called the "systematic paragon of similar knowledge" (Mohs, 1836, Vorrede [Preface])—as the basic idea of his abstractions. All principles seemed to be deducted from natural history. His natural history method was based on an extended scientific system and on a perfect edifice of ideas—this cemented his mineralogy as science and was, in his eyes, progress (Mohs, 1829, 1839).

Mohs wanted to set up his own "school" that would be followed by an elite group of scientists who would spread his methodological views in all teaching institutions of the Habsburg Empire (Klemun, 2004b). Thus, it was paramount to him to have optimal teaching materials available to teach and train high achievers. Mohs believed that the Imperial collection would provide exactly this kind of material, not only for reasons of austerity but also because it was the best institution in Vienna. In addition, Mohs was well aware that referring to the principle of "utility" would be the best way to convince the emperor (Vienna, HHStA, OKäA, B Akten, box 192, 1831, Sternberg to the Hofkämmerer, 11.1.1831). The collection of the Viennese court was still a first-class prestigious object. From 1829, Count Kaspar Sternberg (1761–1838), the founder of the Prague National Museum, tried to bring the Meeting of German Naturalists and Physicians (*Versammlung deutscher Naturforscher und Ärzte*), established by Lorenz Oken in 1822, to Vienna. The successful conferences of this organization in Leipzig and Berlin aimed at promoting and supporting not only natural sciences, but also politics, in particular the idea of German unification. Prior to the preparation of the conference, Sternberg expressed his concerns regarding the bad condition of the court collections. He suggested dusting the enormously rich Imperial collections to avoid scorn and ridicule by the European scholars.

Sternberg also demanded compilation of up-to-date catalogues, which was firmly rejected by von Schreibers, the director of the collections. This example demonstrates the importance attributed to the Imperial collections. According to Sternberg's estimates, more than 20,000 people, mostly travelers from the educated classes, visited the different court collections every year—the Imperial Natural History Collection (including the mineralogical cabinet). The Imperial collections had in fact lost their monopoly position by then. More state museums—referred to as "patriotic"—had been founded in the capitals of the provinces—in Budapest, Prague, and Graz. They outranked the museum in Vienna, not in terms of the number of objects, but because of the exemplary design and structure of the exhibitions. But let us go back to the starting point of the events, to the time of Mohs' appointment in 1826. After Mohs had been ordered to "systematically structure and rearrange the mineralogical collections of the state and the teaching institutions of Vienna," this task was completed within three months. The collections were displayed according to his classification system. Although the decree initially did not mention lessons held in the court collections, the Keeper of the Privy Purse was informed of these plans immediately (Vienna, HKA, Münz- und Bergwesen, Nr. 12544/1826, Laxenburg July 1826 and Vienna, AVA, Studienhofkommission, Medizin, Wien, Kart. 243, Z. 3459/297, July 1826, Emp. Franz I. to Graf Saurau). An agreement was reached to allow lectures when the collection was closed; that is, four times a week during lunchtime. Similar to Schüchs' efforts, the government of Lower Austria demanded "to show consideration for those individuals who are higher civil servants and who have professional reason to use the collections, for instance higher civil servants of the Department of Coinage and Mining" (Vienna, AVA Studienhofkomm. box 243, 277 ex Okt. 1826). Thus, the benefit of the state had to be ensured. Mohs' lectures attracted interest mostly among civil servants, but also young physicians, who were generally interested in natural science and made use of this offer, given the lack of educational opportunities otherwise. The Slovenian pharmacist, Heinrich Freyer, still in education, who would later become the custodian of the state museum in Ljubljana, was one of the interested visitors of the lectures in the mineralogical cabinet (Archive of the Rep. Slovenia, Ljubljana, Freyer, Privat.a. XI, Fasz. 12).

Emperor Franz took further steps to use Mohs and his expertise for the state. He decided that Mohs should embark on exploration trips through the provinces of the "Austrian monarchy" and review proposals submitted to the highest authority for coinage and mining. In this sense, he acted de facto as a governmental consultant for mineralogy and worked for three institutions: he was a professor of the University of Vienna; he gave lectures at the Imperial and Royal Mineralogical Cabinet in the Hofburg, predecessor of the Museum for Natural History today; and served as an expert and reviewer for matters of coinage and mining. Mohs, who presented himself as a theorist in his textbooks, returned in his later years to questions of mining for which he had already shown interest at the beginning of his career. Meanwhile appointed as *Bergrat* (mining counsellor), his *Anleitung zum Schürfen* (Instructions on Prospecting) was distributed to all mining authorities of the provinces of the Monarchy (Vienna, HKA, Münz- und Bergwesen, Zl. 5069 ex 1838). According to the available documents of the Department of Coinage and Mining—even though not all of its holdings have survived (Kostelka and Weiss, 1986)—Mohs was only occasionally asked for help.

When he found incorrect results in studies he reviewed, he always suggested that those were caused by the general difficulty of observing geognostic phenomena. If anything, he criticized the incorrect use of technical terms and a lack of theoretical foundations, which, on the other hand, he did not even expect from mining personnel. Even as a reviewer, Mohs used every opportunity to improve the reputation of mineralogy as an academic discipline toward the general public and to spread his scientific views as a special form of knowledge (Vienna, HKA, Münz- und Bergwesen, 1827, Z. 6767). Apart from the Mohs scale, nineteenth-century historiography evaluated Mohs' scientific achievements rather negatively. According to this view, he did not initiate any progress in Freiberg, and his natural history method had no beneficial effects on scientific research.

His mineralogical nomenclature, building upon his theoretical (crystallographic) principles, which he used to describe several new findings, was never scientifically established. His terminology based on principles was considered too complicated. To give a single example: in 1820, 1822, and 1839, Mohs collected a mineral that is known today as loellingite in Hüttenberg and called it "prismatic arsenic pyrites" (*"prismatischer Arsenik-Kies"*). Chapman implemented the term "Mohsin" in 1843 in recognition of Mohs's achievements, but only a few years later the name "loellingite" became common. Scientists after Mohs did not understand his efforts to reevaluate mineralogy as a science and to improve its reputation toward the general public, because they did not share his objective of a holistic concept of mineralogy. Eduard Suess, the famous Austrian geologist, was quite right when he disparaged Mohs' educational methods as teaching "the shape of letters instead of the ability to read" (Häusler, 1999, p. 27). And yet, Suess' thinking itself was trapped in the dichotomy between theory and practice and subject to the *utility postulate*.

The fact that Mohs' attempt to connect different institutions with each other was taken up again in 1835 illustrates its significance. In 1842, Stephan Endlicher, professor of botany, attributed great scientific importance to museums in his plans for an academy in Vienna. This academy—Endlicher referred to it as an "institution of the sciences"—was supposed to be a coordination center between the library, the court cabinets (museums), and the university (Kadletz-Schöffel, 1992, v. 1, p. 282).

These plans were reminiscent of Mohs' ideas. In institutional terms, Mohs was successful: the improved public reputation of mineralogy in Vienna is linked to his name, and his objective to train and educate outstanding representatives of his subject has been achieved (Mohs, 1839). Wilhelm Haidinger, his most successful student, travel companion, and translator (Mohs, 1825), was one of them. With Haidinger, the heyday of the earth sciences began. Also, the foundation of a museum of coinage and mining ("Montanistisches Museum in Vienna"), which offered classes from 1835 and became the nucleus for recruiting scientists of the second generation, was thanks to Mohs' initiative.

FROM THE COLLECTION OF THE COURT CHAMBER OF COINAGE AND MINING TO THE MINING MUSEUM (1835–1848) IN VIENNA

The Court College of Coinage and Mining (*Hofkollegium des Münz- und Bergwesens*), founded in 1745, was the highest office of the Court Chamber. Its task was to administer and support all state-owned mines of the monarchy. The wise presidents of this office recognized the important role of collections also for this institution and founded a library and a "mineralogical" collection within its framework. Mohs soon learned that not only the Imperial collections but also the collections of the mining authorities he worked for as a mining counselor were suitable for his teaching program.

In 1827, he moved his regular lectures there. On official request, all subordinate authorities could be mobilized to support the extension of the collections, just like in the case of the Joanneum in Graz.

The Imperial and Royal Mint, the part of the Court Chamber permitted to strike coins, got a new building in 1835, and the entire second floor was used for the "mineralogical-geognostic Central Collection" as a museum and teaching space. For the first time, a special collection of the earth sciences was granted independence. Wilhelm Haidinger, a student of Mohs, who had previously taught Mohs' ideas in Edinburgh, used the opportunity to extend this institution in Vienna and turn it into a "Mining Museum" in 1842 (Cernajsek, 1999). More classes and courses in chemistry and mining were offered. Combining teaching and research was in line with the needs of the state to extend the knowledge on mineral deposits by conducting the first systematic survey. The specialization of the collections as a "mining museum" was paralleled with a focus on academically trained civil servants in mining. By referring to the utility principle in favor of the state, Haidinger was able to lay the groundwork for the rise of the earth sciences that thrived between 1835–1848, building upon Mohs' organizational achievements. During these years, knowledge of the earth sciences focused on mining emancipated from Werner's and Mohs' geognosy. At an organizational level, the foundation of the Geologische Reichsanstalt (Geological Survey) that directly emerged from geognosy was proof of its significance. The collections of the mint were taken over by the Survey in 1848, but the lectures were discontinued. In the meantime, the Thun University reforms had made the philosophical faculties independent, and the first chairs of paleontology and soon after of geology were established (Schübl, 2010). The change of name from mining to geology accounted for the expansion of the earth sciences with the rise of geology in the Imperial collections, the Geological Survey, and the universities.

Specialized courses and lectures in earth sciences were exclusively held at the universities, which in turn expanded their collections.

CONCLUSION

Despite the Joanneum's self-professed objective from 1811 to establish teaching courses and to define the purpose of the collections as useful for the country (Anker, 1822), it played not only an important role as a place of teaching but also as a communication space. It was a dynamic space of knowledge for traveling geologists to acquire new insights and exchange different positions. In parallel with the task of focusing on the geognosy of the country, the exhibition concept prioritized rocks and minerals from Styria and those that were "useful." Both publications, such as those by Mathias Anker, and the design and structure of the exhibition he was responsible for, changed from a focus on the object to an interest in stratigraphy. Mohs, as professor of mineralogy at the University of Vienna, on the other hand, was keen to entrench his method of description in all collections. His efforts to link the Imperial collections, the university, and the mining authority did not innovate geology but created an institutional and organizational foundation, from which an independent mining museum could emerge. In 1849, it became the Geological Survey, which would play a pivotal role for geology in the Habsburg Empire. Thus, it was not before the mid-nineteenth century that the sustainable impact of Werner on Mohs and Anker became a thing of the past. Without exception, it was the museum that allowed different ways of following utilitarian thinking at different levels and allowed space for negotiation processes that enabled the shift from the science of specimens to a stratigraphy.

REFERENCES CITED

Unpublished Sources in Archives

[Anonymous], 1811–1848, Jahresberichte des Universalmuseums Joanneum, www.museum-joanneum.at.

Archive of the Geological Survey (today Geologische Bundesanstalt).

German National Museum, 1825 (7 October), Nürnberg, Germany, 449 Autographen, letter, Mohs to Riepl, Freiberg.

Graz, Archiv Meran (by permission of family Harnoncourt), Documents in 11 Volumes.

Vienna, HHStA (Household, Court and State Archive), OKäA (Office of Head Chamberlain).

Vienna, H.K.A. (Archive of the Imperial Chamber), Münz- und Bergwesen.

Vienna, AVA (General Administrative Archive Vienna).

Anker, M., 1822, Uebersicht der aufgestellten vaterländischen, technischen Mineralien-Sammlung im Joanneum nebst allgemeinen Uebersichts-Bemerkungen: Steyermärkische Zeitschrift, no. IV, p. 86–93.

Ash, M., 2002, Wissenschaft und Politik als Ressourcen für einander, *in* Bruch R., ed., Wissenschaften und Wissenschaftspolitik–Bestandsaufnahmen zu Formationen, Brüchen und Kontinuitäten in Deutschland des 20. Jahrhunderts: Stuttgart, Franz Steiner-Verlag, p. 32–51.

Bruch, R., v., 1999, Langsamer Abschied von Humboldt?: Etappen der Universitätsgeschichte 1810–1945, *in* Ash, M., ed., Mythos Humboldt: Wien, Köln, Weimar, Böhlau, p. 29–57.

Cernajsek, T., 1999, Die geowissenschaftliche Forschung in Österreich in der ersten Hälfte des 19. Jahrhunderts, *in* Bachl-Hofmann, C., et al. (Red.), Die Geologische Bundesanstalt in Wien. 150 Jahre Geologie im Dienste Österreichs (1849–1999): Wien, Böhlau Verlag, p. 41–77.

Davis, P., 1996, Museums and the Natural Environment. The Role of Natural History Museums in Biological Conservation: London and New York, Leicester University Press, 286 p.

Dean, D.R., 1979, The word "geology": Annals of Science, no. 36, p. 35–43.

Egglmaier, H.H., 1988, Naturgeschichte. Wissenschaft und Lehrfach, *in* Publikationen aus dem Archiv der Universität Graz no. 22: Graz, Akademische Druck- und Verlagsanstalt, 301 p.

Engelhardt, D. von, 1997, The Romantic Naturforschung around 1800, *in* Teich, M., Porter, R., and Gustafsson B., eds., Nature and Society in Historical Context: Cambridge, UK, Cambridge University Press, p. 195–208.

Fliedl, G., 1992, Das Joanneum – "...kein normales Museum...," *in* Fliedl, G., Muttenthaler, R., and Posch, H., eds., Museumsraum, Museumszeit. Zur Geschichte des Österreichischen Museums- und Ausstellungswesen: Wien, Picus Verlag, p. 11–48.

Flügel, H.W., 2004, Mathias Josef Anker, Arzt, Mineraloge und Geognost der Bidermeierzeit in Graz: Joannea–Mineralogie, no. 2, p. 55–81.

Flügel, H.W., Huber, P., Huber, S., and Machan, A., eds., 2011, Jakob Friedrich van der Nüll. Großbürger und Sammler in Wien an der Wende zum 19. Jahrhundert: Wien, Verlag des Naturhistorischen Museums, 208 p.

Glocker, E.F., 1839, Grundriß der Mineralogie mit Einschluß der Geognosie und Petrefactenkunde, Nürnberg: Johann Leonhard Schrag, 993 p.

Guntau, M., 1984, Abraham Gottlob Werner: Leipzig, Teubner Verlag 1984, 120 p.

Guntau, M., 1996, The natural history of the earth, *in* Jardine, N., Secord, J.A., and Spary E.C., eds., Cultures of Natural History: Cambridge, UK, Cambridge University Press, p. 211–229.

Hamm, E.P., 2000, Goethes Sammlungen auspacken. Die Öffentlichkeit und das Private im naturgeschichtlichen Sammeln, *in* te Heesen, A., and Spary, E.C., eds., Sammeln als Wissen. Das Sammeln und seine wissenschaftsgeschichtliche Bedeutung: Berlin, Wallstein Verlag, p. 85–114.

Häusler, W., 1999, "Bunte Steine." Bildungs- und sozialgeschichtliche Aspekte der österreichischen Erdwissenschaften im Zeitalter der bürgerlichen Revolution, *in* Die Geologische Bundesanstalt in Wien. 150 Jahre Geologie im Dienste Österreichs (1849–1999): Wien, Böhlau Verlag Ges.m.b.H, p. 19–40.

te Heesen, A., 2012, Theorien des Museums: Zur Einführung: Hamburg, Junius Verlag, 200 p.

Hofbauer, G., 2003, Geognosie: Von Füchsels reflektierter Spekulation zu Werners voreingenommener Wahrnehmung, *in* Albrecht H., and Ladwig, R., eds., Abraham Gottlob Werner and the Foundation of the Geological Sciences, Freiberger Forschungshefte D 207: Freiberg, Verlag der Technischen Universität Bergakademie Freiberg, p. 164–176.

Huber, S. und P., 1981, Zur Tradition des Mineraliensammelns im Raume Wien: Vienna, Mitteilungen der Österreichischen Mineralogischen Gesellschaft, no.128, p. 77–86.

Kadletz-Schöffel, H., 1992, Metternich und die Wissenschaften, Dissertationen der Universität Wien: Wien, Lang, 2 volumes, v. 1, 516 p.; v. 2, 240 p.

Keferstein, C., 1828, Teuschtland geognostisch-geologisch dargestellt: Weimar, Verlag des Industrie Comptoirs, 6 vols., 590 p.

Klemun, M., 1998, Zur Geschichte des Naturwissenschaftlichen Vereines für Kärnten, *in* Klemun, M., ed., Werkstatt Natur—Pioniere der Forschung in Kärnten: Klagenfurt, Carinthia II, no. 56. Sonderheft, p. 11–113.

Klemun, M., 2004a, The Royal Natural History Collection in Vienna (18th century): From possessing minerals as treasure towards territorial ambitions as consciousness: Leiden, Scripta Geologica Special Issue no. 4, p. 193–199.

Klemun, M., 2004b, "Die Gestalt der Buchstaben, nicht das Lesen wurde gelehrt." Friedrich Mohs' "naturhistorische Methode" und der mineralogische Unterricht in Wien: Mitteilungen der Österreichischen Gesellschaft für Wissenschaftsgeschichte, no. 23, p. 43–60.

Klemun, M., 2010, Reisen, Aufzeichnen, Beschreiben. Visualisierung erdwissenschaftlichen Wissens und "Mineralogische Reisen" innerhalb habsburgischer Territorien, *in* Lebeau, Chr., and Schmale, W., eds., Images en Capitale: Vienne, fin XVIIe–début XIXe siècles: Bochum, Winkler 2010, p. 95–115.

Klemun, M., 2011, "Geognosie" im Vormärz: AMI BOUÈ (1794–1881) und dessen Mittlerfunktion zwischen den unterschiedlichen europäischen geologischen Wissenskulturen und Kärnten: Carinthia II, no. 201/121, p. 249–268.

Klemun, M., 2015, Geognosie versus Geologie: Nationale Denkstile und kulturelle Praktiken bezüglich Raum und Zeit im Widerstreit: Berichte zur Wissenschaftsgeschichte. Organ der Gesellschaft für Wissenschaftsgeschichte, no. 38, v. 3, p. 227–242.

Klemun, M., 2016, Franz Unger (1800–1870): Multiperspektivische wissenschaftshistorische Annäherungen, *in* Klemun, M., ed., Einheit und Vielfalt. Franz Ungers (1800–1870) Konzepte der Naturforschung im internationalen Kontext: Göttingen, Vandenhoeck und Ruprecht, Vienna University Press, p. 15–92.

Kostelka, L., and Weiss, A., 1986, Friedrich Mohs, Geognostische Reise durch einige Provinzen der k.k. Staaten im Jahre 1836: Klagenfurt [Sonderheft.], Carinthia, v. II, 44 p.

Lentze, H., 1962, Universitätsreform des Ministers Leo Graf Thun-Hohenstein: Wien and Weimar, Böhlau, Sitzungsberichte der österr. Akademie der Wissenschaften, phil. hist. Kl. 239/2, 372 p.

Livingstone, D.N., 2003, Putting Science in Its Place: Geographies of Scientific Knowledge: Chicago and London, University of Chicago Press, 234 p., https://doi.org/10.7208/chicago/9780226487243.001.0001.

Lyell, C., 1830–1833, Principles of Geology. Being an Attempt to Explain the Former Changes of the Earth's Surface, by Reference to Causes now in Operation: London, Murray, 3 vols., v. 1, 511 p.; v. 2, 133 p.; v. 3, 109 p.

MacGregor, A., 1994, The life, character and career of Sir Hans Sloane, *in* MacGregor, A., ed., Sir Hans Sloane: Collector, scientist, antiquary. Founding father of the British Museum: London, British Museum Press, p. 11–44.

Mathé, G., 1989, Friederich Mohs in Freiberg, *in* Postl, W., and Moser B., eds., Friederich Mohs (1773–1839), Graz, Mitteilungen der Abteilung für Mineralogie am Landesmuseum Joanneum, no. 57, p. 15–28.

Mohs, F., 1804, Des Herrn J. F. von der Null Mineralien-Kabinet, nach einem, durchaus auf äussere Kennzeichen gegründeten Systeme geordnet: Wien, Camesinaische Buchhandlung, 1804, 3 vols., 730 p.

Mohs, F., 1807, Die Villacher Alpe, und die dieselbe zunächst umgebenden Gegenden; eine geognostische Skizze: Ephemeriden der Berg- und Hüttenkunde, no. 3, p. 161–228.

Mohs, F., 1812, Versuch einer Elementar-Methode zur naturhistorischen Bestimmung und Erkenntnis der Foßilien: Wien, Camesina 1812, 128 p.

Mohs, F., 1820a, Die Charaktere der Klassen, Ordnungen, Geschlechter und Arten, oder die Charakteristik des naturhistorischen Mineral-Systems: Dresden, Arnoldische Buchhandlung, 226 p.

Mohs, F., 1820b, The Characters of the Classes, Orders, Genera, and Species; or, The Characteristics of the Natural History System of Mineralogy: Edinburgh, W. and C. Tait, 1820, 117 p.

Mohs, F., 1822 and 1824, Grund-Riß der Mineralogie: Dresden, Arnoldische Buchhandlung, 2 vols., v. 1, 630 p.; v. 2, 750 p.

Mohs, F., 1825, Grund-Riß der Mineralogie: Dresden, Arnoldische Buchhandlung (translation by Haidinger, W., 1825, Fr. *Mohs*, Treatise on Mineralogy: Edinburgh and London).

Mohs, F., 1829, Einwürfe des Herrn Prof. Weiss gegen die naturhistorische Methode der Mineralogie beantwortet von Friedrich Mohs: Zeitschrift für Physik und Mathematik no. VI und no. VII.

Mohs, F., 1836, Leichtfaßliche Anfangsgründe der Naturgeschichte des Mineralreichs. Zum Gebrauche bei seinen Vorlesungen über die Mineralogie an dem k.k. Hof- Mineralienkabinete: Vol. 1, Wien, Gerold, 635 p.

Mohs, F., 1839, added by Zippe, Leichtfaßliche Anfangsgründe der Naturgeschichte des Mineralreichs. Zum Gebrauche bei seinen Vorlesungen über die Mineralogie an dem k.k. Hof-Mineralienkabinete Vo. 2: Wien, Gerold, 744 p.

Oldroyd, D.R., 1996, Thinking about the Earth. A History of Ideas in Geology: Cambridge, Massachusetts, Harvard University Press, 410 p.

Pratt, M.L., 1991, Arts of the Contact Zone: Profession, v. 91, p. 33–40.

Raffler, M., 2007, Museum–Spiegel der Nation?: Zugänge zur Historischen Museologie am Beispiel der Genese von Landes-und Nationalmuseen in der Habsburgermonarchie: Wien, Köln, Weimar, Böhlau Verlag, 386 p.

Rahden, W. von, 1992, Ein fast vergessener Aktualist: Georg Christian Füchsel zu Ursprung und Entwicklung der Erde und der Sprache, *in* Naumann, B., Plank, F., and Hofbauer, G., eds., Language and Earth: Elective Affinities between the Emerging Sciences of Linguistics and Geology: Amsterdam, Studies in the History of the Language Sciences, John Benjamins Publishing Company, p. 289–322, https://doi.org/10.1075/sihols.66.15rah.

Rösler, J.H., 1989, Die Hintergründe der Berufung von Friederich Mohs im Jahre 1818 an die Bergakademie Freiberg, *in* Postl, W., and Moser, B., Friederich Mohs (1773–1839): Graz, Mitteilungen der Abteilung für Mineralogie am Landesmuseum Joanneum 57, p. 29–48.

Rudwick, M., 1996, Minerals, strata, fossils, *in* Jardine, N., Secord, J.A., and Spary, E.C., eds., Cultures of Natural History: Cambridge, UK, Cambridge University Press, p. 266–286.

Sandgruber, R., 1980, Indikatoren des Lebensstandards in Wien in der ersten Hälfte des 19. Jahrhunderts, *in* Czeike, F., Wien im Vormärz: Wien, Jugend und Volk, p. 57–74.

Schübl, E., 2010, Mineralogie, Petrographie, Geologie und Paläontologie. Zur Institutionalisierung der Erdwissenschaften an den österreichischen Universitäten, vornehmlich an jener in Wien, 1848–1938: Wien, Grazer Universitätsverlag–Leykam, 304 p.

Wagenbreth, O., 1999, Geschichte der Geologie in Deutschland: Stuttgart Enke im Thieme Verlag, 265 p.

Waidacher, F., 1999, Handbuch der allgemeinen Museologie (third edition): Wien, Köln, Weimar, Böhlau Verlag, 792 p., https://doi.org/10.7767/9783205126904.

Werner, A.G., 1774, Von den äußerlichen Kennzeichen der Fossilien: Leipzig, Siegried Lebrecht Crusius, 302 p.

Wild, R., 1999, Christian Erich Hermann von Meyer (1801–1869)—Der Erforscher der Trias- Saurier, *in* Hauschke, N., and Wild, V., eds., Trias, eine ganz andere Welt. Mitteleuropa im frühen Erdmittelalter: München, Pfeil, p. 587–592.

Wurzbach, C., 1868, Biographisches Lexikon des Kaiserthums Oesterreich, 18 vols.: Wien, Kaiserl.-Königl. Hof- und Staatsdruckerei, p. 443–448.

Manuscript Accepted by the Society 21 November 2017
Manuscript Published Online 25 May 2018

Printed in the USA

The Geological Society of America
Special Paper 535

The museums of Philadelphia

Sally Newcomb*
13120 Two Farm Drive, Silver Spring, Maryland 20904, USA

ABSTRACT

The museums of Philadelphia are noted for both their variety and their longevity. Some of these institutions have been present since before the American Revolution and reflect the continuity within a population for which education was always a predominant goal. Others have arisen over subsequent centuries in response to various needs. Some, such as the American Philosophical Society and the Fairmont Water Works, are eighteenth-century institutions that relatively lately acquired a more formal museum format. Colonial interest in natural history, mineralogy, and natural resources with collateral maps and papers makes them a prime resource for historians of geology. All of these institutions reflect the nature of the museum movement itself from often private or privileged collections to those both welcoming the public and serving as sturdy arms of education. Historians of geology will find maps, instruments, collections, books, and personal and government papers that are of much interest.

INTRODUCTION

When thinking about museums in the United States, probably one thinks first about the iconic museums on the National Mall in Washington, D.C., in New York, or in Boston. In Philadelphia, crowds are drawn to Independence Square and the historic area to learn about the American Revolution in the 1770s, but relatively few of those visitors go elsewhere. However, a closer look reveals a very lively scene for both visual and performing arts, as well as the continuation of more than two and a half centuries of scientific inquiries, supported by institutions that in many cases continue to survive. Philadelphia was the largest city in the new United States until 1790 when it was surpassed by New York City. Washington, D.C., didn't exist until 1790. Yet in Philadelphia, far from the educational and institutional resources of Europe in the eighteenth century, institutions were established to address the scientific and educational needs of the young nation. This legacy is still part of the fabric of our history.

There was much curiosity about the natural history of the Americas on the part of settlers and those who remained in Europe and England. Parrish (2006, p. 6) noted "Because the development of modernity itself and of Enlightenment natural science in particular was so fueled by the European nations' competitive drive to exploit, collect, catalog, and understand the material richness of the Americas, both American nature and the hybrid types of knowledge forged in the colonies were inseparable from that development." The primary demands of economics and resources were and are connected to natural history through natural philosophy and geology, and the libraries and museums of Philadelphia have answered these needs from their inception. It is difficult to focus solely on museums because most are linked to and supported by their extensive libraries, which can be considered archives of human knowledge. Libraries are repositories of manuscripts, maps, memoires, diaries, and government papers, as well as books that complement published scientific papers. Some libraries also have artifact collections.

*senewcomb@earthlink.net

Newcomb, S., 2018, The museums of Philadelphia, *in* Rosenberg, G.D., and Clary, R.M., eds., Museums at the Forefront of the History and Philosophy of Geology: History Made, History in the Making: Geological Society of America Special Paper 535, p. 177–189, https://doi.org/10.1130/2018.2535(11).

What we would now call the intellectual ferment in Philadelphia was very much the product of its time. It differed somewhat from European Enlightenment thought, particularly French, by the marked participation of artisans and the self-educated. The American Revolution itself was part of the mainstream of those ideas. This is not the place for an extended discussion of that enormous subject, but the effect was pervasive. As an example: "The new enthusiasm for scientific attainment voiced by Benjamin Franklin was a reflection of the Enlightenment flooding into the American colonies" (Hindle, 1956, p. 5). Hindle (1956, p. 5–6) also listed John Bartram's (1699–1777) assessment of the population of the Philadelphia area as three groups: those concerned with laying up wealth; those who spent the proceeds; and those for whom it was necessary to work to provide for themselves and their families. It was the latter who had the curiosity to pursue their natural interests. That third group was notably present in Philadelphia, and its interests included education. The "City of Brotherly Love" also demonstrated more tolerance due, some references say, to the Quaker influence (which was less tolerant of deviation in its own members). This was fertile ground for the advancement of museums and other educational institutions.

Museums are collections of tangible objects that evolved from the curio cabinets that Pyenson and Sheets-Pyenson characterized as including the "odd, peculiar, and rare" (Pyenson and Sheets-Pyenson, 1999, p. 126). Collections had a long history by the eighteenth century, associated with the intellectual elites who sought to acquire objects worthy of systematic study or for amusement. Items might be included if they had relation to a larger group or could be fitted into the Great Chain of Being.[1] The inquiring nature of the eighteenth century prompted continued interest in aids to empirical and experimental investigation, including objects and instruments designed for this purpose (Pyenson and Sheets-Pyenson, 1999, p. 128). Discoveries of new lands, with minerals and strange animals and plants, provided much to be examined and questioned, not only among natural philosophers but also by the curious public. Museums, as opposed to curio cabinets, were open to at least some portion of the public on a general basis rather than by private invitation. Philadelphia's institutions and collections were part of the larger trend, but they are unusual in the United States in having arisen early and being persistent. Over the years, the borders between some of the Philadelphia institutions have become a bit blurred as objects have moved between them. All welcome visitors and researchers.

THE LIBRARY COMPANY

Museums and libraries serve not only the high-minded acquisition of knowledge but also the need for practical information or political advantage to be gained by delineating boundaries or documenting the availability of resources. This led to the founding of The Library Company (TLC) in 1731 by Benjamin Franklin and his junto of artisans whose political and intellectual discussions were constrained by lack of books. Private libraries were beyond their means because books were expensive and rare, and new works were slow to arrive from England and the Continent. Bookstores were then nearly nonexistent in Philadelphia. Because each member could afford only a few books, the first idea was for books to be pooled so that others could read them. When this was done, some books disappeared, and some were not treated carefully. A subscription library seemed to be the answer, and The Library Company was established at the instigation of Franklin.[2] Each subscriber paid forty shillings to join and ten shillings annually. Ten directors and a treasurer were elected annually by the subscribers; the directors elected a secretary; and Articles of Agreement were drawn up to be binding on the members for 50 years (Lemay, 2005, p. 94). Franklin hoped for 50 members, but at first only 11 signed up. By fall of 1731, The Library Company reached 50 subscribers. Many of the first members were artisans, but soon wealthier colonists saw the advantages of membership.

There was little fiction in the library. Subjects eventually included history, religion, education, poetry, exploration, science, politics, and philosophy. Soon the highly educated and mathematically literate James Logan (1674–1751), once secretary to William Penn and later resident in Pennsylvania, weighed in with his willingness to consult on books to be purchased. Thomas Godfrey (1704–1749), inventor (at the same time as the British John Hadley [1682–1744]), of the doubly reflecting quadrant[3], went with Franklin to discuss appropriate purchases with Logan. Peter Collinson (1694–1768), a British merchant with a great interest in botany, agreed to act as the agent for The Library Company. Books were purchased, and there were donations, notably from Logan. The collection grew quickly and, over the years, required larger spaces to house it. It served as the Library of Congress from 1774 until 1800 when the national government was often in Philadelphia. The Library Company was the largest public library in America until the 1860s. All the books are still available, and it is still a public library. One example of the riches to be found in The Library Company is the Loganian Library, the complete library of James Logan, which contains books from

[1]This is the idea, advanced in one form or another since antiquity, that nature is arranged hierarchically by small gradations from the simplest matter, through plants, animals, humanity, and angels, to the Creator. The gradation continued through each category with, for example, one more complex plant being "higher" than a simpler one. Dr. R. Thomas (July 2017, personal commun.) has pointed out Mark Dion's installation, which spoofs the Great Chain, at the Wadsworth Atheneum Museum of Art in Hartford, Connecticut.

[2]There is a very extensive literature about Franklin with great detail about founding The Library Company (TLC).

[3]A doubly reflecting quadrant, essentially an octant, is a navigation device that measures the height of the sun or stars above the horizon to find geographic position.

the fifteenth to the eighteenth centuries. From its inception, The Library Company's holdings included works on natural history and extensive reports of exploration as well as virtually all available early books on science. Contact prior to a visit is advised so that materials can be ready.

The Library Company also has a collection of artifacts, but, over the years, many of those that might be expected in a natural history museum have disappeared or been lost, sold, or transferred. It has never strictly speaking been a museum in that it has never made its artifact collection freely open to the public. However, it is the repository of various pieces of equipment used in its members' experiments, including Franklin's and Ebenezer Kinnersley's (1711–1788) apparatus for the famed work on electricity. Its telescope and microscope were frequently borrowed by members; the telescope was employed to observe the transits of Venus in 1761 and 1769. Some apparatus can still be seen in its sixth building on Locust Street, and the artifact collection can be searched online, in the Art and Artifacts section of the Web page, www.librarycompany.org/about. In addition to their ~570,000 books and other items, The Library Company's 550 linear feet of manuscripts are available through the Historical Society of Pennsylvania, located next door. There is an extensive literature about The Library Company both printed and online. A source that places it within the context of the Philadelphia Enlightenment is Chapter 4, The Library Company, of *The Life of Benjamin Franklin: Printer and Publisher, 1730–1747* (Lemay, 2005, p. 93–123).

THE HISTORICAL SOCIETY OF PENNSYLVANIA

In this accounting of resources available in Philadelphia, I would be remiss not to mention the Historical Society of Pennsylvania, although it is not a natural history museum and does not have artifacts. Founded in 1824, it is the repository of ~600,000 printed items and 21 million manuscript and graphic items (www.hsp.org/about-us). There is material from the people associated with early and continuing science in Philadelphia. It includes materials from the seventeenth through the twenty-first centuries, including maps and expedition and travel reports. These include the du Simitière collection (see below) as well as John Bartram's (1699–1777) observations of 1751 on the climate, soils, and rivers of the region. As noted, it is next door to The Library Company and is a resource not to be missed.

THE AMERICAN PHILOSOPHICAL SOCIETY

The American Philosophical Society (APS) is unique in American history. Not an association only of the wealthy, elite, and educated, it was first founded in 1743 by Franklin and his group, who began mostly as artisans, as an outgrowth of their desire to learn from and discuss the ideas gleaned by way of The Library Company, other discussions, and correspondence.[4] Despite its somewhat rocky start, it thrives today as a venerated scholarly organization with regular publications. And as W.J. Bell stated: "That the American Philosophical Society should have had a museum from its inception was implicit in its original character and purpose as an eighteenth-century academy" (Bell, 1967, p. 1). But soon the junto of artisans found it difficult to keep to the over-optimistic schedule of reading and reports, and it languished. The organization was revived in 1767 by some of the original members led by Dr. Thomas Bond (1712–1784), with Franklin as its president. By that time, another society with similar goals had been established; but after some friction, the two united to form the *American Philosophical Society Held at Philadelphia for Promoting Usefull Knowledge* (Orosz, 1990, p. 19). Artifacts and collections were donated from the beginning, members sending all manner of curiosities as well as maps, records, and instruments. Collections were sometimes moved or loaned, especially before Philosophical Hall was built in the period 1785 until 1789, after which the Society finally had sufficient space in which to meet and keep its artifacts.[5] Over the years, members sent minerals, fossils, Indian artifacts, and botanical specimens and drawings to be observed and discussed in the meetings. Scientific instruments were donated, borrowed, and loaned to other institutions. Strictly speaking, the APS holdings didn't qualify as a museum in the sense of being open to the public until long after its founding, at least as sponsored solely by APS. The space was shared by many Philadelphia institutions over the centuries, beginning with the Peale Philadelphia Museum at its inception (see below). The activities of the APS, including publishing, have continued since its 1767 reorganization. The current museum in Philosophical Hall opened as such in 2001. It has rotating exhibits drawing on its own resources and those of other groups such as The Library Company and The Academy of Natural Sciences, featuring scientific instruments, original manuscripts, rare books, works of art, and natural history specimens. In a series of exhibits on Thomas Jefferson's America, the last in that series was titled "Gathering Voices, Thomas Jefferson and Native America." The exhibit of 7 April to 30 December 2017, titled "Curious Revolutionaries: The Peales of Philadelphia," brings the story of Philosophical Hall forward.

For historians of science, the other two collections of APS, held in two other buildings near the museum on Independence Square, are of great interest. One is the comprehensive library with holdings that date to the time of founding of APS (and before), including innumerable papers, correspondence, books, and maps from the personal libraries and donations of members and their families over more than two and a half centuries. The website lists the riches, some of the oldest somewhat unexpected to be in an American library dating from the eighteenth century. Among the notable is the Valentine Darwin collection of ~4800

---

[4]It has been pointed out that by the time the American Philosophical Society (APS) was begun, Franklin, at least, had joined the prior group (Roger Thomas, 2017, personal commun.). Thomas also suggested that the APS was the progenitor of a think tank—an excellent insight.

[5]The story of Philosophical Hall is complex and well covered in many references.

Figure 1. Circumferenter or surveyor's compass, 1796–1798 (Benjamin Rittenhouse and William Lukens Potts, makers). Courtesy of American Philosophical Society, in Philadelphia. Gift from Mrs. Thomas S. Gates, ~1950.

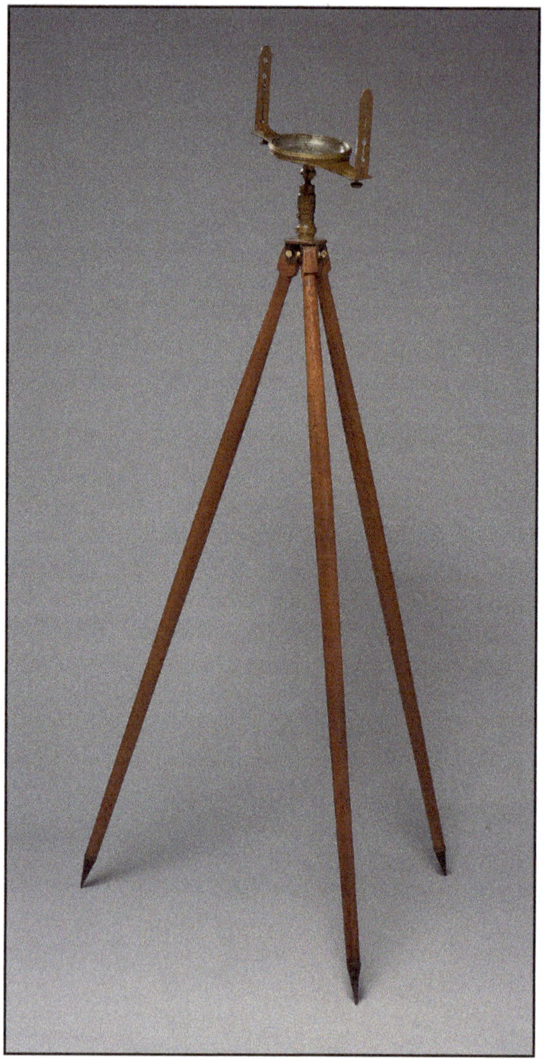

Figure 2. Incomplete Theodolite, ca. 1790. Benjamin Chandlee (maker). Courtesy of American Philosophical Society. Gift of Penrose Hoopes, 23 June 1975.

volumes of the printed works of Charles Darwin in 23 languages. The APS also has the papers of G.G. Simpson. Access to the library is by appointment, but Web access is easy.

The other interest is the artifact collection. The APS museum can show only a small portion of the artifacts still in its possession. The remaining artifacts are kept in their third building at 431 Chestnut Street, with access by appointment. Bell (1967, p. 1–2) stated, "Thus the cabinet grew, without plan or even clear purpose, until in the middle of the nineteenth century it included an astonishing variety of articles, ranging from butterflies to the bones of mastodons, and from Inca pottery to a lock of General Jackson's hair." Now almost all of the paleontological, mineralogical, and ethnic collections have been transferred to other institutions, but APS continues to accept significant donations from members and others. The current collection may be investigated by following the "objects" link on the website. There is still an impressive collection of late eighteenth- and early nineteenth-century instruments that are of interest to historians of geology,[6] particularly those having to do with mapping, locating longitude, and astronomy (see Figs. 1 and 2 for examples). Considering the size and weight of many of these instruments, it is difficult to think of them as portable and suitable for field use. One thing left out of many older reports of mapping and field work is the number of "assistants" required to carry them. Historians of science are most fortunate to have an institution such as APS that has preserved a record stretching from well before the inception of our country itself.

AMERICAN MUSEUM OF DU SIMITIÈRE

Among museums with natural history specimens that vie for status as the "earliest" in the United States was that of Pierre Eugène du Simitière (1737–1784); his museum opened in 1782 and closed in 1785. His paper collection was purchased by The Library Company, where it still resides. Orosz (1990, p. 2) pointed out that there were few models for hopeful museum originators.

[6]I spent a delightful several hours looking at them with the curator, Mary Grace Wahl.

Most museums in Europe were extensions of the cabinets of the wealthy, with access limited to the well born or members of a scientific elite, mainly the same people. Du Simitière was a fascinating character, far better known in the Philadelphia of his day than he is today, a circumstance that has been remedied by the serious study by Gian Domenico Iachini (2012). A native of Switzerland, du Simitière settled permanently in Philadelphia in the early 1770s (Iachini, 2012, p. 135) after an extended stay in the West Indies and various domiciles in the colonies. He was a talented artist, who contributed to the design of the Great Seal of the United States as well as the seals of several states. In his travels, he had acquired very extensive collections of flora and fauna as well as ethnic materials, and he published quite widely on them. He was an astute observer and historian of the early colonies, having collected every book, pamphlet, and news article he could obtain. He became a corresponding member of the American Philosophical Society in 1768, before he moved permanently to Philadelphia (Bell, 1997, v.1, p. 504). He was elected a curator of the APS in 1776.[7]

During this time, while du Simitière tried, usually in vain, to attract financial support, many Philadelphians and other visitors were admitted to see the collection. Due to the large numbers of visitors, in 1782, he established regular hours and began to charge admission (Bell, 1997, p. 512), but he remained under constant financial pressure. The collections and their cataloging were his major interest, but he pursued many avenues to remain solvent. He classified his collections as "natural" and "artificial" as was typical in wünderkammern, or curio cabinets. The natural collection included marine and land productions, petrifications, and botanical specimens. Under artificial, he included antiquities, weapons, ornamental clothing of native peoples, and a collection of paintings (Orosz, 1990, p. 36). Despite publicity, there were never sufficient numbers of visitors. Du Simitière died in 1784, and his museum died quickly thereafter. His executors hoped that the APS would purchase it, but the APS declined, and the contents were put up to public sale. The majority of his books, manuscripts, and broadsides were purchased by The Library Company, while some became part of the collection at the Library of Congress, where they may still be seen (Bell, 1997, p. 513).

THE PHILADELPHIA MUSEUM

The Philadelphia Museum of Charles Willson Peale (1741–1827) is far better known than that of du Simitière for several reasons. Prominent among them is its much longer life as a natural history museum from 1786 until 1845. Peale opened his portrait-only studio to the public for two years before expanding it. It was housed in the APS Philosophical Hall from 1794 until 1810, which at that time was also his family's home. Rather than being the collection of an individual or a small group seeking scientific knowledge, Charles Willson Peale's museum was a commercial venture intended to provide income to support his family. It capitalized on the curiosity of the general public about natural history. Besides his noted portraits, the museum grew from a dried paddlefish and a badly preserved Angora cat to include thousands of specimens including fossils, quadrupeds, birds, reptiles, fish, shells, insects, and minerals (Schofield, 1989, p. 21). Influenced by Georges-Louis Leclerc, Comte de Buffon (1707–1788) and his well-known "Histoire Naturelle," but not agreeing with him on the "degeneracy" of New World organisms, Peale attempted to display the world as organized by Linnaeus. His self-portrait in the museum shows a large room with many cases of specimens (Fig. 3). Schofield (1989, p. 21) stated: "In 1831, the museum contained 250 quadrupeds, 1,310 birds, more than 4,000 insects, 8,000 minerals, 1,044 shells, several hundred fish, more than 200 snakes, lizards, turtles and tortoises, and the major U.S. collection of fossil bones: it had become the primary resource for American natural history."

Peale's objective was to represent the whole world of living things in miniature. It was arranged to show the progression and relations of living beings, the animal, vegetable, and mineral kingdoms, all organized and classified so that the viewer "could come to understand the perfection of the Creator" (Orosz, 1990, p. 48). To this end, Peale and his family collected widely and constantly. Peale's portraits became less popular among his wealthy patrons due to the politics of the American Revolution and those of the egalitarian movement in France, so expansion into a larger museum was likely to be more profitable. The museum had a period of great popularity and success through a number of moves. After C.W. Peale's death, his heirs met with no success in seeking to acquire outside support for the museum, which went bankrupt in 1845. Its collections were sold in 1848. It appeared that "Peale's enlightened notion of a museum as science education and spiritual inspiration for 'everyman' had become archaic" (Schofield, 1989, p. 32). Now museums were becoming the realm of the specialist who used collections to increase scientific knowledge. During the time of the Peale museum's existence, it was a model of public access and education.

THE FAIRMONT WATER WORKS[8]

The Fairmont Water Works Interpretive Center is a different kind of museum but still of much interest to geologists. It is a National Historic Landmark, a Civil Engineering Landmark, and a National Mechanical Engineering Landmark. There are few places that provide a continuous story of such an important facet of city life, dependent on the natural environment. It did not achieve its current form as a repository of history and as an interpretive center until 2003, and it continues to expand. There

[7]His position in the American Revolution was complicated by the fact he was not clearly an American citizen, and he also entertained British officers when they occupied Philadelphia.

[8]The history of the Water Works began as a public utility. It didn't become an educational source until 2003.

Figure 3. *Charles Willson Peale: The Artist in His Museum,* 1822. Courtesy of the Pennsylvania Academy of the Fine Arts, Philadelphia. Gift of Mrs. Sarah Harrison (The Joseph Harrison, Jr., Collection).

must be few such institutions with salaries paid by an entity such as the Philadelphia Water Department itself (Karen Young, director, 2017, personal commun.).

In Philadelphia in the seventeenth century when William Penn's (1644–1718) colony was established, it was noted that Pennsylvania had many sources of good and clean water. This changed as the city grew. Development of the Water Works was begun in the early nineteenth century with a bequest from Franklin himself. He had recognized the role of clean water in combatting yellow fever and other ills as wells and springs became contaminated. Franklin suggested the Wissahickon Creek as a source, but it was early recognized that its volume was insufficient at some times of year. The city turned to the engineer and architect, Benjamin Henry Latrobe (1764–1820), who suggested building a series of spruce and pine pipes to bring Schuylkill River water (then cleaner) from the natural fall line to the city. Even then, the Delaware River was contaminated due to its marshes and waste from sewers and ships. The first Centre Square Works was located in the center of the city, where City Hall is now. By 1801, "63 homes, four breweries, and a sugar refinery" were supplied (www.newjersey.gov/drbc/eduweb/special/fairmont). As the city grew, so did the need for water.

The city's need for water quickly outgrew that first attempt. Latrobe's student, Frederick Graff (1775–1847), was a hydraulic engineer who oversaw expansion and designed the buildings for many years, followed by his son in the same position. The

Figure 4. Fairmont Water Works, 1905. Courtesy of Library of Congress (Historic American Engineering Record) and City of Philadelphia Archives.

dam on the falls of the Schuylkill River had been built in 1819, with water sent to the reservoir up the hill where the Philadelphia Art Museum now stands. From pumps powered by steam to turbines, always expanding, the fascinating history of this, one of the first municipal water systems in the United States (Fig. 4: the Water Works in 1905), as well as that of the entire watershed can be learned at the Fairmont Water Works Interpretive Center. An early map showed the many creeks in the city area, most of which are now sewers. Over the years, the concerns for adequate and uncontaminated water paralleled ours today. The engineering problems were constant. While the Wissahickon Creek (higher in elevation and less flow) was inadequate, the Schuykill River had the problem of high tides that led to slack water that would not turn the turbines (Fig. 5). The Fairmont Water Works stopped operating as such in 1909, but it has remained an integral part of the story of Philadelphia. The subsequent story of the site, including an aquarium, a swimming pool, and various restaurants, may be investigated on Philadelphia websites as well as at the Water Works (see Fig. 6 for a 2016 view). It now serves innumerable school classes and the public with programs that emphasize not only water supply and treatment but also the ecology of the immediate area, the river, and that of the entire watershed. A fish camera (2009) in the fish ladder installed in 1979 fascinates visitors and enables studies of migrating fish.

THE ACADEMY OF NATURAL SCIENCES AT DREXEL UNIVERSITY

The Academy of Natural Sciences of Philadelphia (ANSP), now a component of Drexel University, was founded in 1812 by a group of amateur naturalists who aimed for a "rational disposal of leisure moments" and "the advancement and diffusion of useful, liberal, human knowledge" (www.ansp.org.about/press-room/releases/2012/brief-history-academy-science/, p. 1). Unlike the Peale museum, it was not founded with public education in mind, and it had a long existence before the development of exhibits. It has been called the oldest natural sciences institution in the Western Hemisphere,[9] but clearly that has been challenged by the institutions discussed earlier. The Charleston Museum is probably the earliest (Ewan, 1976, p. 209). The initiation of the Academy of Sciences in 1812 was not unlike the other institutions. The group of men who came together to form it were not from the first families of Philadelphia but were emerging professionals—an apothecary, a dentist, a physician

[9]The American Philosophical Society (APS) embraced science in the tradition of natural philosophy. By the time APS was established, natural science had been formally differentiated as a distinct branch of knowledge.

Figure 5. Fairmont Water Works, interior of New Mill House, showing two of the three Jonval turbines, ca. 1876. Courtesy Fairmont Park Historic Resource Archives, Philadelphia Parks and Recreation.

Figure 6. Fairmont Water Works, August 2016. Courtesy of A. Leonard Pundt.

and chemist, a chemical manufacturer, a distiller, another physician, and an aspiring naturalist (Orosz, 1990, p. 101). At first, the group appeared to have formed more for social than scientific reasons, with requirements for gentlemanly manners. However, they soon bought the extensive mineral collection of Adam Seybert (1773–1825) and began a series of public lectures, open to those who could pay ten dollars (Orosz, 1990, p. 103). The Seybert collection is still in this museum, in its original cabinet. William Maclure's (1763–1840) election to the presidency, from 1817 until 1840, brought his generous financial support and, with his colleagues, much more scientific credibility.

The Academy continued to buy natural history collections, but despite Maclure's support, by the 1820s, it was in debt. In addition, Maclure's and others' commitment to public education led the museum to be opened to the public in 1828 on a limited basis, very few hours of the week, and required paid tickets signed by a member (Orosz, 1990, p. 122). However, in the 1840s, the tide shifted again with the death of Maclure and loss of enthusiasm for his ideal of public education. Emphasis reverted to research by professionals (Orosz, 1990, p. 145). Dr. Joseph Leidy (1823–1891), who practiced medicine but is best known now as a paleontologist, was prominent among the Academy's scientists. Active in many Philadelphia science institutions, he was elected to the Academy in 1845, was soon appointed librarian, served as a curator, then chairman of the board of curators, and finally twice as president of the Academy (Osborn, 1913, p. 343). Leidy's deep paleontological knowledge was demonstrated when he collaborated with Benjamin Waterhouse Hawkins in the first articulated dinosaur skeleton display, the enormous *Hadrosaurus foulkii* in 1868 (Peck and Stroud, 2012, p. 137–140). Leidy brought a high level of science to the institution, but his advancement led to controversy because he embodied professionalism and research, not the public education that Maclure had supported. During his tenure, "the curators were reduced to the status of guards rather than teachers" (Orosz, 1990, p. 148). Leidy must later have extended the scope of his commitment to science, because, in 1885, he took a position with the Wagner Free Institute of Science, where the founding principle was to advance science education—more about that later.

The dichotomy continued to play out at the ANSP, reflecting shifts in the rest of the museum world in the nineteenth and twentieth centuries. As at many current natural history museums, the Academy now supports both a full program of public education and a highly regarded research staff. There are notable collections of fossils, shells, fish, birds, grasshoppers and crickets, and an outstanding library of illustrated natural history volumes and ANSP publications. Special treasures, such as fossils collected by Mary Anning, one of William Smith's iconic geologic maps of 1815, plus maps by John Pinkerton and Maclure himself, as well as the extensive library, are still accessible. The bones of *Hadrosaurus foulkii,* the first nearly complete dinosaur fossil found in the United States, are still in the collection, along with a model. Discovered in New Jersey in 1858, it is notable for Leidy's recognition that it was bipedal, a first for dinosaurs. The library is searchable online, while in the library reading room there are examples from the broad range of materials in storage. Research continues in many areas, and school children and the general public flock to see the dinosaurs and other displays, while also participating in the educational programs.

THE WAGNER FREE INSTITUTE OF SCIENCE

The Wagner Free Institute of Science is a must-see destination for anyone serious about the history and trajectory over time of natural history museums. Perhaps it is the emphasis on education combined with the exhibition of marvels as earlier emphasized by Peale that explains its long history and survival. Located in north Philadelphia near Temple University and accessible from the historic area by the 33 bus, the museum occupies the building constructed for it in 1865 (Fig. 7). William Wagner (1796–1885) collected natural history specimens from his childhood and strongly believed that everyone should have free education in the sciences. To expand the lectures, he first gave from his home. Wagner and his wife founded the Wagner Free Institute as a free adult science education institution in 1855, starting in a public hall with hired lecturers. Later he opened his own hall and museum, the building now a National Historic Landmark that still houses it. Groups of school children come for lectures in the original hall, home to lectures by luminaries such as Joseph Leidy, Edward D. Cope (1840–1897), and Margaret Mead (1901–1978) among many others, before going up to see the exhibits.

More than 100,000 specimens of minerals, fossils, shells, insects, birds, mammals, etc., are displayed in their original cases (Fig. 8). Objects collected by Wagner, Leidy, Cope, and others, remain. When Wagner died in 1885, the board of trustees appointed Joseph Leidy to lead the educational and scientific programs, which were much expanded to include field expeditions. Leidy reorganized the museum systematically so that viewers progressed from examining simpler to more complex organisms. As the educational program flourished, the Wagners opened the first branch of the Philadelphia public library system in 1892, requiring an expansion of the original building. The Transactions of the Wagner Free Institute of Science were published from 1886 until 1944. During that time and for a period afterwards, the library participated in an exchange of publications with many other institutions. The library contains not only William Wagner's book collection dating to the seventeenth century but also a very strong representation of major nineteenth-century science books (Lynn Dorwaldt, librarian, 29 January 2017, personal commun.). This quintessential Victorian natural history museum is open to the public and constitutes a valuable resource for historians of geology and science in general. The exhibits and their associated library embody the history of the natural history museum movement and its relation to education over a period of more than 150 years.

Figure 7. Woodcut, ca. 1865, Wagner Free Institute Museum. Courtesy, Archives, Wagner Free Institute of Science.

THE MUSEUM AND LIBRARY OF THE CHEMICAL HERITAGE FOUNDATION

Of much more recent origin and narrower focus, the museum of the Chemical Heritage Foundation holds much of interest: in addition to displays explaining links between matter, materials, chemical principles, and our daily lives, there are iconic instruments, an eclectic collection of paintings related to chemistry, and a library of famed books from alchemy to those of modern chemistry.[10] The museum was opened in 2008 to document not only the instrumentation involved in making major discoveries but also to show the progress of chemistry through the ages. Major more recent instrumentation occupies the center of the museum, while a sequential exhibit of rare books and apparatus is displayed around the outside walls and progresses up the stairs to show how chemistry advanced through the centuries. This includes some mention of mineralogy. The gallery of alchemical paintings, many rare and old, can best be viewed with a guide who points out the significance of their content. The paintings show many operations and equipment that were instrumental in identifying minerals and their reactions by means of laboratory work and experiments.

MORE MUSEUMS

I have no way of knowing if there are more museums per capita in Philadelphia than in other cities. Even the number to choose from for a population of museum-goers is not clear.[11] Each year that number is swelled by thousands of tourists with an appetite for museums and history. New museums, such as the Museum of the American Revolution, are still opening (19 April 2017). The Mütter Museum of the College of Physicians of Philadelphia was opened in 1863, and it thrives today as a leading

[10]It must be noted that in December 2015, the Chemical Heritage Foundation merged with the Life Sciences Foundation, based in California, and was dedicated to the history and heritage of biotechnology. As of 1 February 2018, they have merged and become the Science History Institute.

[11]There are four major art museums, in each of which there are works with some geological significance.

Figure 8. The Wagner Free Institute of Science Exhibition Hall. Photograph by David Graham. Courtesy, Archives, Wagner Free Institute of Science.

example of Victorian displays of medical specimens, with popular educational programs. It began with Dr. Thomas Dent Mütter's (1811–1859) collection of 1700 anatomical specimens, still in their original display cases despite a move in 1909. The Athenaeum of Philadelphia originally opened in 1814 and has undergone many changes over the years. Its museum includes historical objects, but it is a major repository of architectural information. Another institution with a history that began in revolutionary times is the University of Pennsylvania. Its museum was founded in 1887, emphasizing archaeological exhibits. One of the two David Rittenhouse orreries is exhibited in the Van Pelt-Dietrick Library Center at the University. Founded in 1824, the Franklin Institute has become an excellent, very popular teaching establishment, mainly committed to connecting the public with the mechanical arts but sometimes with the history of science in general. About its inception in 1824, Greene remarked (1976, p. 7): "In no other institution in Philadelphia or elsewhere in the nation was there as strong a working interaction between scientists, technologists, educators, and men of business and industry." The Association of American Geologists was organized in 1840, meeting in the Franklin Institute lecture hall. The Marvin Samson Center for the History of Pharmacy displays artifacts from five centuries of pharmacy. The Philadelphia History Museum was founded in 1938 and offers many exhibits and artifacts relating to the city's history. Besides these larger institutions, there are many smaller venues, often the preserved homes or places of business of eminent people. I'm sure this does not exhaust the list.

CONCLUSIONS

What could have led to the proliferation and persistence of museums and their related libraries and archives in Philadelphia? The largest city in what became the United States of America at the time of the Revolution, its population has long since been surpassed by that of other cities. The national government has been in Washington, the District of Columbia, between Maryland and Virginia, since 1800. In the beginning, the habit of curiosity was embedded in eighteenth-century thought, serviced by the Royal Society and the web of activities it engendered (Parrish, 2006, p. 64) as well as the pervasive ideas of the Enlightenment. There was much interest in natural plants, minerals, and animals, both by themselves and as used by Native American peoples. Pennsylvania and its city, Philadelphia, were founded by William Penn as a Quaker colony tolerant of different modes of worship. Hindle (1976) pointed out the importance of the Quaker influence, with their tradition of self-sufficiency, self-government, and emphasis on education. The commitment to progress was strong. Greene (1976, p. 203) noted the absence of cabinets and libraries of natural history in colonial times, *except* in Boston and Philadelphia. The success of Benjamin Franklin (1706–1790) and others as printers indicates that there were readers—not a universal attribute of colonists on what was the frontier. American education and schooling achieved a higher level of importance than in England, where family and position were premier (Rosencrantz, 1976, p. 346). As mentioned in the note about the Franklin Institute, all classes of citizens came together for mutual education, shifting from traditional elites in favor of free thinkers.

As pointed out, a few Philadelphia museums kept to the "pure" natural history format. But from the establishment of the American Philosophical Society by a group of artisans, not scholars, there was much interest in technology as well. Scientific principles were sought after and valued for explanation and solution of practical problems. With advanced schools unavailable or economically out of reach, many of the "Revolutionary" scientists used self-education to a degree that seems astonishing today. The long success of Peale's museum and the initiation and persistence of many others demonstrate the desire for education. Another example is the story of the Fairmont Water Works, a public utility, as connected to the history of Philadelphia. Franklin and others recognized the relation of pervasive illnesses to the water supply. Technology provided a way to raise the level of river water so that it would flow through pipes in the city. The science came later, linking yellow fever transmission to mosquitos. This story in natural history, illustrating the feedback between science and technology, is encapsulated at the Fairmont Water Works.

The lack of formal education was addressed early. By 1740, there was a movement to establish a charity school. Franklin circulated a proposal for a more comprehensive school in 1749. Unlike other early educational institutions, it would train not only the clergy but would teach the arts as well as practical skills. This became the College of Pennsylvania, which, with additions, became the University of Pennsylvania in 1779. As one would expect, the complex story of its growth is well documented. About Peale's museum, Schofield (1989, p. 22) noted: "Transitions in American popular culture are, for example, to be traced in the museum's changes from a Jeffersonian deistic temple for the religion of humanity, to a Philadelphia resort of family and polite society entertainment, to its ultimate, ambiguous, incarnation as a failing competitor of P.T. Barnum's Jacksonian theater of the absurd." But as time went on, the museums and other institutions of Philadelphia have continued to provide service to the community according to their mandates, both casual and more formal, for the very young through postgraduates, as well as the entire age range of curious people pursuing multiple lines of interest. Today, the museums of Philadelphia continue the "democratization" of education that began before the American Revolution, maintaining the values of Logan, Colden, Franklin, Rittenhouse, Maclure, Wagner, and others of earlier times. Shachtman (2014, p. 188) pointed out that by the time of Thomas Jefferson's (1743–1826) presidency, there was an underpinning of scientific advances and educational resources that allowed a well-educated populace to advance themselves and the country through the application of their intellects. I will let Jefferson, President of the United States and long-time president of the APS, have the last word on museums:

> In the particular enterprises for museums, we have seen the populous and wealthy cities of Boston and New York unable to found or maintain such an institution. The feeble condition of that in each of these places sufficiently proves this. In Philadelphia alone, has this attempt succeeded to a good degree. It has been owing there to a measure of zeal and perseverance in an individual rarely equaled; to a population, crowded, wealthy, and more than usually addicted to the pursuit of knowledge. (Hindle, 1976, p. 210, as quoted from an 1807 Jefferson letter)

ACKNOWLEDGMENTS

Part of the pleasure of working on this paper has been the opportunity to meet the people who work in these institutions. Many thanks to Mary Grace Wahl, associate director for collections and exhibitions for APS, for a wonderful two hours examining eighteenth-century scientific instruments in their storage room. These instruments rarely see the light of day unless they become part of an exhibit in the museum. Similarly, Susan Glassman, director, and Lynn Dorwaldt, librarian, were generous with their time and knowledge about the history and current activities of the Wagner Free Institute of Science. At the Fairmont Water Works, long-time friend and environmental educator, Vivian Williams, showed me all activities, inside and out, and introduced me to the director, Karen Young, and the property manager, Garrett Selby, who filled me in on the current state of both the administration and the physical plant as well as future plans for the Water Works. At the Center for the History of Chemistry, Ann Elizabeth Wiener, museum operations manager, shared her extensive knowledge of the symbolism of the alchemical painting collection. Alexander Till at the Pennsylvania Academy of Fine Arts and Len Pundt and Adam Levine at the Philadelphia Water Department were most helpful in facilitating use of images from their collections. From previous visits, Geological Society of America owes a debt of gratitude to Dr. Ted Daeschler, president and CEO of the Academy of Natural Sciences of Drexel University, as well as to Jim Green, librarian, of The Library Company, for enabling our visits and sharing their expertise. I'm grateful to the two reviewers, Dr. Roger Thomas and an anonymous reviewer, who both provided valued insights and information. And always, thanks to Gary Rosenberg for his editorial abilities and his patience.

REFERENCES CITED

Bell, W.J., Jr., 1967, A Cabinet of Curiosities: Five Episodes in the Evolution of American Museums: Charlottesville, Virginia, University Press of Virginia, p. 1–34.

Bell, W.J., Jr., 1997, Patriot-Improvers: Biographical Sketches of Members of the American Philosophical Society: Philadelphia, Memoirs of the American Philosophical Society, 2 vols., 531 p.

Ewan, J., 1976, The growth of learned and scientific societies in the Southeastern United States to 1860, *in* Oleson, A., and Brown, S.C., eds., The Pursuit of Knowledge in the Early American Republic: American Scientific Societies from Colonial Times to the Civil War: Baltimore, Johns Hopkins Press, p. 208–218.

Greene, J.C., 1976, Science and the public in the age of Jefferson, *in* Hindle, B., ed., Early American Science: New York, Science History Publications, p. 201–213.

Hindle, B., 1956, Pursuit of Science in Revolutionary America, 1735–1789: Chapel Hill, University of North Carolina Press, 410 p.

Hindle, B., 1976, The Quaker background and science in colonial Philadelphia, *in* Hindle, B., ed., Early American Science: New York, Science History Publications, p. 173–180.

Iachini, G.D., 2012, Pierre Eugene du Simitière and the first American national museum: Royal Society of Arts Journal, v. 23, p. 131–159.

Lemay, J.A.L., 2005, The Life of Benjamin Franklin: Printer and Publisher: Philadelphia, University of Pennsylvania Press, v. 2, p. 1730–1747.

Orosz, J.J., 1990, Curators and Culture: The Museum Movement in America, 1740–1870: Tuscaloosa, University of Alabama Press, v. 14, 304 p., https://doi.org/10.2307/2078829.

Osborn, H.F., 1913, Biographical memoir of Joseph Leidy (1823–1891): Washington, D.C., National Academy of Sciences Biographical Memoirs, v. 7, p. 335–396.

Parrish, S.S., 2006, American Curiosity: Cultures of Natural History in the Colonial British Atlantic World: Chapel Hill, University of North Carolina Press, 321 p.

Peck, R.M., and Stroud, P.T., 2012, A Glorious Enterprise: The Academy of Natural Sciences of Philadelphia and the Making of American Science: Philadelphia, University of Pennsylvania Press, 437 p.

Pyenson, L., and Sheets-Pyenson, S., 1999, Servants of Nature: A History of Scientific Institutions, Enterprises and Sensibilities: New York, W.W. Norton and Company, 496 p.

Rosencrantz, B.G., 1976, Early American learned societies as informants on our past, *in* Oleson, A., and Brown, S.C., eds., The Pursuit of Knowledge in the Early American Republic: American Scientific Societies from Colonial Times to the Civil War: Baltimore, Johns Hopkins Press, p. 345–354.

Schofield, R.E., 1989, The science education of an enlightened entrepreneur: Charles Willson Peale and his Philadelphia Museum, 1784–1827: American Studies, with American Studies International, v. 30, no. 2, p. 21–40.

Shachtman, T., 2014, Gentlemen Scientists and Revolutionaries: The Founding Fathers in the Age of Enlightenment: New York, Palgrave/Macmillan, 256 p.

Manuscript Accepted by the Society 21 November 2017
Manuscript Published Online 13 April 2018

The Geological Society of America
Special Paper 535

Carl Akeley's revolution in exhibit design at the Milwaukee Public Museum

Gary David Rosenberg*
Milwaukee Public Museum, Milwaukee, Wisconsin 53202, USA, and
Earth Sciences Department, Indiana University–Purdue University, Indianapolis, Indiana 46202, USA

ABSTRACT

Carl Akeley (1864–1926) started a revolution in museum exhibit design when he created his muskrat diorama for the Milwaukee Public Museum in 1890. It was the first museum exhibit to show an animal in its natural habitat and the first to have a realistic background painted to create the illusion of depth and continuity of the animal's environment. After the Scientific Revolution began and especially during the "Age of the Marvelous" in the sixteenth and seventeenth centuries, collections of natural and human-made objects were displayed in "wünderkammern," rooms of wonder or curiosity cabinets, often arranged in a way that we might now consider strange, without a modern understanding of systematics, environmental, or cultural context. From the Enlightenment of the eighteenth century until Akeley's muskrat diorama, taxonomic groupings dominated museum displays.

Akeley's genius as a taxidermist gave the dioramas an unsurpassed realism. Instead of stuffing animal skins with straw and cotton as taxidermists had done for centuries, Akeley mounted the skins over armatures that he steadily improved as he moved from Milwaukee to the Field Museum in Chicago, and finally to the American Museum of Natural History in New York. He modeled ancillary items such as plants in the diorama with obsessive detail. The results were unprecedented evocations of living animals actively engaged in their ecological niches (herein defined as their place in nature, including their activities).

Akeley's focus was the same as that of the great minds of the Scientific Revolution, and it's the same as scientists' today: the geometry of nature, its structural detail, and spatial relationships. Akeley showed not only the form and structure of animals in his dioramas, but he also defined the environmental space that encompassed them as well. Akeley's focus was more mundane than Descartes's *res extensa* or Stephen Hawking's dark matter between cosmic bodies, but Akeley for the first time showed the general public that they could see the form and structure of animals and the space between in an altogether new way. Today that vision has a name, ecology. And we call its conception in deep time, paleoecology.

**Correspondence:* 1522 N. Prospect Avenue, Unit 502, Milwaukee, Wisconsin 53202, USA; Grosenbe@iupui.edu.

Rosenberg, G.D., 2018, Carl Akeley's revolution in exhibit design at the Milwaukee Public Museum, *in* Rosenberg, G.D., and Clary, R.M., eds., Museums at the Forefront of the History and Philosophy of Geology: History Made, History in the Making: Geological Society of America Special Paper 535, p. 191–202, https://doi.org/10.1130/2018.2535(12).

INTRODUCTION: DISCOVERING CARL AKELEY'S MUSKRAT DIORAMA

Decades ago, I spent my first year of graduate studies at the American Museum of Natural History in New York. When Norman Newell, Curator of Invertebrate Paleontology, learned that I was from Milwaukee, he remarked that the Milwaukee Public Museum was one of the finest museums in the country for exhibits. I did not fully understand the significance of that comment until 2012, when I left my position at Indiana University–Purdue University (IUPUI) in Indianapolis and returned to Milwaukee to volunteer at the museum and to seek new teaching and other opportunities.

Skip Buechel, another volunteer at the museum, pointed out the museum's "Muskrat Diorama" (Fig. 1), which shows a family of muskrats in their pond. Carl Akeley (1864–1926) created it in 1890. It is a coherent view and narrative in the sense that everything in the diorama relates to everything else. A family of muskrats is engaged in their daily activities in a pond where they live. It is a prosaic scene to contemporary eyes, in other words unsurprising, because most everyone has some idea what a local pond or marsh looks like, what kind of wildlife lives there, and how the wildlife interact with each other. Akeley had a lot to do with that awareness. In brief, Akeley started a revolution in exhibit design, and thus in public education, with that display because, until 1890, museum exhibits were simply assemblages of natural (nature-made) or "artificial" (human-made) objects without context. In the decades soon after 1890, human-made artifacts and activities were being shown in their cultural contexts just as wildlife were displayed in their natural environmental contexts.

Stated in philosophical terms, Carl Akeley "filled in the spaces" between objects in museum exhibits. He helped the public visualize a new sense of spatial organization in nature and in their cultural surroundings that had been developing and transforming Western culture since the Renaissance. He engaged a new sense of spatial perception similar to that fostered by the greatest artists of the Renaissance and naturalists and philosophers of the subsequent Scientific Revolution. But Akeley made that perspective visible to the general public for the first time. He facilitated science literacy about natural history. He shared knowledge on a broad scale, and shared knowledge is a cornerstone of democracy. The history and significance of Akeley's achievement are the subject of this report.

WÜNDERKAMMERN COLLECTIONS

Akeley's diorama sits behind a grand assemblage of stuffed animals, skeletons, and human-made, cultural artifacts whose juxtaposition in the same space is puzzling (Fig. 2). It presents unusual natural and artificial (human-made) objects from around the world that have no logical relationship to one another. It is a display in the manner of a "wünderkammer" (plural "wünderkammern"), which is German for a room of curiosities or marvels that were found across Europe in the sixteenth and seventeenth century, "The Age of the Marvelous" (Kenseth, 1991a).

The Milwaukee Public Museum has logically titled the assemblage of exotics, "A Sense of Wonder." It stands at the head of the stairs to the first-floor exhibition hall. It is the first display a visitor sees after climbing the stairs. It confronts the visitor with an unexpected diversity of natural and artificial curiosities that are for the most part foreign to Milwaukee's present-day urban setting and natural environment. It draws the visitor in closer with numerous questions: what are all these strange creatures and human-made artifacts, are they for real, who made the artifacts and what is their purpose, and why are so many kinds of things shown together?

Wünderkammern stood for a different paradigm of nature and perhaps were not intended to educate the general public but were displays of private possession, wealth, and power. Arguably the most famous wünderkammer and the one of greatest impact in the history of geology was Ole Worm's museum (Fig. 3) in seventeenth-century Copenhagen (see also Thomsen, 2009; Tarp, this volume). It is said to have made an impression on Nicholas Steno, a founder of modern geoscience (Thomsen, 2009). Joy Kenseth (1991b, p. 81) described it as "… a bewildering mixture of remarkable naturalia and man-made objects…," an assortment of animals that Europeans at the time (ca. 1654) would have deemed exotic: a narwhal tusk, a stuffed polar bear, a tortoise shell, the saw of a sawfish, a penguin, an armadillo, along with ethnographic objects from the Americas, China, and Turkey, a statue of a male nude, and a miniature replica of Giovanni da Bologna's *Rape of the Sabine Woman,* a large geode, and two stools, "… one with legs in the form of serpents and the other fashioned from the vertebra of a whale." Altogether an "illogical assemblage" to our contemporary eyes (Kenseth, 1991b, p. 81).

The urge to collect and show such novelties arose along with a renewed interest in classical texts, history, and knowledge about nature during the Renaissance (approximately the fifteenth century). The urge was roused during the Age of Exploration (roughly the same time as the Renaissance) and the Scientific Revolution (beginning in 1543 and lasting for ~200 years). It was a momentous period of discovery of previously unknown wildlife, exotic cultures, and natural phenomena.

Marvels were laden with mystical implications. They were held as signs of nature's mistakes or of God's sublime design and wisdom. During the Renaissance and Baroque eras, the Church encouraged the idea that marvels were the transcendent workings of God (Kenseth, 1991a). At the same time, artists held that their exquisite sculptures and extraordinarily detailed artworks, often shown together with natural wonders, rivaled the creations of God and that they, the artists, were custodians of human virtues (Smith, 2004). Exotic objects were associated with the magic powers of alchemy, which a privileged few claimed to control. For example, objects (e.g., mythic "philosophers' stones") were sometimes claimed to enable transformation of base metals into gold and precious gems or to confer immortality. "Putrefactions" and plastic forces within the Earth were said to shape soil into

Figure 1. Carl Akeley's "Muskrat Diorama," created in 1890 at the Milwaukee Public Museum. Photograph courtesy of Matthew Wooten, Milwaukee Public Museum.

Figure 2. "A Sense of Wonder." Milwaukee Public Museum. Photograph courtesy of Matthew Wooten, Milwaukee Public Museum.

Figure 3. Ole Worm's wünderkammer, ca. 1654. Public domain.

"earthy creatures" (Mirollo, 1991, p. 62, what we now know as fossils that are the remains of once living organisms).

Arcane relationships and impossible metamorphoses of objects in wünderkammern competed for attention with the observed characteristics or "scientific detail" of objects. The result was that objects had vague identities. The space between them was poorly defined as well. Inversely, an object cannot be described definitively if the space delimiting it is ambiguous. Consequently, the ambiguity of identity and spatial organization that wünderkammern signified differed from the organization of spatial relationships that was the outcome of the Scientific Revolution.

An interest in observable detail and verifiable structure of nature grew as the Scientific Revolution progressed, and it contributed to the waning interest in wünderkammern by the seventeenth century. Discoveries made with the newly invented microscope and telescope revealed for the first time "infinitely tiny" and "infinitely large" objects, objects at vast distances, and even well-known objects with remarkable characteristics that never before had been seen. These observations transformed the idea of the order of nature and of scale, "profoundly altering previous notions of what constituted size, scale, height, depth, extent, and point of view" (Mirollo, 1991, p. 62).

Moreover, collections of natural objects lost their ability to astonish as they became more common. Their magical properties came into increasing doubt when the discoveries of the Scientific Revolution did not verify their presumed magical properties and they did not fit into the new paradigms of nature. For Protestants during the Reformation, abnormal plants and animals came to signify upheaval and to justify exception to the Catholic doctrine of divine authority (Kenseth, 1991b). The Linnaean classification of animals and plants based on observable similarities in form and structure became hugely influential soon after it was published in the eighteenth century, and it rapidly supplanted the inconsistent and arcane organization of wünderkammer. Finally, the Enlightenment of the eighteenth century and the emergence of Western democracy swept aside the idea that marvels had mysterious powers that only a privileged few could possess. In short, marvels shed their mystique, and nature came to be valued for its ability to reveal new facts about the structure and workings of the cosmos defined by simple rules of geometry. Anyone with a modest education could comprehend these rules, and many did use them to create a new, technological basis of Western civilization (Edgerton, 1991).

SCIENTIFIC DETAIL AND SCIENTIFIC SPACE: RENAISSANCE ANTECEDENTS

Carl Akeley's exhibit is easy to overlook. The diorama almost disappears in the visual commotion of the wünderkammer in front of it. It seems old fashioned, something like a Punch and Judy puppet theater; it is small and unassuming—a handful of people could link hands around it. But its proximity to the wünderkammer assemblage marks the Milwaukee Public Museum's place in a revolution in exhibit design that brought to view the ecological conception of nature that emerged in the 1800s. In brief, Akeley's muskrat and subsequent dioramas succeeded because of Akeley's genius for visualizing and exhibiting wildlife in three-dimensional detail. He had a passion for showing the public wildlife as it looked in nature in order to convey a sense of real animals and plants in real places, evoking a narrative of real activities.

In detail (Fig. 4), it is a display that today's museum visitor would take for granted because the public is at least somewhat familiar with ponds and marshes and the wildlife that inhabits them. Akeley had a lot to do with that; habitat groupings now are commonplace in museums. He showed the public how wildlife looked up close and where and how they lived. In so doing he helped the public understand a basic ecological principle that has come to be universally accepted: every animal lives in its own particular habitat where it interacts with other wildlife and with the physical environment in its own way. That role is called its ecological niche. The muskrat family is busy doing what muskrats do during their day; one at left stands atop a mound of brown rushes that serves as the family's den. A section of the den is exposed below. Another muskrat is curled up within it, sleeping. A muskrat in the water is busy adding rushes to the den. Others to the right are balanced and resting atop vegetation. The pond recedes into the middle ground where one can make out a muskrat mound on the edge of a distant clump of rushes. Birds fly in formation in the sky painted in the background.

Akeley's habitat display was utterly revolutionary, a quantum change from displays of wildlife without context to wildlife within their natural surroundings. It earned Akeley worldwide recognition almost immediately. Shortly after presenting his muskrat diorama, he was offered a position at the British Museum, which he initially accepted; but he was diverted instead to the nascent Field Museum in Chicago.

Akeley's diorama is a window into nature, a view that had its origins in the Renaissance (approximately the fifteenth century).

Figure 4. Carl Akeley's "Muskrat Diorama," created 1890, showing detail. Photograph courtesy of Matthew Wooten, Milwaukee Public Museum.

Renaissance artists such as Leonardo da Vinci (1452–1519) rediscovered Greek and Arabic geometry and used it for the first time to depict nature. That is, they structured landscape images with explicit as well as intuitive aid of geometrically prescribed sight lines, vanishing points, and proportional changes in shape, size, and detail of objects with distance and position to convey an illusionistic, three-dimensional scene in a two-dimensional drawing or painting (Rosenberg, 2001, 2009). An imaginary, gridded window interposed between the artist and nature facilitated the construction of the landscape. The plane of the paper or board on which the landscape is drawn or painted came to signify the window into nature.

Akeley's diorama is viewed through a real window, albeit one that is not gridded. The diorama has a structure that is akin to Leonardo's oldest known work of art, a drawing of a landscape that scholars have variously named, *The Hills of Tuscany,* created in 1473 (Fig. 5A). It is arguably the first work of art in which landscape was the subject—in other words, it marks the beginning of the genre that we now know and take for granted as landscape art. "Landscape" is a scene in which individual objects of nature constitute a whole that is larger than the sum of its parts, just as the human body is considered an entity that is greater than the sum of its individual organs (Rosenberg, 2001, 2009).

Leonardo's scene expresses Steno's principles visually 200 years before Steno published them in writing (Rosenberg, 2001, 2009). The rock layers in the ravine are horizontal and lie atop one another in a definable sequence, logically with the oldest at the bottom and youngest at the top. They can be shown (Rosenberg, 2001) to repeat across the ravine and in the distance (Fig. 5B). Furthermore, the landscape establishes a narrative, which assumes a passage of time: the viewer peers into the distance from which he came or toward which he will go en route to or from the castle, past boats floating upon a distant pond and fields tilled in the distance. More importantly, the spatial organization opens the way for a temporal narrative involving restoration of rock layers that are similar in the outcrops but missing between them as will be discussed.

Figure 5. (A) Leonardo da Vinci, 1473, *View of the Hills of Tuscany.* Scala/Art Resource, New York. Gabinetto dei Disegni e delle Stampe, Uffizi, Florence. (B) Perspectival analysis superimposed on Leonardo da Vinci's *View of the Hills of Tuscany* showing continuity of strata across a ravine and into the illusionistic distance. From G.D. Rosenberg (2001, p. 136, and 2009, p. 21).

This general configuration of the scenery later became a widely adapted formula for Western European landscape depictions, especially in the seventeenth century under the influence of Claude Lorrain (c. 1600–1682). Both Leonardo and Akeley's creations employ a framing device that is used to enhance the illusion of a distant perspective or sense of three dimensions. Leonardo's drawing has a promontory on the left and hill on the right that direct the eye to the center and into the illusionistic distance. The muskrat mound on the left and a clump of rushes on the right serve a similar function in Akeley's diorama. In both works, there is a gap in the middle with detail that lures the eye toward a distant perspective. Leonardo accomplishes this via a ravine or quarry, and Akeley does it with a pool of water. Leonardo draws the eye down the ravine and past a castle sitting on the tip of the promontory at left, into the distance, past boats on a placid lake, and beyond tilled fields that recede proportionally (geometrically) in size toward the distant horizon. Akeley draws the viewer's gaze back into the distance with clumps of floating rushes that progressively diminish in size to the muskrat mound in the middle ground, and then farther into the distance toward a flock of birds flying in the sky painted in the background.

Akeley's crew enhanced the illusion of three dimensions with another artistic technique. It arose and was especially influential in Flemish (Belgian) and other Netherlandish art of the late sixteenth and early seventeenth century. It was an "atmospheric" technique, one involving the use of color and distinctness of objects to evoke distance. Reds and browns, colors that seem to advance toward the eye, were used for objects in the foreground, greens in the middle ground, and blues, which recede from the eye, for the background. In Akeley's diorama (Fig. 4), brown muskrats and rushes are in the foreground, green rushes in the middle ground, and the blue sky is painted in the background. This was, in fact, the first time a painted background was used to suggest depth in a museum display (C. Akeley, 1923, p. 1).

There is no evidence that Akeley was consciously referencing Leonardo's drawing or Netherlandish atmospheric techniques, but it is a fair assumption that his crew of painters had a classical training in art and knew how to use such devices to evoke a three-dimensional landscape. A group of German immigrants known as the "Milwaukee panoramists" were active and influential in the city when Akeley arrived in 1886. They created some of the largest paintings ever made in America. They were known as "cycloramas," dramatic religious or historical narratives that were so large that they were immersive to those who beheld them. They were gradually unrolled to create an emerging drama for enthralled visitors. Or they were mounted on walls surrounding visitors in 360 degrees of an epic scene. Only one cyclorama made in Milwaukee is known to still exist. It is "The Battle of Atlanta," which took place during the Civil War. It has long been a tourist attraction in Atlanta, where it is currently being moved to a new venue for restoration and enhanced display (Jones, 2017b).

Carl Akeley's interest in observable detail manifested itself principally in taxidermy, which had been his calling since he was a young man in Upstate New York, where he grew up. He revolutionized taxidermy as much as he revolutionized exhibit design. He was not satisfied with stuffing animal skins with straw, cotton, rags, and other materials as taxidermists had done for centuries. At their worst, such presentations looked like stuffed pillowcases. He considered himself an artist, and taxidermy was his sculpture (Bodry-Sanders, 1991, 1998; Kirk, 2010; Jones, 2017a). He did produce bronzes, particularly of African animals in action; both the Field Museum in Chicago and the American Museum of Natural History in New York still celebrate these bronzes on their websites (see Web Resources listed below). But his tableaus of stuffed and mounted skins remained his principal focus, and he ceaselessly worked to perfect his methods throughout his life.

He advanced the design of manikins and armatures over which animal skins were skillfully fitted. He was obsessive in his attention to the detail of animal fur, simulation of eyes, even the posture of the mounted animals in order to evoke the animals' "ethos or character manifest by motor behavior" in their natural habitat (D. Akeley, 1930, p. 35).

These objectives were facilitated by his observations and thorough note taking in the field, whether he was hunting muskrats in Wisconsin or big game on African safaris sponsored by his post-Milwaukee employers, the Field Museum in Chicago and later the American Museum of Natural History in New York. He also invented an advanced motion picture camera capable of rapidly swiveling and changing focus so that he could more faithfully record the activities of living animals in their natural habitats and use that information to "bring his animals to life" back in the museum rather than to trust his memory to turn out unconvincing reenactments of animal behavior.

He was notorious for insisting that every blade of vegetation in his dioramas be made to look different from every other blade because that's how plants look in nature. And he pioneered ways to render water, rocks, and soil, as well as the sky with convincing realism. He even applied his techniques of taxidermy to mold buildings and rocks the way that sculptors create molds for statues. He invented a gun to spray cement on the façade of a building from Chicago's Columbian Exposition of 1893 in order to retard spalling. He called the spray "gunite" (which others would develop into a variant now called shotcrete), and the building was the first home of Chicago's Natural History Museum, later to be named the Field Museum. Among other uses, gunite and shotcrete are now sprayed on rock outcrops along highways to retard erosion and to prevent rock falls.

The outcome of his efforts was that everything in his dioramas looked real. The viewer's imagination was stimulated to create a coherent narrative of interactions just as Leonardo's landscape encouraged a story. Muskrats swim in the water and build their dens from the plants that grow in the sediment. The plants help trap sediment so that more plants can grow. Birds flock in the sky and will land in the pond to feed. Sooner or later, rain will fall from the sky to fill the ponds and provide a home for the muskrats. A narrative establishes a sequence of events and passage of time that geologists can appreciate. Prompting the imagination inspires viewers to visualize the stories that nature offers to tell.

SIGNIFICANCE: OUTCOMES OF THE SCIENTIFIC REVOLUTION

Carl Akeley's work grew from the same concept of spatial relationships in nature that occupied the greatest minds of the Scientific Revolution. This is not to say that Akeley was consciously interpreting the arcane writings of such luminaries for the general public, nor that the public was (or is) aware of those connections. But Akeley's objectives were historically consistent, and he made the geometric paradigm of nature visually accessible and intuitively comprehensible for the public. It was an act of generosity that fulfilled one of the promises of the Scientific Revolution—the democratization of scientific knowledge—which has become one of the most important missions of the modern natural history museum.

At first glance, it is remarkable that the shared knowledge that Akeley fostered derives from the same considerations that Leonardo da Vinci's Renaissance drawing of the hills of

Tuscany exemplifies. But it is not surprising, if one considers that perception of spatial relationships is fundamental to all living things[1]. All senses, not simply vision but also hearing, smell, taste, and touch are useful in defining bearing and distance from stimuli, hence help to situate and define spatial relationships[2]. It also follows that important eras such as the Scientific Revolution in Western culture are distinguished by their conceptions of spatial organization (Berger, 1972; Edgerton, 1991; Foucault, 1994).

Every science defines its own space in nature, whether it's the arrangement of organs in the human body (anatomy), the distribution of matter in the cosmos (cosmology), the configuration of the atom (chemistry), or, not least, the structure of our planet and its changes through time (geology). The rediscovery of Greek and Arabic geometry during the Renaissance and use of it for the first time to describe nature made all this possible starting in the Scientific Revolution. Whether use of mathematics is explicit (e.g., employing construction lines or equations) or intuitive (e.g., estimating proportional changes), geometry provides a standard scale for defining three-dimensional objects and their spatial relationships. Anyone with a modest amount of instruction can understand and communicate with others about them (Edgerton, 1991). Shared knowledge is a hallmark of democracy.

Numerous minds debated the nature of object and space over the course of the Scientific Revolution. Immanuel Kant (1724–1804), for example, discussed whether space was a material object or a concept. In 1770, he wrote (Janiak, 2009, p. 2): "Space is not something objective and real, nor a substance, nor an accident, nor a relation; instead, it is subjective and ideal, and originates from the mind's nature in accord with a stable law as a scheme, as it were, for coordinating everything sensed externally."

René Descartes wrote: "… I find [matter] to consist merely in … length, breadth, and depth so that whatever has three dimensions is a part of this matter …." And, "… there cannot be any completely empty space, that is, space containing no matter, because we cannot conceive such a space without conceiving in it these three dimensions, and consequently matter…" (Descartes, Letter to Pierre Chanut, 6 June 1647, from Cottingham, 1992, p. 86). In other words, the geometric parameters that define a material object simultaneously delimit the space that surrounds it, the result being that the surrounding space is also a material object, and the pair function together like a machine.

Few museums would dare plan exhibits of Descartes's or Kant's philosophy for the public. Few would plan a diorama about Descartes's vortices, which he felt compelled to call upon to fill space with materiality (Slowik, 2013). Nor would they try to depict the relationship of Descartes's *res extensa* (extended or corporeal matter) to *res cogitans* (conceptual matter). But inscrutable as such writing might seem, the question of what occupies the space between material objects at all scales from the cosmic to the nano is of concern to scientists yet today. For example, Stephen Hawking and Leonard Mlodinow wrote (2010, p. 110): "There is no such thing as empty space." Contemporary physicists are still on the hunt to define "dark matter" in the "space-time continuum" between objects whether at the cosmological or nano scale (Weatherall, 2016).

Outcrops of rock layers were for Nicholas Steno the operational units of material objects, and the valleys between them in which those rock layers are absent were the operational units of surrounding space. In his *Prodromus* of 1669 (Steno, 1968), he conceptually "filled the spaces" between outcrops with rock layers as functions of time and thereby founded the evolutionary science of geology. He proposed that the discontinuities or spaces between distant outcrops of rock layers were produced by the erosion of once laterally continuous sheets of rock. Thus, Steno the anatomist began the evolutionary narrative of landscape; objects in the landscape and the spaces between them began to constitute an object greater than the sum of its parts (Rosenberg, 2009). Steno recognized the mathematical (geometric) imperative for defining structure and intervening space without the spiritual connotations that Descartes advocated and thus went further than Descartes's concept of nature as machine controlled by a maker (Hansen, 2009; Olden-Jørgensen, 2009). As Steno wrote in his discourse on the brain (translation by Hansen, 2009, p. 163), "… you must also put all the individual parts together in order to understand how they work in coherence." The same held for the landscape, and the outcome was that geoscientists now have a unique perspective on evolutionary changes in the coherence of landforms with the space between them.

Akeley's insistence upon displaying anatomically correct taxidermy and truth to scientific detail of both the animal and of its environment in the intervening space is analogous. Their interface prompts asking how the fit evolved and how was it maintained. This is *natural history*, and a geologist or paleoecologist today would most likely assume that a given spatial arrangement of landforms and/or organisms and their surroundings is but a single frame of a very long sequence of integrated arrangements that originated in deep time and evolved to the present. Simply stated, geologists and paleoecologists hold that spatial organization has a history and, implicitly, that space and time are a continuum (inseparable) in a way that is compatible with Einstein's conception of them[3].

How the debate about dark matter and the space-time continuum will play out is subject to conjecture. But 1890 was the right time for Akeley's ecological rendition. Ecology and the polemics of conservation were gaining attention by the mid-1800s with the writings of Muir, Thoreau, Emerson, Pinchot, and others in response to the increasing industrialization and urbanization of

[1]Plants included. Although Peter Wohlleben's (2015) book, *The Hidden Life of Trees*, is sometimes uncomfortably anthropomorphic, the author cites several scientific papers that study the ability of plants to respond to spatial cues.

[2]Taste may seem to provide no spatial clues for humans, but consider organisms in our evolutionary past that probably derived spatial information from tasting seawater or sediment, as many animals still do today.

[3]But not as Newton conceived space and time, i.e., as backdrops to the material cosmos. I thank Vic Baker, University of Arizona, for this insight.

the young American nation. And by 1913, Lawrence Joseph Henderson's "The Fitness of the Environment" discussed the integration of organism and environment head-on.

The Milwaukee landscape was rapidly transitioning at the time of Akeley's arrival in 1886. The population of Milwaukee, then a rival to Chicago, was burgeoning with ~200,000 residents. Until the mid-1800s, southeastern Wisconsin was heavily forested, and maps of the 1890s still showed large areas of swamp (Gurda, 1991, p. 125–126). The forest was rapidly being depleted by agriculture (especially wheat) and logging. Wetlands were being filled in for housing and industries such as tanneries, breweries, and iron works, which had become central to the area's economy. Milwaukee was then known as "The Machine Shop of the World" (Gurda, 1991).

Thus, Akeley did not have to go far to study muskrats in their natural habitat—nor encroachments upon it. Increase Lapham (1811–1875) was warning that the loss of forests from around Milwaukee would lead to increased flooding along the Milwaukee and Menomonee Rivers (Gurda, 1991, p. 144). Akeley's original objective was to observe, collect, and create dioramas of all mammals of Wisconsin in addition to muskrats; so it is safe to assume he was as sensitive to loss of the region's natural resources and pre-European habitat as Lapham was.

Akeley was not a polemicist who advocated for science literacy in print, but the immediate recognition that his muskrat diorama brought him indicated that it resonated with Milwaukee's progressives, who placed a high priority on education. Immigrants or their first-generation, American-born children then constituted 86% of Milwaukee's population, in contrast to an average of 33% in other American cities (Gurda, 1991). Most of these came from the German states beginning in the mid-1800s, and these included a few thousand German socialists who influenced the development of Milwaukee beyond their numbers. The term "socialist" had different connotations than it does today. The socialists were largely "German Forty-Eighters," progressives who lost a battle to reform the German states in 1848, a year when revolution swept with varying success across Europe. The autocracy was the winner in the German states, and the progressives fled into exile, many to American cities where German immigrant communities were already established, particularly Milwaukee but also Cincinnati and St. Louis.

The Forty-Eighters were social organizers and free-thinking, anti-Catholic "secular humanists" who studied the "teachings of nature and history" as an approach to harmony with nature. In 1851, they founded the German-English Academy in Milwaukee to educate the public for the world's work. They were led by Peter Engelmann (1823–1874) (see also Coorough Burke and Sheehan, this volume), who believed in hands-on learning and whose personal collection of fossils, minerals, and plants formed the core of what is now the Milwaukee Public Museum's natural history collections (Gurda, 1991).

Another Forty-Eighter of note was Margarethe Meyer-Schurz (1833–1876), who started America's first kindergarten in the town of Watertown near Milwaukee in 1856. Hers was a private kindergarten; the first public kindergarten was later established in St. Louis. The rationale was the same for both: recognition that public education from an early age was vital for personal, local community, and national advancement.

Margarethe's husband, Carl Schurz (1829–1906), became a general in the Union forces during the Civil War. Subsequently, he was elected the first German-born senator (by then, from Missouri) and became Secretary of the Interior under Rutherford B. Hayes. He was a conservationist who argued for forest preservation, and he pushed, with mixed results, for reforms in the Civil Service. The Republican Party had been founded in 1854 in Ripon, Wisconsin, on an anti-slavery and immigrants' rights platform. Schurz was an active member, but throughout his later career, his opinions and efforts on behalf of civil rights for African Americans and Native Americans were mixed.

There is no doubt that Akeley recognized the ecological significance of his works. He was as ardent and arguably as accomplished a conservationist as John Muir, who grew up in Wisconsin but left in 1864, well before Akeley arrived. The two never met, but both were influential in expanding or establishing national parks, Muir in the United States and Akeley in Africa.

There was only one degree of separation between them. Both met Teddy Roosevelt. Muir influenced Roosevelt to expand Yosemite National Park and to set in motion action in Washington that ultimately led to federal administration of all national parks. Akeley went with Roosevelt on safari in Africa, where both hunted big game. And like Roosevelt, Akeley achieved almost mythic status for his death-defying heroics and for the collections he accumulated on his safaris (M.L.J. Akeley, 1929, 1940; C.E. Akeley and M.L.J. Akeley, 1946; Bodry-Sanders, 1991, 1998; Kirk, 2010). In Akeley's case, it was with active participation of both his first and second wives. But Akeley was conflicted about killing and rationalized it for his museum exhibits while deploring hunting as a sport for trophies to hang on the wall. He wrote (C. Akeley, 1923, p. 57): "Men have spoken of darkest Africa, but the dark chapters of African history are only now being written by the inroads of civilization." And (C. Akeley, 1923, p. 114): "I have found but little enjoyment in shooting any kind of animal ..." but shooting animals that do not provide a challenge "... made me feel a great deal like a murderer."

Although his mountain gorilla diorama, "The Old Man of Mikeno" at the American Museum of Natural History in New York, was one of his proudest achievements and remains one of his most famous, he put his conservation ethic into practice when he convinced King Albert of Belgium to establish Virunga National Park in the Congo in 1925, Africa's first national park, in order to protect the mountain gorilla, which he saw as increasingly endangered by big game hunters and civil conflicts in the region.

Akeley died in 1926 at age 62 from illness on safari in Africa, and a few years later, William Morton Wheeler eulogized Akeley's accomplishments. He was Akeley's mentor when both worked at Ward's Scientific in New York and, subsequently, when Wheeler was Director of the Milwaukee Public Museum, where

Akeley was his employee. Wheeler wrote (D. Akeley, 1930, p. 35) that Akeley's displays were "... prophetic of a change which through the influence of the ethologists, behaviorists, psychologists, and physiologists, has now pervaded the whole field of the biological sciences, so that we have come to see that *an organism cannot be isolated, even conceptually, from the peculiar environment to which it has become adapted during aeons of geologic time without a serious misunderstanding of its true nature*" (italics added).

Wheeler was not the first to view Akeley's accomplishments through the lens of deep time. Henry Fairfield Osborn, Director of the American Museum of Natural History, adopted a similar perspective in his preface to Akeley's (1923) book, *In Brightest Africa.* Akeley did not leave behind an extensive written record, but his book chronicled in detail one of his African safaris sponsored by the American Museum of Natural History. Osborn (1923, preface) wrote: "Through sculpture, Carl E. Akeley is recording the vanishing greatness of the natural world of Africa. *We paleontologists alone realize that in Africa the remnants of all the royal families of the Age of Mammals are making their last stand,* that their backs are up against the pitiless wall of what we call civilization. Human rights are triumphing over animal rights, and it would be hard to determine which rights are really superior or most worthy to survive" (italics added).

In his reference to adaptation over the course of geologic time, Wheeler's tribute presages another outcome of Akeley's work. Akeley was aware of Darwin's theory of evolution, and he fully accepted its tenets. He cited the evolutionary relationships between apes and humans several times in his book, *In Brightest Africa,* and he understood that animals were adapted to and had evolved along with their environments. Although Akeley was not formally educated as a paleontologist, at the very least his dioramas fostered the realization that animals in the distant past also occupied their own ecological niches.

The science of reading such relationships from the fossil record would become formalized as paleoecology by the mid-1900s. It is reasonable to assume that Norman Newell had Carl Akeley in mind when he praised the quality of the Milwaukee Public Museum's exhibits to me, for Norman Newell worked daily among the American Museum of Natural History's galleries distinguished by Carl Akeley's dioramas and, along with John Imbrie, was one of the founders of paleoecology (Imbrie and Newell, 1964).

CONCLUSIONS

There is something to be said for wünderkammern. Assemblages of strange objects arouse curiosity even if—or perhaps because—they have not been explained. People's imaginations facilitate their participation in searching for understanding, which is certainly a vital aspect of science. But science does not stop there, and natural history museums evolve accordingly. Science fits wonders into concepts that expand the understanding of the object, help generate more facts, and raise more questions. Natural history museums strive to involve the public in those developments.

The influence of Carl Akeley's muskrat diorama spread from Milwaukee to Chicago, then to New York, and onward. Natural history exhibits became more immersive, the Milwaukee Public Museum's "Silurian Reef Diorama" (Fig. 6) almost literally so (see also Coorough Burke and Sheehan, this volume). The diorama seems to surround the viewer in a Silurian sea much the way the Milwaukee panoramists sought to immerse the viewer in the history of religion and human warfare. As stated, museums began to put not only animals and plants in the context of

Figure 6. Diorama of a Silurian Reef. Milwaukee Public Museum. Photograph by G.D. Rosenberg.

Figure 7. Diorama of a Buffalo Hunt. Milwaukee Public Museum. Photograph by G.D. Rosenberg.

their natural environment but also human artifacts and people in their cultural context. Barriers between exhibits and visitors were eliminated. Some examples from the Milwaukee Public Museum include dioramas showing museum curators hunting for fossil Dromaeosaurs in the Hell Creek Formation; a Costa Rican rain forest through which visitors can walk and study wildlife; and a "Buffalo Hunt" with Native Americans on horseback chasing bison that are heading for a real-life little boy, who is obviously convinced he is part of an important event in early American history (Fig. 7).

In retrospect, it is likely that Carl Akeley was blessed to have lived in a progressive city that embraced all four dimensions of scientific spatial relationships and their ecological and evolutionary significance. Today, the value of the scientific perspective is being questioned by political forces; this is manifest in the deterioration of funding for natural history museums (Conniff, 2016). The challenge is not merely to maintain but also to expand space for science on behalf and with the participation of the general public. It is what our democracy is all about.

ACKNOWLEDGMENTS

I thank James Kelley and Alan Young (Milwaukee Public Museum) and Penelope Bodry-Sanders (Lemur Conservation Foundation and American Museum of Natural History) for extensive discussion about Carl Akeley and his legacy at the Milwaukee Public Museum. Patricia Coorough Burke (Milwaukee Public Museum) kindly reviewed the manuscript. Matthew Wooten (Milwaukee Public Museum) kindly provided images of "A Sense of Wonder" and of Carl Akeley's "Muskrat Diorama" at the Milwaukee Public Museum. Paul Mayer (Field Museum, Chicago) kindly provided images of Akeley's diorama, "The Four Seasons," at the Field Museum for use in my talk at the GSA Annual Meeting, September 2016, on Carl Akeley. I also thank Victor Baker, University of Arizona, and William Brice, University of Pittsburgh, Johnstown, for extensive review.

WEB RESOURCES

The web resources cited in this paper were last accessed 21 November 2017 and are at:

American Museum of Natural History, New York. Akeley Hall of African Mammals: http://www.amnh.org/exhibitions/permanent-exhibitions/mammal-halls/akeley-hall-of-african-mammals.

Field Museum of Natural History, Chicago. Web Page for Carl Akeley: https://www.fieldmuseum.org/about/history/carl-akeley.

REFERENCES CITED

Akeley, C.E., 1923, In Brightest Africa: Garden City, New York, Doubleday, Page and Co., 267 p.

Akeley, C.E., and Akeley, M.L.J., 1946, Adventures in the African Jungle: New York, Dodd Mead and Company, 275 p.

Akeley, D., 1930, Jungle Portraits: New York, Macmillan, 398 p.

Akeley, M.L.J., 1929, Carl Akeley's Africa: The Account of the Akeley-Eastman-Pomeroy African Hall Expedition of the American Museum of Natural History: New York, Dodd Mead and Company, 321 p.

Akeley, M.L.J., 1940, The Wilderness Lives Again: Carl Akeley and the Great Adventure: New York, Dodd Mead and Company, 398 p.

Berger, J., 1972, Ways of Seeing: London, British Broadcasting Corporation and Penguin Books, 165 p.

Bodry-Sanders, P., 1991, Carl Akeley: Africa's Collector, Africa's Savior: New York, Paragon House, 298 p.

Bodry-Sanders, P., 1998, African Obsession: The Life and Legacy of Carl Akeley: Jacksonville, Batax Museum Publishers, 298 p. (Note: This is a reprint and revision of Bodry-Sanders, 1991.)

Conniff, R., 2016, Our natural history, endangered: New York Times, 1 April 2016: http://www.nytimes.com/2016/04/03/opinion/ournatural-history-endangered.html?emc=eta1&_r=0 (accessed May 2017).

Coorough Burke, P., and Sheehan, P.M., 2018, this volume, Museums at the intersection of science and citizen: An example from a Silurian reef, *in* Rosenberg, G.D., and Clary, R.M., eds., Museums at the Forefront of the History and Philosophy of Geology: History Made, History in the Making: Geological Society of America Special Paper 535, https://doi.org/10.1130/2018.2535(17).

Cottingham, J., 1992, Descartes: Oxford, UK, Blackwell, 167 p.

Descartes, R., 1647, Letter to Pierre Chanut, 6 June 1647, *in* Cottingham, J., 2010, Descartes: Oxford, UK, Blackwell, p. 86.

Edgerton, S., 1991, The Heritage of Giotto's Geometry: Art and Science on the Eve of the Scientific Revolution: Ithaca, Cornell University Press, 313 p.

Foucault, M., 1994, The Order of Things: The Archaeology of the Human Sciences: New York, Vintage Books, 387 p.

Gurda, J., 1991, The Making of Milwaukee: Milwaukee, Milwaukee County Historical Society, 446 p.

Hansen, J.M., 2009, On the origin of natural history: Steno's modern, but forgotten philosophy of science, *in* Rosenberg, G.D., ed., The Revolution in Geology from the Renaissance to the Enlightenment: Geological Society of America Memoir 203, p. 159–178, https://doi.org/10.1130/978-0-8137-1203-1-203.0.159.

Hawking, S., and Mlodinow, L., 2010, The Grand Design: New York, Bantam Books, 208 p.

Henderson, L.J., 1913, The Fitness of the Environment: New York, Macmillan, 312 p.

Imbrie, J., and Newell, N., 1964, Approaches to Paleoecology: New York, John Wiley, 422 p.

Janiak, A., 2009, Kant's views on space and time: Stanford Encyclopedia of Philosophy, http://plato.stanford.edu/entries/kant-spacetime/#Aca (accessed May 2017).

Jones, M., 2017a (3 Feb.), Akeley revolutionized museum taxidermy: Milwaukee, Milwaukee Journal Sentinel, Friday, p. C1.

Jones, M., 2017b (11 Feb.), Art on a Grand Scale. Massive, 130-year-old Civil War cyclorama, with Milwaukee signature, carefully moved across Atlanta: Milwaukee, Milwaukee Journal Sentinel, p. A01.

Kenseth, J., 1991a, The age of the marvelous: An introduction, *in* Kenseth, J., ed., The Age of the Marvelous: Hanover, New Hampshire, Hood Museum of Art, Dartmouth College, p. 25–59.

Kenseth, J., 1991b, A world of wonders in one closet shut, *in* Kenseth, J., ed., The Age of the Marvelous: Hanover, New Hampshire, Hood Museum of Art, Dartmouth College, p. 81–101.

Kirk, J., 2010, Kingdom under Glass: New York, Henry Holt and Company, 376 p.

Mirollo, J.V., 1991, The aesthetics of the marvelous: The wondrous work of art in a wondrous world, *in* Kenseth, J., ed., The Age of the Marvelous: Hanover, New Hampshire, Hood Museum of Art, Dartmouth College, p. 61–79.

Olden-Jørgensen, S., 2009, Nicholas Steno and Rene Descartes: A Cartesian perspective on Steno's scientific development, *in* Rosenberg, G.D., ed., The Revolution in Geology from the Renaissance to the Enlightenment: Geological Society of America Memoir 203, p. 149–158, https://doi.org/10.1130/978-0-8137-1203-1-203.0.149.

Osborn, H.F., 1923, Preface, *in* Akeley, C.E., In Brightest Africa: Garden City, New York, Doubleday, Page and Company, 267 p.

Rosenberg, G.D., 2001, An artistic perspective on the continuity of space and the origin of modern geologic thought: Earth Sciences History, v. 20, no. 2, p. 127–155, https://doi.org/10.17704/eshi.20.2.n845766556q63672.

Rosenberg, G.D., 2009, The measure of man and landscape in the Renaissance and Scientific Revolution, *in* Rosenberg, G.D., ed., The Revolution in Geology from the Renaissance to the Enlightenment: Geological Society of America Memoir 203, p. 13–40, https://doi.org/10.1130/978-0-8137-1203-1-203.0.13.

Slowik, E., 2013, Descartes' Physics: Stanford Encyclopedia of Philosophy, http://plato.stanford.edu/entries/descartes-physics/ (accessed May 2017).

Smith, P.H., 2004, The Body of the Artisan: Art and Experience in the Scientific Revolution: Chicago, University of Chicago Press, 352 p., https://doi.org/10.7208/chicago/9780226764269.001.0001.

Steno, N., 1968, The Prodromus, *in* Winter, J.G., ed., The Prodromus of Nicolaus Steno's Dissertation Concerning a Solid Body Enclosed by Process of Nature within a Solid: New York, Haffner, p. 205–277.

Tarp, L., 2018, this volume, Museum Wormianum: Collecting and learning in seventeenth-century Denmark, *in* Rosenberg, G.D., and Clary, R.M., eds., Museums at the Forefront of the History and Philosophy of Geology: History Made, History in the Making: Geological Society of America Special Paper 535, https://doi.org/10.1130/2018.2535(06).

Thomsen, E., 2009, Niels Stensen—Steno, in the world of collections and museums, *in* Rosenberg, G.D., ed., The Revolution in Geology from the Renaissance to the Enlightenment: Geological Society of America Memoir 203, p. 75–91, https://doi.org/10.1130/2009.1203(05).

Weatherall, J.O., 2016, Void: The Strange Physics of Nothing: New Haven, Connecticut, Yale University Press, 187 p.

Wohlleben, P., 2015, The Hidden Life of Trees: Vancouver, Greystone Books, 260 p.

Manuscript Accepted by the Society 21 November 2017
Manuscript Published Online 13 April 2018

Printed in the USA

The Geological Society of America
Special Paper 535

Geology and paleontology at the California Academy of Sciences, 1895–2016: A brief overview

Alan E. Leviton
Michele L. Aldrich*
California Academy of Sciences, Golden Gate Park, San Francisco, California 94118, USA

ABSTRACT

Too often we view museums as display centers, not places where basic research takes place. But that is unfortunate. Museums bridge the gap between scientists and public outreach and, on balance, manage well. This is true for many of the world's major museums, including the California Academy of Sciences, where geology plays a significant role in both basic research and outreach programs.

In 1889, the Academy began construction of its building on Market Street in downtown San Francisco to replace the ramshackle church that had been its home for some years. It would now have space for public exhibits and its core research activities. For some of its exhibits, spectacular materials were purchased from Ward's Natural Science Establishment, but for others, it drew upon its own collections, notably rocks and minerals. The new Academy opened its doors to the public in early 1892, but not long afterward both buildings and contents were lost in the 1906 San Francisco earthquake. The Academy rebounded and undertook construction of new facilities in Golden Gate Park where, for many years, its exhibits, apart from the magnificent Steinhart Aquarium and its displays of living fishes, consisted mostly of habitat dioramas but also included a mineralogy hall and, in time, a fossil hall, *Life through Time.* The mineral hall played to the human urge to collect and classify, embodied in the "rock-hound" approach of youngsters who fill their rooms with natural materials. And basic research by curators, e.g., F.M. Anderson, GD. Hanna, L.G. Hertlein, A.G. Smith, and, more recently, P.D. Roopnarine, R. Mooi, and G.C. Williams, resulted in many publications on stratigraphy, paleontology, paleocommunities, and evolution and extinctions.

In the late 1980s, confronted by seismic concerns, the Academy replaced all but one of its buildings. And, because of exciting developments, the new geology exhibits, introduced by an eye-catching globe and text about the Earth's crust, panels on

*Michele L. Aldrich passed away on 23 November 2016 following a brief illness. She was coauthor of this work, which was presented at the 2016 GSA Annual Meeting in Denver in session T-95 sponsored by GSA's History and Philosophy of Geology Division. This presentation is a modestly enlarged version of the oral presentation given at the meeting, and it is here with much affection dedicated to her memory.

Leviton, A.E., and Aldrich, M.L., 2018, Geology and paleontology at the California Academy of Sciences, 1895–2016: A brief overview, *in* Rosenberg, G.D., and Clary, R.M., eds., Museums at the Forefront of the History and Philosophy of Geology: History Made, History in the Making: Geological Society of America Special Paper 535, p. 203–215, https://doi.org/10.1130/2018.2535(13).

earthquakes (California's "bugaboo") and on continental drift and plate tectonics and their impact on Earth history, reinforce the educational aspects of the Academy's public outreach. Also given Kociolek and Fourtanier's studies of diatoms, and more recently paleobiodiversity and biogeography by Mooi, Roopnarine, and students, and with fossil and mineral collections, managed by Jean DeMouthe, now numbering in the millions of specimens, outreach and basic research continue unabated.

1853 FOUNDING OF THE ACADEMY TO 1906 SAN FRANCISCO EARTHQUAKE AND INITIAL RECOVERY

The California Academy of Sciences' involvement with geology and paleontology dates from its founding in 1853. In that year, Andrew Randall (1819–1856), formerly a field assistant to David Dale Owen (1807–1860) during the latter's survey of the Wisconsin territory, and John Boardman Trask (1824–1879), who had an abiding interest in the Sierra foothills, its geology and mining districts, joined five other San Franciscans in founding the California Academy of Natural Sciences. Also in 1853, William Phipps Blake (1826–1910), of the Pacific Railroad Survey, published the first geologic map of the San Francisco region. Thus, from its founding, geology played a significant role in Academy activities, and one that has continued throughout its 165-year history.

It should be kept in mind in perusing what follows that the California Academy of Sciences is not solely dedicated to research or public outreach in the earth sciences. It has a much broader outlook harboring several other world-class research collections, among them being botany, fishes, reptiles and amphibians, entomology, invertebrate zoology, and anthropology. And while not all have an extended relationship to geology and/or paleontology, insofar as the research activities of specific curators are concerned, there is considerable interplay, especially so in invertebrate zoology. Without belaboring the matter, in the recent past in anthropology, both Nina Jablonski and Zeray Alemsagrad dealt with the Koobi Fora fauna of the East African Plio-Pleistocene of the Lake Turkana region, whereas others drew heavily on Tertiary geology in the Philippines, Gondwanan geology of South Africa, Cenozoic history of Madagascar, in discussing matters of historical and biogeographical significance.

Returning to the geological studies at the Academy, and specifically with respect to its early years, roughly 1853–1907, Aldrich and Leviton covered those years of Academy activities in earlier presentations and publications (1982, 1987, 1993, 1997; Aldrich et al., 1986; Leviton et al., 2006, 2010; Rodda and Leviton, 1983), so that our starting point for this historical synopsis will be the mid-1890s with the report on submarine canyons off the California coast, published in 1897 in the Academy's *Proceedings* series by Academy associate and former Academy President George Davidson (1825–1911) (Fig. 1A) and Frank Marion Anderson (1863–1943) (Fig. 1B), who was elected a resident member of the academy in 1899, and then in 1903 as curator of geology and invertebrate paleontology, to serve without salary.

George Davidson (1825–1911; Dickie et al., 1914), who served as head of the U.S. Coastal Survey (later the U.S. Coast and Geodetic Survey) for the Pacific Coast until his summary dismissal in 1894, was both an avid oceanographer and astronomer. He was elected to membership in the Academy in 1869, though he attended Academy meetings in the earlier 1860s and occasionally gave presentations on his work in Alaska and along the Pacific Coast.

Frank Marion Anderson's (1863–1945) research focused on Cretaceous and Lower to Mid-Tertiary stratigraphy and paleontology of California, Baja California (e.g., 1902, 1905, 1908, 1911, 1927, 1928, 1929; Anderson and Martin, 1914; Anderson and Hanna, 1925, 1935), and Colombia (1927, 1928, 1929). He was also an avid collector of fossils and, in 1902, he began to transfer his invertebrate fossils to the Academy. Furthermore, as with the other Academy curators of his time, Anderson's research office and the department's collections were located in the building situated behind and connected by an enclosed bridge to the Academy's new Market Street building (Fig. 2), which on the public ground floor (Fig. 3) had on display several excellent paleontology exhibits, including a reconstructed mammoth and a giant sloth, the two obtained from Ward's Natural Science Establishment. There was also a considerable display of rocks and minerals from its own collections, many dating to the Whitney California geological survey period of the 1860s, during which time,

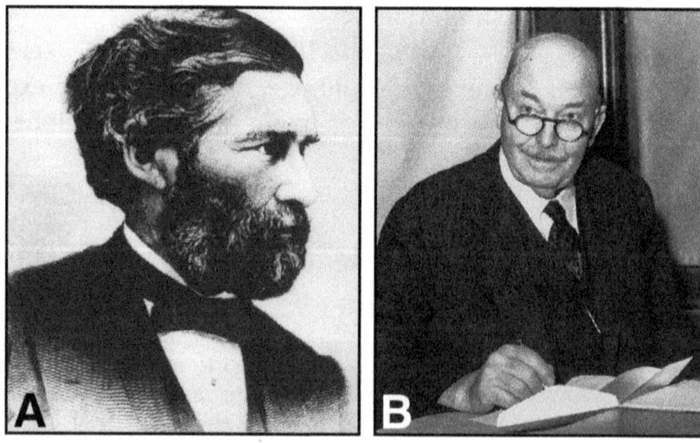

Figure 1. (A) George Davidson (California Academy of Sciences Archives). (B) Frank Marion Anderson (California Academy of Sciences Archives).

Figure 2. California Academy of Sciences, Market Street building, 1890–1906 (California Academy of Sciences Archives).

Josiah Dwight Whitney (1819–1896), William Henry Brewer (1827–1910), and other members of the survey, participated in the activities of the fledgling California Academy of Natural Sciences (the Academy's formal name prior to 1868).[1]

Unfortunately, as a result of the 18 April 1906 San Francisco earthquake and fire (Fig. 4), the Academy's two buildings (Figs. 5 and 6) and nearly all of its prized possessions, research collections included, were lost. And though the event impacted much of the Academy's activities, its curators nonetheless continued to pursue their research, mostly in temporary quarters located elsewhere in the city (Figs. 7 and 8; Gough Street, California Academy of Sciences), as did the Academy's staff, who participated in the Academy's 1905–1906 Galapagos expedition.

Figure 3. Public exhibits on ground floor of Academy museum (California Academy of Sciences Archives).

With respect to the Galapagos expedition, and more specifically with respect to geology and paleontology, 26-year-old Washington Ochsner (1879–1927) (Figs. 9A, 9B, and 10B) served as expedition geologist. Furthermore, as a result of his and his expedition companions' efforts in the Galapagos, a substantial number of fossils were brought back, thereby initiating a new paleontological collection. Although Ochsner's subsequent association with the Academy was limited, nonetheless, in collaboration with Academy life member and former curator, William Healey Dall (1845–1927) (Fig. 10A), he prepared an extensive review of Tertiary and Pleistocene mollusks of the Galapagos Islands, which was published posthumously for both in 1928.

Figure 4. Fire that swept downtown San Francisco, 18 April 1906 (California Academy of Sciences Archives).

[1]"... January 13 [1868], when the Academy's new constitution was slightly amended and again adopted for a while. The new constitution provided that the name of the society should be the 'California Academy of Sciences' dropping the word 'Natural' of the old title. Its object was to be 'the promotion of science,' and 'this was to be accomplished by the holding of meetings for scientific intercourse and discussion, by the reading and publication of papers containing original contributions to science, by the establishment of a museum and library, and by other suitable means." (from the *Minute Books* [of the Academy meetings], 6 January 1868–2 January 1872; see page 95 in Leviton and Aldrich, 1997, *Theodore Henry Hittell's The California Academy of Sciences, 1853–1906*, Memoirs of the California Academy of Sciences, 22.

Figure 5. Devastation wrought by the earthquake to the Academy's Market Street buildings (the Academy's museum and research quarters, left rear).

Figure 6. Interior of the Academy's museum building following the earthquake and fire that engulfed the building (California Academy of Sciences Archives).

POST–1906 EARTHQUAKE RECOVERY AND THE HANNA/HERTLEIN ERA TO 1969

The Academy (Fig. 11) relocated to Golden Gate Park, and in 1916 the first of several new buildings opened to the public. The building, which was named North American Hall and which housed public exhibits on the ground floor, also provided facilities for research and collections. At this time and for several decades thereafter, and apart from its new Steinhart Aquarium, which opened in 1923, academy public exhibits focused on dioramas of animals and their habitats, mostly North America and, in time and with the addition of African Hall, Africa. Otherwise, there was little relating to either fossils or geology except one wing in North American Hall (Figs. 12A and 12B) that was devoted exclusively to the display of hundreds of species of gems, minerals, and rocks, a near must in California considering its mining history, but one that also allowed visitors, with their own specimens in hand, to attempt identification of their precious holdings.

During the rebuilding period, research continued to flourish. In 1910, Anderson brought in Roy Ernest Dickerson (1877–1944; see Harris, 1944) (Fig. 13A), who served as curator of invertebrate paleontology and, for a brief time, as director of the Academy's museum (1914), until his departure in 1919 for the Philippines and a future in petroleum geology. Dickerson was an active researcher with an interest in Tertiary faunal units on the Pacific Coast as well as the influence of early Tertiary climate on the faunas (1917a, 1917b, 1921, 1922).

But before leaving the Academy in 1919, Anderson and Dickerson attracted G Dallas Hanna (1887–1970) (Figs. 13B and 14A), who had diverse research interests that ranged from mammals of the Pribilof Islands, recent and fossil mollusks (e.g., Hanna, 1923, 1966), descriptive geology (e.g., Hanna and Hertlein, 1927; Hanna, 1930, 1952) and stratigraphy, the great Alaskan earthquake (1964), and oil seeps (Hanna, 1963), indeed even insects of the Pribilof Islands (Hanna et al., 1921), but with an overall abiding interest in microfossils, notably diatoms, their diversity as well as their use in establishing stratigraphic relationships (Hanna, 1927, 1932, 1970, et seq.; Hanna and Brigger, 1966). Hanna married Margaret Hughes (Fig. 14B), née Moore, who had in the late 1920s and early 1930s served

Figure 7. The Academy's temporary administrative quarters on Gough Street following the earthquake and fire (California Academy of Sciences Archives).

Figure 8. Rebuilding the Academy's temporary library housing materials donated by museums and other institutions worldwide (California Academy of Sciences Archives).

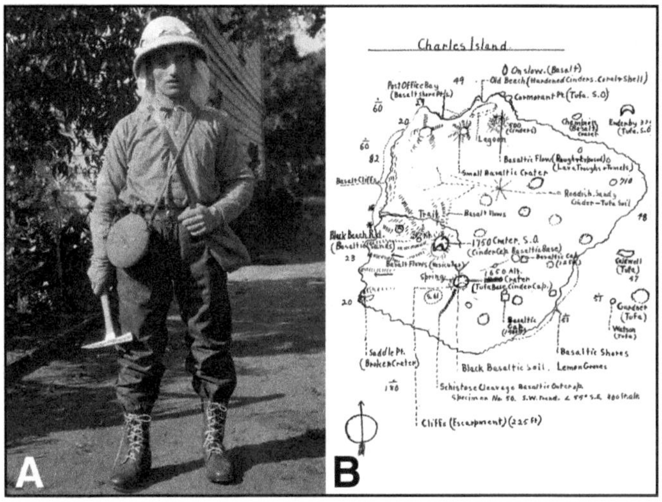

Figure 9. (A) Washington Ochsner before leaving for the Galapagos in 1905 (California Academy of Sciences Archives). (B) Page from Ochsner's field notes, titled "Charles Island" (California Academy of Sciences Archives).

as Joseph Cushman's lead foraminifera artist. The two Hannas served the Academy for many years, and Margaret, not surprisingly, was most supportive of Dallas Hanna's micropaleontological and stratigraphic studies as well his interests in such diverse matters as producing binocular and periscope lenses for the U.S. Navy during World War II, in-house construction of the Academy's planetarium projector, and in-house offset color printing for Academy scientific publications.

In 1929, Hanna and his then new departmental associate, Leo George Hertlein (1898–1972; see Addicott, 1970) (Fig. 15A), launched an aggressive program of research in micropaleontology and invertebrate paleontology and stratigraphy, though

Figure 10. (A) William Healy Dall (portrait in *Proceedings of the California Academy of Sciences*, 1928). (B) Washington Ochsner (portrait in *Proceedings of the California Academy of Sciences*, 1928).

Figure 11. Entrance to the first new building of the California Academy of Sciences that opened to the public in 1916 (California Academy of Sciences Archives).*

Figure 12. (A) Mineral Hall wing in the newly opened North American Hall (California Academy of Sciences Archives). (B) Updated Mineral Hall, ca. 1990s (California Academy of Sciences Archives).

Figure 13. (A) Roy Ernest Dickerson (American Association of Petroleum Geologists ©1944, *American Association of Petroleum Geologists Bulletin*, v. 28, p. 888, reprinted by permission of the American Association of Petroleum Geologists, whose permission is required for further use). (B) G Dallas Hanna, ca. 1919 (California Academy of Sciences Archives).

Figure 14. (A) G Dallas Hanna, ca. mid-1950s (California Academy of Sciences Archives). (B) Margaret Hanna (California Academy of Sciences Archives).

Hertlein's attention was mostly directed to Tertiary marine mollusks of California and the Galapagos Islands (e.g., 1925, 1939, 1963, 1968, 1972) and regional paleontology (e.g., Hertlein and Jordan, 1927).

But both Hanna and Hertlein were supportive of regional geology investigations, and in this they attracted colleagues who signed on as research associates in the then Department of Paleontology, notably geologist and civil engineer, George Gester (1884–1959; see Moody, 1960) (Fig. 15B), at the time chief geologist for Standard Oil of California, but who continued his Academy association and explorations as a geologist after his retirement (e.g., Gester, 1962 [posthumously published]), Carl Grunsky (1855–1934), and Allyn Smith (1893–1976; see Emerson, 1977) (Fig. 16A), then head of personnel for the Pacific Telephone Company.

Apart from his research associate appointment in geology and paleontology, Grunsky served as Academy president from 1914 to 1931, and then in 1932, he succeeded Barton

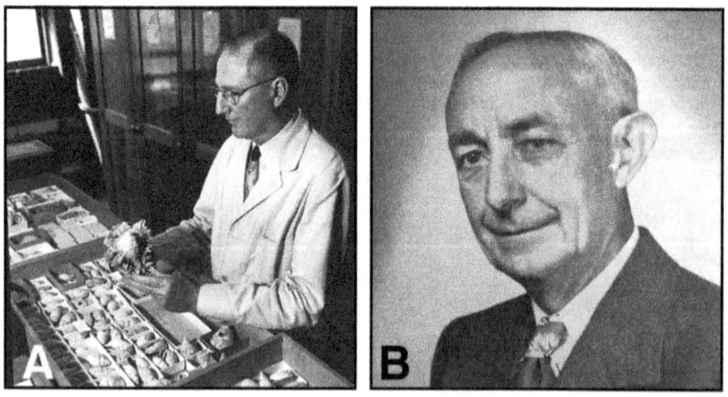

Figure 15. (A) Leo George Hertlein (California Academy of Sciences Archives). (B) George Gester (American Association of Petroleum Geologists ©1960, *American Association of Petroleum Geologists Bulletin*, v. 44, p. 374, reprinted by permission of the AAPG, whose permission is required for further use).

**Erratum:* A different image of the building appeared in the version of this chapter published online on 13 April 2018.

Warren Evermann (1853–1932) as director of the Academy's museum and Steinhart Aquarium. According to Hanna, although Grunsky's scientific contributions were negligible, he did preside over meetings with sensitivity and a good understanding of both people and matters being discussed.

Allyn Smith (see Miller, 1977; Kellogg, 1986), on the other hand, was appointed research associate in malacology in 1939, and following his retirement from the telephone company in 1954, he came to the Academy as research malacologist and assistant to the Academy's director and marine biologist, Robert C. Miller (Fig. 16B). Then, in 1960, he was appointed associate curator of invertebrate zoology and paleontology, and from 1963 to 1972, department chair. Smith managed to publish more than 100 papers covering his studies on recent and fossil snails and chitons, among them the section on "Amphineura" in the *Treatise on Invertebrate Paleontology* (1960), as well as contributions on Paleozoic and Mesozoic mollusks (e.g., Smith et al., 1968, 1971), including two summary catalogs with bibliographies, one in 1973, Fossil Chitons from the Mesozoic, and a second in [Smith and Hoare] 1987, Paleozoic Polyplacophora.

At this point, we must emphasize that for Academy curators in paleontology, geology, and invertebrate zoology, the distinction among the fields had long since blurred; indeed they were seen as interdependent, and, as we well know today, they collectively provide the breadth of understanding required to deal with past and present phylogenetic, ecological, and biogeographic relationships as well as, indeed, historical geological events. Thus, as one peruses their publications, one finds that the Hanna, Hertlein, and Smith paleontological studies are interspersed with treatments of extant invertebrates, mostly molluscan, and microfossils.

Furthermore, and although implied but not dealt with specifically, the research curator staff of geology and paleontology engaged in field work throughout their careers at the Academy. Much was done locally and throughout California, but more lengthy excursions took place too, notably to Mexico, as for example the California Academy of Sciences expeditions to the Gulf of California in 1921, the Revillagigedo Islands in 1925, the 1953 six-week Orca Expedition sponsored by the Sefton Foundation to investigate the faunas, floras, and geology of the islands in the Gulf of California, and not long afterward, the Bechtel Foundation sponsored at least two field excursions to Baja California, one of two months duration in 1958, to Baja California del Sur, the second in 1961 to the Vizcaiño Desert region. In each case, a member or members of the Academy's geology and paleontology group participated, e.g., Galapagos, Hertlein; Gulf of California, Hanna; Revillagigedo Islands, Hertlein; Orca Expedition, Hanna; Bechtel Foundation, Baja California del Sur, Smith. There were other travels, including Alaska, which held a special place in Hanna's heart for while in his youthful prime, he traveled by dogsled roundtrip from Bristol Bay to Iditarod.

In 1967, Victor Zullo (1936–1993; see Newman et al., 1994) (Fig. 17A) joined the Academy's staff as curator of invertebrate zoology and geology during the administration of botanical cactaceaologist George Lindsay (Fig. 17B), a strong supporter of basic research at the Academy. Zullo's fascination with Cenozoic barnacles and biostratigraphy resulted in at least 15 publications during his Academy tenure, capped by the coauthored treatment of Cirripedia (Newman et al., 1969) in the *Treatise of Invertebrate Paleontology*. But Zullo's tenure at the Academy was short-lived, for in 1971, he left to join the faculty of the University of North Carolina.

POST–G DALLAS HANNA AND LEO G. HERTLEIN PERIOD, 1970–2016

In 1970, G Dallas Hanna died, and he was soon followed by the passing of Leo Hertlein in 1972. But before the latter event, and shortly after Victor Zullo's departure, the Academy acquired the services of Peter U. Rodda in 1971 (Fig. 18A), who at the time was with the University of Texas in Austin. A year later, in 1973, Welton Lee (1973) (Fig. 18B), an invertebrate zoologist, was brought in, and he and Rodda were destined to play

Figure 16. (A) Allyn G. Smith (California Academy of Sciences Archives). (B) Robert C. Miller (California Academy of Sciences Archives).

Figure 17. (A) Victor Zullo (left) appearing on the TV program *Science in Action*, ca. 1969, with program host Earl S. Herald (right), then superintendent of the Academy's Steinhart Aquarium (California Academy of Sciences Archives). (B) George E. Lindsay, Academy director (California Academy of Sciences Archives).

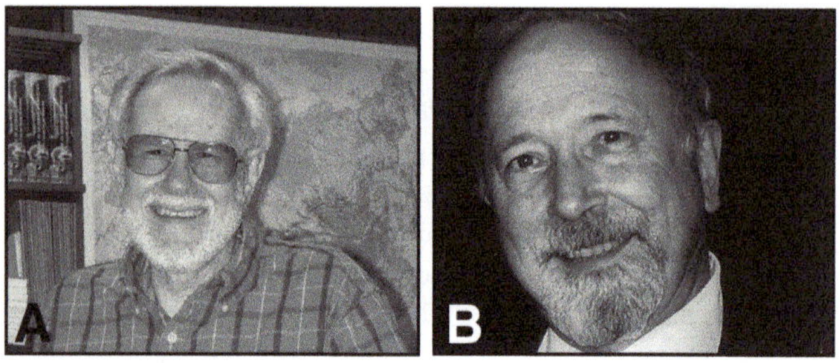

Figure 18. (A) Peter U. Rodda (photograph by A. Leviton, 2000). (B) Welton Lee (California Academy of Sciences Archives).

significant roles in greatly expanding the Academy's invertebrate research collections, recent and fossil. Additionally, Rodda became deeply committed to the Academy's public outreach activities in both geology and paleontology. And in this he would be strongly supported by Ian Campbell (1899–1978) (Fig. 19A), who, having served a lengthy term as professor of geology at California Institute of Technology, became not only California State Geologist but also an Academy research associate, trustee, and, in due course, its president. Campbell was not unappreciative of the parts played by Rodda and others in advancing the Academy's stature in research and public outreach, including temporaries, such as Sam Van Landingham (1935–2013) (Fig. 19B), who had come to the Academy as research diatomist, ca. 1973–1978. At the time, Sam had already published the first of his eight-volume catalogue of fossil and recent genera and species of diatoms, to which he added the additional volumes at a steady pace (1967–1979).

Returning to Peter Rodda, he played a seminal role in conceptualizing and overseeing the development of a new Academy Hall, *Life through Time,* with its remarkable, almost living exhibits (Fig. 20). Regretfully, Rodda's time was so thoroughly subsumed by this that his own research with Mesozoic ammonites suffered. But the new hall turned out to be a stunning success. It must be noted that in the planning of the new facility, Rodda did have the support of departmental staff, one of

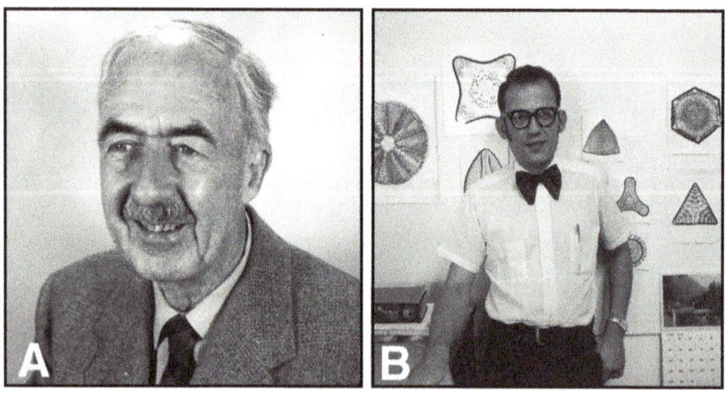

Figure 19. (A) Ian Campbell (California Academy of Sciences Archives). (B) Sam Van Landingham (California Academy of Sciences Archives).

Figure 20. Exhibit in *Life Through Time* hall, ca. 1990 (California Academy of Sciences Archives).

whom, mineralogist Jean DeMouthe (1949–2017) (Fig. 21A), had come to the Academy in 1973 as a curatorial assistant but rapidly rose to become senior collections manager of the Academy's extensive and ever-growing geology and paleontology collections (Fig. 21B). DeMouthe was also a registered geologist in California and acting county geologist for San Mateo County as well as adjunct professor at San Francisco State University. As one might expect for a mineralogist, her publications and public outreach activities strongly favor minerals (e.g., DeMouthe, 1989, 1990, 1991, 2005) but not exclusively because she was extremely interested in reaching out to the public of all ages through publication (DeMouthe, 1994, 1999) and lectures, as well as her eclectic interests (DeMouthe, 1977).

Also, not long after the loss of G Dallas Hanna and the departure of Van Landingham, and in continued recognition of the hugely important research potential of the Academy's collection of microfossils, notably diatoms, in 1989, the Academy

California Academy of Sciences
Earth Sciences Collections ca 1985
(Source: From compilation notes by Peter U. Rodda)

Invertebrates (Recent and Fossil)
Recent 150,000 lots (approximately 1,000,000 specimens)
Fossil 50,000 cataloged lots (approximately 700,000 specimens, world-wide but primarily Western North America and the Pacific region, Mesozoic and Cenozoic)
Types 5,500 fossil and 5,000 living invertebrates

Micropaleontology
Foraminifera (no data on collection size but extensive)
Diatoms 45,000 mostly center-mount slides, including 274 holotypes (NB: diatom library contains 6,200 volumes and reprints)

Rocks and Minerals
B Approx.. 15,000 representing 1,670 mineral species and 1,000 gem stones

Figure 21. (A) Jean DeMouthe (photograph by A. Leviton, 2016). (B) Earth Sciences Collections at the California Academy of Sciences, ca. 1985.

Figure 22. (A) John Patrick Kociolek (photograph by A. Leviton, 2004). (B) Elisabeth Fourtanier and public visitors to diatom laboratory at the Academy (California Academy of Sciences Archives).

attracted John Patrick Kociolek (Fig. 22A), Pat to his friends and colleagues, and his erstwhile associate and postdoctoral fellow, Elisabeth Fourtanier (Fig. 22B) (1993–1997), who thoroughly enjoyed reaching out to the public and talking about diatoms. Pat was especially interested in the taxonomy and phylogeny of diatoms, with notable interest in living freshwater diatoms but also in the historical events that impacted them in both space and time. Apart from taxonomic and broader biological studies, the two indulged in a collaborative effort to produce a massive catalogue of diatom names, bibliography, and associated data, worldwide and across time. Regretfully, only the first two volumes of the catalog were completed and published (Fourtanier and Kociolek, 2009a, 2009b) because of other distractions, namely that following an initial several years of research and an assortment of relatively light administrative duties, in 1998 Kociolek was asked to take on the responsibility of executive director of the Academy, in which capacity he served until his departure in 2007 for the University of Colorado.

With Patrick's commitment to administration, and with the loss of two curators in invertebrate zoology and paleontology, the Academy sought a couple of rising stars to fill the vacancies. Thus, in 1990, it acquired two talented scientists, Gary C. Williams (Fig. 23A) and Richard Mooi (Fig. 23B). Williams' research centered on the systematics, biogeography, and evolutionary biology of octocorallian coelenterates, but he also had a penchant for the history of science (Williams, 1997, 2007, 2008) as well as for travel to exotic places, not the least of which have been his excursions in the Andes of South America, backpacking from the Andean crest into the Amazon Basin of Bolivia, also traveling the length of the Andes to Tierra del Fuego, but also studies in South Africa, the Philippines, and elsewhere. Richard J. Mooi's echinoderm research spans the range of paleontology, zoology, and ecology (e.g., Mooi, 1989, 1990, 2001; Mooi and David, 1997; Mooi et al., 2000), but he is also a dedicated teacher. Indeed, in the mid-1990s, Mooi became involved in the direction of the Academy's "Summer Systematics Institute" (Figs. 24A and 24B), now a National Science Foundation–sponsored program, in which students, selected by application from across the United States, spend a summer in study of the fundamentals of systematics, including methodologies, for both fossil and Recent organisms, and the most current applications for data gathering and analysis, thus providing a unique research participation experience.

Then, two years after Peter Rodda's retirement in 1997, the paleontology and geology staff was further enhanced by the 1999 arrival of Peter D. Roopnarine (Fig. 24C), whose interests in paleobiogeography, mass extinctions, paleocommunity dynamics, evolutionary ecology, and related topics, as well as teaching, have resulted in a significant output of research as well as an enthusiastic cluster of students who participate in research activities in invertebrate zoology and paleontology. Of particular interest are Roopnarine's mathematical modeling of paleocommunity collapse during mass extinctions, morphological response of bivalve taxa to paleocommunity composition, ecological modeling of paleocommunity food webs, all impacting community structure, survival and/or collapse across time, and how deep-time web reconstructions can serve as analogs for the present (e.g., Roopnarine, 2001a, 2001b, 2006; Roopnarine

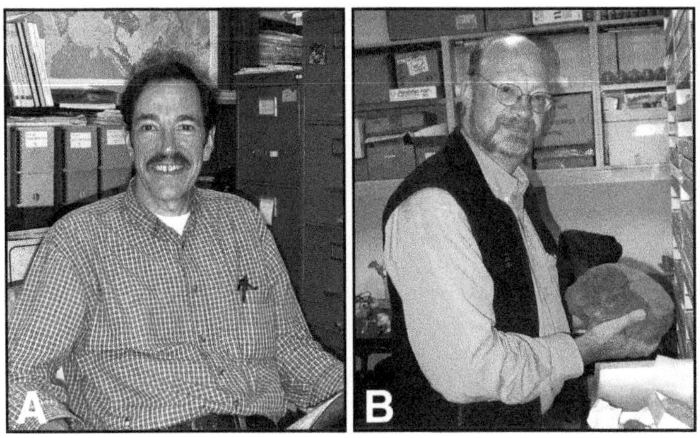

Figure 23. (A) Gary C. Williams (photograph by A. Leviton, 2002). (B) Richard Mooi (photograph by A. Leviton, 2016).

Figure 24. Students at the Summer Systematic Institute, (A) in the classroom working on an echinoderm phylogeny and (B) in the invertebrate zoology and geology collection room (photograph by R. Mooi). (C) Peter Roopnarine (photograph by A. Leviton, 2016).

Figure 25. Newly constructed building of the California Academy of Sciences in Golden Gate Park, 2008 (California Academy of Sciences Archives).

et al., 1998; Roopnarine and Angielczyk, 2015; Roopnarine and Dineen, 2017).

Like other major natural history museums throughout the world that have responded to the exciting developments in the earth and other natural sciences, the Academy has seen a rebirth of its public outreach programs. But there was another major interruption in the Academy's physical being. In 1989, following California's Loma Prieta earthquake, its buildings were deemed seismically unsafe; thus a second round of demolition and new construction began. In 2008, a new "Academy" building (Fig. 25) opened to the public, and except for one building of the older complex, African Hall, the new facility incorporated a different genre of exhibits, which, aside from the dome housing a living tropical rainforest, a greatly expanded aquarium, and

Figure 26. (A) Globe delimiting plate boundaries on public exhibit (photograph by A. Leviton, 2016). (B) Fault exhibit at one time on public display at the Academy (California Academy of Sciences Archives).

Figure 27. Exhibit displaying items to have in the home in the event of an area-wide earthquake occurrence (photograph by A. Leviton, 2016).

Figure 28. Exhibit displaying food and related items to have in the home in the event of an area-wide earthquake occurrence (photograph by A. Leviton, 2016).

Figure 29. Design drawing of newly opened gem and mineral hall at the Academy (2016) (California Academy of Sciences Archives).

its ever popular planetarium, now allowed for rapid response to new events in the natural sciences. In the area of earth sciences, the Academy is committed to informing its visitors about earthquakes, California's biggest fear, addressing such matters as global plate distributions, the underlying tectonics of plate movements (Figs. 26A and 26B), and the surficial paleobiological impacts thereof, as well as how to prepare for and respond to the ever-present likelihood of serious local earthquakes (Figs. 27 and 28). There is even a special rumble room for visitors to sense the intensity of the shaking that can occur.

Of course, main-floor exhibits tend to be less stable nowadays so that in due course it can be expected that the Academy's earthquake exhibits will give way to other exciting but not necessarily geological presentations.

But all is not lost because though initially seemingly deemphasized in its recent exhibits, in 2016 the Academy opened a completely redesigned hall (Fig. 29) devoted to rocks, minerals, and gems, here too benefiting from the knowledgeable input of geology collections manager, Jean DeMouthe. Once again the public's rock-hound enthusiasts will also be accommodated.

SUMMARY

The research role of museums, whether in the biological or geosciences, is not often readily apparent to the public. Yet, natural history museums, worldwide, like the Academy, draw upon their research scientists and associates to create exhibits and other programs that convey current understandings of the background science in their public outreach and education programs. With respect to the Academy, from its formative years in the nineteenth century to the present day, it has reached out to the public with vital information about earthquakes, biodiversity, evolution, habitat degradation, and climate, much of which is based upon research done by its staff and close associates. Also, not immediately apparent are the financial aspects of the public outreach programs that enable the public to make credible decisions relating to family and other activities, in the Academy's case, dating back to the 1868 Hayward earthquake when Academy President James Blake served as chair of the San Francisco Chamber of Commerce's Earthquake Committee, which was established to understand the nature of the event and its future implications. But equally important, museums such as the Academy foster collaboration among the neontologists, geologists, and paleontologists, in field and laboratory research, as well as in the development of exhibits, thus helping to blur the boundaries among the subdisciplines of science and to exemplify the value of interdisciplinary research leading to improved educational experiences in our schools as well as to greater public awareness of the place of science in our lives. Indeed, the Academy's integration of a living tropical rainforest in close proximity to its paleontological and geological exhibits as well as its novel citizen scientists program in field research, both locally and in foreign countries, serve as important role models for the future development of museums.

ACKNOWLEDGMENTS

We take this opportunity to thank our friends and colleagues who are or have been associated with the California Academy of Sciences for sharing their knowledge of Academy activities in geology and paleontology with us. We especially want to take note of Kelly Jensen, digitization specialist in the Academy's Library and Archives, Yolanda Bustos, former Academy archivist, and Rebekah Kim, current Academy archivist, for ferreting out a number of the images we present here, also Dr. Peter U. Rodda, who provided a copy of an unpublished memorial he had written relating to Margaret Hanna, and Gary C. Williams, who had a file of manuscript notes by Dr. Rodda and others relating to the history of the Department of Invertebrate Zoology and Geology, and Paula Sillman of the American Association of Petroleum Geologists, who most kindly gave permission to reproduce two portraits published in the *American Association of Petroleum Geologists Bulletin*. We also acknowledge with appreciation Dr. Peter Roopnarine, who kindly critiqued this manuscript, as did Drs. Gary Williams and Jean DeMouthe. We also express our appreciation to two anonymous reviewers for their comments that assuredly helped improve the overall presentation and to Dr. Gary Rosenberg, who not only offered specific recommendations for the improvement of our manuscript but who also patiently guided us through the GSA editorial maze. Despite the best intentions of our colleagues mentioned above, the authors remain fully responsible for all errors of commission or omission herein.

At this point, we must also offer a word of explanation relating to the absence of additional details about our individual investigators and their accomplishments. For the past three years, the Academy archives have for the most part been unavailable because of a mold infestation in the library and specimen collection rooms throughout the institution. The process of cleaning up and rendering materials safe from deterioration required much material to be moved offsite while structural and other changes are being made to the onsite storage areas. We deeply regret having to limit our discussions to memories and materials at hand, but we do hope there are adequate indications included herein to excite the imaginations of our history-oriented colleagues.

REFERENCES CITED

Addicott, W.G., 1970, Biographical sketch of Leo G. Hertlein [with full bibliography]: The Nautilus, v. 84, p. 37–421.

Aldrich, M.L., and Leviton, A.E., 1993, Directory of historians of American science: History of Science in America, News & Views, v. 10, no. 3 (Special Issue), p. 1–59.

Aldrich, M.L., and Leviton, A.E., 2000, West and east: The California Academy of Sciences and the Smithsonian Institution, 1853–1906, *in* Ghiselin, M., and Leviton, A., eds., Cultures and Institutions of Natural History: California Academy of Sciences Memoir 25, p. 183–202.

Aldrich, M.L., Bolt, B., Leviton, A.E., and Rodda, P.U., 1986, The "report" of the 1868 Haywards Earthquake: Bulletin of the Seismological Society of America, v. 76, p. 71–76.

Anderson, F.M., 1902, Cretaceous deposits of the Pacific Coast: Proceedings of the California Academy of Sciences, ser. 3 (Geology), v. 2, p. 1–154, pls. 1–12.

Anderson, F.M., 1905, A stratigraphic study in the Mount Diablo Range of California: Proceedings of the California Academy of Sciences, ser. 3 (Geology), v. 2, p. 157–248, pls. 13–35.

Anderson, F.M., 1908, A further stratigraphic study in the Mount Diablo Range of California: Proceedings of the California Academy of Sciences, ser. 4, v. 3, p. 1–40.

Anderson, F.M., 1911, The Neocene deposits of Kern River, California, and the Temblor Basin: Proceedings of the California Academy of Sciences, ser. 4, v. 3, p. 73–148, 9 pls.

Anderson, F.M., 1927, The marine Miocene deposits of north Colombia: Proceedings of the California Academy of Sciences, ser. 4, v. 16, p. 87–95, 2 pls.

Anderson, F.M., 1928, Notes on Lower Tertiary deposits of Colombia and their molluscan and foraminiferal fauna: Proceedings of the California Academy of Sciences, ser. 4, v. 17, p. 1–29, 1 pl.

Anderson, F.M., 1929, Marine Miocene and related deposits of north Colombia: Proceedings of the California Academy of Sciences, ser. 4, v. 18, p. 73–213, 16 pls.

Anderson, F.M., and Hanna, GD., 1925, Fauna and Stratigraphic Relations of the Tejon Eocene at the Type Locality in Kern County, California: San Francisco, Occasional Papers of the California Academy of Sciences, no. 11, 249 p., 16 pls.

Anderson, F.M., and Hanna, GD., 1935, Cretaceous geology of Lower California: Proceedings of the California Academy of Sciences, ser. 4, v. 23, p. 1–34, 11 pls., 2 figs.

Anderson, F.M., and Martin, B., 1914, Neocene record in the Temblor Basin, California, and Neocene deposits of the San Juan District, San Luis Obispo County: Proceedings of the California Academy of Sciences, ser. 4, v. 4, p. 15–112, 10 pls.

DeMouthe, J., 1977, Horse Markings and Coloration: Cranbury, New Jersey, A.S. Barnes & Co., 101 p.

DeMouthe, J., 1989, Natural materials Part I: Minerals other than quartz: Journal of International Snuff Bottle Society, v. 21, p. 4–13.

DeMouthe, J., 1990, Natural materials Part II: Minerals other than quartz: Journal of International Snuff Bottle Society, v. 22, p. 9–15.

DeMouthe, J., 1991, Natural materials Part III & IV: Minerals other than quartz: Journal of International Snuff Bottle Society, v. 23, p. 12–26.

DeMouthe, J., 1994, An introduction to the geology of California: Rocks and Minerals, v. 69, p. 360–365, https://doi.org/10.1080/00357529.1994.9925617.

DeMouthe, J., 1999, Geology of the Fritzgerald Marine Reserve, San Mateo County, California, *in* The Natural History of the Fritzgerald Marine Reserve, published By Friends of Fritz[g]erald, p. 131–140.

DeMouthe, J., 2005, Natural Materials: Sources, Properties & Uses: Oxford, UK, Elsevier Publishing, 204 p.

Dickerson, R.E., 1917a, Climate and its influence upon the Oligocene faunas of the Pacific Coast, with descriptions of some new species from the *Molopophorus linolnensis* zone: Proceedings of the California Academy of Sciences, ser. 4, v. 7, p. 157–182, 5 pls.

Dickerson, R.E., 1917b, Climate zones of Martinez Eocene time: Proceedings of the California Academy of Sciences, ser. 4, v. 7, p. 193–196.

Dickerson, R.E., 1921, Notes on a fauna of the Vigo group and its bearing on the evolution of marine molluscan faunas: Proceedings of the California Academy of Sciences, ser. 4, v. 11, p. 1–26.

Dickerson, R.E., 1922, Tertiary and Quaternary history of the Petaluma, Point Reyes and Santa Rosa quadrangles: Proceedings of the California Academy of Sciences, ser. 4, v. 11, p. 527–601, 34 pls.

Dickie, G.W., Harrison, R., and Christy, S.B., 1914, II. George Davidson: Proceedings of the California Academy of Sciences, ser. 4, v. 4, p. 1–13.

Emerson, W.K., 1977, Obituary: Allyn Goodwin Smith, 1893–1976 [portrait]: The Nautilus, v. 91, p. 37.

Fourtanier, E., and Kociolek, J.P., 2009a, Catalogue of Diatom Names. Part I: Introduction and Bibliography: Occasional Papers of the California Academy of Sciences 156, part I, x + 156 p.

Fourtanier, E., and Kociolek, J.P., 2009b, Catalogue of Diatom Names. Part II: *Abas* through *Bruniopsis*: Occasional Papers of the California Academy of Sciences 156, part II, xii + 231 p.

Hanna, GD., 1923, Expedition of the California Academy of Sciences to the Gulf of California in 1921. Land and freshwater mollusks: Proceedings of the California Academy of Sciences, ser. 4, v. 12, p. 483–527, 4 pls.

Hanna, GD., 1927, Cretaceous diatoms from California: Occasional Papers of the California Academy of Sciences 13, 48 p., 5 pls.

Hanna, GD., 1930, Geology of Sharktooth Hill, Kern County, California: Proceedings of the California Academy of Sciences, ser. 4, v. 19, p. 65–83.

Hanna, GD., 1932, The diatoms of Sharktooth Hill, Kern County, California: Proceedings of the California Academy of Sciences, ser. 4, v. 20, p. 161–263.

Hanna, GD., 1952, Geology of the continental slope off central California: Proceedings of the California Academy of Sciences, ser. 4, v. 27, p. 325–358, 8 pls.

Hanna, GD., 1963, Oil seepages on the Arctic coastal plain, Alaska: Occasional Papers of the California Academy of Sciences 38, 18 p., 1 pl.

Hanna, GD., 1966, Introduced mollusks of Western North America: Occasional Papers of the California Academy of Sciences 48, 108 p., 4 pls.

Hanna, GD., 1970, Fossil diatoms from the Pribilof Islands, Bering Sea, Alaska: Proceedings of the California Academy of Sciences, ser. 4, v. 37, p. 167–234, 2 maps.

Hanna, GD., and Brigger, A.L., 1966, Fossil diatoms from southern Baja California: Proceedings of the California Academy of Sciences, ser. 4, v. 30, p. 285–308.

Hanna, GD., Van Dyke, E.C., et al., 1921, Insects of the Pribilof Islands, Alaska: Proceedings of the California Academy of Sciences, ser. 4, v. 11, p. 153–195.

Hanna, GD., and Hertlein, L.G., 1927, Expedition of the California Academy of Sciences to the Gulf of California in 1921: Geology and paleontology: Proceedings of the California Academy of Sciences, ser. 4, v. 16, p. 137–157, 1 pl.

Harris, R.C., 1944, Memorial: Roy Ernest Dickerson (1877–1944) [portrait.]: Bulletin of the American Association of Petroleum Geologists, v. 28, p. 888–890.

Hertlein, L.G., 1925, Pectens from the Tertiary of Lower California: Proceedings of the California Academy of Sciences, ser. 4, v. 14, p. 1–35, 6 pls.

Hertlein, L.G., 1939, Marine Pleistocene mollusks from the Galapagos Islands: Proceedings of the California Academy of Sciences, ser. 4, v. 23, p. 367–380, 1 pl.

Hertlein, L.G., 1963, Contributions to the biogeography of Cocos Island, including a bibliography: Proceedings of the California Academy of Sciences, ser. 4, v. 32, p. 219–289.

Hertlein, L.G., 1972, Pliocene fossils from Baltra (South Seymour) Island, Galapagos Islands: Proceedings of the California Academy of Sciences, ser. 4, v. 39, p. 25–46.

Hertlein, L.G., and Jordan, E.K., 1927, Paleontology of the Miocene of Lower California: Proceedings of the California Academy of Sciences, ser. 4, v. 16, p. 605–647, 5 pls.

Kellogg, M.G., 1986, A bibliography and catalog of taxa for Allyn Goodwin Smith (1983–1976): Occasional Papers of the California Academy of Sciences, no. 145, p. 1–12.

Leviton, A.E., and Aldrich, M.L., 1982, John Boardman Trask: California's physician-geologist, 1853–1856, *in* Leviton, A.E., Rodda, P.U., Yochelson, E., and Aldrich, M.L., eds., Frontiers of Geological Exploration of Western North America: San Francisco, California, Pacific Division, American Association for the Advancement of Science, p. 37–69.

Leviton, A.E., and Aldrich, M.L., 1987, James Blake, M.D.: President of the California Academy of Sciences, 1868–1872: Fellows Newsletter, v. 7, p. 4–8.

Leviton, A.E., and Aldrich, M.L., 1997, Theodore Henry Hittell's The California Academy of Sciences, 1853–1906, A Narrative History: San Francisco, California, California Academy of Sciences, 623 p.

Leviton, A.E., Aldrich, M.L., and Elsbernd, K., 2006, The California Academy of Sciences, Grove Karl Gilbert, and photographs of the 1906 earthquake mostly from the Archives of the California Academy of Sciences: Proceedings of the California Academy of Sciences, ser. 4, v. 57, p. 1–34.

Leviton, A.E., Aldrich, M.L., Williams, G.C., and Rodda, P.U., 2010, Geology at the California Academy of Sciences, 1853–1907: Proceedings of the California Academy of Sciences, ser. 4, v. 61, p. 547–573.

Miller, R.C., 1977, Obituary: Allyn G. Smith [portrait.]: Malacological Review, v. 10, p. 87–88.

Moody, G.B., 1960, George Clark Gester (1884–1959) [portrait.]: Bulletin of the American Association of Petroleum Geologists, v. 44, p. 374–376.

Mooi, R., 1989, Living and fossil genera of the Clypeasteroidea (Echinoidea: Echinodermata): An illustrated key and annotated checklist: Smithsonian Contributions to Zoology 488, 51 p.

Mooi, R., 1990, Paedomorphosis, Aristotle's lantern, and the origin of the sand dollars (Echinodermata: Clypeasteroidea): Paleobiology, v. 16, p. 25–48, https://doi.org/10.1017/S0094837300009714.

Mooi, R., 2001, Not all written in stone: Interdisciplinary syntheses in echinoderm paleontology: Canadian Journal of Zoology, v. 79, p. 1209–1231, https://doi.org/10.1139/z00-217.

Mooi, R., and David, B., 1997, Skeletal homologies of echinoderms: The Paleontological Society Papers, v. 3, p. 305–335.

Mooi, R., Martinez, S., and Parma, S.G., 2000, Phylogenetic systematics of Tertiary monophorasterid sand dollars (Clypeasteroidea: Echinoidea) from South America: Journal of Paleontology, v. 74, p. 263–281.

Newman, W., Zullo, V.A., and Withers, T.H., 1969, Cirripedia, *in* Treatise on Invertebrate Paleontology, v. R:4:1: Boulder, Colorado, and Lawrence, Kansas, Geological Society of America and University of Kansas Press, p. 206–296.

Newman, W., Buckeridge, J.S., Portell, R.W., and Spivey, H.R., 1994, Victor August Zullo (24 July 1936–16 July 1993) [portrait]: Journal of Crustacean Biology, v. 14, p. 399–405, https://doi.org/10.1163/193724094X00371.

Rodda, P.U., and Leviton, A.E., 1983, Nineteenth century earthquake investigations in California: Earth Sciences History, v. 2, p. 48–56, https://doi.org/10.17704/eshi.2.1.576065779g862057.

Roopnarine, P.D., 2001a, The development and classification of evolutionary mode in stratophenetic series: C computational approach: Paleobiology, v. 27, p. 446–465, https://doi.org/10.1666/0094-8373(2001)027<0446:TDACOE>2.0.CO;2.

Roopnarine, P.D., 2001b, A history of diversification, extinction, and invasion in tropical America as derived from species-level phylogenies of chionine genera (Family Veneridae): Journal of Paleontology, v. 75, p. 644–657, https://doi.org/10.1017/S0022336000039706.

Roopnarine, P.D., 2006, Extinction cascades and catastrophe in ancient food webs: Paleobiology, v. 32, p. 1–19, https://doi.org/10.1666/05008.1.

Roopnarine, P.D., and Angielczyk, K.D., 2015, Community stability and selective extinction during the Permian–Triassic mass extinction: Science, v. 350, no. 6256, p. 90–93.

Roopnarine, P.D., and Dineen, A.A., 2018, Coral reefs in crisis: The reliability of deep-time food web reconstructions as analogs for the present, *in* Tyler, C., and Schneider, C., eds., Marine Conservation Paleobiology: New York, Springer Verlag, https://doi.org/10.1101/113795 (in press).

Roopnarine, P.D., Fritzgerald, P., Byars, G., and Kilb, K., 1998, Coincident boron profiles of bivalves from the Gulf of California: Implications for the calculations of paleosalinities: Palaios, v. 13, p. 395–400, https://doi.org/10.2307/3515327.

Smith, A.G., 1960, Amphineura, *in* Moore, R.C., and Pitrat, C.W., eds., Treatise on Invertebrate Paleontology. Part I, Mollusca 1: Boulder, Colorado, and Lawrence, Kansas, Geological Society of America and University of Kansas Press, p. I12, I41, I49, I149.

Smith, A.G., 1971, The Carboniferous genus Glyptochiton de Koninck, 1883 (Mollusca: Polyplacophora): Proceedings of the California Academy of Sciences, ser. 4, v. 37, p. 567–574.

Smith, A.G., 1973, Fossil chitons from the Mesozoic—A checklist and bibliography: Occasional Papers of the California Academy of Sciences, no. 103, 30 p.

Smith, A.G., and Hoare, R.D., 1987, Paleozoic Polyplacophora—A checklist and bibliography: Occasional Papers of the California Academy of Sciences, no. 146, 71 p.

Smith, A.G., Sohl, N.F., and Yochelson, E.L., 1968, New Upper Cretaceous Amphineura (Mollusca): Contributions to Paleontology. Description of chitons from the Gulf Coastal Plain and Puerto Rico: U.S. Geological Survey Professional Paper 593G, p. G1–G8, pl. 1.

Williams, G.C., 1997, Preliminary assessment of the phylogeny of Pennatulacea (Anthrozoa: Octocorallia), with a reevaluation of Edicaran frond-like fossils, and a synopsis of the history of evolutionary thought regarding sea pens, *in* Proceedings of the Sixth International Conference on Coelenterate Biology, Leiden, 1995: Leiden, The Netherlands, Nationaal Natuurhistorisch Museum, p. 497–509.

Williams, G.C., 2007, History of Invertebrate Zoology at the California Academy of Sciences: Proceedings of the California Academy of Sciences, ser. 4, v. 58, p. 197–239.

Williams, G.C., 2008, William Dampier: Science, exploration, and literary influence, including his Hydrographic Treatise of 1699: Proceedings of the California Academy of Sciences, ser. 4, v. 59, p. 533–663.

Manuscript Accepted by the Society 21 November 2017
Manuscript Published Online 13 April 2018
Revised Manuscript Published Online 20 July 2018

Printed in the USA

The Geological Society of America
Special Paper 535

The Rutgers Geology Museum: America's first geology museum and the past 200 years of geoscience education

Lauren Neitzke-Adamo*
Rutgers Geology Museum, Rutgers University, Geology Hall, 85 Somerset Street, New Brunswick, New Jersey 08901, USA, and *Department of Earth and Planetary Sciences, Rutgers University, Wright-Rieman Laboratories, 610 Taylor Road, Piscataway, New Jersey 08854, USA*

A.J. Blandford
Department of History, Rutgers University, 111 Van Dyck Hall, 16 Seminary Place, New Brunswick, New Jersey 08901, USA

Julia Criscione
Rutgers Geology Museum, Rutgers University, Geology Hall, 85 Somerset Street, New Brunswick, New Jersey 08901, USA, and *Department of Earth and Planetary Sciences, Rutgers University, Wright-Rieman Laboratories, 610 Taylor Road, Piscataway, New Jersey 08854, USA*

Richard K. Olsson
Department of Earth and Planetary Sciences, Rutgers University, Wright-Rieman Laboratories, 610 Taylor Road, Piscataway, New Jersey 08854, USA

Erika Gorder
Special Collections and University Archives, Rutgers University Libraries, Alexander Library, 169 College Avenue, New Brunswick, New Jersey 08901, USA

ABSTRACT

The Rutgers Geology Museum is America's first geology museum. Rocks, fossils, and minerals had been collected into "cabinets of curiosities" since first contact between Europeans and Native Americans, and beginning in the late eighteenth century, many of these small personal cabinets were expanded, organized, and made available to the public at natural history and philosophical societies in Philadelphia and Boston. In the first decades of the nineteenth century, geology was widely recognized as an important new science that influenced the organization of collections on display at the growing number of colleges, academies, societies, lyceums, and museums that began popping up all over the United States, but it was not until 1872 that the first museum dedicated specifically to geology was built at Rutgers College. Rutgers University, known as Rutgers College until 1924, is itself one of the oldest colleges in America. Originally chartered as Queens College in the British colony of New

**Corresponding author:* lauren.adamo@rutgers.edu.

Neitzke-Adamo, L., Blandford, A.J., Criscione, J., Olsson, R.K., and Gorder, E., 2018, The Rutgers Geology Museum: America's first geology museum and the past 200 years of geoscience education, *in* Rosenberg, G.D., and Clary, R.M., eds., Museums at the Forefront of the History and Philosophy of Geology: History Made, History in the Making: Geological Society of America Special Paper 535, p. 217–236, https://doi.org/10.1130/2018.2535(14).

Jersey in 1766, Rutgers, along with Harvard (1636), William and Mary (1693), Yale (1701), University of Pennsylvania (1740), Columbia (1754), Princeton (1755), Brown (1764), and Dartmouth College (1769), was among the nine colonial colleges founded before the American Revolution.

Since its inception, the Rutgers Geology Museum's primary mission has been to educate the public on natural history–related topics. How this was accomplished has varied greatly through the years and originated with the "cabinet" of minerals that was displayed to students, alumni, and professors in the days of Dr. Lewis Beck, the first geology professor at Rutgers College. Through the efforts of Dr. George Cook, professor and vice president of Rutgers College, Geology Hall was erected in 1872 as the permanent home for the collections, and the museum and its collections became the focal point of the natural history courses taught at the time. The many professors and curators who tended to the museum and its collections over the next half century helped shape and dictate the future of science and geology education at the university, and with the creation of the Department of Geology in 1931, the museum became a center of leading geologic research and the outlet to present the results to the community. Today, the museum strives to connect with the local K–12 and university communities to inspire the next generation of geoscientists to continue building upon the legacy that the many Rutgers University geologists worked so hard to build.

INTRODUCTION

The study of geology developed in Europe during the late eighteenth century and involved the uniting of three historically distinct fields: long-debated theoretical ideas about the history of the Earth—natural philosophy; the collection and classification of minerals, fossils, and other curiosities from underground—natural history; and the management of mines—practical mining (Porter, 1977; Laudan, 1987; Rudwick, 2007). Although a largely speculative mineralogical map of North America was presented in Paris in the 1750s, it wasn't until the end of the American Revolution that the first deliberately "geological" surveys were conducted and published in the 1780s (Merrill, 1964; Cailleux, 1972; Eyles, 1972). While Americans had always been extremely curious about their mountains and soils, these first geological reports were not widely available, and the spread of specifically "geological" knowledge and practices was initially limited to a handful of individuals who had received training in Europe (Hindle, 1956; Merrill, 1964; Greene, 1984; Newell, 1993; Parrish, 2006).

In 1787, Samuel Latham Mitchill, a Long Island native, published his *Geological Remarks on the Maritime Parts of the State of New-York* in a pamphlet primarily concerned with anatomy (Mitchill, 1787). Mitchill had just returned from completing a medical degree at the University of Edinburgh, and he had been a member of the recently formed Natural History Society where James Hutton's new geological theories were avidly discussed (Hall, 1934). Mitchill became one of the most influential early American geologists and naturalists, and his pamphlet is probably the first time the term "geological" was printed in America. That same year, naturalist Johann David Schoepf included numerous geological descriptions of America in his German language travel narrative, *Travels in the Confederation, 1783–1784* (Merrill, 1964; Schoepf and Morrison, 1968; Greene, 1984). Schoepf was a German physician and surgeon who had been stationed with Hessian and British troops during the Revolutionary War. He sailed to America with specific instructions to observe the mountains and soils of America, which previous travelers, such as the itinerant Linnaean naturalist, Peter Kalm (author of a 1755 travel account), had overlooked (Kremers, 1907; Merrill, 1964; Spieker, 1971). Although this work was not fully translated into English until the twentieth century, the sections on mining in New Jersey and Pennsylvania were published by a Philadelphia periodical in 1790, reflecting both the awareness of Schoepf's work among Pennsylvania naturalists (many of whom were literate in German) and the growing interest in mineralogical and geological subjects (Hoffman, 1901; Bates, 1965; Baatz, 1985; Reid-Maroney, 2001).

This small community gradually increased in size until it could support the publication of the foundational texts of American geology: William Maclure's *Geological Map of the United States,* in 1809; the release of the first American geological textbook, Parker Cleaveland's *Elementary Treatise on Mineralogy and Geology* in 1816; and the national distribution of geological articles in the *American Journal of Science and Arts,* commonly referred to as Silliman's Journal, beginning in 1818. These publications helped to extend and promote interest in geological surveying so that by the 1830s, geology was considered one of the most important and popular sciences in the country (Merrill, 1964; Kohlstedt, 1976; Greene, 1984; Daniels, 1994).

In the decades that followed, the number of geological texts and images, including those for children, multiplied. Attendance at geological lectures became so numerous that colleges began to integrate geology into their curriculum, and state legislatures began allocating thousands of dollars for the production of geological surveys and maps. These state geological survey projects represent the first examples of publicly funded scientific research in the United States. Because every state underwrote some measure of geological surveying, the surveys collectively helped to create an informal and decentralized corps of paid scientists. Eager to compare notes and formulate a national picture of American strata that could compete with and be related to European formations, these survey geologists created the first truly national scientific society, the Association of American Geologists and Naturalists. The society continues today as the American Association for the Advancement of Science and still holds their annual meeting in a different city each year. American geology finally received a national direction in 1879 with the creation of the United States Geological Survey (Merrill, 1964; Kohlstedt, 1976; Greene, 1982; Greene, 1984; Newell, 1993, Daniels, 1994; Gerstner, 1994; Lucier, 2008; Lewis, 2011; Spanagel, 2014).

The rapid expansion of the geological public during the nineteenth century was a global phenomenon that was driven by innovations in theory and practice developed in European centers and through the footwork and formulations of local observers from all over the world. Geology both fueled and was fueled by the global expansion of industrialization and its attendant thirst for coals, iron, and other ores. In America, the study of geology and the extraction of these resources was integrated with the nation's unique political structure of federal and state governments and the activism of local associations, which one foreign observer described as ubiquitous, "Everywhere that, at the head of a new undertaking, you see the government in France and a great lord in England, count on it that you will perceive an association in the United States" (de Tocqueville, 2002, p. 489–490). America did not have a large class of independently wealthy gentlemen scientists as in Europe or state-funded mining schools like those of France and Germany; therefore, the growth of geology during the first half of the nineteenth century was instead a mixture of the top-down lobbying efforts of early industrialists and the bottom-up interests of local groups of "citizen scientists" (Greene, 1982; Newell, 1993; Larson, 2001; Lucier, 2008; Lewis, 2011; Spanagel, 2014). College students, farmers, doctors, postal officers, judges, and teachers all wanted to understand their local soils, minerals, and artifacts, which led to a demand for lectures by experts, public "cabinets" to share and display collections, and initiatives to fund local and regional geological surveying. In the second half of the nineteenth century, geological work became professionalized, and specialized scientific instruction was more widely available at the new land-grant colleges and through scientific schools developed within many of the older schools, many of which also developed their own institutional museums (Hendrickson, 1961; Guralnick, 1975; Kuritz, 1981; Dupree, 1986; Kohlstedt, 1988; Daniels, 1994; Lewis, 2011).

The development of geology at Rutgers closely follows this larger national pattern, and the Rutgers Geology Hall and Museum are among the few remaining architectural legacies of this history. The still largely original and intact spaces of the museum display not only the historical collections but provide us a way of seeing nineteenth-century geology in its native habitat. Here, the preservation of the physical spaces where geological and natural historical specimens were collected, organized, displayed, and chemically and instrumentally analyzed also embodies the social organization of the human and financial capital—the students, professors, schools, states, and companies—that produced and paid for geological knowledge in the nineteenth century (Alpers, 1991).

EARLY AMERICAN GEOLOGY AS GEOLOGICAL EDUCATION

While Samuel Latham Mitchill published his geological remarks as early as 1787 (Mitchill, 1787), it was not until he had become integrated into community institutions such as Columbia College in New York City, the New York Society for the Promotion of Agriculture, Arts, and Manufactures, and the short-lived American Mineralogical Society (1798) that Mitchill's interest in geology found a wider audience. Mitchill began teaching natural history, chemistry, and geology at Columbia College in 1792, and many of these lectures were open to subscription by the public (Hall, 1934; Spieker, 1971; Greene, 1984). Similarly, Benjamin Silliman's first lectures on geology and natural history at Yale, beginning in 1807, extended beyond the school itself to include what he referred to as "volunteer amateurs" and inspired a new cohort of native-trained geologists such as Amos Eaton and Edward Hitchcock (Hall, 1934; Greene, 1984; Brown, 1989; Herbert, 2012; Spanagel, 2014).

Crucial to Silliman's success in New Haven specifically, and geological instruction in America everywhere, was access to a chemical laboratory for mineralogical analysis and the availability of comprehensive collections of European minerals, rocks, and fossils that were so essential for learning how to "read the rocks." Americans could make their own local collections, but mineral classification was based on European specimens, and without access to material examples, it was impossible to determine "even *the names* of the most common stones and minerals" (Silliman, 1818, p. 36). At the turn of the century, there were only a handful of collections available—Adam Seybert's in Philadelphia, Samuel Mitchill's in New York, and a collection David Hosack had donated to Princeton. As these collections increased, they also became more available to the public because some were donated to large institutions. Seybert, for example, transferred his collection to the Academy of Natural Sciences in Philadelphia, and the independently wealthy Rhode Island Colonel George Gibb eventually donated his famously "huge" collection to Yale, where it was available to any curious traveler. However, a collection alone was not sufficient. These immutable rocks needed to be enlivened by inspiring instruction and their meanings made mobile through networks of exchange. Although Harvard College had a well-catalogued collection of hundreds of specimens from around the world, they did not have a Silliman like Yale or a Parker Cleaveland like Bowdoin College in Maine (Greene and Burke, 1978; Latour, 1987; Brown, 1989).

During the 1820s, geology grew in popularity as it became increasingly associated with the successful pursuit of internal improvements, such as the Erie Canal, the excavation of which unearthed exciting new fossil evidence about the vastness of Earth's history and the extinct creatures that had lived in these "former worlds" (Guralnick, 1972; Rudwick, 1992; Newell, 1993; Larson, 2001; Spanagel, 2014). However, there were still relatively few places where one could receive experienced training in subjects that would soon be thought of as "science," such as natural history, geology, chemistry, and engineering. In 1824, the British ended a restriction on the emigration of "artisans" with mechanical and engineering knowledge, allowing geologically trained surveyors such as Robert C. Taylor and John Finch to supplement the small but growing ranks of geological surveyors in America in the 1820s and 1830s (Torrens, 1990; Spanagel, 2014). The demand for trained surveyors, engineers, mechanics, chemists, and geologists and the popularity of these subjects led to a wave of new institutions such as the Rensselaer Institute, now known as Rensselaer Polytechnic, and educational reforms in older colleges such as Rutgers (Guralnick, 1972; Greene, 1984; Torrens, 1990; Geiger, 2000).

John Finch arrived in America in 1823, almost directly from a debtor's prison in London, and he hoped to benefit from these reforms and to obtain a position as a professor at one of the colleges that were adding natural history, chemistry, and geology to their curriculum (Torrens, 1990). Finch quickly integrated himself into the American geological community by giving lectures at local societies and publishing several articles in Silliman's Journal. In 1829, Finch was given permission to use the College building at Rutgers to deliver a series of 15 geological lectures, the first of its kind at the college. Like those at Columbia and Yale, these lectures took place physically at the College, but they were by subscription and open to the public (Torrens, 1990). What is especially interesting about this lecture series is that it was initiated and organized by a group of students in direct response to the public demand for instruction in geology. This same group of students also petitioned the board of trustees to allow them to form a Natural History Society, and the following year they succeeded in soliciting donations for a specimen cabinet and the purchase of several books. This mineral cabinet, originally displayed in Van Nest Hall (Fig. 1), for which there are very few records, can be considered the origins of the Rutgers Geology Museum (Rutgers College Faculty Minutes, Rutgers University Special Collections and University Archives, 1829–1831). Most likely Finch brought a small collection of his own to illustrate the lectures and spent time helping the residents name and classify their own local finds. In 1830, when the board decided to create a permanent professorship of natural history, they hired an American naturalist and chemist, Lewis C. Beck, instead of Finch (McCormick, 1966; Torrens, 1990).

THE FIRST GEOLOGY PROFESSOR AT RUTGERS COLLEGE, LEWIS CALEB BECK

The first geology professor at Rutgers College, Lewis Caleb Beck (1798–1853; Fig. 2) had institutional and personal ties to both Samuel Latham Mitchill and Amos Eaton, one of the founders of Rensselaer Institute, who helped inspire and shape the members of the first generation of professional American geologists (Kelly and Burrage, 1920; Fenton and Fenton, 1945; Wilkerson, 1963; Spanagel, 2014). Born in Schenectady, New York (1798), Beck graduated from Union College in 1815 and pursued a medical degree at the College of Physicians and Surgeons, the medical department of Columbia University located in New York City, and he began practicing medicine in 1818 (Gross, 1861; Kelly and Burrage, 1920). His older brother, T. Romeyn Beck, had assisted Eaton in the publication of *A Geological Survey of the County of Albany* in 1818 as part of a geological survey of Albany and Rensselaer County, New York, commissioned by the owner of one of the last remaining Dutch patroonships, Stephen Van Rensselaer. In 1821, Beck himself assisted Eaton in his next survey, *A Geological and Agricultural Survey of Rensselaer County, in the State of New-York,* and the results of these surveys became the standard for subsequent geological surveys (Spanagel, 2014). When the Rensselaer Institute opened in 1824, Eaton hired Beck as a junior professor of botany, mineralogy, and zoology (Gross, 1861; Kelly and Burrage, 1920; Wilkerson, 1963; Aldrich, 2000).

Like many of the other well-known geologists of the time who started out as lawyers, clergymen, surveyors, zoologists, and chemists, for Beck, geology was one pronounced aspect of a career as a "naturalist" that included publications in botany, chemistry, and mineralogy as well as a *Gazetteer of Illinois and Missouri.* There were few paying jobs for geologists, and those that existed did not pay well. While some geologists, such as Benjamin Silliman, had permanent positions, most early geology instructors, like Beck, had to piece together a living through multiple appointments. Even permanent professorships, like Silliman's, paid less than similar positions in the traditional fields that we now refer to as the humanities, and teaching work was often supplemented by private consulting work for wealthy individuals, mining companies, or the government geological surveys. Beck found teaching jobs in Vermont, New York, and New Jersey, and in 1836, he was hired as the mineralogist for the New York Natural History Survey. Later on, he received an appointment at the Patent Office to carry out some of the first chemical analyses of agricultural products as part of what would eventually become the Food and Drug Administration (Wilkerson, 1963; Newell, 1993; Aldrich, 2000; Lucier, 2008).

At Rutgers College, Beck taught geology and mineralogy as part of the natural history and chemistry courses and was heavily involved in studying and publishing on the local geology and

Figure 1. View of Van Nest Hall with students gathered in front, ca. 1850. Van Nest Hall was originally used to house and display the mineral and fossil "cabinet" during the days of Lewis C. Beck's tenure before the Rutgers College Geology Museum was built. Photograph courtesy of the Rutgers University Archives.

mineralogy of New Jersey and New York (Beck, 1842; Gross, 1861; Wilkerson, 1963). During his travels on foot and horseback over the years, Beck took copious notes and collected local geologic, mineralogical, and botanical specimens. These journeys, along with his chemistry background, fostered a lifelong pursuit into the examination and understanding of the chemical composition of minerals. Beck conveyed these practices and his extensive knowledge of chemistry to his students by combining field observations with chemical analyses performed in the laboratory. He amassed an extensive personal collection in tandem with the nearly 3000 mineral specimens gathered for the people of the State of New York (Gross, 1861). The Rutgers Geology Museum did not have a permanent building during the tenure of Lewis Caleb Beck, and Beck most likely used his own personal collection to supplement the mineral "cabinet" for instructional purposes. Shortly after his death, his collection was purchased by friends of Rutgers College and placed in the custody of the student Natural History Society. This collection would serve as a key component of the museum's mineralogical and teaching collections in subsequent years, and the Beck collection remains the oldest scientific collection still on display today at Rutgers University (Wilkerson, 1963).

GEORGE HAMMELL COOK, THE MORRILL ACT, AND THE FOUNDING OF THE GEOLOGY MUSEUM

In 1853, George Hammell Cook (Fig. 3), one of Rutgers University's most revered and well-known figures in its 250-year history, was hired by Rutgers President Theodore Frelinghuysen and the Board of Trustees to replace Beck as full-time professor of chemistry and natural science. Cook was urged to pay an informal and secret visit to discuss this position. With the promise of a higher salary, less arduous duties, the chance to do more advanced scientific research, and US$2000 to purchase laboratory equipment, he decided to accept the position at Rutgers College (Sidar, 1976). His appointment and accomplishments at the college led to the building of Geology Hall, the permanent home of the Rutgers Geology Museum, and the significant improvement of Rutgers College through the attainment of land grant status (McCormick, 1966; Sidar, 1976).

Figure 2. Portrait of Lewis Caleb Beck, professor of chemistry and natural science (1830–1853) and first professor of geology at Rutgers College. Photograph courtesy of the Rutgers University Archives.

Figure 3. Portrait of George Hammell Cook, professor of chemistry, geology, agriculture (1853–1889), vice president (1864–1889), and founder of the Geology Museum (1872) at Rutgers College. Photograph courtesy of the Rutgers University Archives.

Cook was born in 1818, raised in Hanover, New Jersey, and later moved to Troy, New York, to attend college at Rensselaer Polytechnic Institute between the years of 1838 and 1840 (Wilkerson, 1963; Sidar, 1976). He enrolled at Rensselaer with a keen interest in engineering but quickly changed his focus to geology and natural history due to the influence, once again, of Professor Amos Eaton, whose energy and enthusiasm, despite his declining health, continued to exert as strong an influence on his students as always (Sidar, 1976). Like Beck before him, Cook was also offered a position at Rensselaer Polytechnic Institute and strung together a series of tutoring and teaching jobs to support himself (Wilkerson, 1963; Sidar, 1976).

In 1846, he left teaching to try to pursue his fortunes in the glass-making business, but after a series of financial issues and the folding of the company (Sidar, 1976), he returned to education in 1848 as a teacher of mathematics and natural philosophy at Albany Academy (Wilkerson, 1963). Cook was unhappy teaching at a school with so few students; so he began lecturing at Albany Medical College in 1852, where he co-taught several courses in chemistry with Dr. Lewis Beck (Sidar, 1976). In 1854, shortly after Cook arrived at Rutgers College, he was named the State Geologist by the New Jersey Geological Survey and retained this title until his death in 1889 (Wilkerson, 1963; Sidar, 1976). In 1856, Cook was granted an honorary Ph.D. from New York University, a measure of his academic distinction during this period because there were no earned doctorates in the country until 1863 (Brubacher and Rudy, 1976; Sidar, 1976; Sheets-Pyenson, 1987; Bruce, 1988; Angulo, 2009). Over the years at Rutgers College, Cook's position was altered from professor of chemistry and natural history to include agriculture in 1867, and it was finally changed to professor of geology and agriculture in 1882. Additionally, he was named vice president of Rutgers College in 1864 and remained in this position until his death (Sidar, 1976).

Cook's simultaneous activities at Rutgers and with the government-run New Jersey Geological Survey during this time exemplify the dual appointments that many state geological survey geologists held (Merrill, 1978; Newell, 1993). As a professor at Rutgers, he helped establish the Bachelor of Science program and one of the first experimental agricultural stations in the country, and as the State Geologist of New Jersey, his research efforts encouraged the economic development of the clays, soils, iron, zinc, and water supplies throughout the state (Wilkerson, 1963; Sidar, 1976). The Morrill Land Grant Act, signed by President

Abraham Lincoln on 2 July 1862, allowed for the funds raised by the sale of public lands to be applied to the endowment and support of colleges in the study of military tactics, agriculture, and engineering (Vesey, 1965; Geiger, 2000). In the early 1860s, the college was struggling with low student enrollments and lack of funds for professors' salaries; therefore, Cook along with Professor David Murray, presented a proposal to the Board of Trustees in December 1863 to obtain funds from the Federal government under the Morrill Act. The board immediately accepted this proposal and embarked on an intensive campaign to secure the grant from the state legislators (McCormick, 1966; Sidar, 1976).

Cook himself undertook a majority of this task, personally seeking out legislators and other important and influential men of the time (McCormick, 1966; Sidar, 1976). Events culminated when the state legislators voted against Princeton University and the State Normal School at Trenton and voted in favor of Rutgers College (McCormick, 1966). In order to comply with the terms of the grant, the College needed to make some improvements to their facilities without the use of state funds. A rundown farm that would eventually become the experimental agricultural station was purchased, and a well-equipped observatory was built, largely through the gift of Daniel S. Schanck (Fig. 4). After a year-long fundraising campaign that brought in over US$121,000, it was decided that half of these funds would be designated to "build a geological hall" (McCormick, 1966). Henry Janeway Hardenbergh, great-great-grandson of the first Rutgers College president and the architect of the Kirkpatrick Chapel and Van Nest Hall on campus (Fig. 1) and the Waldorf-Astoria Hotel and Dakota building in New York City, was hired to build the large brownstone building. Construction on Geology Hall was completed in 1872, with an armory in the basement, a lecture hall and facilities for the physical sciences on the first floor, and the two-story geology museum with its large windows and open gallery occupying the central floors of the building (Fig. 5; McCormick, 1966).

The Rutgers Geology Museum within Geology Hall opened at this time with the purpose of displaying and making the ever-growing mineral and fossil collections compiled by Beck, Cook, and others more accessible to the public and for use in the instruction of geology and natural history classes (McCormick, 1966). The construction and opening of a permanent home for the Rutgers Geology Museum represented the culmination of work Cook had been pursuing since 1853 when he took over from Lewis Beck, and it represents both Cook's outsized influence within the College and the role that geology played in this period as one of the most preeminent and popular sciences.

Figure 4. View of the Schanck Observatory sometime in the 1890s. This facility, built in 1866, was made possible with the funds donated by Daniel S. Schanck, for whom the Observatory is named. Photograph courtesy of the Rutgers University Archives.

Figure 5. Exterior view of Rutgers College Geology Hall sometime in the late 1800s. Geology Hall was built by Henry Janeway Hardenbergh and completed in 1872. The basement of the building was used as an armory, while the first floor contained a sizable lecture hall that was used to hold physical sciences classes. The Rutgers Geology Museum was built on the second and third floors of this building and resides here to this day. Photograph courtesy of the Rutgers University Archives.

In 1857, Cook reinstated the Natural History Society to stimulate and satisfy interest in natural history. The objective of the society was to expand the "cabinet" or museum of minerals that was started by Beck, and originally on display in Van Nest Hall on campus (Fig. 1), with Cook himself serving as president.

This collection was the most visible indicator of the entire sciences program at Rutgers College, and it was proudly displayed and toured by students, faculty, and alumni. Cook kept up extensive correspondence with colleagues around the world, even throughout the Civil War, to acquire and exchange mineral, fossil, and other natural history specimens to expand the collection (Sidar, 1976). By the end of his career at Rutgers College, Cook had collected and organized more than 4000 mineral and fossil specimens and acquired several other important artifacts, such as a mastodon skeleton from Mannington, New Jersey (Fig. 6) and several large slabs of Triassic sandstone from Morris County, New Jersey; these sandstone slabs contained footprints from seven different species of ancient reptiles including dinosaurs (Fig. 7; Wilkerson, 1963).

The origins of the Rutgers Geology Museum can be traced back to the antebellum era, when the student-led Society for Natural History organized the very first "cabinet" of minerals, just one of many of the numerous mineralogical and geological collections held by local colleges and academies, agricultural and natural history societies, and lyceums all over the country. Some of the most notable examples include the collection of the American Philosophical Society—which Charles Willson Peale displayed as part of his American Museum (1790), Yale College's mineral gallery (1811), the cabinet at Philadelphia's Academy of Natural Sciences (1812), and the Lyceum of Natural History of New York (1817). However, as geological collecting and observation increasingly shifted from the hands of "citizen scientists" to paid labor done by state surveyors, many of these smaller local societies sold off or transferred their collections to new institutional homes dedicated to natural history, as was the case with the many collections transferred to the Boston Museum of Natural History in 1830 (Bates, 1965; Greene, 1984; Preston, 1986; Conn, 1998; Yanni, 1999).

Figure 6. View of the mastodon skeleton from Mannington, New Jersey, on display in the Rutgers Geology Museum with William S. Valiant. Valiant was the assistant curator of the museum from 1893 to 1903 and the curator from 1903 to 1923. Photograph courtesy of the Rutgers University Archives.

Figure 7. View of the large slabs of Triassic sandstone from Morris County, New Jersey, acquired by George H. Cook. These slabs contain the trackways of seven different species of dinosaurs and reptiles. Photograph courtesy of the Rutgers University Archives.

Scientific research became more specialized during the middle decades of the nineteenth century, and the work of collecting, displaying, and explaining natural historical phenomena became the domain of larger government and university museums, such as the New York "State Cabinet of Natural History" (1845), the Smithsonian Institution (1846), Harvard's Museum of Comparative Zoology (1859), Yale's Peabody Museum (1866), and the American Museum of Natural History in New York City (1869). All of these museums had exhibits and collections featuring rocks, minerals, and fossils; however, the Rutgers Geology Museum is unique in that when it opened in 1872, it was the first museum dedicated exclusively to geology and can be considered America's first geology museum (Preston, 1986; Sheets-Pyenson, 1987; Winsor, 1991; Conn, 1998; Yanni, 1999; Chaplin, 2001; Miller and Buhay, 2007).

Shortly thereafter, several other geology museums opened up around the country, including the Colorado School of Mines Geology Museum (1874; Colorado School of Mines, 2017), the Arizona Mining and Mineral Museum (1884; Arizona Mining and Mineral Museum, 2015), the Orton Geological Museum at Ohio State University (1893; Ohio State University, 2017), the University of Wisconsin–Madison Geology Museum (~1877; University of Wisconsin Geology Museum, 2017), the South Dakota School of Mines and Technology Geology Museum (1885; South Dakota Mines, 2017), and the Fryxell Geology Museum in Illinois (ca. late 1880s; Augustana College, 2017).

CHESTER, VALIANT, AND THE FIRST MUSEUM CURATORS

In 1891, two years after Cook's death, Albert Huntington Chester was hired as a professor of general and applied chemistry, and then reappointed as a professor of chemistry and mineralogy and as the first official curator of the Rutgers Geology Museum in 1892 (Fig. 8; Rutgers College, 1916). Chester was born in 1843 in Saratoga Springs, New York, and began his career in mineralogy when he graduated with a degree in Mining Engineering from Columbia School of Mines in 1868, an M.A. from Union College in 1872, and a Ph.D. from Columbia University in 1878. He then went on to teach chemistry, mineralogy, and natural history at Hamilton College as a professor of mining and metallurgy from 1870 to 1891 (Paine, 1924; Aiken, 1945; Wilkerson, 1963). During his tenure at Hamilton College, not only was Chester a well-respected lecturer, but he also oversaw the Department of Chemistry and Mineralogy as the department head (Aiken, 1945). He also obtained, examined, and classified thousands of mineral specimens from type localities around the world, without the help of modern mineral identification techniques, in order to publish a comprehensive catalogue of mineral names and compositions (Chester, 1886). Chester also undertook the rearranging and relabeling of the State of New York's large mineral collection, a large portion of which was collected by Lewis Beck during his time at the New York Geological Survey, as it had fallen into disrepair with nearly all the labels misplaced or lost (Chester Collection Hamilton College, retrieved 2017). Chester worked diligently to restore order to this collection, and this experience curating and classifying mineral collections is most likely one of the reasons he was offered the position at Rutgers College and the Geology Museum.

When Chester arrived at Rutgers College, the Geology Museum already possessed a large collection of items that covered many topics within geology, including a collection of rock samples from across Europe and North America, a broad range of paleontological samples that covered the major topics in the field but with special emphasis on New Jersey, a large collection of shells, many stone tools and artifacts, thousands of mineral specimens, and the recently mounted mastodon skeleton (Fig. 6). He continued to personally accumulate mineral specimens as more samples were sent to him for analysis, and this ~4800-piece collection, later donated to Rutgers after Chester's death, remained a valuable research and teaching set at Rutgers University. Despite Chester's strong interest in mineralogy, little time was spent putting the Geology Museum's vast collections in order prior to the mid-1890s, a daunting task because of the many different systems of mineralogical and geological classification the nineteenth-century labels would have reflected (Greene, 1982; Newell, 1993;

Figure 8. Portrait of Albert H. Chester, ca. 1880s, courtesy of the Hamilton College Archives. Chester became the first official curator of the Rutgers Geology Museum in 1892 and retained that title until his death in 1903.

Gerstner; 1994; Aldrich, 2000; Lucier, 2008). William S. Valiant was hired in 1893 to help Chester undertake the task of reorganizing the museum's collection as the assistant curator and held this position until Chester's death in 1903 (Figs. 6 and 9; Wilkerson, 1963). Valiant was then named museum curator and continued in this capacity until 1923, when his health would no longer allow him to complete his duties (Hayes, 1927).

Valiant was born in Rome, New York, in 1845, and he developed an interest in geology, fossils, and minerals at a very early age. He started exploring the local geology and fossil hunting around the age of 13 and educated himself by studying geologic reports and corresponding with State Geologist, James Hall (Hayes, 1927; Wilkerson, 1963). Valiant developed a keen eye for observation and became an expert at classifying minerals despite having no university training, and he actually helped catalogue minerals for the Rome Academy in 1875 (Wilkerson, 1963). He is also responsible for one of the most important paleontological finds of the century when he discovered a new species of trilobite, *Triarthrus becki,* which was the first specimen recovered with preserved antennae and appendages intact (Hayes, 1927). Later on in his life, he worked closely with the state geologists as a contributor to the Annual Reports of the New Jersey Geological Survey (Kümmel, 1909).

Under the guidance of Chester and Valiant, much progress was made in cleaning up the Geology Museum's collections and making improvements to the physical space. They continued to acquire more natural history specimens from around the world, including a giant Japanese spider crab exoskeleton and a right whale skeleton, which were both mounted in 1899 (Figs. 9 and 10, respectively; Wilkerson, 1963). Contrary to what one might assume about the limited scope of responsibility for an assistant curator, Museum records show that Valiant actually did the majority of the work organizing, re-labeling, and recording the thousands of fossil and mineral specimens in the collection. As Chester's health declined, Valiant took on more of the responsibilities of the museum and was performing the majority of the curator duties long before Chester's death (Chester Collection, Rutgers University Special Collections and University Archives, 1871–1903; Valiant, 1899). Valiant kept diligent records of the museum visitors and encouraged all the visitors to ask questions

Figure 9. View of a giant Japanese spider crab with William S. Valiant. This spider crab was gifted to Rutgers College from the Japanese government as a gesture of good will for accepting many Japanese students at the College in the late 1800s. It was originally mounted in 1899 and remained on the walls until the 1980s, when it started to fall into disrepair. This giant spider crab specimen was returned to its historic location in the Rutgers Geology Museum in 2015 after a three-year restoration project. Photograph courtesy of the Rutgers University Archives.

Figure 10. Interior of the Rutgers Geology Museum with a view of the right whale, mounted in 1899, hanging from the ceiling. The whale bones were later given to Yale University where they were discarded. The floor of the museum is covered with cabinets filled with mineral and fossil specimens that would be visually inspected by students taking mineralogy and natural history courses at Rutgers College during the late 1800s. Photograph courtesy of the Rutgers University Archives.

about the exhibits. He took great pride in the educational value of the museum, and he was constantly looking for ways to increase its impact to the general public (Rutgers College Geology Museum Records, 1860–1919; Valiant, 1899).

MODERN MUSEUM ERA

Geology Courses and the Creation of the Geology Department

Many other prominent geologists stepped into the role of Museum Director during the next several decades, including Dr. Joseph Volney Lewis from 1911 to 1927 (Curator from 1904 to 1911), Dr. Albert Orion Hayes from 1927 to 1944, Dr. Helgi Johnson from 1944 to 1961, Dr. James H.C. Martens from 1961 to 1967, and Dr. Raymond C. Murray from 1967 to 1977 (Wilkerson, 1963). Until the 1920s, much of Geology Hall was occupied by the Department of Physics, and the geology and mineralogy professors taught courses in the dedicated classroom, mineralogy laboratory, and museum space within the building. From the early 1900s until the late 1920s, the activity amongst the geology professors was largely centered around classroom instruction and working on cataloging and displaying the vast rock, mineral, and fossil collections of Beck, Cook, Chester, and others (R.K. Olsson, 2016, personal commun.).

Geology at Rutgers began to take on its more modern form when Albert Orion Hayes (Fig. 11) was hired in 1926. This same year, the Department of Physics vacated the premises in Geology Hall, allowing the geology professors to completely take over all four floors of the building, with the museum and offices on the second and third floors and classrooms, laboratories, and faculty lounges in the basement and first floor (R.K. Olsson, 2016, personal commun.). At this time, Rutgers College received a considerable donation willed by Mr. Fred Canfield to be used toward the furtherance of geology at the college (Wilkerson, 1963). This donation helped geology to officially become a major department when the first B.S. degree in geology was awarded in 1931. The first earned doctorate and master's degrees in geology followed shortly after in 1938 and 1939, respectively (R.K. Olsson, 2016, personal commun.).

The Rutgers College Geology Museum and Geology Department continued to grow through the mid-1900s as new faculty were hired and more collections were acquired. Many new courses in geology were offered between 1926 and 1945, including paleontology, geomorphology, human origins, sedimentology, stratigraphy, and crystallography (Wilkerson, 1963). As the focus of geological research and instruction shifted increasingly to these new sub-fields, the historic mineral collections were less and less frequently used in lectures and coursework. Despite this, the museum's collections continued to grow as additional acquisitions were made, most significant of which was the Rowe collection donated in 1940 (R.K. Olsson, 2016, personal commun.). George Rowe was the mining captain for the New Jersey Zinc Company in Franklin, New Jersey, from 1906 to 1934, and the ~2400-piece collection contains an array of minerals from around the world but prominently features fluorescent minerals from New Jersey (Wilkerson, 1963; Rowe, 1999). Many of these and other historic minerals remain on display today at the Geology Museum, and the collection remains a valuable reference collection with many samples from all over the world and from some localities in New Jersey and New York that are no longer accessible (Fig. 12; Beck Papers, Rutgers University Special Collections and University Archives, 1839).

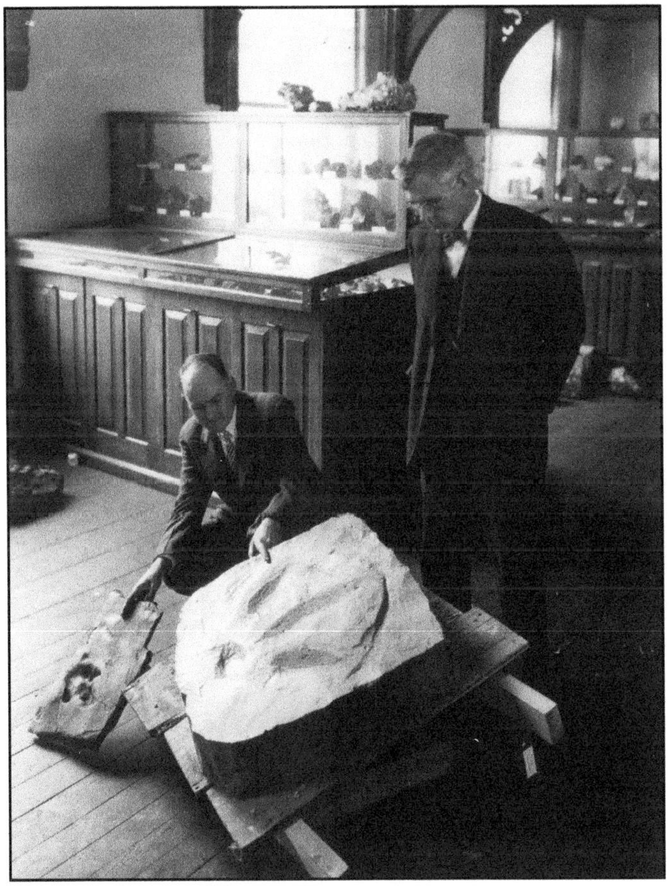

Figure 11. Professor Albert Orion Hayes (standing; 1926–1951) and Professor Alfred C. Hawkins (squatting; 1927–1933) examining a dinosaur footprint that was discovered in the clay pits in Woodbridge, New Jersey, in 1930. Photograph courtesy of the Rutgers University Archives.

In 1977, the Geology Department moved from the Geology Museum within Geology Hall to renovated quarters in a building on another campus, thus completing the relocation of all Rutgers College science departments across the Raritan River to Busch Campus (Wilkerson, 1963; R.K. Olsson, 2016, personal commun.). Under the direction of Dr. Richard K. Olsson (Fig. 13), hired in 1957 and department chairman from 1977 to 1996, the Geology Department increased its research focus by hiring many new faculty members and laying the groundwork for the strong geologic research institution it is today (R.K. Olsson, 2016, personal commun.). This move marked the end of active research activities within the museum and the expansion of the museum's education and outreach mission despite the commencement of the annual open house program prior to this time.

Outreach and Public Education

The Geology Museum first expanded its public outreach initiative with the establishment of its annual open house at the end of January in 1968 (Fig. 14). This event, first suggested by museum technician, Al Bliss, and supported by Dr. Murray, entailed a series of faculty-organized lectures, exhibits, rock and mineral identifications, and movies in Geology Hall for Rutgers students, the public, high school teachers, and students. The sale of various mineral collections to fundraise for the museum began in 1975 and has since been a major activity of the open house. In the early years of the open house, Rutgers faculty performed most of the lectures, and, in subsequent years, invited speakers from other universities, research centers, government departments, and academies all over the country. The annual open house has been host to several major scientific discoveries; for example, the Moon rocks collected during the Apollo 11 lunar landing (1969) were exhibited in 1971, and the first public presentation of the rediscovery of the *Titanic* was presented by Dr. Robert Ballard of Woods Hole Oceanographic Institute in 1986 (R.K. Olsson, 2016, personal commun.). The lecture program included the latest research results from various fields of geology, paleontology, mineralogy, archeology, oceanography, evolution, extinction, glaciology, seismology, physics and astronomy, meteorites, etc., from around the world. Records of attendance were not kept prior to 2010, but with an average of well over 2000 people in attendance every year, this event remains the largest and most well-attended public event of the year, and 2018 marked the 50th anniversary of the first open house.

From 1977 until 2009, the head of the Department of Earth and Planetary Sciences, known as the Geology Department until 2007, also served as the director of the Geology Museum. The Geology Department heads who also filled the role of museum director include Dr. Richard K. Olsson (1977–1996), Dr. Michael J. Carr (1996–1999), and Dr. Kenneth G. Miller (2000–2009). In 1978, Dr. Olsson appointed William Selden (Fig. 15) as curator of the museum, a position he held until his

June 24th 1836.

Visited Staten Island in Steam Boat Bolivar- Presented a letter of introduction from the Gov. to Dr Rockwell Health Officer- Called with him upon Major Howard. at Howard Place. about ½ a mile from the Quarantine. Major H's house is about 600 feet above the level of the sea- He has a collection of minerals & curiosities- Toad Hill is 50 or 60 feet higher & is thought to be the highest land on the Island. Major H. some time since dug a well. 85 feet below the surface & retained specimens of the strata- some of them very beautiful. He has some duplicates which he promised me-

July 1. 1836. Rode to the village of Richmond the Seat of Justice for the County. It is said to be about 7½ miles about S.W from the Quarantine, Four miles from the Quarantine found some fine specimens of White Talc of several varieties It occurs on the West side of the road to Richmond near the house of Mr Garretson. It is imbedded in detached pieces in the earth. the rock below is similar to that found at Quarantine.

In the afternoon visited Major Howard & examined his collection of minerals from Howard Place & the Island generally- The following is a catalogue.

Radiated quartz in boulders presenting considerable variety- Howard place.

Hematitic Iron Ore. in small masses. Howard Place.

Carbonate of Magnesia compact & crystallized. Howard Place-

Talc fine specimens. Do—

Figure 12. Specimens and historic documents from the mineral and archive collection of Dr. Lewis Beck. (A) Notes from Beck's original journal describing some of the minerals still in the collections today. Images courtesy of the Rutgers University Archives. (B, C) Quartz specimens from Diana, Lewis County, New York. (D) Quartz specimen with talc collected near the "Quarantine" Hospital in Tompkinsville, Staten Island, New York. Stephenson (2004) had previously stated that this mineral locality, which was once small farmland, is now built-up and no longer accessible. This study was able to match the historic mineral specimens and journal entries to this locality that is no longer a viable collection site.

Figure 13. Professor Richard K. Olsson (left; 1957–present), Professor Raymond Murray (1973–1977), and John Beardsloe III (right) of the New Brunswick Theological Seminary; they are examining the remains of Egyptian priestess Iset-Ha on display at the Rutgers Geology Museum, ca. 1969 (photograph courtesy of the Rutgers University Archives).

retirement in 2009. Under his guidance, new exhibits were created, and the annual open house continued to grow. A significant effort was made by Selden and a few dedicated volunteers to inventory the many thousands of rock, mineral, and fossil specimens in the historic Rowe, Chester, Cook, and Beck collections, which had been moved out of Geology Hall in the 1960s due to the growing geology faculty (R.K. Olsson, 2016, personal commun.). However, hundreds of these historic minerals can still be seen on display within the museum today. In the 1990s, the Friends of the Geology Museum was established to create a stronger connection between the museum and its patrons with the purpose of sponsoring museum activities that enhance the services to the public and enrich the educational environment of the community. Selden was also responsible for first introducing guided tours of the museum for K–12 school and special interest groups, a program that is still in effect today and brings thousands of children to the museum each year (Fig. 16; R.K. Olsson, 2016, personal commun.).

During a Rutgers University transition in 2009, control of the Geology Museum was transferred from the Department of Earth and Planetary Sciences, to the unit of Undergraduate Academic Affairs under the dual directorship of Dr. Kathleen Scott from Biological Sciences and Dr. Carl Swisher from Earth and Planetary Sciences. The museum's outreach programs were greatly expanded during this time to include new interactive family-friendly programs on weekends and week nights, guided field trips to local geologic sites, and increased programming to facilitate K–12 interaction at the annual open house event (Fig. 16). Faculty and staff with science and educational backgrounds enhanced the guided tour programs with hands-on learning activities that addressed the Core Curriculum Standards, now called Next Generation Science Standards, and they started offering a wider variety of tours, such as bilingual tours for the English as a Secondary Language classrooms, Boy Scout and Girl Scout groups, etc. Today these programs reach an average of 8000–10,000 people a year and more than

Figure 14. Photographs from the first Rutgers Geology Museum open house in 1968. (A and B) The mineral sale; (C) Professor Steven Knowlton Fox (1948–1981) talking with visitors and volunteers; and (D) a view of the interior of the museum. Photographs courtesy of Richard K. Olsson.

100 different schools and groups within the Rutgers University and New Jersey communities.

In 2015, oversight of the Rutgers University Geology Museum was moved back to the School of Arts and Sciences, and Dr. Lauren Neitzke-Adamo and Dr. Patricia Irizarry-Barrato were named co-directors. An advisory committee, consisting of museum and university staff, undergraduate and graduate students, faculty members, and active museum members, was created with the hope of developing a mission for the museum that will further cement the commitment to enhance the Geology Museum's impact on Rutgers University and its surrounding communities. The museum continues to be committed to providing varied, dynamic, and scientifically relevant exhibits and offering stimulating K–12 outreach opportunities while preserving the historic aesthetics and cultural integrity of the museum. New efforts will be made to expand the role of the museum in Rutgers University education, and additional internship and place-based learning opportunities will be created and offered to university students to enhance their learning and professional experiences. Significant improvements have been made to the types and quality of the programs offered at the Rutgers Geology Museum in recent years, and the museum staff hopes that these new efforts will help establish the museum as a leading outreach and education center at Rutgers University and throughout New Jersey.

THE MUSEUM AS A RUTGERS RESOURCE AND CONCLUSIONS

With the increase in fiscal constraints in recent years to state and university natural history museums, many museums have been forced to shutter their doors, and the priceless collections have been forcibly relocated to other institutions, dismantled, or sold for profit (Allmon, 1994; Dalton, 2003; Froelich, 2003; Gropp, 2003; Conniff, 2016). The value of natural history museums and their collections as resources for research and public education is commonly misunderstood by the general public (Allmon, 1994). Even though active research is no longer conducted at the Rutgers Geology Museum, it still provides value to the research community as a means to document the past, a repository of reference material, and as a center for disseminating current research to the public through hands-on education (Allmon, 1994; Suarez and Tsutsui, 2004). Throughout the museum's history, members of the Rutgers University faculty have taken an active role in the creation of new and exciting exhibits

Figure 15. Museum Curator William Selden giving a tour to a group of schoolchildren at the Rutgers University Geology Museum. Photograph courtesy of the Rutgers University Archives.

based on the latest natural history–related discoveries. In the days of Cook, Chester, Valiant, etc., the professors took the lead in this endeavor; whereas, today, the museum directors, staff, and educational specialists oversee the majority of the exhibit renovations and consult with university professors to ensure the scientific accuracy and relevance of the new displays. Furthermore, many Rutgers University undergraduate and graduate students assist the museum staff with exhibit design and other museum activities, through internships and place-based course requirements, providing valuable professional development experience for the next generation of educators and researchers.

In addition to the immense historic value of the museum and the building, the Rutgers Geology Museum remains a valuable resource as a place of learning. The popularity of science museums has significantly increased over the past 50 years (Friedman, 2007), and research has shown that people of all ages learn from museum visits through the memories they create and transformative experiences (Ramey-Gassert et al., 1994; Rennie and McClafferty, 1995; Weil, 1995; Rennie and Johnston, 2004; Soren, 2009; and others). Visitors can be transformed by gaining new appreciation of a topic they knew little about, having their previous views or beliefs challenged, or when they are provided with opportunities to explore new ideas and concepts (Soren, 2009). These experiences may have profound life-long effects; for example, a museum visitor might decide to become a paleontologist after watching scientists work on excavating fossils at a museum. Change can also be as seemingly insignificant as when a student correctly recalls a concept or answer based on a visit to a museum (Rennie and McClafferty, 1995). Thousands of school-aged children from all over New Jersey and the surrounding states visit the Rutgers Geology Museum each year, and in some cases, these visits might be their only exposure to science topics outside of the classroom. As a free museum, the Rutgers Geology Museum serves many underprivileged areas including New Brunswick, New Jersey, the town in which the museum is located. Courses across a broad range of subjects, including geology, ecology, evolution, anthropology, history, art history, creative writing, art, American studies, education, etc. bring hundreds of Rutgers University students to the Geology Museum each year to utilize the resources and exhibits available in order to enhance the university students' learning experiences.

Figure 16. The Rutgers Geology Museum today. (A) A view of the inside of the museum in 2017 showing the Mannington mastodon and the newly restored spider crab that was remounted in 2015. (B) Visitors participate in activity stations during an annual open house. (C) Visitors collect fossils on a field trip to Big Brook Preserve in Colts Neck, New Jersey. (D) A group of Girl Scouts pose in front of the mastodon as they explore the exhibits at a "Late Night at the Museum" event.

Much more time and effort would need to be invested to transform the museum into one of the leading outreach and educational centers in New Jersey; however, the Rutgers Geology Museum remains a treasure within the Rutgers University and New Jersey communities. As the first geology museum in America, the long geologic, architectural, mineralogical, and educational history surrounding the Rutgers Geology Museum deserves to be preserved so that many future generations can experience this museum first hand. The museum staff is embarking on a new chapter in the history of the Rutgers Geology Museum, and as part of this effort, they hope to enhance the many exhibits, tours, and programming available to the public. By forging new relationships across New Jersey, the museum hopes to expand its mission and to continue to meet the changing educational needs of the times, while keeping the Rutgers and New Jersey communities, which were so pivotal in demanding the need for a public space for geologic education in the first place, as its top priority.

ACKNOWLEDGMENTS

Thank you to Gary Rosenberg and the anonymous reviewers for providing many helpful and insightful comments that helped improve this manuscript. A big thanks goes to Dr. Carla Yanni for her helpful discussions on the history of Geology Hall, to Robert Selden for his information on the inner workings of the Geology Museum, and to Dr. Marie-Pierre Aubry for allowing us access to the historic mineral collections. Thank you to Dr. Ronald Ransome and the Dean's Office of the School of Arts and Sciences at Rutgers University for providing funding to present this research. This paper is dedicated to the memory of

Christopher J. Lombardi, a friend and colleague who was proud to be part of the history and tradition of geology at Rutgers.

REFERENCES CITED

Aiken, W.A., 1945, 1895 Class Annalist's Letter [speech]: Clinton, New York, Hamilton College.

Aldrich, M.L., 2000, New York State Natural History Survey, 1836–1845: A Chapter in the History of American Science: Ithaca, New York, Paleontological Research Institution, 288 p.

Allmon, W.D., 1994, The value of natural history collections: New York, Curator, v. 37, p. 83–89.

Alpers, S., 1991, The Museum as a Way of Seeing, *in* Karp, I., and Lavine, S.D., eds., Exhibiting Cultures: The Poetics and Politics of Museum Display: Washington, D.C., Smithsonian Institution Press, p. 25–32.

Angulo, A.J., 2009, William Barton Rogers and the Idea of MIT: Baltimore, Johns Hopkins University Press, 240 p.

Arizona Mining and Mineral Museum, 2015, History: http://www.miningmineralmuseum.com/history (accessed June 2017).

Augustana College, 2017, Fryxell Geology Museum: https://www.augustana.edu/locations/fryxell-geology-museum (accessed June 2017).

Baatz, S., 1985, Venerate the Plow, A History of the Philadelphia Society for Promoting Agriculture 1785–1985: Philadelphia, Philadelphia Society for Promoting Agriculture, 124 p.

Bates, R.S., 1965, Scientific Societies in the United States: Oxford, UK, Pergamon Press, 246 p.

Beck, L.C., Beck Papers, Travel Journals, Survey of New York, May 9, 1839–August 22, 1839: Rutgers University Special Collections and University Archives, Collection AC-2146, Box 1, Folder 2 of 3.

Beck, L.C., 1842, Mineralogy of New-York: Albany, W. & A. White & J. Visscher, 536 p.

Brown, C.M., 1989, Benjamin Silliman: A Life in the Young Republic: Princeton, New Jersey, Princeton University Press, 400 p.

Brubacher, J.S., and Rudy, W., 1976, Higher Education in Transition: A History of American Colleges and Universities, 1636–1976: New York, Harper and Row, 564 p.

Bruce, R.V., 1988, The Launching of Modern American Science, 1846–1876: Ithaca, New York, Cornell University Press, 446 p.

Cailleux, A., 1972, The geological map of North America (1752) of J.E. Guettard, *in* Schneer, C.J., ed., Two Hundred Years of Geology in America: Hanover, New Hampshire, University Press of New England, p. 43–52.

Chaplin, J.E., 2001, Subject Matter: Technology, the Body, and Science on the Anglo-American Frontier, 1500–1676: Cambridge, Massachusetts, Harvard University Press, 428 p.

Chester, A.H., 1886, Catalogue of Minerals Alphabetically Arranged with Their Chemical Composition and Synonyms: New York, John Wiley and Sons, 52 p.

Chester, A.H., Albert Huntington Chester Papers Collection, 1871–1903. Rutgers University Special Collections and University Archives, Collection No. R-MC 010. Chester Collection Newspaper Clipping: Faculty and Staff Biographical Material Files, Hamilton College Archives, Accession #0000.106, retrieved February 2017.

Colorado School of Mines, 2017, Geology Museum: http://www.mines.edu/Geology_Museum (accessed June 2017).

Conn, S., 1998, Museums and American Intellectual Life, 1876–1926: Chicago, Chicago University Press, 305 p.

Conniff, R., 2016 (3 April), Our natural history, endangered: New York Times, Sunday Review: https://www.nytimes.com/2016/04/03/opinion/ournatural-history-endangered.html (accessed June 2017).

Dalton, R., 2003, Natural history collections in crisis as funding is slashed: Nature, v. 423, p. 575, https://doi.org/10.1038/423575a.

Daniels, G.H., 1994, American Science in the Age of Jackson: Tuscaloosa, University of Alabama Press, 304 p.

de Tocqueville, A., 2000, Democracy in America [Translated from De La Démocratie en Amérique by Mansfield, H.C., and Winthrop, D.]: Chicago, University of Chicago Press, 722 p.

Dupree, A.H., 1986, Science in the Federal Government: A History of Policies and Activities: Baltimore, Johns Hopkins University Press, 488 p.

Eyles, V.A., 1972, Mineralogical maps as forerunners of modern geological maps: The Cartographic Journal, v. 9, no. 2, p. 133–135, https://doi.org/10.1179/caj.1972.9.2.133.

Fenton, C.L., and Fenton, M.A., 1945, The Story of Great Geologists: Freeport, New York, Doubleday and Company, 301 p.

Friedman, A.J., 2007, The extraordinary growth of the science-technology museum: New York, Curator, v. 50, p. 63–75, https://doi.org/10.1111/j.2151-6952.2007.tb00250.x.

Froelich, A., 2003, Smithsonian science: First class on a coach budget: Bioscience, v. 53, p. 328, https://doi.org/10.1641/0006-3568(2003)053[0328:SSFCOA]2.0.CO;2.

Geiger, R.L., 2000, The American College in the Nineteenth Century: Nashville, Tennessee, Vanderbilt University Press, 368 p.

Gerstner, P., 1994, Henry Darwin Rogers, 1808–1866 American Geologist: Tuscaloosa, University of Alabama Press, 328 p.

Greene, J.C., 1984, American Science in the Age of Jefferson: Ames, Iowa State University Press, 484 p.

Greene, J.C., and Burke, J.G., 1978, The science of minerals in the age of Jefferson: Transactions of the American Philosophical Society, v. 68, no. 4, 113 p., https://doi.org/10.2307/1006294.

Greene, M.T., 1982, Geology in the Nineteenth Century: Changing Views of a Changing World: Cornell History of Science Series: Ithaca, New York, Cornell University Press, 324 p.

Gropp, R.E., 2003, Are university natural science collections going extinct?: Bioscience, v. 53, p. 550, https://doi.org/10.1641/0006-3568(2003)053[0550:AUNSCG]2.0.CO;2.

Gross, S.D., 1861, Lives of Eminent American Physicians and Surgeons of the Nineteenth Century: Philadelphia, Lindsay and Blakiston, 836 p.

Guralnick, S.M., 1972, Geology and religion before Darwin: The case of Edward Hitchcock, theologian and geologist (1793–1864): Isis, v. 63, no. 4, p. 529–543, https://doi.org/10.1086/351001.

Guralnick, S.M., 1975, Science and the Ante-Bellum American College: Philadelphia, American Philosophical Society Memoir 109, 227 p.

Hall, C.R., 1934, A Scientist in the Early Republic: Samuel Latham Mitchill, 1764–1831: New York, Columbia University Press, 162 p.

Hayes, A.O., 1927, William S. Valiant: Science, v. 65, p. 415, https://doi.org/10.1126/science.65.1687.415.

Hendrickson, W.B., 1961, Nineteenth-century state geological surveys: Early government support of science: Isis, v. 52, no. 3, p. 357–371, https://doi.org/10.1086/349475.

Herbert, R.L., 2012, The complete correspondence of Edward Hitchcock and Benjamin Silliman, 1817–1863: The American Journal of Science and the Rise of American Geology: https://www.amherst.edu/media/view/425361/ (accessed February 2017).

Hindle, B., 1956, The Pursuit of Science in Revolutionary America, 1735–1789: Chapel Hill, University of North Carolina Press, 410 p.

Hoffman, F., 1901, Fragmentary notes from the reports of two naturalists in North America: Pharmaceutical Reviews, v. 16, p. 260–265.

Kelly, H.A., and Burrage, W.L., 1920, American Medical Biographies: Baltimore, Norman, Remington Company, 1320 p.

Kohlstedt, S.G., 1976, The Formation of the American Scientific Community: The American Association for the Advancement of Science 1848–1860: Urbana, University of Illinois Press, 264 p.

Kohlstedt, S.G., 1988, Curiosities and cabinets: Natural history museums and education on the antebellum campus: Isis, v. 79, no. 3, p. 405–426.

Kremers, E., 1907, Johann David Schoepf: Pharmaceutical Reviews, v. 21, p. 156–158.

Kümmel, H.B., 1909, Geological Survey of New Jersey: Annual Report of the State Geologist, For the Year 1908: Trenton, New Jersey, MacCrellish and Quigley, State Printers, 159 p.

Kuritz, H., 1981, The popularization of science in nineteenth-century America: History of Education Quarterly, v. 21, no. 3, p. 259–274, https://doi.org/10.2307/367698.

Larson, J.L., 2001, Internal Improvement: National Public Works and the Promise of Popular Government in the Early United States: Chapel Hill, University of North Carolina Press, 352 p.

Latour, B., 1987, Science in Action: How to Follow Scientists and Engineers through Society: Cambridge, Massachusetts, Harvard University Press, 288 p.

Laudan, R., 1987, From Mineralogy to Geology: The Foundations of a Science, 1650–1830, Science and Its Conceptual Foundations: Chicago, University of Chicago Press, 285 p., https://doi.org/10.7208/chicago/9780226924755.001.0001.

Lewis, A.J., 2011, A Democracy of Facts: Natural History in the Early Republic: Early American Studies: Philadelphia, University of Pennsylvania Press, 216 p.

Lucier, P., 2008, Scientists and Swindlers: Consulting on Coal and Oil in America, 1820–1890: Baltimore, Johns Hopkins University Press, 448 p.
McCormick, R.P., 1966, Rutgers: A Bicentennial History: New Brunswick, New Jersey, Rutgers University Press, 336 p.
Merrill, G.P., 1964, The First One Hundred Years of American Geology: New York, Hafner Publishing Company, 773 p.
Merrill, G.P., 1978, Contributions to a History of American State Geological and Natural History Surveys: New York, Arno Press, 549 p.
Miller, R.F., and Buhay, D.N., 2007, Gesner's Museum of Natural History, An Early Canadian Geological Collection [Online]: Geoscience Canada, v. 34, no. 1, https://journals.lib.unb.ca/index.php/GC/article/viewArticle/10232/10611 (accessed June 2017).
Mitchill, S.L., 1787, Observations Anatomical, Physiological, and Pathological, on the Absorbent Tubes of Animal Bodies. To Which Are Added, Geological Remarks on the Maritime Parts of the State of New-York: New York, J. McLean and Co., 16 p.
Newell, J.R., 1993, American geologists and their geology: The formation of the American geological community, 1780–1865 [Ph.D. thesis]: Madison, University of Wisconsin, 742 p.
Ohio State University, 2017, Orton Hall: https://ortongeologicalmuseum.osu.edu/orton-hall (accessed June 2017).
Paine, N.E., 1924, 1874 Class Annalist's Letter [speech]: Clinton, New York, Hamilton College.
Parrish, S.S., 2006, American Curiosity: Cultures of Natural History in the Colonial British Atlantic World: Chapel Hill, University of North Carolina Press, 344 p.
Porter, R., 1977, The Making of Geology: Earth Science in Britain, 1660–1815: New York, Cambridge University Press, 300 p.
Preston, D.J., 1986, Dinosaurs in the Attic: An Excursion into the American Museum of Natural History: New York, St. Martin's Press, 244 p.
Ramey-Gassert, L., Walberg, H.J., III, and Walberg, H.J., 1994, Reexamining connections: Museums as science learning environments: Science Education, v. 78, no. 4, p. 345–363, https://doi.org/10.1002/sce.3730780403.
Reid-Maroney, N., 2001, Philadelphia's Enlightenment, 1740–1800: Kingdom of Christ, Empire of Reason, Contributions to the Study of World History: Westport, Connecticut, Greenwood Press, 199 p.
Rennie, L.J., and Johnston, D.J., 2004, The nature of learning and its implications for research on learning from museums: Science Education, v. 88, S1, p. S4–S16, https://doi.org/10.1002/sce.20017.
Rennie, L.J., and McClafferty, T., 1995, Using visits to interactive science and technology centers, museums, aquaria and zoos to promote learning in science: Journal of Science Teacher Education, v. 6, p. 175–185.
Rowe, G.H., III, 1999, Autobiography of George H. Rowe: Hobe Sound, Florida, 69 p.
Rudwick, M.J.S., 1992, Scenes from Deep Time: Early Pictorial Representations of the Prehistoric World: Chicago, University of Chicago Press, 294 p.
Rudwick, M.J.S., 2007, Bursting the Limits of Time: The Reconstruction of Geohistory in the Age of Revolution: Chicago, University of Chicago Press, 732 p.
Rutgers College, 1916, Catalogue of the Officers and Alumni of Rutgers College from 1766–1916: Trenton, New Jersey, State Gazette Publishing Co., 500 p.
Rutgers College Faculty Minutes, 1829: Rutgers University Special Collections and University Archives, Collection No. R-Minutes.
Rutgers College Faculty Minutes, 1830: Rutgers University Special Collections and University Archives, Collection No. R-Minutes.
Rutgers College Faculty Minutes, 1831: Rutgers University Special Collections and University Archives, Collection No. R-Minutes.
Rutgers College Geology Museum Records, 1860–1919 (Unprocessed): Rutgers University Special Collection and University Archives, Collection No. 1017.
Schoepf, J.D., and Morrison, A.J., 1968, Travels in the Confederation, 1783–1784 [Translated from Reise durch einige der mittlern und südlichen vereinigten nordamerikanischen Staaten nach Ost-Florida und den Bahama-Inseln, unternommen in den Jahren 1783 und 1784: Erlangen, 1788]: New York, Bergman Publishers, 344 p.
Sheets-Pyenson, S., 1987, Cathedrals of science: The development of colonial natural history museums during the late nineteenth century: History of Science, v. 25, p. 279–300, https://doi.org/10.1177/007327538702500303.
Sidar, J.W., 1976, George Hammell Cook: A Life in Agriculture and Geology, 1818–89: New Brunswick, New Jersey, Rutgers University Press, 282 p.
Silliman, B., Sr., 1818, Review of an Elementary Treatise on Mineralogy and Geology, Being an Introduction to the Study of these Sciences, and Designed for the Use of Pupils; for Persons Attending Lectures on These Subjects; and as a Companion for Travellers in the United States of America: American Journal of Science, v. 1, p. 35–52.
Soren, B.J., 2009, Museum experiences that change visitors: Museum Management and Curatorship, v. 24, no. 3, p. 233–251, https://doi.org/10.1080/09647770903073060.
South Dakota Mines, 2017, Museum of Geology: http://www.sdsmt.edu/Academics/Museum-of-Geology/About/History/ (accessed June 2017).
Spanagel, D.I., 2014, DeWitt Clinton and Amos Eaton: Geology and Power in Early New York: Baltimore, Johns Hopkins University Press, 288 p.
Spieker, E.M., 1971, Schöpf, Maclure, Werner, and the earliest work on American geology: Science, v. 172, no. 3990, p. 1333–1334, https://doi.org/10.1126/science.172.3990.1333.
Stephenson, K., 2004, The Quarantine war: The burning of the New York Marine Hospital in 1858: Public Health Chronicles, v. 119, p. 79–92.
Suarez, A.V., and Tsutsui, N.D., 2004, The value of museum collections for research and society: Bioscience, v. 54, no. 1, p. 66–74, https://doi.org/10.1641/0006-3568(2004)054[0066:TVOMCF]2.0.CO;2.
Torrens, H.S., 1990, The transmission of ideas on the use of fossils in stratigraphic analysis from England to America 1800–1840: Earth Sciences History, v. 9, p. 108–117, https://doi.org/10.17704/eshi.9.2.er06u08533g47457.
University of Wisconsin Geology Museum, 2017, About: http://geoscience.wisc.edu/museum/about/ (accessed June 2017).
Valiant, W.S., 1899, Rutgers Geology Museum Brochure: William S. Valiant Collection (Unprocessed): Rutgers University Special Collections and University Archives, Collection No. R-MC 156.
Vesey, L.R., 1965, The Emergence of the American University: Chicago, University of Chicago Press, 520 p.
Weil, S.E., 1995, A Cabinet of Curiosities: Inquiries into Museums and Their Prospects: Washington, D.C., Smithsonian Institution Press, 264 p.
Wilkerson, A.S., 1963, An Abbreviated History of Geology at Rutgers, the State University, from 1830 to 1963: New Brunswick, New Jersey, Rutgers University Press, 96 p.
Winsor, M.P., 1991, Reading the Shape of Nature: Comparative Zoology at the Agassiz Museum: Chicago, University of Chicago Press, 324 p. https://doi.org/10.7208/chicago/9780226902081.001.0001.
Yanni, C., 1999, Nature's Museums: Baltimore, Johns Hopkins University Press, 199 p.

Manuscript Accepted by the Society 21 November 2017
Manuscript Published Online 6 June 2018

Printed in the USA

The Geological Society of America
Special Paper 535

The role of the Dunn-Seiler Museum, Mississippi State University, in promoting public geoliteracy

Renee M. Clary*
Amy Moe-Hoffman*
Department of Geosciences, Mississippi State University, Box 5448, Mississippi State, Mississippi 39762, USA

ABSTRACT

Mississippi State University's Dunn-Seiler Museum is a small, traditional museum housed in the Department of Geosciences. Founded in 1946, the museum preserves the geologic specimens collected by faculty, students, and donors and includes holotypes and extensive Cretaceous collections. Throughout its seven decades, the Dunn-Seiler Museum has transitioned in its role as a repository and displayer of geological objects to a facility that uses best practices in informal education to promote visitor understanding of important geological concepts, including biodiversity, evolution, geologic time, and sustainability. The first small public gallery had limited access and displayed specimens with identification cards but little interpretive signage. Beginning in the early 2000s, best research practices were utilized to build interpretive displays to enhance the visitor's educational experience. In the past decade, the Dunn-Seiler participated within national organizations to expand educational opportunities with resources at the national level. Throughout this transition to increased public education, the museum still curated and maintained research collections. Today the Dunn-Seiler Museum bridges the community of researchers and the general public. While the museum's role at the forefront of geology was initially in its storage and *display* of research specimens, its role has expanded to *communicate* geology for improved public geoliteracy.

INTRODUCTION

The Dunn-Seiler Museum is a small, traditional museum housed in the Department of Geosciences at Mississippi State University (MSU) (Fig. 1). Founded in 1946 to preserve and care for the extensive geologic collections made by the faculty and students, the museum reposits more than 50,000 fossils, including holotypes and important Cretaceous specimens that were primarily collected and identified by university geology faculty. The public museum exhibits a small percentage of the collections, with displays focusing upon minerals, rocks, karst specimens and models, and meteorites. The majority of displays are dedicated to fossils, from Precambrian through Pleistocene, primarily arranged in order of geologic time. Admission is free, and the museum is publicly accessible during normal university operating hours. Geology graduate students staff the museum during the

*E-mail: RClary@geosci.msstate.edu; AMHoffman@geosci.msstate.edu.

Clary, R.M., and Moe-Hoffman, A., 2018, The role of the Dunn-Seiler Museum, Mississippi State University, in promoting public geoliteracy, *in* Rosenberg, G.D., and Clary, R.M., eds., Museums at the Forefront of the History and Philosophy of Geology: History Made, History in the Making: Geological Society of America Special Paper 535, p. 237–248, https://doi.org/10.1130/2018.2535(15).

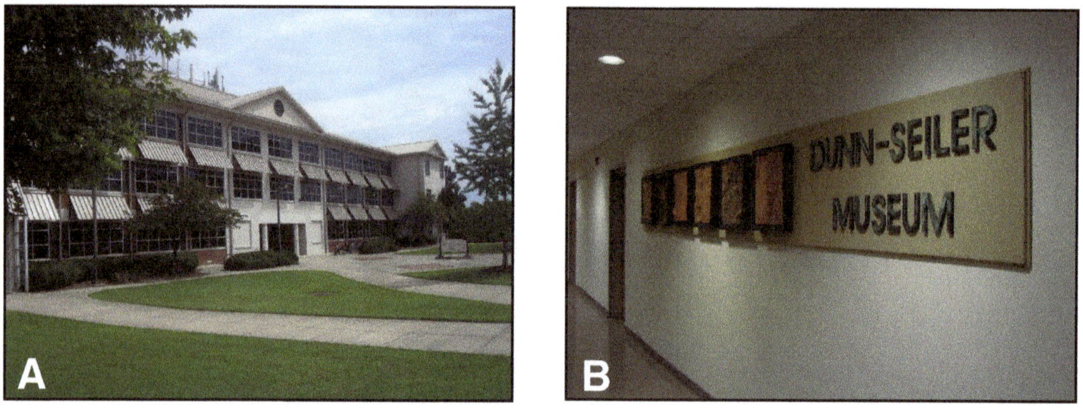

Figure 1. Dunn-Seiler Museum (B) is housed in the Department of Geosciences, Hilbun Hall, Mississippi State University (A). Photographs by Athena Nagel and Renee Clary.

fall and spring semesters. In addition to tours of the Dunn-Seiler Museum's displays, the museum also conducts fossil-collecting opportunities for visitors at one of the MSU Cretaceous outcrops (Fig. 2).

The Dunn-Seiler Museum houses one of only two paleontological collections in Mississippi, with the other display held at the Mississippi Museum of Natural Science (MMNS) in Jackson. Located in Oktibbeha County in northeastern Mississippi, the Dunn-Seiler is the nearest geology and Earth science museum for several school districts. Other than the Jackson, Mississippi, museum, which is located ~130 miles (~210 km) from the Dunn-Seiler, the other nearest museums displaying geological specimens are the McWane Science Center (Birmingham, Alabama; ~140 miles [~225 km]) and the Pink Palace Museum (Memphis, Tennessee; ~180 miles [~290 km]).

More than 1000 visitors' signatures are recorded in the Dunn-Seiler Museum visitor's book annually. The majority of Dunn-Seiler Museum visitors are enrolled university students and students participating in field trips from schools and organizations within ~2 h traveling distance to the campus. In 2013, Cable News Network (CNN) identified Oktibbeha County, Mississippi, within the top 25 income-disparate counties in the United States (Sutter, 2013). Additionally, Mississippi has the highest proportion of African Americans in any state in the United States (currently 37.4% by the 2010 U.S. Census), the lowest per capita personal income, and the highest poverty rate in the nation (Nave, 2017). In public schools, 68.3% of Mississippi students are eligible for free or reduced lunch (AECF [Anne E. Casey Foundation], 2011). Therefore, the Dunn-Seiler Museum serves a primarily rural region and high-needs school districts. In addition to socioeconomic challenges, the area served by the Dunn-Seiler Museum has a large number of residents who identify as creationists.

Throughout its seven-decade existence, the Dunn-Seiler Museum played different roles within the university and greater community. In the beginning, the Dunn-Seiler Museum's purpose was to store the research specimens of faculty and students and display some of the geological objects of the greater

Figure 2. In addition to museum tours, the Dunn-Seiler provides fossil-collecting excursions for visitors at a campus Cretaceous outcrop. Visitors can keep the 70-m.y.-old fossils they collect. In these photographs, participants in the Brickfire Project collect fossils at the outcrop. Photographs by Renee Clary.

museum collections within the university setting. In the early 2000s, the museum transitioned to display specimens using best practices in informal learning for effective geoscience education and improved geoliteracy of visitors. New displays incorporate many of the Earth Science Literacy Principles (ESLI [Earth Science Literacy Initiative], 2010) while focusing upon the exposed Cretaceous landscape of northeastern Mississippi with which most viewers are familiar. Within the past decade, the museum further expanded its presence by partnering with national outreach efforts. While the museum's role at the forefront of geology was initially in its storage and display of research specimens as geological curiosities, its role has expanded to *communicate* geology for improved public understanding. This small university museum transitioned from its early wünderkammer presentations (Rosenberg, this volume) to contextual displays for effective informal education of the local underserved populations.

OPPORTUNITIES TO LEARN IN MUSEUM SETTINGS

Informal or Free-Choice Learning

Most people find it intuitive that adults learn the majority of their science outside the classroom, but many are surprised to learn that even our K–16 school-age students acquire more of their scientific knowledge through informal venues compared with traditional science classroom settings (Falk and Dierking, 2002). Informal learning can include television documentaries, articles on social media, or visits to museums and outdoor learning centers. Researchers demonstrated that this "free-choice" learning is effective for geoscience education and can occur in multiple settings (Falk, 2001; Falk and Dierking, 2002; McComas, 2006; Wandersee and Clary, 2006). After-school programs, workshops, and designed environments (Bell et al., 2009) can be used effectively for science learning.

Museums as Informal Learning Sites

Informal museum displays and the educational opportunities contained within them are not static. New designs and venues incorporate the latest research-based practices and advances in technology. Science museums in particular have evolved since their inception (Friedman, 2010). The first museums consisted of private collections turned over to private agencies. Technology collections then joined the displays in the nineteenth century. In the United States, Charles Willson Peale (1741–1827), founder of one of the first American museums, organized displays of specimens to provide a better understanding of nature for his visitors, although the specimens' labels were to "speak for themselves" (i.e., only the specimen cards that accompanied each object provided identifications and information) (Kohlstedt and Brinkman, 2004, p. 7). At the beginning of the 1900s, science and technology museums began to include informal education as a goal instead of an unintended result. By the mid-twentieth century, a third generation of museums had education as the *only* goal (Friedman, 2010). Importantly, many traditional museums remained, and several of them repurposed their exhibits to improve visitors' educational experiences. Since the goal to optimize educational potential in museums should be embraced by the entire geoscience community (Pasquaré Mariotto and Venturini, 2017), many museums, including the Dunn-Seiler, sought opportunities for greater inclusivity of geologists, educators, and the general public.

Once visitor education became a museum goal, displayed objects with specimen labels only were no longer sufficient. Collections must be accompanied with appropriate text and illustration (Hooper-Greenhill, 1999) to improve geoliteracy. Signage should be succinct, written at an appropriate reading level, use active voice, include illustrations where possible, and focus upon one major construct (Wandersee and Clary, 2007). The learning objectives for the viewer should inform the signage content and design.

For the Dunn-Seiler Museum, the small traditional museum must balance its role between optimizing public learning in displays and programming, while performing the critical university tasks of archiving, conserving, and researching. Pinna (2000a) observed that schizophrenia exists between research and public display since the culture of the museum still belongs in the hands of the researchers. However, the Dunn-Seiler Museum's philosophy differs from the typical research university approach (Pinna, 2000b) in that scientific research *informs* scientific communication, which is optimized within the public display using best practices in geoscience education for improved geoliteracy. Scientific results are incorporated into Dunn-Seiler exhibits at a level for effective public learning, using researched techniques in signage and informal education (e.g., following Wandersee and Clary [2007] signage guidelines). Therefore, visitors to the museum can find displays on cave formations that include Emeritus Professor John Mylroie's karst research and Mississippi's hadrosaur bones, displayed among Cretaceous marine organisms that are interpreted through the "bloat and float" hypothesis.

Earth Sciences Literacy Initiative: Template for Geoscience Learning

The National Science Foundation funded the Earth Science Literacy Initiative (ESLI) to identify the critical geoscience knowledge necessary for all Americans to achieve scientific literacy (ESLI, 2010). ESLI distilled this content into nine "Big Ideas" that include (1) science as a way of knowing (Earth scientists use repeatable observations to explain our planet); (2) geologic time (Earth is 4.6 billion years old); (3) system science (Earth is a complex multivariate system); (4) dynamic processes (Earth is continuously changing); (5) the role of H_2O (Earth is the water planet); (6) evolution and organismal impact (life evolves and continuously modifies Earth); (7) natural resources (humans depend on Earth for resources); (8) natural hazards (natural hazards pose risks to humans); and (9) human impact (humans significantly alter Earth) (ESLI, 2010). Informal learning environments that include these Big Ideas in their displays and outreach have the potential to improve the geoliteracy of their visitors (Clary and Wandersee, 2014a).

DUNN-SEILER MUSEUM COLLECTIONS

The museum collections include rocks, minerals, and ~50,000 fossil vertebrate, invertebrate, and plant specimens. Cores are also reposited. The vast majority of specimens are marine invertebrate fossils, followed by vertebrate, and then plant fossils.

The geology of Mississippi is well represented within the Dunn-Seiler collections. Mississippi formations include the oldest Paleozoic strata (Devonian and Mississippian shales, sandstones, and limestones) that are minimally exposed in the state's extreme northeastern corner. These are geologically followed by Cretaceous strata (limestones, chalk, and sandstones), Paleogene sandstones, limestones, and clays, and Quaternary alluvium in the Delta region (western Mississippi). The strata dip westward toward the synclinal axis of the Mississippi Embayment (Dockery and Thompson, 2016). Mississippi has yielded several "famous" fossil specimens, some of which are reposited outside the state, including the holotype jaw of *Panthera* (which the statue of Joseph Leidy holds in his hand outside the Academy of Natural Sciences of Drexel University) and famous Paleocene and Eocene mammal fossils from the Red Hot Truck Stop locality in Meridian, Mississippi. Dunn-Seiler collections contain many commonly found Cretaceous marine invertebrates (e.g., oysters, cephalopods, and gastropods) and reptiles (mosasaurs, turtles, and crocodiles), Paleogene marine invertebrates and vertebrates (including Mississippi's state fossil [Eocene whales, *Basilosaurus* and *Zygorhiza*]), a plethora of petrified wood (Mississippi's state rock), and Eocene plant fossils.

Of the museum's Mississippi specimens, a majority of the fossils are Cretaceous in age since the local area contains multiple Cretaceous chalk outcrops, which have been systematically collected by MSU researchers for more than 100 years (Fig. 3). The extensive Cretaceous collections document transgressive and regressive phases and paleoclimates. The Cretaceous typed and figured specimens include barnacles and crocodile (skull and postcranial) holotypes.

In addition to Mississippi specimens, the Dunn-Seiler Museum includes Mazon Creek plant fossils, Cretaceous marine invertebrates from the United Kingdom, and a fairly diverse graptolite collection.

Museum collections are used by faculty, students, and researchers for research, education and outreach, and display in the small public museum. Recently loaned research specimens include Cretaceous dinosaur remains and ammonites. Researchers have also performed on-site investigations of the Cretaceous crocodile, *Eothoracosaurus*. Graduate students currently utilize Dunn-Seiler collections for their thesis investigations involving Pleistocene megafauna of the southeastern United States and bioimmuration of *Exogyra* shells.

Only a small percentage of the museum's specimens are exhibited in the public gallery. While the Dunn-Seiler Museum continues to improve research access to its collections, the public impact is optimized through the museum displays and outreach activities.

EARLY HISTORY OF THE DUNN-SEILER MUSEUM

The Dunn-Seiler Museum (DSM) was established in 1946 as "Franklin Seiler Museum" to preserve and care for the extensive geologic collections made by the faculty and students of Mississippi State University's (MSU's) Department of Geosciences. The name originally honored Franklin Carl Seiler (1916–1945), former MSU student and then geology instructor from 1937 until his death in the European theater during World War II. In 1962, the museum's name was changed to recognize significant achievements of Dr. Paul Heaney Dunn (1895–1967), a University of Chicago–educated paleontologist who served as department head for almost three decades (1934–1962). In 1962, university administration granted permission to officially create the museum in Hilbun Hall as both a memorial to Seiler and in honor of Dunn (Meyer-Arendt, 1994). Therefore, through Dunn's leadership and efforts, the museum became a permanent university and community resource on Mississippi State University's campus.

Throughout its tenure, the Dunn-Seiler existed in various locations and stages of accessibility. Museum specimens, some collected during the late 1800s, were displayed at McCain Hall at Mississippi State University, which first housed the Department of Geology and Geography before it moved to Hilbun Hall (Hearn, 2003). Hilbun Hall currently houses both the departments of Geosciences and Physics and Astronomy.

The DSM resided on the second floor of Hilbun Hall near paleontology professor Ernest Russell's (1923–2013) office. Russell collected Cretaceous fossils and cores (mostly from the state of Mississippi) and managed the collections until his retirement. In 1984, paleontologist Christopher Dewey took over the museum curation. When the building's renovation began in 1999, the Dunn-Seiler's collections were dispersed across campus to biology and anthropology departments, the university library, and research and storage buildings on campus. Upon Hilbun's renovation in 2003, the museum relocated on the first floor of Hilbun Hall in a 900 ft^2 public gallery, and a ribbon-cutting ceremony was held (spearheaded by geography instructor, Joan Mylroie [Hearn, 2003]). However, not all collections and specimens returned to the smaller allocated space.

The 2003 Dunn-Seiler Museum displayed minerals, rocks, and fossils in locked glass-fronted cabinets. Fossils were arranged chronologically from Precambrian through the Pleistocene. Traditional labels with scientific names, and sometimes collection locality, were included with the specimens, but very little interpretation signage was contained within the displays (Fig. 4). The museum's showpiece was considered to be a *Triceratops* cast, mounted above children's eye level and partially surrounded

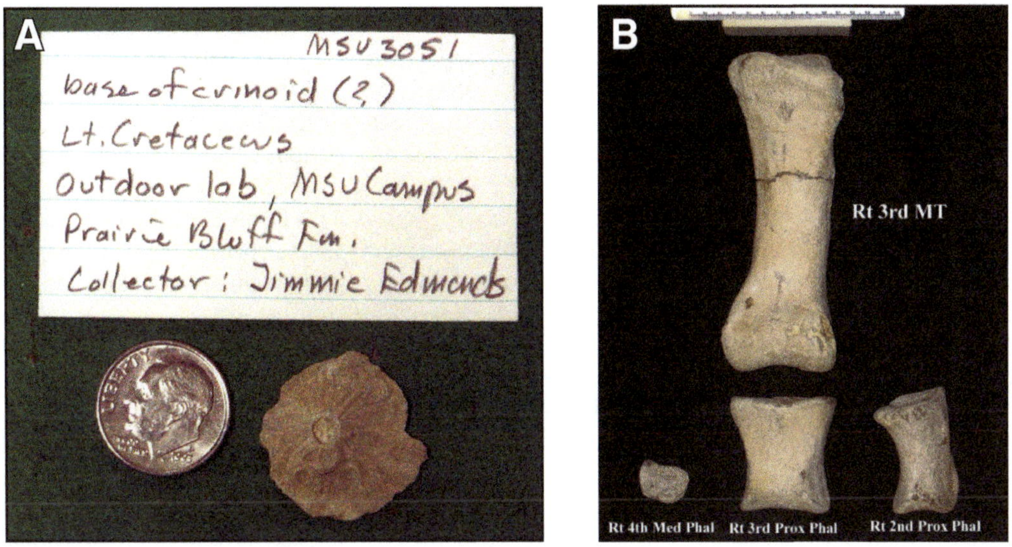

Figure 3. The Dunn-Seiler Museum's research collections contain holotypes, an irreplaceable Cretaceous core, and extensive Cretaceous fossils from the local landscape, including a crinoid holdfast (A) collected at an outcrop beyond the biology building; and unusual dinosaur fossils (B). Photographs by George Phillips.

with protective Plexiglass. Accessibility to the public museum space was also limited: a retired professor remarked that he often walked past the first-floor museum, with its doors closed and its lights off.

DUNN-SEILER'S TRANSITION TO IMPROVED INFORMAL LEARNING

Several events combined toward the end of the first decade of the twenty-first century to broaden the accessibility and educational effectiveness of the Dunn-Seiler Museum. Renee Clary became director of the museum in 2008, and Amy Moe-Hoffman soon followed as collections manager. Moe-Hoffman brought her previous experience from Paleontological Research Institute and University of Colorado Boulder to formally define the Dunn-Seiler Museum's mission statement, goals, and curation and loan policies. Without a budget, the Dunn-Seiler Museum became a service project as funding opportunities and staffing mechanisms were explored.

Retrieving the Collections

Some of the collections were still scattered across campus from their relocation during building renovation. Collections housed in biological sciences were "donated" to a state museum by a biology professor upon his retirement. Fortunately, the curator in the state museum recognized the Dunn-Seiler specimens' numbers and alerted the museum director. Multiple boxes were

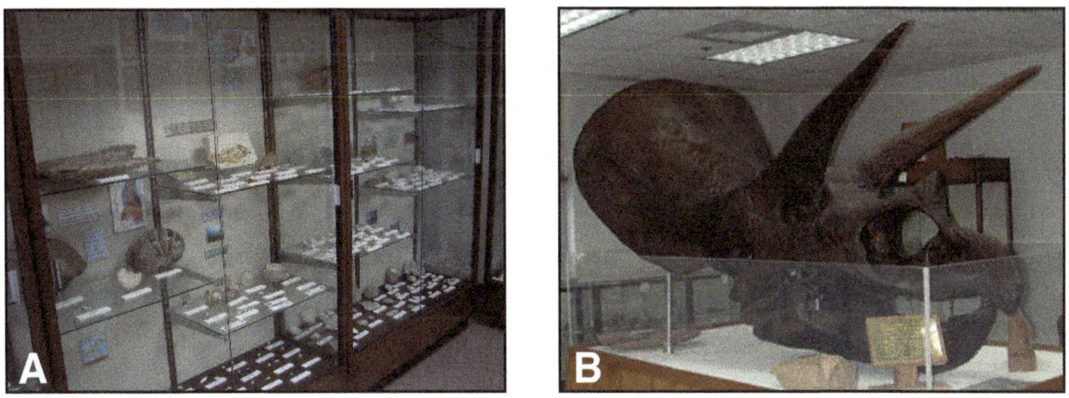

Figure 4. When the Dunn-Seiler Museum re-opened in 2003, specimens were displayed with identification labels but very little interpretive signage (A). The show piece of the museum was a *Triceratops* cast, displayed behind Plexiglas and above the eye level of children (B). Photographs by Renee Clary.

retrieved from the Department of Anthropology and Middle Eastern Cultures' climate-controlled research storage building. A small secured room and a larger basement space in Hilbun Hall became the storage facilities for the collections.

Early Challenges for a Public Dunn-Seiler Museum

Between 2007 and 2010, outreach and tours for the Dunn-Seiler were arranged individually with instructors. The general museum description on the departmental Web page led to early conflicts and awkward situations when some groups arrived to see labeled specimens older than 6000 years and to hear a history of the planet that they felt was contrary to their Biblical interpretations. The director quickly edited the website so that it alerted potential guests that the collections "help visitors visualize and understand the 4.6 billion year history of our planet." Groups requesting information about the museum and outcrop tours were directed to the website, and they were advised that the fossils at the campus outcrop were 70 million years old. While some groups elected not to visit upon learning this, the discomfort on campus that resulted from religious conflicts with geologic time was minimized.

Keeping the museum open during regular hours was challenging. In 2007, a U.S. Department of Education grant funded students who majored jointly in a science and science education. Three students were employed through a U.S. Department of Education grant to work 10 h/wk during regular fall and spring semesters in the museum through 2011, and this, combined with directed individual studies courses, allowed some regular public hours for the Dunn-Seiler Museum. The Dunn-Seiler Museum staff next advocated for and received assistance from geology graduate students. Beginning in the fall 2011 semester, graduate students, who were funded as teaching assistants, staffed the museum to fulfill their 20 h/wk assistantship obligations. While staffing the museum, the graduate students are also available to provide tutoring assistance to undergraduate students in general geology courses. The following year, the Dunn-Seiler Museum partnered with the MSU Visitors Center to offer tours to both the museum and the campus Cretaceous outcrop, where guests could search for and keep the 70-m.y.-old fossils they found. The Visitors Center provides Maroon VIP students[1] to escort groups to different stations (museum, outcrop, and other waypoints) during campus visits.

The Cretaceous outcrop is located adjacent to a large commuter parking lot within walking distance of the museum. The eroded chalk within the Prairie Bluff Formation has topographic relief and provides abundant collecting opportunities. Developed as a geocache site by the museum director and retired extension professor, Michael Williams, the outcrop yields several species of Cretaceous oysters, gastropod steinkerns, ammonites, belemnites, and an occasional shark's tooth or mosasaur vertebrae. Visitors are encouraged to compare their fossil finds with identified fossil displays at the Dunn-Seiler, and tours address Mississippi's sea-level and climate changes, extinct organisms, fossilization, and geologic time through fossils at the site.

[1]The Maroon VIPs are student volunteers who serve the Mississippi State University (MSU) Welcome Center, Office of the President, and MSU colleges and departments. They primarily serve as greeters for the university. With the Dunn-Seiler Museum, the Maroon VIPs escort visiting groups to ensure that they successfully navigate MSU's campus and arrive at their scheduled tours and waypoints on time. The Dunn-Seiler Museum has partnered with the MSU Welcome Center and Maroon VIPs from 2012 to the present.

IMPROVING GEOLITERACY THROUGH DUNN-SEILER DISPLAYS

Retrieving the collections and keeping the Dunn-Seiler Museum publically available on a regular schedule were not the only challenges faced by the museum. The format of the museum displays was not conducive to informal learning, since display case upon display case contained numerous specimens lined in rows on the glass shelves, each with its individual identification card. Just as in Peale's museum, the specimens were expected to "speak for themselves." With time and funding constraints, the museum staff searched for opportunities to modify the museum's exhibit format to actively engage visitors and to present geological content to facilitate viewers' "big picture" understanding of the Earth system.

The specimens mistakenly "donated" to the state museum contained a large compressed fossilized trunk section of *Lepidodenron*, and a table was pulled into the museum to hold this and a very early collected Triassic section of *Eubrontes* dinosaur tracks. These became the first touchable exhibits for the visitors, many of whom are K–12 students.

Addressing the Earth Science Literacy Initiative: Mississippi's Geologic Timeline

An opportunity emerged to modify the existing displays. Until 2010, a geology professor had displayed his personal specimens of lithified sedimentary structures in the museum, but he decided he needed them for his courses. This emptied two and a half display cases. The Mazon Creek plant fossils that remained in the third case were relocated to the Paleozoic display in the Pennsylvanian cabinet. The three empty display cases provided an opportunity to better engage the museum visitors, improve the informal learning experience, address Big Ideas of Earth Science Literacy (ESLI, 2010), and connect to the geographic affiliation and sense of place of the local visitors. The result was the "Mississippi's Geological Timeline" exhibit, which relates to visitors beyond simple specimen display since it focuses upon the local landscape and the state's geologic history (Farnum, 2004).

Undergraduate Jessica Kelly, funded under the Teachers for a Competitive Tomorrow grant, began to research the geology of the state and select specimens for the cabinets. Using the state geological map as a guideline, Jessica identified specimens from

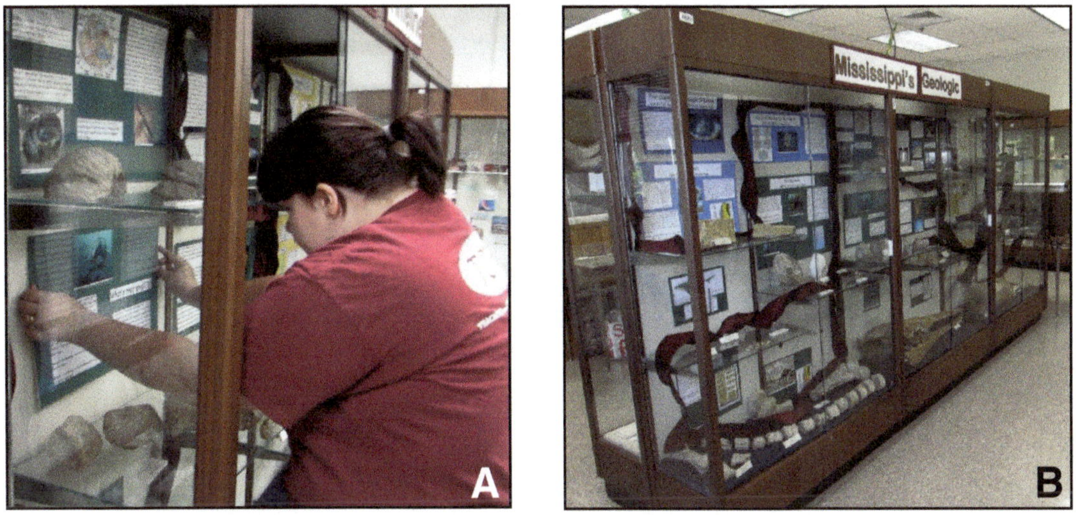

Figure 5. Mississippi's Geologic Timeline includes fossils found within the state (B). The display is color-coded with signage to correspond with the colored strata on the state's geologic map. The maroon ribbon guides the viewer's eye throughout the display. Student Jessica Kelly (A) worked on the exhibit. Photographs by Renee Clary.

the Devonian, Cretaceous, Paleogene (Eocene and Oligocene), and Neogene periods, with special focus on the Cretaceous environment that underlies the university campus. Kelly, supervised by Clary, included the remains of a hadrosaur dinosaur that were uncovered in marine limestone, presumably a result of "bloat and float," as well as an interesting crocodile skull, *Eothoracosaurus*, which was recovered locally. Kelly incorporated a fossil vertebrae from the state whale fossil, *Basilosaurus*, and samples of the state rock, petrified wood. Clary directed Kelly to use the Earth Science Literacy Principles (ESLI, 2010) to develop the signage, which addresses the age of the planet (Big Idea 2), the dynamic character of the Earth (Big Idea 4), and the evolution of life forms within the state (Big Idea 6). Each period is color-coded with signage that corresponds to the geologic map of Mississippi (e.g., green for Cretaceous fossils), and the visitor's eye is guided through the display with a large maroon (MSU's school color) ribbon, which courses throughout the cases (Fig. 5).

Tectonic and evolutionary milestones of Mississippi's geological history are addressed with signage, including major changes in life forms, from trilobites through dinosaurs to mammoths and mastodons, and changes in the local environment since the mosasaur-inhabited Cretaceous seas. By tapping into visitors' geological sense of place (Clary and Wandersee, 2006) and including fossils that are observed locally, the display extends the campus Cretaceous outcrop experience and promotes a more scientifically literate view of the complexity of the Earth system.

Newer Opportunities: Interpreted Displays and Interactive Displays

In 2014, retired teacher Nora Gainey donated multiple rock, mineral, and fossil specimens that she and her late husband had amassed for their classrooms to the Dunn-Seiler Museum through the MSU Foundation. Moe-Hoffman designed new displays to educate viewers about the new specimens and differentiate between items that are collected and items used for research (Fig. 6). The Gainey exhibits were unveiled at the 2014 National Fossil Day event on campus. When hall display cabinets became available, Moe-Hoffman designed another exhibit for the 2017 Giverny Award[2], given annually to the best children's science storybook (Fig. 7). While the Giverny book does not always focus upon geology, the majority of the books, awarded by the 15 Degree Laboratory since 1998, address some of the Big Ideas of Earth Science Literacy (ESLI, 2010) and support the overall educational mission of the Dunn-Seiler Museum.[3] Clary and Moe-Hoffman also teamed up with MSU geologist Adam Skarke to propose a MSU Schillig Special Teaching grant project[4] that constructed an augmented reality topographic sandbox. The sandbox is the latest hands-on interactive addition to the

[2]The Giverny Award was established by the late James Wandersee (1946–2014) and Elisabeth Schussler in 1998, both at Louisiana State University at that time. Wandersee administered the award as the director of the 15 Degree Laboratory. Upon Wandersee's death in 2014, Renee Clary became the director of the 15 Degree Laboratory. Clary and Wandersee collaborated for 12 years, and co-founded the EarthScholars Research Group.

[3]Past Giverny winners include the 2002 *Rare Treasures* (by Don Brown), which focuses on the life of Mary Anning; the 2004 *Lonesome George* (by Francine Jacobs and Jean Cassels), which addresses extinction and human impact on our planet; and the 2017 winner, *The Tree Lady* (by H. Joseph Hopkins and Jill McElmurry), which discusses Kate Sessions and trees' environmental adaptations.

[4]The internal MSU Ottilie Schillig Special Teaching Projects grants provide funding for improving undergraduate education. The grants are competitive and awarded annually.

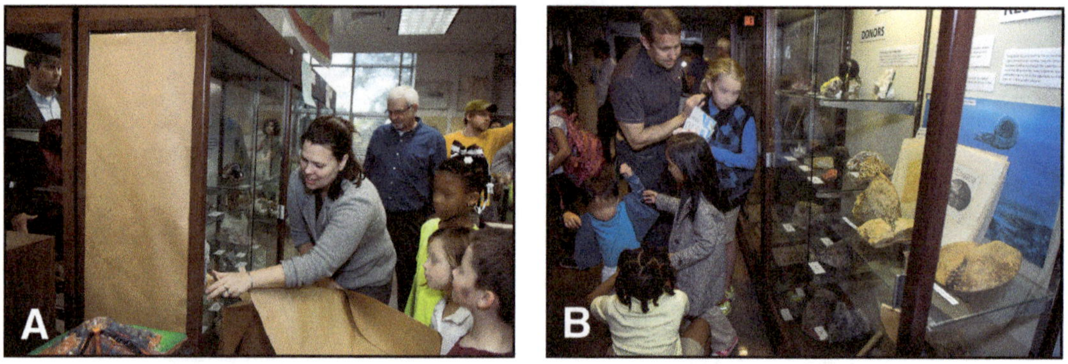

Figure 6. When Nora Gainey donated specimens she and her husband collected throughout their years as teachers, Amy Moe-Hoffman designed displays to showcase the new arrivals (B). Moe-Hoffman unveiled the exhibit at the 2014 National Fossil Day celebration (A). Photographs by Megan Bean/Mississippi State University.

Dunn-Seiler Museum when it is not being used in university classrooms (Fig. 8).

IMPROVING GEOLITERACY THROUGH COLLABORATION

Within the past decade, the museum further expanded its presence beyond its display space to include additional informal learning opportunities. The Dunn-Seiler Museum's participation with U.S. National Fossil Day, Earth Day Network, Darwin Day, and Cargo for Conservation[5] ensures that the museum capitalizes on the resources available at the national level. In collaboration with these national events, the Dunn-Seiler holds competitions to engage Mississippi's K–12 students and hosts public museum events to involve the local community. Although these national venues and resources are integrated, the Dunn-Seiler adapts events to focus upon the local geology and environment and to engage the public in recognizing their role in the Earth system.

National Fossil Day in Earth Science Week

To improve public understanding of the geosciences, the Dunn-Seiler Museum in 2010 initiated outreach competitions for Mississippi's K–12 students. The museum participated in the first U.S. National Fossil Day, part of Earth Science week and sponsored by the U.S. National Park Service. Since the hadrosaur bones recovered in Mississippi were being returned to the Dunn-Seiler Museum after their loan to researchers in Mississippi and Florida, the competition invited Mississippi's school children to write a creative story about the Mississippi dinosaur or artistically draw or create a model of it. For National Fossil Day, a public outreach event, "Fossil Extravaganza," was also held and included "Adopt-a-Fossil," creating imprint "fossils" by pressing fossils into soft modeling clay, tours of the museum, and searching for fossils in Paleozoic gravel (Fig. 9). Although the focus of the Fossil Art and Story contest has changed over the years to include trilobites, mosasaurs, and mammoths, the event has been held annually to correspond with National Fossil Day. In order to effectively engage the local population and draw attention to the geology of the state, the Dunn-Seiler modifies and develops the annual National Fossil Day theme to address *Mississippi's* geology. In 2017, we expanded the theme of the National Fossil Day Paleozoic fish logo to include all of Mississippi's prehistoric fish, since Paleozoic strata were not extensively represented in the state. The focused attention on Mississippi's geology ensures that K–12 students and the general public find local, relevant evidence supporting some of the major constructs identified by the Earth Science Literacy Initiative: geologic time (Big Idea 2),

Figure 7. Although constrained by the existing cabinets, exhibits are now designed with informational signage. Renee Clary, as Director of the 15 Degree Laboratory, announced the 2017 Giverny Award winner at Mississippi State in April 2017; Amy Moe-Hoffman designed the displays. Photograph by Beth Wynn/Mississippi State University.

[5]Cargo for Conservation (sometimes called "Suitcase for Survival") is a program through the U.S. Fish and Wildlife Service. An assortment of illegal wildlife products are permanently loaned to schools, universities, and organizations to teach about the illegal wildlife trade.

Figure 8. Visitors to the Dunn-Seiler Museum's 2017 Fossil Extravaganza enjoyed the augmented reality topographic sandbox. Young visitors manipulated sand to create landforms. Photograph by Renee Clary.

a dynamic, changing planet (Big Idea 3), and evolution of life forms (Big Idea 6).

The Earth Day Network

In the spring of 2011, the Dunn-Seiler Museum extended its K–12 competitions by becoming a member of the Earth Day Network. The Creative Recycling competition challenged Mississippi's K–12 students to find a solution for a plastic bottle that would otherwise be discarded. The competition was a huge success and has been repeated annually, with Gaining Ground Sustainability Institute of Mississippi (GGSIM) partnering with the museum to co-sponsor the event from 2012 onward. The winning projects are announced at the MSU Earth Fair and displayed in the Dunn-Seiler Museum (Fig. 10). Each year the challenge focuses upon items that are typically discarded and end up in landfills, including Styrofoam, cardboard, water bottles, bottles caps, and telephone books. Although the challenge coincides with national Earth Day events, the Dunn-Seiler Museum ties the event to Mississippi's recycling efforts, sanitary landfills, and sustainable practices to address the Earth Science Literacy Initiative's Big Ideas that humans depend on the Earth for natural resources (Big Idea 7), and humans significantly alter the planet (Big Idea 9). Middle school teachers attending a professional development program in the natural sciences (2010–2013) encouraged their students to actively participate in the competition. Their feedback helped in the dissemination of sustainability activities to other middle school teachers (Clary and Wandersee, 2014b).

Darwin Day and Cargo for Conservation

The university sporadically acknowledged Darwin Day in the early 2000s, but in 2013 Amy Moe-Hoffman teamed with Diana Outlaw in the MSU Department of Biological Sciences to develop a thorough, inclusive outreach focused upon biodiversity and evolution. Since 2013, Darwin Day hosts multiple events across the community and involves the Dunn-Seiler Museum in fossil tours and fossil digs at the Cretaceous outcrop (Fig. 11). When a local school superintendent made the museum director aware of the Cargo for Conservation opportunity, Moe-Hoffman submitted a request for specimens. The Dunn-Seiler Museum received two boxes of confiscated items as a permanent loan

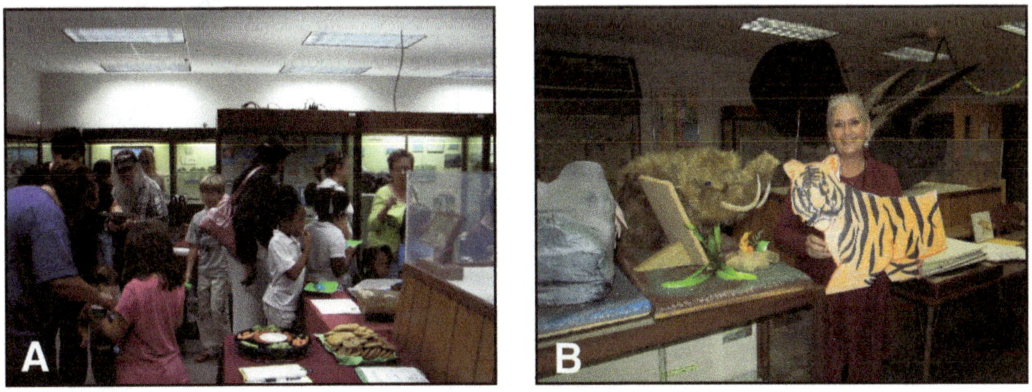

Figure 9. The Dunn-Seiler Museum hosted Fossil Extravaganza for the first National Fossil Day in 2010 (A). The museum implemented a K–12 art and story contest for National Fossil Day, although the focus of the competition has changed over the years to include a variety of Mississippi's extinct Paleozoic, Mesozoic, and Cenozoic animals. Teacher Tammie Bright, Yazoo County Middle School, brings student entries for the 2016 Pleistocene competition (B). Photographs by Renee Clary.

Figure 10. Beginning in the spring 2011, the Dunn-Seiler Museum became a member of the Earth Day Network, and sponsored K–12 recycling competitions for Mississippi's K–12 students. Each year, students are asked to develop a creative solution for an item that would otherwise be thrown away. In 2015, Athena Nagel judged Styrofoam entries (A), and in 2016, Maurice Testa and Amy Moe-Hoffman staffed the Mississippi State University Earth Day booth that displayed the winning water bottle entries (B). Photographs by Renee Clary.

from the U.S. Fish and Wildlife Service. The items, including an ocelot and mink fur coat, Asian leopard fur panel, elephant skin hiking boot, caviar face cream, and an ivory bracelet, have added to the Darwin Day celebration and have been used in classrooms and within various campus displays. Darwin Week events and Cargo for Conservation events and displays are designed to include several of the major concepts within the Earth Science Literacy Initiative (2010), specifically focusing on natural selection, biodiversity, and sustainability to educate that science is a way of knowing (Big Idea 1), the planet is a complex system (Big Idea 3), life evolves and modifies the planet (Big Idea 4), and humans depend upon Earth resources (Big Idea 7) and also significantly impact the planet (Big Idea 9).

Further Leveraging the University Community

Opportunities also presented themselves for the Dunn-Seiler Museum to participate in community engagement within the greater university Museums and Galleries committee. In 2009, a joint exhibit was planned within the MSU Visitors Center gallery space, adjacent to the Davis Wade football stadium. Gameday visitors could learn more about the various museums on campus and what tours and facilities were available within the "Treasures from the Vault" exhibit. In 2014, the university committee, now directed by Moe-Hoffman, designed MSU Museums and Galleries brochures to alert visitors to the opportunities on campus, including the Dunn-Seiler Museum, and the committee offered a

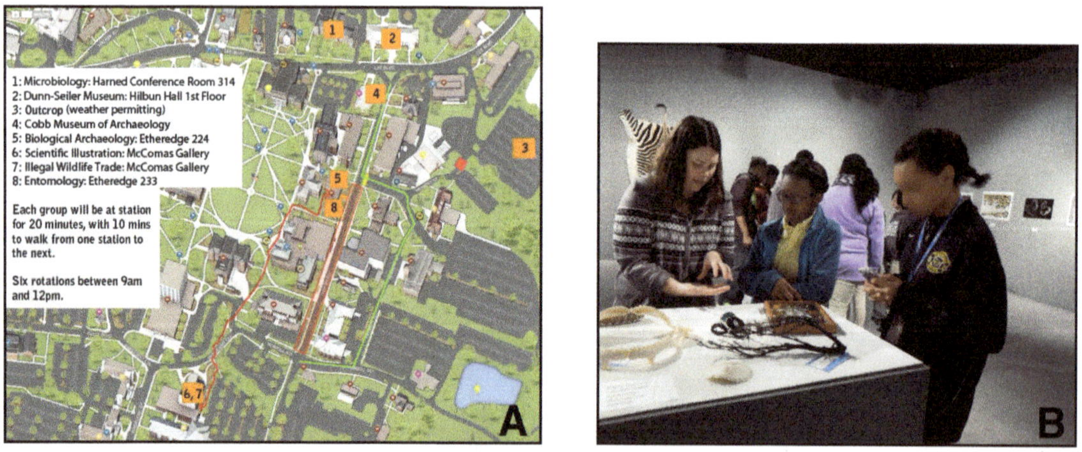

Figure 11. Darwin Day events began in earnest in 2013, when Amy Moe-Hoffman teamed with Biological Sciences to showcase biodiversity and natural selection at various stations across campus (A). When the Cargo for Conservation specimens arrived from the U.S. Fish and Wildlife Service in 2015, the items were included in Darwin Day outreach as well as in classrooms and other public events (B). Map by Amy Moe-Hoffman; photograph by Megan Bean/Mississippi State University.

TABLE 1. THE AVERAGE NUMBER OF VISITORS IMPACTED BY DUNN-SEILER MUSEUM EDUCATIONAL OUTREACH, 2013–2017

Outreach event	Number of events	Number of people impacted
Museum and outcrop tours	25	1300 visitors
Teacher professional development workshops	4	85 visitors
Informal speaking engagements and/or demonstrations	25	600 visitors
K–12 outreach competitions (National Fossil Day and Earth Day)	2	300 project submissions
Informal public outreach events (Fossil Extravaganza, Earth Day, Darwin Day, and Night at the Museum)	4	600 visitors

new "Science Night at the Museum" event for the general public in March 2017. The Dunn-Seiler Museum was a focal point of this new event.

DISCUSSION

What do the Dunn-Seiler Museum's improved displays, K–12 competitions, and activities mean in terms of outreach to the general public, and does this make a difference? The Dunn-Seiler Museum's tours and outreach events have touched thousands of children and adults over the past decade (Table 1). In addition to hosting competitions, school visits, and public events, the Dunn-Seiler Museum receives requests for speakers, demonstrations, and science fair judges and accommodates student camps and teacher professional development workshops within the summer months. Teachers from distant schools (i.e., out-of-state or schools greater than a ~2 h driving distance) requesting resources are directed to the Dunn-Seiler Museum's loaner kits and electronic presentations and resources hosted on our website. Since a chaperone or adult will often sign the visitor's book for a large group, and some people choose not to sign in to record their visit, the visitor numbers are likely an underestimation of the actual number of people who came through the Dunn-Seiler. However, the numbers only reveal part of the impact from the small public gallery. Geologist and Professor Emeritus John Mylroie noted that the role of this small university museum in northeast Mississippi cannot be underestimated, since the museum combats the young Earth creationism position held by a large percentage of the local population.

The changes are necessarily incremental with limited funding and time resources. Cumulatively, the changes add up, although there is still much to be done. The museum staff still typically avoids the "e" word in local discussions and museum tours, while maintaining scientific accuracy. Renee Clary observed that some groups became unreceptive—the "curtain comes down"—once evolution is mentioned, and she accepts that the Dunn-Seiler Museum cannot overcome years of religious education within a short tour. (The staff and graduate students talk with visitors about change over time and how some organisms became extinct.) However, the Dunn-Seiler can begin the scientific discussion and relate radiometric age-dating methods for Earth, for example, to more accepted practices of chemotherapy with radioactive elements.

Through the Dunn-Seiler Museum displays, public events, and K–12 art, story, and recycling competitions, the public is educated on biodiversity, extinction, geologic time, evolution, and conservation—promoting public geoliteracy in language respectful to the local community. In the midst of increased outreach, the research has continued within museum collections, too. While reinventing itself toward public outreach, the museum remembers its mission to serve as a research repository, and it accepts the university's mission to include service to the state (Allmon, 2004). Therefore, the Dunn-Seiler Museum now successfully bridges two communities—the research of faculty scientists and the public's general understanding of earth products, events, and concerns. The changes within this small university museum's displays and outreach parallel those within larger U.S. museums and document the shift from science *objects* toward science *education* (Rader and Cain, 2014). In the case of the Dunn-Seiler Museum, the informal education focuses upon the local geology and important concepts within the Earth Science Literacy Initiative for improved geoliteracy of the local underserved population. The Dunn-Seiler Museum has transitioned from an initial role of collection storage and "speak for themselves" specimen displays to a research collections facility with interpretive displays that improve public awareness of our planet's history and its resources.

ACKNOWLEDGMENTS

We thank Gary Rosenberg and our two anonymous reviewers for their thorough review and helpful feedback, which greatly improved the focus and presentation of our research.

REFERENCES CITED

AECF (Anne E. Casey Foundation), 2011, 2010 Kids Count Data Book: State Profiles of Child Well-Being: Baltimore, Maryland, http://datacenter.kidscount.org/ (November 2017).

Allmon, W., 2004, Opening a new natural history museum in twenty-first century America: A case study in historic perspective, *in* Leviton, A., Williams, G., and Aldrich, M., eds., Museums and Other Institutions of Natural History: Past, Present, and Future: San Francisco, California, California Academy of Sciences, p. 251–274.

Bell, P., Lewenstein, B., Shouse, A.W., and Feder, M.A., eds., 2009, Learning Science in Informal Environments: People, Places, and Pursuits: Washington, D.C., The National Academies Press (National Research Council Committee on Learning Science in Informal Environments, Board on Science Education, Center for Education, Division of Behavioral and Social Sciences and Education), 348 p.

Clary, R.M., and Wandersee, J.H., 2006, A writing template for probing students' geological sense of place: Science Education Review, v. 5, no. 2, p. 51–59.

Clary, R.M., and Wandersee, J.H., 2014a, Lessons from US fossil parks for effective informal science education: Geoheritage, v. 6, no. 4, p. 241–256, https://doi.org/10.1007/s12371-014-0116-x.

Clary, R.M., and Wandersee, J.H., 2014b, How low can you go?: Interdisciplinary student-impact investigations for environmental awareness and sustainability: Science Scope, v. 37, no. 7, p. 19–26, https://doi.org/10.2505/4/ss14_037_07_19.

Dockery, D., and Thompson, D., 2016, The Geology of Mississippi: Jackson, University Press of Mississippi, 692 p.

ESLI (Earth Science Literacy Initiative), 2010, Earth Science Literacy Principles: The Big Ideas and Supporting Concepts of Earth Science: National Science Foundation. Retrieved from http://www.earthscienceliteracy.org/ (accessed May 2017).

Falk, J.H., 2001, Free-Choice Science Education: How We Learn Science Outside of School: New York, Teacher's College Press, Columbia University.

Falk, J.H., and Dierking, L.D., 2002, Lessons without Limit: How Free-Choice Learning Is Transforming Education: Walnut Creek, California, Alta Mira Press, 208 p.

Farnum, J., 2004, In search of relevance: The museum of the twenty-first century, *in* Leviton, A., Williams, G., and Aldrich, M., eds., Museums and Other Institutions of Natural History: Past, Present, and Future: San Francisco, California, California Academy of Sciences, p. 244–250.

Friedman, A., 2010, The evolution of the science museum: Physics Today, v. 63, no. 10, p. 45–51, https://doi.org/10.1063/1.3502548.

Hearn, P., 2003, Dinosaur replica among displays at refurbished MSU museum: Mississippi State University Newsroom. Retrieved from http://www.msstate.edu/newsroom/article/2003/11/dinosaur-replica-among-displays-refurbished-msu-museum/ (accessed May 2017).

Hooper-Greenhill, E., ed., 1999, The Educational Role of the Museum (2nd edition): London, Routledge, 368 p.

Kohlstedt, S.G., and Brinkman, P., 2004, Framing nature: The formative years of natural history museum development in the United States, *in* Leviton, A., Williams, G., and Aldrich, M., eds., Museums and Other Institutions of Natural History: Past, Present, and Future: San Francisco, California, California Academy of Sciences, p. 7–32.

McComas, W.F., 2006, Science teaching beyond the classroom: Science Teacher (Normal, Illinois), v. 73, no. 1, p. 26–30.

Meyer-Arendt, K., 1994, The Department of Geosciences at Mississippi State University—A Brief History: Retrieved from https://www.geosciences.msstate.edu/about/ (accessed November 2017).

Nave, R., 2017 (14 Sept.), Mississippi still has worst poverty, household income: Mississippi Today: Retrieved from https://mississippitoday.org/2017/09/14/mississippi-still-worst-poverty-household-income-u-s/ (accessed November 2017).

Pasquaré Mariotto, F., and Venturini, C., 2017, Strategies and tools for improving Earth Science education and popularization in museums: Geoheritage, v. 9, no. 2, p. 187–194, https://doi.org/10.1007/s12371-016-0194-z.

Pinna, G., 2000a, Scientific research versus public exhibits: A schizophrenic aspect of natural history museums, *in* Ghiselin, M., and Leviton, A., eds., Cultures and Institutions of Natural History: Essays in the History and Philosophy of Science: San Francisco, California, California Academy of Sciences, p. 329–332.

Pinna, G., 2000b, A philosophy for natural history museums, *in* Ghiselin, M., and Leviton, A., eds., Cultures and Institutions of Natural History: Essays in the History and Philosophy of Science: San Francisco, California: California Academy of Sciences, p. 333–337.

Rader, K., and Cain, V., 2014, Life on Display: Revolutionizing U.S. Museums of Science and Natural History in the Twentieth Century: Chicago, University of Chicago Press, https://doi.org/10.7208/chicago/9780226079837.001.0001.

Rosenberg, G.D., 2018, this volume, Carl Akeley's revolution in exhibit design at the Milwaukee Public Museum, *in* Rosenberg, G.D., and Clary, R.M., eds., Museums at the Forefront of the History and Philosophy of Geology: History Made, History in the Making: Geological Society of America Special Paper 535, https://doi.org/10.1130/2018.2535(12).

Sutter, J., 2013, Change the list: The most unequal counties in America. Cable News Network (CNN). Retrieved from http://www.cnn.com/2013/10/29/opinion/sutter-list-income-inequality/index.html (accessed November 2017).

Wandersee, J.H., and Clary, R.M., 2006, Fieldwork: New directions and examples in informal science education research, *in* Mintzes, J., and Leonard, W., eds., NSTA Handbook of College Science Teaching: Theory, Research, & Practice: Arlington, Virginia, National Science Teachers Association (NSTA Press), p. 167–176.

Wandersee, J.H., and Clary, R.M., 2007, Learning on the trail: A content analysis of a university arboretum's exemplary interpretative science signage system: The American Biology Teacher, v. 69, no. 1, p. 16–23.

Manuscript Accepted by the Society 13 December 2017
Manuscript Published Online 13 April 2018

Printed in the USA

The Geological Society of America
Special Paper 535

Twenty-first-century natural history: Planetary geology in natural history museums

J.C. Aubele
L.S. Crumpler
New Mexico Museum of Natural History and Science, 1801 Mountain Road NW, Albuquerque, New Mexico 87104, USA

ABSTRACT

The traditional model of the natural history museum developed during an age of exploration. In the twenty-first century, natural history museums can demonstrate the excitement of science and enhance geoscience education by using the space-age exploration of our solar system and incorporating the geoscience subdiscipline of planetary geology. Natural history museums reach a self-selected, self-directed, and multigenerational audience. This audience can choose to pursue a range of exhibits and programs in various sciences offered by a museum. The public may be interested in geoscience but often has limited knowledge or understanding of the science. Planetary geology offers an effective way to add content and technology to the traditional natural history museum and a new way to interest museum visitors in basic geoscience. Over the past decade, the New Mexico Museum of Natural History and Science (NMMNHS) has successfully used planetary geology to enhance geoscience education by incorporating the following techniques: (1) geoscience related to a specific planet or planetary mission; (2) geoscience related to a planetary problem; and (3) planetary geology related to art. Use of these techniques has allowed the NMMNHS to reach multiple and underrepresented audiences, to encourage interest in basic geoscience, and to better serve the science education needs of the state of New Mexico. The addition of planetary geology to the traditional range of science topics enables natural history museums to continue their evolution as relevant sources of geoscience and provides them with an additional and effective way to teach geoscience in the twenty-first century.

INTRODUCTION

Over the past decade, the New Mexico Museum of Natural History and Science (NMMNHS) has effectively used the topic of planetary exploration to teach fundamental geoscience and the geology of New Mexico. The NMMNHS is a public museum run by the state of New Mexico. It is located in the state's largest urban metropolitan area and is mandated to serve the entire state. Our annual visitor attendance averages 200,000 visitors per year, including school groups, in a state with a population of only 2 million residents. The museum education department serves 30,000 additional students, teachers, and members of the public annually through targeted science, technology, engineering, and mathematics (STEM) education programs. New Mexico is a minority-majority state (populated by 52% underrepresented minorities) and a largely rural and economically disadvantaged

Aubele, J.C., and Crumpler, L.S., 2018, Twenty-first-century natural history: Planetary geology in natural history museums, *in* Rosenberg, G.D., and Clary, R.M., eds., Museums at the Forefront of the History and Philosophy of Geology: History Made, History in the Making: Geological Society of America Special Paper 535, p. 249–262, https://doi.org/10.1130/2018.2535(16).

state. New Mexico students in STEM rank lower than average on national and international benchmarks at all grade levels. The state ranks near the bottom for student performance, scoring among the lowest in the nation in mathematics and reading. It ranks second to last in the United States for high school graduation rates. In the Albuquerque Public School system, seven out of ten students qualify for the federal meals program and attend high-poverty Title I schools that receive national funding. The NMMNHS recognizes the great need for effective STEM education in New Mexico and has worked toward making the museum a positive force for this effort. In New Mexico, with its iconic geological landscape, teaching geoscience effectively can be a portal to opening the door to science for our visitors.

Role of Natural History Museums in Education

Public natural history museums, defined in this paper as those not affiliated with a university, serve their communities and regions as resources that are accessible and non-threatening for pre-kindergarten (pre-K) through senior citizens, including underrepresented audiences. Visitors to natural history museums are a self-selected, self-determined, and multigenerational audience who choose to spend their discretionary time and money at the museum and who choose the exhibits, information, and programs they wish to peruse. This type of science education is referred to as informal science education or free-choice learning (Falk and Dierking, 1997). The average person spends only 5% of their life in a formal classroom, including four years of college (Falk and Dierking, 2010). Therefore, natural history museums and other informal science education institutions, including zoos, aquariums, botanical gardens, visitors centers for parks and monuments, and educational television, play an increasingly important role in educating the public about science by reaching them during the other 95% of their lives.

Museums are recognized as a trusted source for science information for the public and a resource for science education for pre-K–12 students. In 2001, a national survey sponsored by the American Alliance of Museums showed that 87% (±3%) of 1000 randomly selected adult respondents viewed museums as "one of the most trustworthy sources of objective information" and, in fact, as "the most trustworthy source of information in America, rated higher than local newspapers, nonprofits researchers, the U.S. government, or academic researchers" (Lake Snell Perry and Associates Research, 2001). Another important study on informal science institutions and trust asked specifically about information related to global warming, and 72% of those surveyed expressed trust in science and natural history museums for information related to that particularly contentious topic (Leiserowitz, et al., 2010). Informal science education institutions have additionally become a trusted resource for content-driven teacher professional development workshops. Pre-K–12 classroom, science, and homeschool teachers participate in museum workshops on a variety of science topics or make use of museum-written curricula, loaner kits, or other materials, and then become the resources for their own students and colleagues.

Natural history museums support "the lifelong learning continuum that begins with early childhood and is based in science … creating the building blocks for cognitive, social, regulatory and emotional development" (Institute for Museum and Library Services, 2003). One of the key educational strengths of museums is object-based learning (OBL)—that is, the use of real objects to facilitate interest and learning. OBL is particularly effective in learning about science and technology. While natural history museums have collections of objects and display them in exhibits that are designed to accommodate various learning styles and educational standards, the average time during which a museum visitor explores an individual exhibit is less than 10 min (Hooper-Greenhill, 1996). In order to increase visitor appreciation and understanding of the exhibit, natural history museums incorporate volunteer docents or guides in the exhibit, supplemental materials, additional short-term exhibits highlighting new discoveries, and associated programs including on-site and off-site lectures, classes, and workshops for all age groups. Museums can use these associated programs to make a face-to-face connection between the public and their local scientists in informal and comfortable venues including evening events, short talks or presentations, relaxed Q&A sessions, outreach presentations to local civic organizations, family days with activities for multigenerational groups, and special events (Aubele et al., 2011). A way to visualize this package of informal education is a pyramid (Fig. 1). The associated

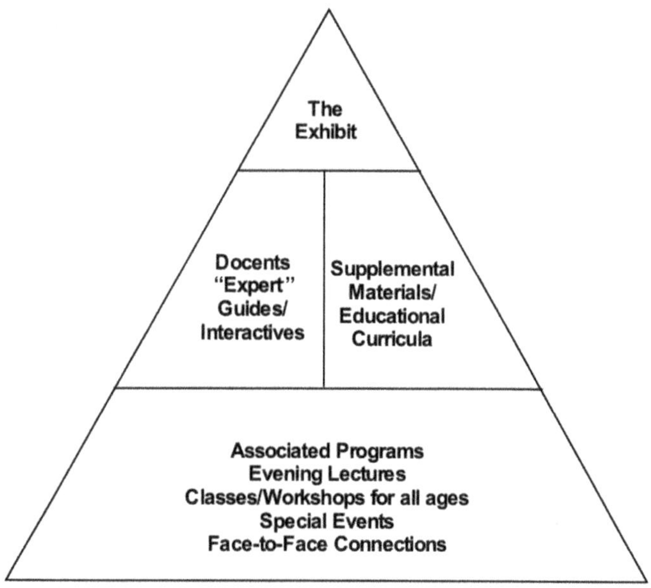

Figure 1. The pyramid of informal science education typical of that offered by a public natural history museum. The exhibit provides the initial science content information that is then enhanced by external information via docents and supplemental documents as well as associated age-appropriate educational programming targeted to specific audiences. The exhibit is only a part of the learning experience.

educational programming enables the museum to leverage the science content, visuals, and interactives of an exhibit in order to increase the science content available to the average visitor with more information-intense approaches.

However, it is difficult to measure museum learning from these methods. Many adult scientists have anecdotal stories about their early experience (in a museum or other informal science education setting) that shaped their desire to become a scientist, but the efficacy of museum education is difficult to quantify. Currently the positive anecdotal evidence of learning in museums is primarily supported by data acquired using the following techniques: museum attendance numbers and their increase or decrease during specific exhibits; numbers of attendees for associated educational programs; the amount of time spent by visitors viewing a particular exhibit; written surveys returned by randomly selected visitors on specific topics; and evaluations collected by pre-K–12 teachers tracking the performance and interest in science of students who visited the museum.

Natural History Museums and Geoscience Education

Every natural history museum has geoscience-related objects, exhibits, and programming. Geoscience impacts every aspect of human society. For example, the unique and beautiful landscape created by geology is a major component of tourism and recreation and the creation of local and national parks, monuments, and museums. Most moderate-sized communities have a rock and mineral club, and there is significant interest in, and appreciation of, geology. However, *being interested and having knowledge about a science are not equivalent.* In a survey of randomly selected adult visitors to the New Mexico Museum of Natural History and Science (NMMNHS), 53% of the total respondents selected Earth Sciences/Geology as the science topic that they felt they knew the least about. The same survey found that only 15% of the respondents selected the Earth Sciences/Geology exhibits in the museum as those that most captured their attention when visiting the museum. Earth Sciences/Geology ranked third after Life Sciences/Fossils and Space Sciences exhibits (Fig. 2). In two decades of educating the general public through exhibits, classes, lectures, and other programming at the NMMNHS, we have found that even in a land where geology is a dominant element of the surrounding landscape, the physical and temporal concepts of geology and fundamental geologic processes are difficult for those untrained in geology to assimilate and understand. This may make it not only difficult for the average museum visitor to enjoy and learn from traditional geoscience exhibits, but the visitor may be unable to make use of the information to better understand the local geology and landscape. Lack of exposure to and training in geology, as compared with other sciences, is likely an important factor. To overcome this inherent deficiency, a way to make geology exciting and relevant is desirable.

Natural History Museums and Planetary Geology

The geoscience subdiscipline of planetary geology provides a new technique for natural history museums to present geoscience education (Aubele and Crumpler, 2016a). Planetary geology incorporates the topics of "outer space" (space science) and "exotic new worlds" (planetary science) while using the techniques and principals of terrestrial geoscience to explore, sample, analyze, and understand the geological terrain, materials, processes, and evolution of the other planets and natural satellites of our solar system (Short, 1975; Greeley and Carr, 1976; Wilhelms, 1993). The concept is essentially comparative planetary geology, a science investigation approach that uses discoveries and interpretations of other planets and is proven to be useful in understanding the geology of Earth (Aubele, 1996). The comparative

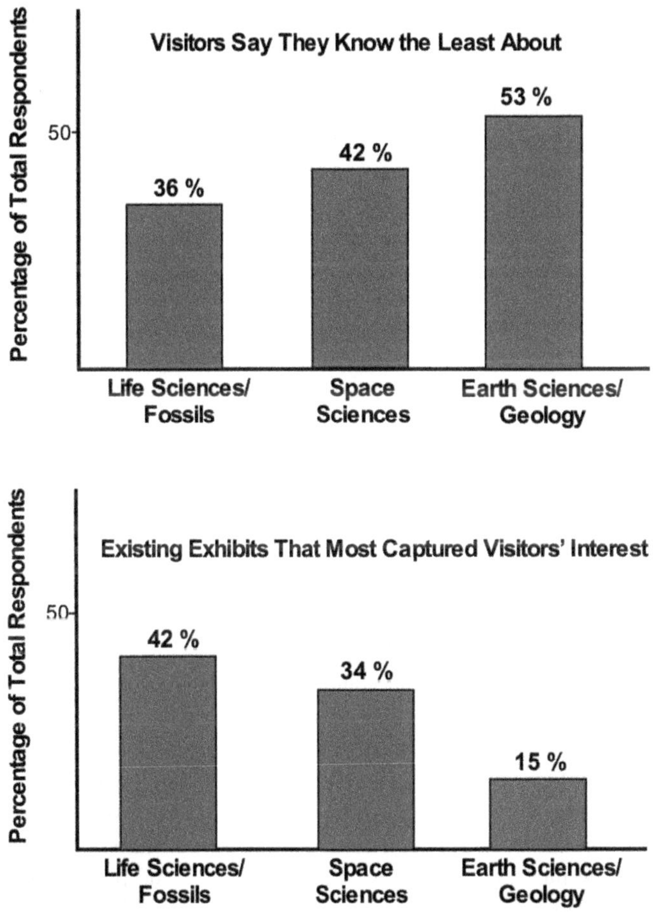

Figure 2. Results from a survey of 100 randomly selected adult visitors to the New Mexico Museum of Natural History and Science (NMMNHS) in Fall 2017. They were asked to choose the one topic, from the three major science topics covered by the NMMNHS, that they felt they knew the least about. Fifty-three percent of visitors self-identified Earth Sciences/Geology as being the topic they knew the least about. In the same survey, only 15% of visitors said that existing exhibits themed to Earth Sciences/Geology captured their attention, the lowest percentage of the three major science topics.

planetology approach also encourages thinking about the study of Earth as a global system, providing a larger-scale, exterior view of geoscience on our own planet. Traditional geosciences developed from the outcrop and evolved to a regional view. With the advent of satellites and planetary mission technology, new global data sets have been acquired by numerous Earth-observing mission satellites. Geoscience on our own planet can be presented using a planetary geology approach of the "big picture" followed by selected local, detailed "in situ" investigations. This approach includes environmental geology topics that have not been included in traditional natural history museums in the past but are important for the twenty-first century, such as changes in weather patterns, land use, ocean temperature, sea level, and the areal extent of rainforests, deltas, and glaciers.

In addition to the simple concept of comparative planetology, the concept of basic geographic exploration and discovery can be important in developing excitement about the geosciences. The exploration of our solar system is the twenty-first-century equivalent of the great early voyages of discovery and scientific explorations. Natural history museums developed in association with the great European voyages of exploration and discovery that began in the late fifteenth century and continued to the eighteenth century. Renaissance "cabinets of curiosities" were private collections of exotic geological and biological specimens. When these privately owned collections were donated to universities or other institutions, early natural history museums, with no accessibility for the general public, began to be established in the mid-sixteenth to mid-seventeenth centuries. In 1683, the Ashmolean Museum, at Oxford University, was opened to the public (https://www.ashmolean.org/history-ashmolean) in a decision that was considered to be controversial in the seventeenth century. It contained a variety of donated collections, including geological, biological, and archeological and/or anthropological specimens discovered and collected by travelers. This became the standard model for today's public natural history museum.

Planetary exploration today holds the attention of the public in a way that only a few other areas of science can. In many ways, the discovery, observation, and exploration of our solar system captivates the public in this century just as those great ocean voyages or discoveries of previously unknown continents and species did in past centuries. Planetary exploration is also a newsworthy event. A featured story or headline in print or digital media about a new planetary discovery can provide "free advertising" and a ready-made audience for public geoscience education using an existing museum exhibit. For STEM educators and professional geoscientists in the United States, planetary research presents a unique opportunity to ignite passion and interest in geoscience in an entirely new group of young minds, as well as introducing them to other areas of science or engineering (Fig. 3).

Natural history museums have an advantage in presenting and interpreting the "new geology" on other planets since they are already experienced sources of terrestrial geoscience information. Most natural history museums already include a planetarium and/or observatory as well as astronomy or space science exhibits; a growing number of museums have curators or research scientists with ongoing space science research programs that can inform and impact exhibits and educational programming. Although pre-K–12 formal curriculum or standards do not specifically include "planetary geology," they do include Earth and Space Science. Since planetary geology is a very broad science that incorporates many different disciplines, tools, and data sets and uses primary themes such as observation, comparison, and making connections across disciplines, it can easily be incorporated into all of the basic sciences, mathematics, engineering, literacy, history, and careers and used with the Common Core State Standards Initiative (http://www.corestandards.org/) and the Next Generation Science Standards (National Academy of Sciences, 2012) that have been adopted by most states.

A key educational strength of museums, object-based learning (OBL), can be an impediment to the planetary approach. For example, many natural history museums focus on the biological and paleontological sciences or the mineralogical subdiscipline of geoscience because these topics use collections as a resource both for materials and expertise. Models or replicas of spacecraft or samples such as lunar rocks or meteorites are usually the only planetary component to be part of a physical collection. Because it is impossible to collect and display categories of geological morphologies or physical geological processes in a museum's

Figure 3. Science on a Sphere, on loan from National Aeronautics and Space Administration Goddard Spaceflight Center, on exhibit at the New Mexico Museum of Natural History and Science (NMMNHS) in 2013, and one of the museum's volunteer "Junior Docents," a group of teens (ages 13–17) who learn science and public presentation skills. Credit: NMMNHS.

collection of physical objects, these subjects have languished without help from educators and museums even in regions such as New Mexico where the geology is exposed in dramatic scenery. It is especially true in regions where rock outcrops are not evident, such as those covered by forests or swamps or grasslands and where, without such guidance, the general public may have a hard time visualizing the connection between geologic processes and their effect on the local scenery. This may be changing. During the past few decades, natural history museums have sought to enhance their traditional collections and exhibits by adding hands-on interactives and new technology to previously static exhibits filled with objects on display. Both the data sets and the technology associated with modern planetary geology exploration provide a technological component to basic geoscience education. For example, we have found that images of geology on other planets, such as volcanic edifices, tectonic scarps, and alluvial debris fans, returned from a planetary mission, are easily displayed on monitors and manipulated by visitors at interactive stations. This allows for comparison between geology on various planets, including local planetary analogs. These types of modern interactives offer an important technological adjunct to presenting geoscience in natural history museums.

While the planetary connection allows museums to impart the enthusiasm and excitement of science and new scientific discoveries, it can also be used to relate to local or regional geoscience. The science can become relevant through: (1) connections with local geoscientists who are involved in planetary missions; and (2) local or regional planetary analogs in the surrounding landscape. The connection with local scientists involved in planetary missions is important in humanizing science. Elementary and middle school students frequently believe that science is too difficult for them, that they are not "smart enough," that they do not know enough, or that scientists and engineers can be only one specific type of person. Through contact with local scientists they soon learn that large-scale, highly technological, modern voyages of planetary exploration require people from many different disciplines and career paths and from many different backgrounds, nations, cultures, and heritage. The National Aeronautics and Space Administration–National Science Foundation (NASA-NSF) Passport to Knowledge educational activity called Marsapalooza made use of this technique (https://www.nasa.gov/audience/forstudents/5-8/features/F_Marsapalooza.html) to raise public literacy and present local scientists and engineers involved in Mars missions as role models. The fact that there are local or regional scientists or engineers from their own city or state who are involved in newsworthy planetary missions is usually a surprise to the public. Meeting these people face-to-face in an informal, relaxed setting within their public museum promotes a personal connection that enhances interest and learning (Aubele and Stanley, 2007).

Presentations and activities on local geoscience topics that have value as analogs to other planets are equally important. Previously, geologists could only use examples from Earth when teaching geology, but there is now more than one data point in our understanding of the spectrum of geological processes and evolution (Crumpler et al., 2007). This was the innovative approach taken by the Cleveland Museum of Natural History (CMNH) in an exhibition that opened in 1997 (Hannibal, 1998). The exhibition, called *PLANET e,* was the opening installation of the CMNH Reinberger Hall of Earth and Planetary Exploration. The exhibit described and discussed basic geologic processes such as volcanism, tectonism, and glacial, aeolian and lacustrine deposition and erosion, using not just Earth but multiple planets and moons in our solar system as examples. One related concept that is particularly important, and often overlooked, is the fact that no matter where students grow up, they often feel that other places are more interesting. Teaching geoscience by using strong local analogs to another planet can make the local geology seem much more exotic and interesting. We have an advantage in the southwestern United States. As a semiarid region with volcanoes, dry riverbeds called arroyos, canyons, and stratigraphically layered mesas, our state of New Mexico does resemble Mars. However, regardless of the region and the local geology, there is always a comparison of some kind with another planet in our system that can be utilized (Greeley, 2013). Natural history museums are especially effective if they involve the general public in understanding the local or regional geology; and if there are strong local analogs to another planet, it can make the local geology more relevant and suddenly seem much more interesting to students of all ages.

As an example, the New Mexico Museum of Natural History and Science (NMMNHS) created a small exhibit of rocks and minerals found in New Mexico that were also identified on Mars. The concept was to use the planetary analog in what would normally be a typical rock and mineral display. We observed that many visitors spent more time reading and discussing this exhibit than the average time spent at other similar-sized, comparable rock and mineral exhibits (Fig. 4). Observation and analysis based on the technique of compare and contrast are useful as educational tools because they assist non-geologists in understanding the geological principles and processes that have produced their surrounding stratigraphy and landscape. There is abundant information available for these analogs, even if a museum has no in-house research in planetary geology. NASA and NASA-related research facilities, European Space Agency (ESA), and international agreements on missions flown by Japan, India, and China all provide abundant information and data of all kinds, including collected images and data sets available in the public domain as well as specifically designed lesson plans and curricula for formal and informal science educators (https://www.nasa.gov/audience/foreducators/index.html).

EXAMPLES OF MUSEUM PROGRAMS USING PLANETARY GEOLOGY TO ENHANCE GEOSCIENCE EDUCATION

In order to describe the use of planetary geology in geoscience education provided by natural history museums, three

Figure 4. Minerals found in New Mexico that were also identified on Mars. This exhibit at the New Mexico Museum of Natural History and Science (NMMNHS) focused on the subdiscipline of mineralogy and used planetary geology to enhance visitor's interest in the minerals and rocks of New Mexico. Credit: NMMNHS.

successful approaches are delineated in the following section. Detailed examples from the New Mexico Museum of Natural History and Science are provided followed by similar programs from other institutions.

Geoscience Related to a Specific Planet or Planetary Mission

Mars in New Mexico: A Decade of Exhibits, Education, and Public Programming

The fact that a research curator at the New Mexico Museum of Natural History and Science (NMMNHS) is a team member on the Mars Exploration Rover (MER) mission has been used over the mission lifetime to allow the public to participate in the ongoing discoveries while also learning something about the natural history of New Mexico. In early 2004, the NMMNHS began a program of museum exhibits and education specifically using Mars and the MER mission. The permanent exhibit entitled, *Making Tracks on Mars* (conceived and produced in-house), has been on display since December 2003. The full-scale, detailed MER mission rover replica was built in-house by museum volunteers with funding from local companies and donors and by permission of the Jet Propulsion Laboratory (Fig. 5). The mural and rockscape were produced by local artists to model the proposed views and surface features of Mars. For example, the type of rocks (basalt) and their size distribution were selected from local occurrences and placed in the exhibit landscape to illustrate the size and frequency of rocks associated with a rock population covering 8% of the surface, the expected rock population at the Spirit rover landing site in Gusev crater (Golombek et al., 2005). This characteristic of the site is then used to engage visitors in a discussion of the basaltic rocks of New Mexico and their similarity to those seen on Mars and the fact that rocks in any landscape are distributed in size according to an exponential distribution. The exhibit also includes an interactive panoramic camera that encourages children to stop and explore the way we see the surface from a rover and think about how we learn new things from in situ remote sensing. There are

Figure 5. A young visitor enjoying the interactive panoramic camera on the rover replica in the permanent exhibit *Making Tracks on Mars* at the New Mexico Museum of Natural History and Science (NMMNHS). The rover is a full-scale replica. The interactive kiosk visible on the left is one of three created by the NMMNHS and allows the visitor to explore the science instruments on the rover, the mission, and the geology of Mars. The mural and rockscape model the views and surface features of Mars. The exhibit is designed to be a learning opportunity. Features within the exhibit enable visitors to observe the color of the sky and ground, the rock population density at the rover site, the nature of volcanic rocks, the implication of visible tracks leading from the lander pad, and the obvious direction of the wind based on sand wind shadows associated with rocks. The exhibit was installed in December 2003, just prior to the rover's landing, and it remains a popular part of the museum space science wing. Credit: NMMNHS.

also three interactive kiosks (created and produced in-house) that encourage visitors to explore the mission, instrumentation on the rovers, and the geology of the planet Mars. Each of these offers the opportunity for a docent to build on the visitor's interest with additional verbal information.

Associated educational programs using Mars and the MER mission were created for pre-K–12 students and their teachers, families, and the general public. These programs included outreach presentations to schools and adult clubs and civic organizations, after-school programs, Mars-themed summer science camps, Mars family days, public evening lectures, adult classes, field trips to New Mexico–Mars geology analog sites, and special events for bilingual (Spanish and English) participants. The Museum produced a complete teachers' guide and curriculum entitled *Making Tracks on Mars: Teacher Resource and Activity Guide* (Aubele et al., 2006a, 2006b; Aubele et al., 2012) designed for formal and informal science education classrooms and events (Fig. 6). Mars-related workshops for teachers have been offered each year at the Museum and at sites throughout the state of New Mexico. The curriculum was distributed nationally through a link with the Lunar and Planetary Science Institute in Houston. Making use of social media as an educational tool, MER science team member and co-author of this paper, L.S. Crumpler, began and continues to author a blog entitled, *Rover Field Notes from Mars,* which is accessible under the Planetary Society's (www.planetary.org/) guest blogs section related to the Mars Exploration Rover Mission. (Go to http://www.planetary.org/blogs/guest-blogs/2017/crumpler-sol4857.html.)

The museum formed a strong partnership with a regional Native American junior-senior high school (Figs. 7A–7C). The concept of place-based geoscience is effective and powerful, particularly in the education of Native American students (Semken, 2005). This place-based approach has been extremely successful as part of the Mars educational programming provided

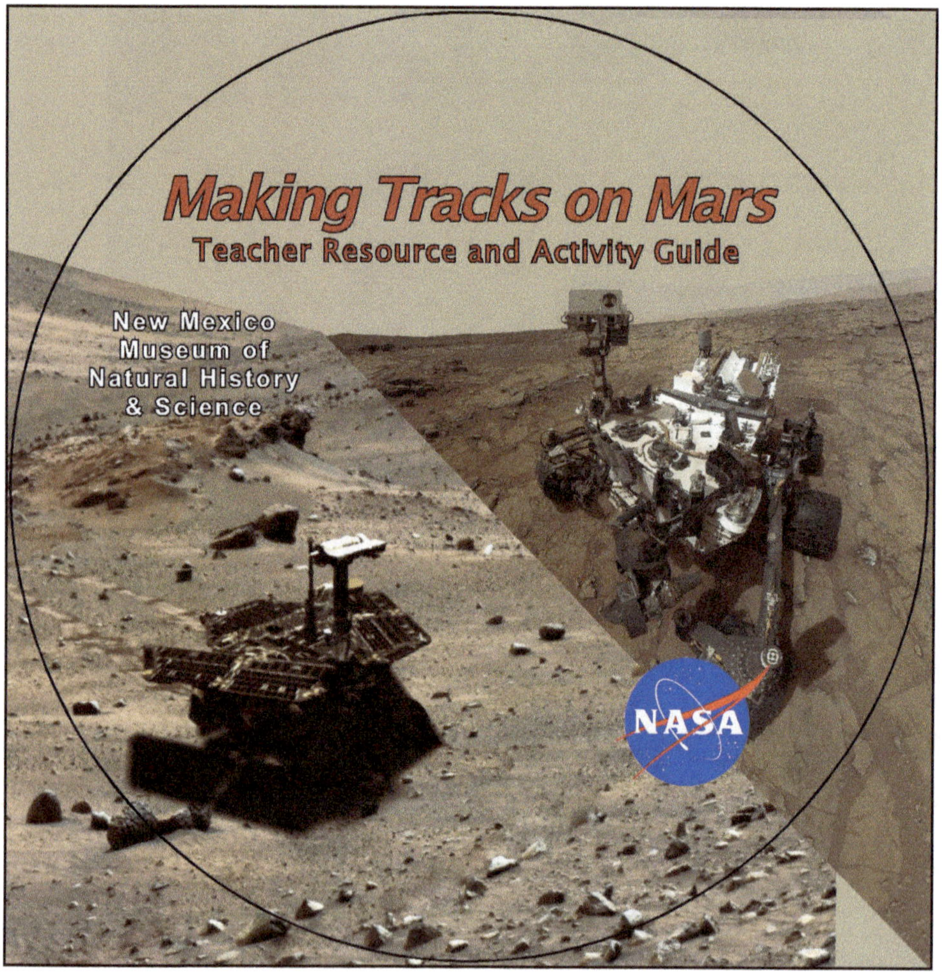

Figure 6. Cover for the *Making Tracks on Mars Teacher Resource and Activity Guide,* produced by the New Mexico Museum of Natural History and Science (NMMNHS). The Teachers' Guide consists of 20 tested new or adapted hands-on activity lesson plans and additional accompanying information about Mars and rover missions. It is available in a second edition, 508-compliant, version on DVD, and is designed for classroom and science teachers, home school parents, and informal science educators working with museums or after-school programs. Credit: NMMNHS.

to Native American students in New Mexico. The NMMNHS worked closely with Laguna-Acoma Junior-Senior High School, which is located on the Pueblo of Laguna Indian Reservation, 45 miles west of Albuquerque, New Mexico. The school serves the Native American communities of the Pueblo of Acoma and the Pueblo of Laguna as well as the traditional Hispanic mountain communities of Cubero and Seboyeta. The school is ~93% minority (Native American and Hispanic) students. High school students, with their teacher, carried out a series of field trips and measurements with guidance from museum scientists and educators. They measured rock type, size, and distribution in a variety of field areas in their own "backyard" on the reservation. When they analyzed some of the Mars rover images, they were amazed to discover similarities to their local geology; and they presented their research in a series of school outreach programs to students at other New Mexico middle and high schools. All of the students who participated in this project continued their educations and became the first in their families to graduate from college.

The overarching theme of this ongoing educational program uses a newsworthy planetary mission to excite and interest the public; and it uses Mars to teach local geology by linking New Mexico and Mars (Aubele et al., 2004; Aubele and Crumpler, 2016b). The exhibit initially opened in January of 2004 at the same time as the rover landings on Mars. Visitor attendance for 2004 showed a substantial increase of 12,000 additional visitors over 2003. This occurred the same year that a new science center opened nearby and museum attendance was projected to decrease. Since 2004, a minimum of 3 million visitors have visited the Mars-related exhibits, and participation in our Mars mission–specific educational programming has totaled more than 20,000 adults and children during the past decade (Aubele, 2013a, 2013b). One special evening event was attended by more than 1800 people. The most frequent request for lecture topics from the public is for our periodic Mars mission update lectures, which routinely fill a 300-seat theater. Our curriculum and teacher professional development workshops have reached ~6000 classroom and science teachers during the past decade. The Mars programming has been one of our most successful thematic geoscience education efforts.

Other Examples

A similar approach, using a specific solar system mission, is exemplified by the Royal Ontario Museum (ROM) (Mader et al., 2017) in Toronto, which has a link to the current OSIRIS-REx

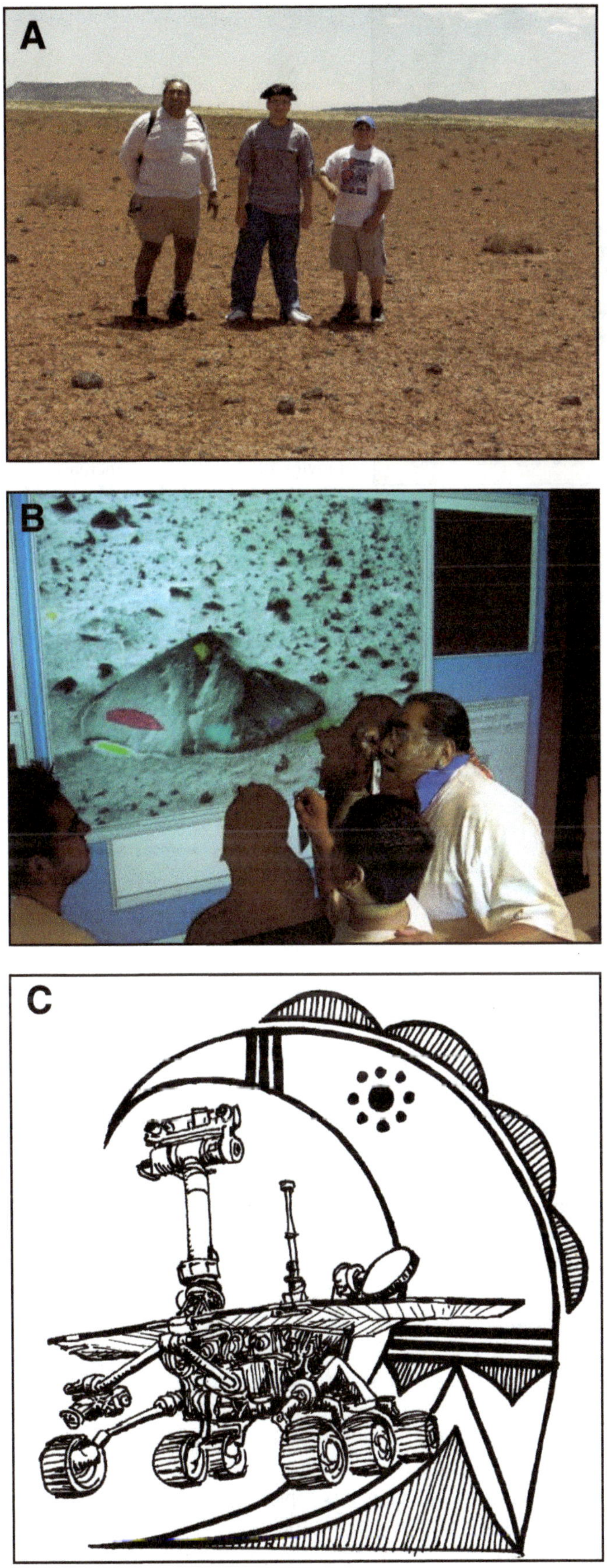

(Origins, Spectral Interpretation, Resource Identification, Security–Regolith Explorer) mission launched by NASA in 2016 (sample return from an asteroid) through their Endowed Chair of Mineralogy, who is a collaborating scientist on the mission. The ROM has worked closely with the Canadian Space Agency and U.S. planetary scientists to educate the public about the mission and the science through exhibits, lecture and panel events, outreach, summer camps, school-visit programs, and participation in the OSIRIS-Rex Ambassador training. Similarly, the Rose Center at the American Museum of Natural History and Science (AMNH) publicized a direct connection to the James Webb Space Telescope through an instrument component that was conceived and designed at the museum by scientists in the museum's department of astrophysics (American Museum of Natural History, 2010). When the James Webb Telescope launches, the AMNH is expected to install an exhibit about the space telescope and present the science to their visitors with their personal "hometown" connection.

Geoscience Related to a Planetary Problem

Mars Meteorites, Geochemistry, and Geochronology

Because of the success of our Mars in New Mexico program in enhancing geoscience education in the state, the New Mexico Museum of Natural History and Science (NMMNHS) began to experiment with other ways of using planetary geology to teach basic geoscience. Using planetary geology to introduce the general public to the geological subdisciplines of geochemistry and geochronology was the intent of a short-term special exhibit and associated programming entitled, *It Came from Mars,* produced by the NMMNHS and on exhibit from May 2015 to May 2016. The exhibit was designed to take advantage of popular culture, specifically the 1950s space invader movie theme (Fig. 8A) and Sherlock Holmes mysteries. Using meteorites from Mars and a Holmes-style series of clues and deductions, the exhibit presented detailed information about mineralogy and petrology, age dating, and atmospheric composition (Fig. 8B). Linking these topics to

Figure 7. (A) As part of their geoscience research in partnership with the museum, the Laguna-Acoma Junior-Senior High School team measured rock population density in a Mars analog area on Laguna Pueblo reservation land. Laguna-Acoma Junior-Senior High School in New Mexico serves the Native American communities of the Pueblo of Acoma and the Pueblo of Laguna as well as the traditional Hispanic mountain communities of Cubero and Seboyeta. (B) The Laguna-Acoma Junior-Senior High School team study a rover image. From right to left: science teacher Mr. Joseph Aragon (past lieutenant governor of Acoma Pueblo), high school junior, Brandon Herrera (Laguna Pueblo), and high school senior, Mark Vallejos (Seboyeta, New Mexico). (C) Using their cultural heritage and references, the Laguna-Acoma students produced their own logo for their participation in the mission as a partner to the NMMNHS. The logo consists of an eagle arching over, and protecting, the rover and includes traditional design elements that symbolize landscape and water. Credit: NMMNHS for A–C.

Figure 8. (A) The title panel of the exhibit, *It Came from Mars,* used popular culture, in this case the space invader movies of the 1950s, to increase visitor interest. Credit: New Mexico Museum of Natural History and Science (NMMNHS). (B) In order to answer the natural question, "how do we know the meteorites are from Mars?" part of the exhibit cast the problem as a Sherlock Holmes mystery, thus encouraging visitors to re-discover the scientific method. The exhibit was on display from May 2015 to May 2016 at the NMMNHS. Credit: NMMNHS.

Mars allowed the museum to educate visitors about complicated concepts in a way that encouraged their interest in these subjects. Visitor attendance increased significantly during the May 2015 to May 2016 period that it was on display, with an additional 16,000 more visitors than during the previous May 2014 to May 2015 time period. Many of the visitors stated that they were specifically coming for this exhibit when entering the museum. Associated lectures and special events included programs targeted toward K–6 students, teens, adults, and families, and these also experienced higher than average participant numbers. The year-long run of the exhibit culminated in a special "Mars Fiesta" offering an opportunity for several hundred museum visitors to interact one-on-one with local Mars scientists. The success of this special exhibit has led to plans for the development of other ways to use planetary geology to enhance the presentation of basic geoscience information.

Other Examples

The American Museum of Natural History's short-term exhibit, *Exoplanets* (conceived and produced in-house as part of NASA's Navigator Program) and its permanent exhibit, *Why Is Earth Habitable?* in the Gottesman Hall of Planet Earth both reflect this approach to presenting geoscience subdiscipline topics by linking them with planetary topics (Natural History Museum, 2017). Many other geoscience topics and subdisciplines could be presented in a similar way, including economic resources and ore minerals on Earth and predicted on other bodies in the solar system, atmosphere, climate, and global warming on Earth contrasted with Venus' greenhouse atmosphere (where the term was originally coined) (Grosfils et al., 2000), and the cave, ocean, and thermal springs environments for extremophiles on Earth compared with the planetary conditions necessary for the development of life and assumptions for looking for life on other planets.

Planetary Geoscience Related to Art

Teachers of science and technology have recently begun to incorporate the arts as a useful and effective way to enhance creativity and understanding of basic science in students. This is referred to as science, technology, engineering, arts, and mathematics (STEAM). Geoscience was an early adopter of this idea in a publication from the Geological Society of America on the art of geology (Moores and Wahl, 1988). Planetary geology is an excellent way to present this confluence of art and science. The images taken by planetary spacecraft are beautiful, often color enhanced, and present scenes that are frequently evocative in shape, texture, or patterns. They are easily presented as art at the same time that they can be used via associated text and docents to present the science behind the beauty. NASA has acknowledged that fact with two recent books, *Earth as Art* (Friedl, 2012) and *Mars: The Pristine Beauty of the Red Planet* (McEwan et al., 2017). Every full-page image in both books is accompanied by easily understandable geologic descriptions. In his introduction to the latter book, co-author and outreach education lead for the Mars Reconnaissance Orbiter HiRISE (high-resolution imaging and spectrometer experiment) team, Ari Espinosa, writes, "Space exploration needs engineers, physicists, mathematicians, geologists, and astronomers. But it also needs poets, thinkers, and dreamers." To present science to those who might be more comfortable with the arts and humanities is the intent of adding art to science.

Space Photography

In the museum environment, the presentation of images acquired for scientific purposes is perhaps one of the simplest ways to engage visitors in the science of exploration while enveloping them in the familiar environment of an art gallery, complete with open views and seating. An example of this approach is the photo exhibit *25 Years of Space Photography,* curated by Jay Belloli, Director of the Baxter Art Gallery (California Institute of Technology), and made possible by a grant from the IBM Corporation. It is a compilation of 140 pictures chronicling the first images to be acquired by the U.S. unmanned space program of the bodies of our solar system from 1965 through 1989. These images were matted and framed with explanatory text and traveled the globe to more than 50 art galleries and museums in the 1990s. In 2015, the complete collection was granted to the NMMNHS, and some of the images, along with additional first-time planetary exploration images (from 1990 to 2016) selected by the museum's space science research curator, were exhibited for one year as the museum's special anniversary celebration exhibition (Fig. 9). This use of planetary data as science, art, and history surreptitiously enhanced science education for those visitors who might find art or history more accessible or interesting than science (Sudbury, 1996). *Los Angeles Times* art critic, Christopher Knight, wrote in the original catalog (Belloli and Dean, 1985; Knight 2012) "... framed and hanging on the walls of a gallery, space photographs are radically re-contextualized: they are transformed *from offerings of visual (scientific) evidence into aesthetic objects for examination*." Knight also stressed the scientist-as-artist in linking these photographs with the tradition of nineteenth-century photographers who charted and first depicted America's wilderness through paintings or photographs.

Other Examples

The NMMNHS has made additional use of this technique. To celebrate the 10th anniversary of the Mars Exploration Rover mission in 2014, the museum created a short-term exhibit, entitled, *Mars Landscape Art by Spirit and Opportunity*. This exhibit was designed to blend art and science in the same way as the *25 Years* exhibit and to immerse the museum visitor in Mars with an art gallery–style exhibit of wall-size panoramic images (Fig. 10). The exhibit was reviewed locally by the arts editor (not the science writer) of our metropolitan newspaper and drew increased visitor numbers to the museum throughout its entire run (Gomez, 2014). Similarly, a short-term exhibit in 2016 at the Natural History Museum, London (UK) entitled, *Michael Benson's Other Worlds: Visions of Our Solar System* (Natural History Museum, 2017), presented an artist's viewpoint concerning images of the planets taken by spacecraft. The associated website produced by the museum included a video feature, *Portrait of a Planetary Scientist,* profiling the work that a planetary scientist does and successfully linking the science to the art (http://www.nhm.ac.uk/visit/exhibitions/otherworlds-visions-solar-system-exhibition.html). An example of using the beauty of planet Earth as a teaching tool in museums is the Smithsonian Institution Traveling Exhibition Service (SITES) exhibition entitled, *Earth from Space,* curated by Andrew K. Johnston. The exhibit toured nationally from 2006 to 2015 and consisted of 40 free-standing panels showing images of places on Earth from space and describing what we have learned about our planet from these global views and data sets. It is currently an online exhibition on the SITES website (Smithsonian Institution, 2006, http://earthfromspace.si.edu/online_exhibition.asp).

THE CHALLENGE FOR NATURAL HISTORY MUSEUMS IN THE TWENTY-FIRST CENTURY

The contemporary exploration of space has brought new discoveries that, like those made during the European Age of Exploration, have the potential for great consequences. We are the first generation of humans to see all the classical planets and their moons in our solar system, including our own planet from space, and natural history museums such as the NMMNHS are uniquely

Figure 9. (A) View of the exhibit *25 Years of Space Photography* on display in 2016 in celebration of the 30th anniversary of the New Mexico Museum of Natural History and Science (NMMNHS). The exhibit was designed in the style of an art gallery with serene wall colors and benches as an effort to encourage visitors to treat space images as art as well as science. (B, C). Two examples of the images in the exhibit *25 Years of Space Photography*. (B) The false-color rendition of the surface of Mercury is reminiscent of an impressionist painting. (C) The black and white radar image of the terrain on Venus focuses your attention on the patterns of lines as in an abstract piece of art. Credit: NMMNHS.

qualified to excite the public with this new geology frontier. In the twenty-first century, natural history museums can enhance their traditional role as providers of informal geoscience education by using the excitement of the space-age "voyages" of natural history explorations of new worlds. Based on the experience of the New Mexico Museum of Natural History and Science, and that of the other museums described here, planetary geology is a demonstrated method for presenting geoscience topics to a museum audience that might otherwise never explore the significance of geology. The inclusion of planetary geology also provides a way for natural history museums to continue a relevant role in informal science education as the science of geology continues to evolve and new geological discoveries are made on other planets. This will allow natural history museums to continue to be new, fresh, and relevant for future generations.

ACKNOWLEDGMENTS

We thank the following: NASA, the MER Mission Science Team, and Jet Propulsion Laboratory; Gary Rosenberg and Renee Clary, co-conveners of the GSA Annual Meeting theme session on "Museums at the Forefront of the History and Philosophy of Geology" and co-editors of this volume, for their invitation to make an oral presentation at the meeting and for

Figure 10. View of the exhibit, *Mars Landscape Art by Spirit and Opportunity,* celebrating the 10th anniversary of the Mars Exploration Rover (MER) Mission. Created in-house and on exhibit in 2014 at the New Mexico Museum of Natural History and Science (NMMNHS). The exhibit focused on the beauty of the landscape as though the two rovers were photographers on an exotic assignment on Mars. Credit: NMMNHS.

their encouragement of this paper; Deb Novak, Director of Education, New Mexico Museum of Natural History & Science (NMMNHS), for her support of space science education at the museum; Jessica Newell, Cleveland Museum of Natural History, M.M. Mader, Royal Ontario Museum, and Andrew Johnston, Smithsonian Institution, for photographs, information, and interesting discussions relevant to this paper; Margie Marino, Executive Director of NMMNHS, Jim Greenhouse, Space Science Director of NMMNHS, the New Mexico Museum of Natural History Foundation, and our NMMNHS colleagues (especially Education and Exhibits Departments) who helped make possible this museum's decade-long program of Mars-related exhibitions, education, and outreach.

REFERENCES CITED

American Museum of Natural History, 2010, Rose Center to go to space: http://www.amnh.org/our-research/science-news/2010/rose-center-to-go-to-space/.

Aubele, J.C., 1996, Planetology, Comparative, *in* Dasch, J., ed., Encyclopedia of Earth Sciences, Vol. 2 (M–Z): New York, Macmillan Publishing Company, p. 853–859.

Aubele, J.C., 2013a, Ten years of exhibits, education, and public outreach presenting the Mars exploration rover mission at the New Mexico Museum of Natural History and Science: Houston, Texas, Lunar and Planetary Institute, Lunar and Planetary Science XLIV, abstract 2926.

Aubele, J.C., 2013b, The impact of Mars exploration on teaching, outreach, and student education: A decade of education and public outreach at the New Mexico Museum of Natural History and Science using the Mars exploration rover mission: Geological Society of America Abstracts with Programs, v. 45, no. 7, p. 470.

Aubele, J.C., and Crumpler, L.S., 2016a, 21st century natural history: Planetary geology in museums: Geological Society of America Abstracts with Programs, v. 48, no. 7, https://doi.org/10.1130/abs/2016AM-286924.

Aubele, J.C., and Crumpler, L.S., 2016b, It's like Mars: Using planetary analogs to teach place-based geoscience at the New Mexico Museum of Natural History and Science: Geological Society of America Abstracts with Programs, v. 48, no. 7, https://doi.org/10.1130/abs/2016AM-286806.

Aubele, J.C., and Stanley, J., 2007, Informal science education + planetary missions = increased public support and science literacy: Houston, Texas, Lunar and Planetary Institute, Lunar and Planetary Science XXXIX, abstract 2334.

Aubele, J.C., Stanley, J., Aragon, J., Grochowski, A., Jones, K., and Crumpler, L., 2004, Increasing science literacy and public support for planetary science: MER museum exhibits, educational programming and public outreach at the NM Museum of Natural History and Science: Houston, Texas, Lunar and Planetary Institute, Lunar and Planetary Science XXXVI, abstract 2343.

Aubele, J.C., Stanley, J., Grochowski, A., Jones, K., and Aragon, J., 2006a, Making Tracks on Mars: Teacher Resource and Activity Guide, CD-ROM or DVD: Albuquerque, New Mexico Museum of Natural History and Science.

Aubele, J.C., Stanley, J., Grochowski, A., Jones, K., and Aragon, J., 2006b, Using Mars and the MER mission to teach science: A curriculum designed for teachers and their students: San Francisco, California, American Geophysical Union, Abstract 6218 presented at 2006 Fall Meeting.

Aubele, J.C., Stanley, J., Grochowski, A., Jones, K., and Aragon, J., 2012, Mars curriculum for K-12 science education, 2nd edition, Making Tracks on Mars Teacher Resource and Activity Guide: Houston, Texas, Lunar and Planetary Institute, Lunar and Planetary Science XLIII, abstract 1659.

Aubele, J.C., Shinabery, M., and Stanley, J., 2011, Increased public support and planetary geology literacy through informal science education: Lessons from 30 years of museum and science center education and exhibits: Geological Society of America Abstracts with Programs, v. 43, no. 5, p. 244.

Belloli, A.P.A., and Dean, L., eds., 1985, 25 Years of Space Photography, Exhibition Catalog: Pasadena, Baxter Art Gallery, California Institute of Technology in association with New York and London, W.W. Norton, 120 p.

Crumpler, L.S., Aubele, J.C., and Zimbelman, J.R., 2007, Volcanic features of New Mexico analogous to volcanic features on Mars, *in* Chapman, M., ed., The Geology of Mars: Evidence from Earth-Based Analogs: Cambridge, UK, Cambridge University Press, p. 95–125, https://doi.org/10.1017/CBO9780511536014.005.

Falk, J.H., and Dierking, L.D., 1997, The Museum Experience: Washington D.C., Whalesback Books, p. 97–114.

Falk, J.H., and Dierking, L.D., 2010, The 95% solution: American Scientist, v. 98, p. 486–493, https://doi.org/10.1511/2010.87.486.

Friedl, L., ed., 2012, Earth as Art: NASA, Washington, D.C., U.S. Government Printing Office, 158 p.

Golombek, M.P., Arvidson, R.E., Bell, J.F., III, Christensen, P.R., Crisp, J.A., Crumpler, L.S., Ehlmann, B.L., Fergason, R.L., Grant, J.A., Greeley, R., Haldemann, A.F.C., Kass, D.M., Parker, T.J., Schofield, J.T., Squyres, S.W., and Zurek, R.W., 2005, Assessment of Mars Exploration Rover landing site predictions: Nature, v. 436, p. 44–48.

Gomez, A., 2014 (2 Feb.), Mars: A Lot Like Home: The Sunday Journal, Albuquerque Journal Publishing Company, p. A1 and A6.

Greeley, R., 2013, Planetary morphologic processes, *in* Greeley, R., Introduction to Planetary Geomorphology: New York, Cambridge University Press, p. 34–58.

Greeley, R., and Carr, M., eds., 1976, A Geological Basis for the Exploration of the Planets: National Aeronautics and Space Administration, Scientific and Technical Information Office, NASA SP-417, 109 p.

Grosfils, E., Aubele, J.C., Crumpler, L.S., Gregg, T., and Sakimoto, S., 2000, Volcanism on Earth's seafloor and Venus, *in* Gregg, T., and Zimbelman, J., eds., Environmental Effects on Volcanic Eruptions: From Deep Oceans to Deep Space: New York, Kluwer Academic/Plenum Publishers, p. 113–142, https://doi.org/10.1007/978-1-4615-4151-6_5.

Hannibal, J.T., 1998, Rocks, minerals, planets, and more: The Cleveland Museum of Natural History's Reinberger Hall of Earth and Planetary Exploration, *in* Hansen, M.C., ed., Ohio Geology: A Quarterly Publication of the Division of Geological Survey, Summer 1998, p. 6.

Hooper-Greenhill, E., 1996, Museums and Their Visitors: London and New York, Routledge, p. 140–170.

Institute for Museum and Library Services, 2003, Symposium Overview: Washington, D.C., Report of the 21st Century Learner Symposium, September 2003, p. 4.

Knight, C., 2012 (23 April), Review: History of Space Photography is out of this world: Los Angeles Times: http://articles.latimes.com/keyword/business.

Lake Snell Perry and Associates Research, 2001, Americans' Perceptions of Museums: Key Findings: Nationwide Public Opinion Survey: American Alliance of Museums, unpublished Information Center Report available to member museums.

Leiserowitz, A., Smith, N., and Marlon, J.R., 2010, Americans' knowledge of climate change: New Haven, Connecticut, Yale University, Yale Project on Climate Change Communication: http://environment.yale.edu/climate/files/ClimateChangeKnowledge2010.pdf.

Mader, M.M., Tait, K.T., and Ireland, D., 2017, Museums and planetary missions: Houston, Texas, Lunar and Planetary Institute: Lunar and Planetary Science, XLVIII, abstract 2804.

McEwan, A.S., Hansen-Koharcheck, C., and Espinoza, A., 2017, Mars: The Pristine Beauty of the Red Planet: Tucson, University of Arizona Press, 448 p.

Moores, E.M., and Wahl, F.M., 1988, The Art of Geology: Geological Society of America Special Paper 225, 140 p., https://doi.org/10.1130/SPE225-p1.

National Academy of Sciences, 2012, A Framework for K-12 Science Education: Practices, Crosscutting Concepts, and Core Ideas: Washington, D.C., The National Academies Press, 400 p., https://doi.org/10.17226/13165.

Natural History Museum, London, UK, 2017, Discovering Earth: Science meets art (video), from the exhibit Otherworlds: Visions of our solar system: http://www.nhm.ac.uk/visit/exhibitions/otherworlds-visions-solar-system-exhibition.html.

Semken, S., 2005, Sense of place and place-based introductory geo-science teaching for American Indian and Alaskan Native: Journal of Geoscience Education, v. 53, no. 2, p. 149–157, https://doi.org/10.5408/1089-9995-53.2.149.

Short, N.M., 1975, Planetary Geology: Englewood Cliffs, New Jersey, Prentice-Hall, p. 13–15.

Smithsonian Institution, 2006, Earth from Space online exhibit: http://earthfromspace.si.edu/online_exhibition.asp.

Sudbury, P., 1996, Science by stealth, *in* Hooper-Greenhill, E., ed., The Educational Role of the Museum: London and New York, Routledge, p. 166–174.

Wilhelms, D.E., 1993, To a Rocky Moon: Tucson, University of Arizona Press, p. 1–6.

Manuscript Accepted by the Society 3 January 2018
Manuscript Published Online 6 June 2018

The Geological Society of America
Special Paper 535

Museums at the intersection of science and citizen: An example from a Silurian reef

Patricia Coorough Burke*
Peter M. Sheehan*
Geology Department, Milwaukee Public Museum, 800 West Wells Street, Milwaukee, Wisconsin 53233, USA

ABSTRACT

Milwaukee Public Museum has been presenting scientific concepts to audiences for 134 years. The exhibit methods have moved beyond specimen display and beyond the museum walls. Today, museums bring science to the public and scientific community through searchable collections databases, contextual websites, and social media. The "Silurian Reef" exhibit, with associated collections, website, and online database are examples of the ways science, public audiences, and museums interact and how these interactions have evolved over the past 130 years.

The Schoonmaker Reef in Wauwatosa, Wisconsin, was the first fossil reef recognized in North America and attracted geologists such as James Hall (1811–1898), T.C. Chamberlin (1843–1928), and I.A. Lapham (1811–1875) to recognize its importance. Fossils from this locality and others in SE Wisconsin form the large Milwaukee Public Museum (MPM) collection. The MPM Silurian diorama, reconstructing one of these reef ecosystems, was created in 1985 as part of the "Third Planet" exhibit. This was one of the earliest museum exhibits to present plate tectonics and the evolution of life as one story. MPM's Silurian collections were central to published research on the biodiversity and ecology of Silurian communities (Watkins, 1993). The collections, research, and exhibit were the foundation for an innovative website; The Virtual Silurian Reef (VSR) was developed in 1997. The VSR is an educational outreach website that explores the significance of Silurian reefs and concepts of evolution, plate tectonics, and biodiversity. More recently, MPM, in partnership with the Field Museum, digitized our Silurian collections and created a searchable online database housed on a redesigned VSR.

The collections, website, and searchable database have given academic researchers better access to specimens in both MPM and Field Museum of Natural History (FMNH) in Chicago collections. A larger impact has been the broad audiences reached. The website and exhibit have been used in National Science Foundation–funded educational outreach, by educators, artists, and fossil-hunting kids. Images and text from the website and database are found on interpretive signs in local and state parks and on trails that overlook historic collecting localities. The fossil specimens have even been used to model bronze fossil play sculptures for a city park.

*E-mails: coorough@mpm.edu; sheehan@mpm.edu.

Coorough Burke, P., and Sheehan, P.M., 2018, Museums at the intersection of science and citizen: An example from a Silurian reef, *in* Rosenberg, G.D., and Clary, R.M., eds., Museums at the Forefront of the History and Philosophy of Geology: History Made, History in the Making: Geological Society of America Special Paper 535, p. 263–271, https://doi.org/10.1130/2018.2535(17).

INTRODUCTION

Natural history museums provide the public with an informal entry to science. Observation of the natural world is guided by interpretive dioramas and curated displays of specimens. The Milwaukee Public Museum (MPM) has a long history of being at the nexus of science, education, innovative exhibition, and community. The institution's collections and exhibits began as a small educational tool and grew to become one of the country's top public museums, by changing and adapting across the decades. In this article, we will examine the changing nature of the interaction between museums and citizens. We will trace how one local fossil collection grew from a single drawer in a school room to a 20,000 specimen collection used as a tool for research, public exhibits, and education. *The Silurian Reef* collections began with the initial collecting of Peter Engelmann (1823–1874) and his students in 1851. Over a span of 166 years, the collections have been the subject of scientific collecting and research, an immersive ancient reef habitat in a curved diorama shell, educational outreach, a website, a major digitization project, a searchable online database, and an artistic rendering of fossil organisms as playground equipment.

THE BEGINNING OF A MUSEUM

The beginning of MPM collections and educational tradition started with a Prussian political refugee who came to America in 1849. Peter Engelmann was educated at the University of Heidelberg as well as Humboldt University in Berlin. Engelmann made his way to Milwaukee, a city heavily populated by German immigrants. Engelmann's background and education led him to teaching, and, in 1851, he was hired to head a newly formed school, the German-English Academy. Engelmann embraced Alexander von Humboldt's tradition of learning from nature through observation and experience. As part of his curriculum, Engelmann led his students on excursions into the countryside collecting and observing. The initial collecting focused on botany and geology specimens. The collections became part of the classroom's teaching resources and were housed at the school. The specimens were first located in a cupboard, then filled an entire classroom, and eventually, the whole assembly hall on the top floor of the school became the museum. These natural history specimens would later form the nucleus of collections used to establish the Milwaukee Public Museum (Fig. 1).

The field excursions began as educational experiences for students but generated great interest in parents and other adults. Soon Engelmann, with other curious-minded adults, formed a natural history society to pursue interest in the sciences and to maintain an object collection for the school. The society officially organized on 6 May 1857, under the name *Der Naturhistorische Verein von Wisconsin* (The Natural History Society of Wisconsin). The society collected specimens and maintained a reference library with up-to-date publications in the various sciences. In 1874, the society's annual report indicated an income of $540.99, mostly from membership, and expenses of $414.16, which included expenditures for display, books, and museum cleaning. By 1879, the organization received a state charter and incorporated under the name, the Natural History Society of Wisconsin. In 1902, the charter was amended, and the name changed to the Wisconsin Natural History Society (Thomson, 1908). The stated mission of the organization was "To arouse and sustain an interest in the study of nature and especially of facts and material pertaining to the natural history and ethnology of Wisconsin."

The society members were serious in their pursuit of knowledge. They formed curatorial committees in botany, geology, zoology, and archaeology. They held monthly meetings that often included presentations by members in their area of expertise. The society published proceedings and special papers, first in German and later in both German and English. More significant for Wisconsin's paleontology, the society encouraged a group of knowledgeable, enthusiastic amateur collectors.

The society members were actively collecting fossils at a golden moment in Milwaukee history, as the city's population tripled between 1845 and 1860. Rapid urban development created a building boom that kept local quarry operations active and exposed abundant fossils. The fossil collections amassed at this time were distributed to museums across the country. F. Holbrook Day's collection is at Harvard University's Museum of Comparative Zoology; Dr. Thomas Greene's collection is at the University of Wisconsin–Milwaukee's Greene Gallery; Edgar E. Teller's collection is at the Field Museum of Natural History (FMNH) in Chicago; C.E. Monroe's collection is at the Milwaukee Public Museum; and Increase Lapham's collection of over 10,000 geology specimens went to the University of Wisconsin–Madison (unfortunately, those were lost in the 1884 Science Hall fire).

THE NATURAL HISTORY SOCIETY AND THE SCIENTISTS

The activities of the society and its members also drew the attention of professional geologists to the area. Increase Lapham, known as the first scholar of Wisconsin for his wide-ranging studies of area botany, geology, and archaeology, corresponded with James Hall concerning fossil identifications. James Hall, head of The New York Geological Survey, as well as Robert Whitfield (American Museum of Natural History), Stuart Weller (University of Chicago), and T.C. Chamberlin studied the regional paleontology. In 1844, Lapham first recognized the Wauwatosa, Wisconsin, Schoonmaker Quarry as an area of geologic interest. He informed Hall, who recognized it as the first North American Silurian fossil reef in 1862 (Lapham's papers). Chamberlin (1877) published descriptions of the reef and inter-reef lithologic facies and their fauna in *The Geology of Wisconsin Survey of 1873–1879.* All these scientists utilized and depended upon the collections and observations made by the local natural history society and corresponded with the society members on the significance of the fossils.

A PUBLIC MUSEUM

By 1881, the society's collection had grown beyond the storage capacity of the German-English Academy. An innovative solution, not seen in other cities, was to donate the entire collection to municipal government with a declared purpose to "increase the usefulness of the museum and bring it into closer touch with the public." The City of Milwaukee willingly accepted the donation once state-enabling legislation passed. On 13 April 1882, the city was empowered to accept the donation and establish a free public museum for the "exhibition of objects of Natural History and Ethnology, and for the purpose to receive, hold and manage the collections so donated" (Laws of Wisconsin, 1882). The Milwaukee Public Museum opened to the public in 1884 with a mission to educate the public and to be open to citizens without charge.

The move from private to public and from classroom to cabinet display required some adjustments. Attendance soared, but the objects displayed without a teacher or guide highlighted the need for interpretive materials. Initially, tours led by knowledgeable guides and labeled cabinet displays helped to put the specimens in context.

Innovation in museum display was needed. At the Milwaukee Museum, that innovation began with the display of a Lapland sled collection in 1887. The sled came as part of a lot including reindeer skins and antlers, a sled and harness, and some costume items. Traditional display methods of the time would place the taxidermied reindeer on a display base grouped with other natural history specimens, and the sled along with the other pieces would be displayed with ethnological collections. Instead, Carl Akeley (1864–1926) (see Rosenberg, this volume) (Fig. 2), the taxidermist and artist commissioned to produce the exhibit, harnessed the reindeer to the sled with a fur-clad mannequin holding the reins on a base made to simulate snow.

In 1889, Carl Akeley became a full-time museum artist and began work on his first habitat display. Akeley used a muskrat

Figure 1. Children viewing mastodon on display at Milwaukee Public Museum in the early 1900s. Courtesy of Milwaukee Public Museum (https://www.mpm.edu/).

Figure 2. The young exhibit innovator, Carl Akeley; photograph taken in Milwaukee at Stein Studios, circa 1886. Courtesy of Milwaukee Public Museum (https://www.mpm.edu/).

colony as a demonstration model to show museum trustees his exhibit vision. The display, completed in 1890, was a muskrat group in a curved diorama shell with three-dimensional materials in the foreground blending into a painted background. Akeley did not stop with just the surface view. He created a cross section of the muskrat lodge against the glass front of the exhibit to display the interior structure of the lodge and the subsurface entryway. The exhibit went beyond object display, presenting the interaction between plant, animal, and environment. No one who has seen the muskrat diorama could ever again look at a muskrat lodge in nature without imagining the underwater entry. This landmark display is still on view at the MPM (Fig. 3).

Habitat displays became a signature style for the Milwaukee Public Museum. Utilizing artistic dioramas to help in scientific understanding of the natural world was a successful educational tool throughout the museum. The exhibition technique was deployed in 1977 when the MPM began planning a major renovation to its 1940s Geology Hall. The organizing theme of the new exhibit was the recent but still somewhat controversial theory of plate tectonics. The vision was to create

Figure 3. Milwaukee Public Museum muskrat diorama made by Carl Akeley in 1890. The display is approximately 2 meters wide by 1.5 meters high. Courtesy of Milwaukee Public Museum (https://www.mpm.edu/).

Figure 4. "Silurian Reef" exhibit in Milwaukee Public Museum's "Third Planet" exhibit was created in 1985 and is 7 meters wide and 4 meters high. Courtesy of Milwaukee Public Museum (https://www.mpm.edu/).

time-specific habitat displays of life through geologic time, including a Silurian reef, a Carboniferous forest, a Cretaceous riverine habitat, and lastly a Pleistocene glacier. The Silurian reef was particularly important because the bedrock under most of Milwaukee is Silurian and has been the target of quarrying operations and local fossil collecting for more than 160 years. The reef diorama shows Wisconsin 430 million years ago when the state was covered by a shallow, tropical sea that extended across much of what is now the Midwestern United States. The organisms and the reef habitat are based on locally collected fossil evidence. The exhibit uses the "Milwaukee style" (Lurie, 1983, p. 16), pioneered by Akeley, in which background murals seamlessly blend into a three-dimensional foreground with life-like plastic and wax animal models based on fossils in the museum's collection. On display beside the reconstructed reef diorama are the fossils used to model the exhibit animals (Fig. 4).

The Silurian model reef organisms, produced by exhibit fabricator, Chase Studios, are based largely on fossils from the Middle Silurian Racine Formation of southeast Wisconsin. The reef exhibit contains 57 different models representing 13 major groups of reef organisms, including stromatoporoids, tabulate and rugose corals, bryozoans, brachiopods, bivalves, gastropods, cephalopods, cystoids, crinoids, and trilobites. The exhibit places animal models in a shallow marine habitat and projects water-wave ripples–effect stage lighting to produce an underwater atmosphere (Fig. 5). The visual effect draws visitors to pause and invites them to study both the diorama and the fossils. The most ferocious predators of the time—four-foot-long orthoconic nautiloid cephalopods—can be observed swimming and feeding on

Figure 5. Close-up of "Silurian Reef" exhibit. Courtesy of Milwaukee Public Museum (https://www.mpm.edu/).

trilobites, providing the audience with ideas about the paleobiology of these 430-m.y.-old fossils. The exhibit image has become iconic. It can be seen in trail signage and park exhibits around the Great Lakes and has been incorporated into educational websites and wiki pages.

Paleoecological Study on a Historic Collection

Shortly after the "Third Planet" exhibit opened, Dr. Rod Watkins arrived as a curator in the geology section (Fig. 6). Watkins brought his expertise in the Silurian paleoecology of the Welsh Borderlands faunas (Watkins, 1979) to the study of MPM's Silurian collection. He began the monumental task of curating the Silurian fossil collection. Over the decades, large limestone blocks and slabs had been collected and deposited in the museum's collection with just a note of where they had been collected. Watkins saw the potential for a paleoecological study of the Wisconsin Silurian reefs in these dusty overlooked boulders. During the next seven years, he would split these limestone blocks down into more than 11,000 Silurian reef and inter-reef individual fossil specimens that he identified and placed in the collection. Using this curated collection, Dr. Watkins wrote more than 25 scientific articles, including "The Silurian (Wenlockian) Reef Fauna of Southeastern Wisconsin" (Watkins, 1993), which summarized and interpreted the fossils found at localities throughout the area (Fig. 7).

Bringing the Museum to the Public Electronically

The Silurian reef and inter-reef fossils were well documented in scientific journals (Watkins, 1993, Watkins and Coorough, 1997; Watkins et al., 2007), and Milwaukee area school children enjoyed the immersive reconstructed reef on visits to the museum, but the exposure beyond the museum walls was limited. The World Wide Web changed that. In 1996, MPM's geology section applied for and received a grant from the Wisconsin Advanced Technology Fund to create an educational outreach site around the Silurian reef diorama and associated fossil collection. The project was a model for 3D, interactive, distance-learning environments at an early stage of the Internet. Using the Silurian fossil collections, current research, local rock outcrops, and the immersive Silurian reef diorama, geology staff created *The Virtual Silurian Reef* (VSR) (https://www.mpm.edu/content/collections/learn/reef/). The site provided online resources and activities for teachers, and most importantly, it opened the doors of a building and let the world stroll through digitally.

Natural history museums face a constant dilemma in presenting exhibit content, with the need to balance specimen display with interpretive material and explanatory text. The average visitor spends 15–30 seconds at an exhibit, and this time is coveted both by curatorial staff wanting to share detailed information and display artists wanting to protect the visual experience with minimal signage. The VSR website freed the geology staff from the constraints of exhibit text to create a content-rich resource. In the educational tradition of MPM, the creators took the opportunity to produce a site that went beyond pretty images. The VSR uses reefs as a vehicle to introduce geologic concepts, including plate tectonics, biodiversity, and environmental change. The site has more than 50 themes to follow with images and animations to aid in comprehension.

The "Silurian Reef" exhibit, website, and collection provided an excellent platform for the 1998 National Science Foundation (NSF)–supported *Science Explorations: An Enrichment Program for Middle School Girls*. The program targeted young women at risk of dropping out of math and science with after-

Figure 6. Dr. Rodney Watkins working on the Silurian collection in 1986. Courtesy of Milwaukee Public Museum (https://www.mpm.edu/).

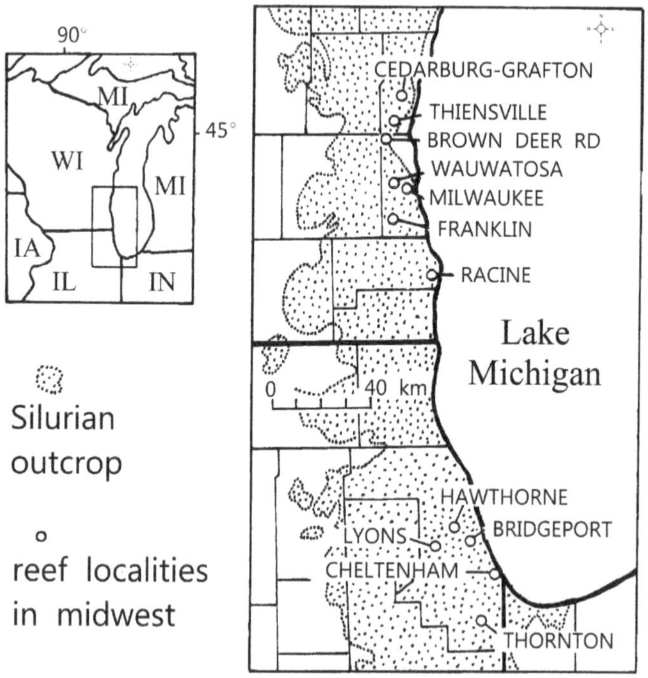

Figure 7. Milwaukee area Silurian localities map. Courtesy of Milwaukee Public Museum (https://www.mpm.edu/).

school enrichment programming. As part of the geology study, the students conducted a paleodiversity study using the reef diorama as their study area. In one activity, the students were tasked with determining if the "Silurian Reef" exhibit's diversity reflected collection data results found in Watkins' 1993 paper. The students studied the biodiversity pages on the VSR website to establish a method to measure the diversity in the reef exhibit. One afternoon was spent counting the animals in a gridded exhibit window. The data were compiled and then compared with results in Watkins' research. The process allowed the students to understand biodiversity by collecting their own data. The young women presented their results as posters in the museum at the completion of the program.

Handwritten Paper Records to Digital Database

Ever since the first catalog entry, technology has been changing museums and the way collections are utilized. The Silurian reef fossil collection records at MPM began as a classroom collection cataloged by shelf locations or cabinet numbers. Later, each discipline maintained its own catalog book, and subsequently, each discipline established subdisciplines. The geology collections have handwritten catalog books for minerals and rocks, meteorites, paleobotany, vertebrate paleontology, and invertebrate paleontology. A series of National Science Foundation (NSF) grants, from 1983 through 1991, improved collection facilities and access to collection information by developing in-house computer cataloging.

The Silurian reef fossil collection is the first of MPM's geology collections to be brought into the twenty-first century through an Institute of Museum and Library Services (2010–2013)–supported digitization grant. The project was a cooperative venture with The Field Museum of Natural History (FMNH) to convert handwritten paper-based records for more than 31,000 specimens to electronic form. The conversion makes the information about the fossil specimens more easily searchable and preserves associated data. The handwritten labels and fossils were photographed. The images along with the specimen data were put into a specialized museum collection management system (EMu). The digitization effort has made it possible to share collection data with aggregators that publish and share biodiversity data sets, such as Integrated Digitized Biocollections (iDigbio) and Global Biodiversity Information Facility (GBIF).

As part of the digitization effort, the Virtual Silurian Reef pages were given a much needed redesign. The limits of 1990s computers and bandwidths restricted image resolution and interactivity on the original site. The site was static, but the content remained accurate. In planning the rebuilding of the site, the focus was to maintain content, improve image resolutions, make the content easily updatable, and keep maintenance requirements minimal.

We were able to utilize a built-in feature of the EMu collections management system—the narratives module—to populate web page templates. The process integrated the collection management system and the educational web content for efficient use of digital assets and collection's staff time. The update brought a new layout, higher-resolution images, greater interactivity, and new scientific illustrations. The refreshed Virtual Silurian Reef (http://silurian-reef-dev.fieldmuseum.org/narrative/392) hosts a searchable online database that allows any visitor to view MPM and FMNH Silurian collections data and images from within the content-rich website.

OUT OF THE MUSEUM AND INTO THE NEIGHBORHOOD

As a tool for making people aware of the natural world, the Silurian reef fossils have been utilized for more than 130 years. Several recent efforts have highlighted and preserved the history of scientific discovery related to these fossils. A unique project involves development of a park on a portion of a 19 acre former quarry site. The Hartung Quarry neighborhood straddles the west side of Milwaukee and east side of Wauwatosa and includes the former Hartung Quarry.

Frederick Hartung began a stone quarry operation on his farm in 1910, and, during the next 54 years, he removed approximately three million tons of rock for use in buildings and road construction. The stone in the quarry came from two fossiliferous Silurian formations. MPM holds a large collection of Hartung Quarry fossils that were collected during the years of stone removal. Quarry operations stopped in 1964, and, for the next 40 years, the quarry pit was used as a landfill for City of Milwaukee construction debris. By 2005, the landfill had reached capacity, and the site was closed. During the same 95 year period, Milwaukee and Wauwatosa had developed westward, and residential homes surrounded the quarry.

As the municipalities contemplated what to do with the filled quarry site, the neighbors organized, forming the Hartung Park Community Association. The group successfully lobbied the local governments to work cooperatively to create a park and preserve local history on the former Hartung Quarry site. As the park planning began, the community association wanted to make sure the connection between the quarry's ancient past was not buried by the modern park. A vision for something more interactive than a plaque led to the idea of a play area that included fossils in some form.

In 2008, MPM geologist, Dr. Peter Sheehan, was contacted by a former student and Hartung Park neighborhood resident, Michael Cape. Mr. Cape wanted to bring Laura Priebe, a neighbor and local sculptor, to meet Sheehan and get some background on Hartung fossils for a public art project. The artist was planning to create oversized playground sculptures of the animals that lived in the shallow seas of the area 430 million years ago. She needed to learn what they looked like and how they lived. During the next two years, MPM geology staff assisted the endeavor by sharing their knowledge of the Hartung fauna, answering questions, and loaning fossil specimens and images to the artist.

Figure 8. Hartung Park's "Colossal Fossil" play area. Courtesy of Milwaukee Public Museum (https://www.mpm.edu/).

The "Colossal Fossil" play area was unveiled in 2010 and includes seven oversized bronze models of Silurian fossils ranging in size from two feet to five feet across. The sculptures include a trilobite, a brachiopod, a crinoid, a coral, and a cephalopod sitting on limestone slabs. Along with the sculptures, three educational information boards tell the story of the fossils, utilizing images and text from the MPM Virtual Silurian reef website (Figs. 8–11).

Another local Silurian site highlighted by area preservationists is the original Schoonmaker reef quarry, identified by James Hall as a fossil reef. The site gained National Historic Landmark status in 1998. Until recently, the site remained a brushy bluff, strewn with rubble and under private ownership. In a cooperative

Figure 10. Hartung Park's "Colossal Fossil" trilobite sculpture. Courtesy of Milwaukee Public Museum (https://www.mpm.edu/).

Figure 9. Hartung Park's "Colossal Fossil" crinoid sculpture. Courtesy of Milwaukee Public Museum (https://www.mpm.edu/).

Figure 11. Hartung Park's "Colossal Fossil" eurypterid sculpture. Courtesy of Milwaukee Public Museum (https://www.mpm.edu/).

initiative between the City of Wauwatosa and a local developer, the site to the south of the bluff has been developed into a housing complex—appropriately named "the Reef"—with a public trail skirting the northern exposure of the old quarry wall, including exposures of the Schoonmaker reef.

CONCLUSIONS

Though the original German founders of MPM might not recognize their creation today, they would be pleased to know that the endeavor grew into a major museum that produced an enormous body of scientific work, informed millions of visitors, and can now reach curious students and scientists across the globe. The museum has adapted to changing technology and visitor demographics but has remained true to the original mission of arousing and sustaining an interest in study of the natural world. The Silurian reef collection and its many uses chronicle how museum collections remain relevant and useful through decades. Scientific papers document the collection with analysis, while the "Silurian Reef" exhibit presents fossil evidence with visual art, and the "Colossal Fossil" playground brings the past into playful everyday life.

Museums and the collections at their foundation remain a trusted information source in a world of alternative facts. The MPM's mission remains pertinent today even as the tools to achieve the mission have expanded beyond exhibits to include websites, searchable databases, and more recently social media tools. We will continue to be a place to see authentic objects and get help understanding them.

REFERENCES CITED

Annual Report for The Natural History Society of Wisconsin, 1874, https://books.google.com/books/ Bericht der Naturhistorische Vereins von Wisconsin 1874 (July 2017).

Chamberlin, T.C., 1877, Geology of eastern Wisconsin, *in* Chamberlin, T.C., ed., Geology of Wisconsin Survey of 1873–1877, v. 2: Madison, printed by authority of the Legislature of Wisconsin, p. 93–405.

Hall, J., 1862, Physical geography and general geology, *in* Hall, J., and Whitney, J.D., Report on the Geological Survey of the State of Wisconsin, v. 1: Albany, printed by authority of the Legislature of Wisconsin

Increase A. Lapham Papers, Wisconsin Historical Society Archives; http://content.wisconsinhistory.org/cdm/landingpage/collection/p15932coll7 (March 2017).

Laws of Wisconsin, Chapter 328, published 14 April 1882.

Lurie, N.O., 1983, A Special Style: The Milwaukee Public Museum 1882–1982: Milwaukee, Milwaukee Public Museum, 150 p.

Rosenberg, G.D., 2018, this volume, Carl Akeley's revolution in exhibit design at the Milwaukee Public Museum, *in* Rosenberg, G.D., and Clary, R.M., eds., Museums at the Forefront of the History and Philosophy of Geology: History Made, History in the Making: Geological Society of America Special Paper 535, https://doi.org/10.1130/2018.2535(12).

Thomson, J.V., 1908, Handbook of Learned Societies and Institutions: Carnegie Institution of Washington, Publication Number 39, p. 418.

Watkins, R.W., 1979, Benthic community organization in the Ludlow Series of the Welsh Borderland: British Museum of Natural History (Geology) Bulletin, v. 31, p. 178–277.

Watkins, R.W., 1993, The Silurian (Wenlockian) reef fauna of southeastern Wisconsin: Palaois, v. 8, p. 325–338, https://doi.org/10.2307/3515264.

Watkins, R., and Coorough, P.J., 1997, Silurian sponge spicules from the Racine Formation, Wisconsin: Journal of Paleontology, v. 71, no. 2, p. 208–214, https://doi.org/10.1017/S0022336000039147.

Watkins, R., Mayer, P.S., and Coorough, P.J., 2007, Silurian of the Great Lakes Region, Part 5: Sedimentology and paleontology of the Silurian Waubakee Formation, Milwaukee and Ozaukee counties, Wisconsin, Milwaukee Public Museum: Contributions in Biology and Geology, no. 103, p. 68.

Manuscript Accepted by the Society 21 November 2017
Manuscript Published Online 25 May 2018

Printed in the USA

The Geological Society of America
Special Paper 535

The Children's Museum of Indianapolis: A history of leveraging field expeditions and lab work to enhance public engagement

Dallas C. Evans
The Children's Museum of Indianapolis, 3000 North Meridian Street, Indianapolis, Indiana 46208, USA

Abstract

Any child that has been to a museum, gone stomping through a creek, or gazed at the stars knows that science learning isn't confined to a classroom.

Children are eager to explore the wonders of the natural world, and parents and teachers value the importance of science education—thus, The Children's Museum of Indianapolis (TCMI) has collected science objects and conducted fieldwork since it was first established.

During the 1930s, museum staff members drove equipment-laden Model T cars on expeditions known as Prairie Treks. Indianapolis schoolchildren were given the chance to venture to the western United States and investigate the plants and wildlife of the region. Campers learned to identify birds and animals, pan for gold, make plaster casts of dinosaur footprints, and collect fossils and rocks to add to the museum's collection.

The natural science collection at TCMI is composed of more than 10,000 unique objects that help foster both curiosity and enthusiasm for the sciences. Science is an intensely hands-on and investigative endeavor, and this is reflected in the scope and the use of the objects in the collection. Items related to zoology, botany, and geology provide core materials that are utilized throughout the museum in exhibits, programs, and interpretation.

TCMI is closing in on its 100-year history. Its unique mission, as the world's largest children's museum, helps it to provide public engagement with the geosciences. Today each year more than 1.2 million visitors can experience programs ranging from self-guided discovery to active participation with scientists and their current research. Thousands have joined dinosaur excavations in the rocks of the Hell Creek Formation of South Dakota, prepped fossil materials in the Paleo Prep Lab, and even assisted in collection-based research.

Evans, D.C., 2018, The Children's Museum of Indianapolis: A history of leveraging field expeditions and lab work to enhance public engagement, *in* Rosenberg, G.D., and Clary, R.M., eds., Museums at the Forefront of the History and Philosophy of Geology: History Made, History in the Making: Geological Society of America Special Paper 535, p. 273–287, https://doi.org/10.1130/2018.2535(18).

EXPLORING THE HELL CREEK FORMATION AT THE RUTH MASON QUARRY

The adventure begins in Faith, South Dakota—population 489. Less than a dozen miles north of this small town, among the prairie grass and pockets of heavily eroded badlands is our field site, the Ruth Mason Quarry. This remote dig site contains dinosaur bones by the thousands, and almost all of them are from the same type of animal, the duck-billed dinosaur known as *Edmontosaurus*.

It's a site that is well suited for science education; the bones are sturdy, and the mudstone matrix is soft. This site is primarily monospecific; almost every fossil recovered is from an *Edmontosaurus annectens*. However, there is also the scattered evidence of theropods, and the occasional serrated shine of tyrannosaur teeth may be spotted among the disarticulated duck-billed dinosaur bones. For well over a decade, The Children's Museum of Indianapolis (TCMI) has excavated this locality, bringing children as young as eight years old and their families to unearth and study the fossils buried there.

Interestingly enough, this fossil site was first discovered by an even younger child, a seven-year-old girl exploring her family ranch in the early 1900s. While climbing among the steeply cut bluffs that overlook the meandering Moreau River, a young Ruth Mason spotted what she thought might be fossilized bones.

The property where she made her discovery lies in the northwestern portion of South Dakota and borders the Cheyenne River Sioux Reservation. In this area, the rocks of the Maastrichtian-age Fox Hills Formation and those of the Hell Creek Formation lie exposed. Here the dinosaur-bearing outcrops of the Hell Creek consist of colorful bands of planar and crossbedded sandstones, mudstones, and lignites. Unbeknownst to the young Ruth Mason, the family ranch was situated on one of the most prolific Late Cretaceous dinosaur strata in North America.

While Ruth Mason was keenly interested in finding out more about her discovery, it proved difficult to get others to share her enthusiasm. Through the years, she made several attempts to verify her find, but she couldn't locate anyone willing to take the time to visit the remote ranch. It wasn't until after her 80th birthday, during a chance conversation with Peter Larson of the Black Hills Institute for Geological Research, that she found someone with the knowledge and willingness to examine the site (Fiffer, 2000).

When Larson and the staff from Black Hills Institute started to investigate, the size and scope of the discovery began to be revealed. It was obviously a scientifically significant site, a large bone bed of disarticulated hadrosaur remains extending over many acres and containing hundreds if not thousands of animals. Subsequent research revealed the origins of the depositional environment to be that of a meandering freshwater fluvial system (Christians, 1992).

As the friendship between Mason and Larson developed, an agreement was made to conduct extensive excavations at the site. Larson along with his staff and volunteers would dig at this quarry from 1979 to 1991, efficiently excavating thousands of dinosaur bones. Through their efforts, composite hadrosaur skeletons from the Ruth Mason Quarry would grace museums in the United States, Japan, Wales, Ireland, and the Netherlands (Larson and Donnan, 2002).

After the death of Ruth Mason and with other projects looming, Larson ended his time at the quarry. (Larson was also deeply embroiled in legal actions linked to the excavation of the *Tyrannosaurus rex* "Sue" from an adjacent property.) Others would dig at Ruth Mason Quarry, including crews from the University of Wisconsin at Madison and the University of Chicago. However, it wasn't until 2001 that staff members from The Children's Museum of Indianapolis would first venture to the site, meet the private landowners (the heirs of Ruth Mason), and evaluate the site for its paleontological and educational opportunities.

Museums must cautiously proceed when committing to active field programs. There is much to consider, such as legal due diligence, relevance to mission, transportation, accommodations, staffing, safety, and security. These issues take on a greater importance when participation by children and their families enters into the mix. Luckily, TCMI had a long history as both a collecting institution and in creating family programs, or in the parlance of South Dakota—"this wasn't our first rodeo."

MARY STEWART CAREY AND THE ORIGIN OF THE MUSEUM

When The Children's Museum of Indianapolis was first established in 1925, it was due to the work and dedication of Mary Stewart Carey (1859–1938). She was influential, civic minded, and a staunch advocate for childhood education.

Mary Stewart Carey was well known for her involvement in community leadership and her work with several local educational and charitable organizations. In 1922, she donated a 40 acre orchard as a site to establish a private school, giving it the name, The Orchard School. Carey was also recognized as the driving force behind the development and legislative adoption of the Indiana state flag (Boddenhammer and Barrows, 1994). But it was a trip to the Brooklyn Children's Museum that provided Carey with the inspiration to seek the establishment of a new museum dedicated to the children of Indianapolis.

At that time, children's museums were relatively new on the cultural landscape. Established in 1899, the Brooklyn museum was the nation's first, followed by one in Boston and another in Detroit. While most "adult" museums were dedicated to the active collection and preservation of objects, their exhibitions primarily consisted of passive viewing. It was in Brooklyn where youth were first encouraged to pursue their own interests, to discover and learn, and even actively manipulate items from the museum collection (Din, 1999). After witnessing this interaction, Carey was motivated to create another such museum, the nation's fourth.

With the support of the community, The Children's Museum of Indianapolis opened in a two-story carriage house belonging to the Propylaeum, a local civic club. Soon a board of trustees was established, and Carey became the first museum president.

Later, her Meridian Street mansion would become the site of the museum from 1927 through 1946.

Carey and other board members sought to create a hands-on space that was intellectually engaging and with access to artifacts and objects to inspire creativity. The first objects donated to the museum were from the Carey household, but the board members soon made an appeal for donations from the community. When the museum first opened its doors, they had a collection of over 600 science and cultural objects donated by area schoolchildren and their families. Many of the items still belong to the museum's collection today and include such natural objects as zoological specimens, minerals, locally collected fossils, and marine shells.

HILLIS HOWIE AND THE PRAIRIE TREK EXPEDITIONS

In 1926, a 23-year-old craft and nature study instructor at The Orchard School in Indianapolis began developing plans for a summer program for school children. Hillis Howie (1903–1982), having read about the proposal to pave Route 66, had the grand and ambitious vision to organize annual expeditions to the American West. The plan was to camp, live off the land, and explore the nature and cultures of the area before they were inexorably changed by the completion of the paved highway.

As both a school teacher and scoutmaster for Boy Scout Troop 18, Howie was able to find students eager to participate that first year. He recruited former Children's Museum employee, Stuart Springer, as his assistant, and together they acquired two Model T cars for the journey. Laden with gear and luggage, and with added canvas water bags hanging from the back, the cars helped transport the two adults and nine boys on a roundabout adventure of nearly 6000 miles.

That was the beginning of a series of trips that Howie would call the Prairie Treks.

At first, these trips were primarily an extension of Howie's scout troop, but in 1930, as a board member of the museum, he proposed to continue the expeditions under the banner of The Children's Museum to "seek and secure archaeological and natural history exhibits for the museum" (Kriplen, 1982, p. 43).

As an instructor, Howie was strongly influenced by the writings of the early twentieth-century philosopher and educational reformer, John Dewey (1859–1952) (Shields, 2011). Arguably one of the most influential educational thinkers of his era, Dewey's core concepts on formal education and childhood learning emphasized that people learn best through experiential and inquiry-based education (Dewey, 1910) and the importance of actively involving students in real-life challenges (Radu, 2011). Thus, for Howie, these trips would take on a progressive and pragmatic approach, emphasizing the particular interests of the children (Fig. 1).

Those who participated in the Prairie Treks were each asked to pick a specific subject of interest and study—with a focus in geology, mammalogy, ornithology, herpetology, archaeology, ethnology, photography, or journalism. So while they were busy camping, hiking, and canyoneering, the children were also busy cooking, cleaning, and setting up camp and then pursuing their areas of interest by studying, journaling, and collecting. Soon items were being brought back from the trips, added to the museum's collections, and used in programs and exhibits. Those who chose journalism were responsible for writing articles and updates that would be published in the local newspaper, *The Indianapolis Star*.

In the summer of 1934, Howie acquired a sizeable ranch among the foothills of the Zuni Mountains of New Mexico; he named it Cottonwood Gulch. This property would act as a base camp for the Prairie Treks and would later become the home of an educational foundation sponsoring wilderness expeditions and outdoor programs.

Many decades later, it can be difficult to grasp the impact that Howie's work may have had on those young participants. Luckily, some indications of his influence remain. One of the students on the Prairie Trek of 1938 would later become a novelist and pen over a dozen books including the iconic *Slaughterhouse-Five* and *Galápagos*. When asked about the trip during an interview, Kurt Vonnegut Jr. recalled, "It took me a long time to realize what a great man Hillis Howie was. That's part of the American experience to suddenly come across a truly great person who never becomes rich or famous, but who is enormously beneficial just to those near him. Hillis Howie was such a person, a great naturalist…" (Nuwer, 1987, p. 18).

Vonnegut would later dedicate the novel *Galápagos* to Hillis Howie (Vonnegut, 1985) and in an interview indicated that Howie was the inspiration behind the character, Mary, the schoolteacher heroine of that novel.

Figure 1. Hillis Howie (standing, third from right) and members of the 1930 Prairie Trek expedition readying to depart the museum for their western adventure. This and all subsequent photos courtesy of The Children's Museum of Indianapolis.

Howie led the museum-sponsored trips from 1930 to 1942, when World War II intervened, and then one final Prairie Trek for the museum took place in 1946. Howie would then move on to graduate school at Yale University and have a long and successful career in teaching, school administration, and outdoor education (Winslow, 1982).

Through these summer expeditions, Hillis Howie was able to create an adventure on a grand scale, transform the lives of the children involved, and set the tone for science education for an entire institution. While the museum would continue with public engagement in the sciences, outdoor programs, and local collecting trips, it would take several decades until the museum was once again ready to embark on month-long expeditions to collect geological materials in the western United States.

Much has changed at The Children's Museum of Indianapolis since the days of Mary Stewart Carey and Hillis Howie, but the core principles and practices that guided their vision continue to this day.

THE MUSEUM TODAY

Today, the Children's Museum of Indianapolis is a 482,950-square-foot facility situated on 29 acres. Its mission is to create extraordinary learning experiences across the arts, sciences, and humanities that have the power to transform the lives of children and families. With over 1.2 million visitors a year, The Children's Museum of Indianapolis is the world's largest museum dedicated to children and their families (Fig. 2).

Just as in the days of Mary Stewart Carey, the museum seeks to educate, enlighten, and engage through innovative exhibits and programs. TCMI has continued to selectively add to its collection, using objects to inspire imaginations and connect generations. The bulk of the collection, more than 120,000 artifacts, represents the arts and humanities, while an additional 10,000 specimens are curated within the natural science collection.

Visitors enjoy exhibits in 11 major galleries, which include *Fireworks of Glass* featuring the work of contemporary glass artist Dale Chihuly; *The Power of Children: Making a Difference,* which shares the extraordinary stories of Anne Frank, Ruby Bridges, and Ryan White; and *National Geographic Treasures of the Earth,* an immersive exhibit exploring three famous archaeological sites—the tomb of ancient Egyptian pharaoh Seti I, the site of China's Terracotta Warriors of Xi'an, and the Caribbean shipwreck of Captain Kidd's Cara Merchant. The museum also creates the *Take Me There* exhibit series allowing families to discover the history, art, traditions, and family life of other cultures. The first was *Take Me There: Egypt* showing the mosques, museums, marketplaces, archaeological digs, and homes of Egypt. The next was *Take Me There: China,* one of the largest exhibits on contemporary China ever produced in the United States. Additionally the museum hosts international traveling exhibits, such as *Tutankhamun: The Golden King and the Great Pharaohs* and *Terracotta Warriors: The Emperor's Painted Army.*

Science exhibits within the museum include *Dow AgroSciences ScienceWorks,* where children and their families can explore the work of real scientists such as hydrologists, biologists, and geologists. They can also visit *Beyond Spaceship Earth with the Schaefer Planetarium and Space Object Theater,* where they discover the story of space exploration from the National Aeronautics and Space Administration's Project Mercury program, which sent the first Americans into space to the International Space Station. The museum has an exhibit devoted to paleontology called, *Dinosphere: Now You're in Their World,* a one-of-a-kind exhibit designed to immerse visitors into the Late Cretaceous period.

Dinosphere: Now You're in Their World

As visitors walk down the gentle ramp into *Dinosphere,* they hear the muffled noise of large animals in the distance, the sound of humming insects, and crashing footsteps. The use of dramatic sound, light, and theming helps to transport visitors back in time to the Late Cretaceous.

Peals of thunder and brief flashes of lightening cause visitors to glance up to see a thunderstorm roll across the domed ceiling of *Dinosphere*. Also projected on the dome are vivid sunrises, a darkened night with a full moon, and the occasional low-flying Pteranodon. Against this backdrop, the dinosaurs are presented not as a static skeletal display, but in a lush environment of pines, ferns, cycads, and magnolias. *Dinosphere* is multisensory and convincingly immersive, giving visitors a feel for the Late Cretaceous environment.

In an article for *Curator* magazine, Charity Counts discusses exhibits in terms of "the spectacle" or even "spectacular design," where museums incorporate design principles often associated with theater productions and theme parks to create a greater depth of experience. "Spectacular Design is a method for producing an exhibition that is intended to transport visitors to new locations, engage them in a story being told, and have lasting impact" (Counts, 2009, p. 281).

The *Dinosphere* gallery is in a building particularly well suited as a dramatic setting to engage the senses, although the space started out as something much different. When it first opened in 1996, it was a 300-seat, large-format theater called the Cinedome. However, just a few years after completion of the Cinedome, attendance was already lagging; the construction of another IMAX theater just a few miles away at the Indiana State Museum, coupled with the costly maintenance of the projection equipment, left museum staff searching for better uses for the space. This was the impetus for the creation of *Dinosphere: Now You're in Their World.* Repurposing that domed theater to create a multi-level, multisensory environment was no small task; it took four years of planning and two years of construction at a price tag of US$23 million. *Dinosphere* opened in June of 2004 (Fig. 3).

Despite the theatrical use of setting, light, and sound, the dinosaurs are the true stars of the show. In this space the actual fossils, unearthed from the Cretaceous rocks of North America are placed on exhibit. Children and their families can see real dinosaur bones up close and personal.

Figure 2. The Children's Museum of Indianapolis as it stands today.

"Is it real?" is an often-heard question asked by children as they enter *Dinosphere*. During the initial planning stages of the gallery, variations of that same question were apparent during early surveys, interviews, and with focus groups. Front-end evaluations revealed that visitors voiced a consistent desire: they wanted to see real dinosaur bones. In response to this desire, whenever possible, the actual fossil specimens are on display. Next to each dinosaur is a sign clearly denoting whether the dinosaur is cast material or fossil. The ability of museums to provide authentic experiences is an important factor in visitor satisfaction. Rebbert (2007, p. 1322) observed that "In a world where virtual experiences are ever increasing, museums provide tangible encounters with real objects."

To create *Dinosphere,* the museum sought the assistance of a group of international advisors with backgrounds in vertebrate paleontology, exhibition, and informal education. In the

Figure 3. An adult *Alamosaurus* with two juveniles burst out of the dome. These sculptures, by artist Brian Cooley, helped dramatically signify the transition from large format theater to *Dinosphere* gallery.

role of senior advisor was Robert Bakker, working with fellow paleontologists Phil Currie, Paul Sereno, Pete Larson, Neal Larson, and Dong Zhiming. Key among the advisors were also paleobotanist Eva Koppelhus; paleoartist Michael Skrepnick; paleoart collector John Lanzendorf; museum educator David Cassady; and renowned experts in informal education Lynn Dierking and John Falk.

The advisors provided critical evaluations and label review and offered their expertise with program and interactive development. They indicated institutions that were leaders in the field in terms of paleontology exhibits, public programs, public prep labs, and citizen science. They helped to arrange contacts and site visits with such iconic institutions as the American Museum of Natural History, Royal Tyrrell Museum of Paleontology, The Field Museum, the Page Museum, and the Denver Museum of Nature and Science. The advisors also helped to direct specimen acquisition and assisted the museum in navigating the often tricky relationships within academic, commercial, and amateur paleontology. They helped to ensure that specimens came with the required metadata (such as site maps, stratigraphic data, legal due diligence, etc.) and aided in bringing these scientifically important fossils into the public trust. Years after the gallery opening, the advisory board continues to guide the development of *Dinosphere*.

The design and interpretive strategies utilized in *Dinosphere* help generate opportunities for discovery and conversation (Fig. 4). The floor plan is nonlinear, allowing families to freely roam and take the path that most gains their interest. Labels are brief and fact based, and there are "dig deeper" touch screens available for those wanting to explore more about the dinosaurs, their discovery, preparation, and display. The gallery is collections rich, providing opportunity for visitors to touch both fossils and casts.

Three main dinosaur scenes serve as the focal point for the exhibit underneath the dome:

"T. rex *Attacks*"

This display contains a classic confrontation of the Late Cretaceous—two hungry tyrannosaurs, an adult and juvenile, are stalking a *Triceratops*. What could be the possible outcomes from this scene? Two against one seems like unfavorable odds, but the size of the ceratopsian is formidable. Here visitors are encouraged to investigate and come to their own conclusions.

The main attraction in this scene is the subadult *Tyrannosaurus rex* known as Bucky. It's named after rodeo cowboy Bucky Derflinger, who discovered the remains of the dinosaur when driving cattle across his father's ranch. Found among the bones of this *T. rex* was a furcula, a small boomerang-shaped element formed by the fusion of two collarbones—also known as a wishbone. The skeleton is 40% complete, which is quite good for a theropod from the Hell Creek Formation, but visitors often find it shocking when they begin to understand the paucity of the fossil record (Fig. 5).

The display offers several starting points for discussion among visitors. Conversations centered on preservation in the fossil record, the curious link between birds and dinosaurs, and how paleontology is a unique science in which anyone can participate and perhaps make a major contribution.

"Predator or Scavenger"

This scene features dinosaurs excavated at a single site from the 70-m.y.-old rocks of the Two Medicine Formation. A *Gorgosaurus, Maiasaura,* and *Bambiraptors* were discovered in western Montana by the Linsters, a family of amateur paleontologists (Fig. 6).

Figure 4. The design of the *Dinosphere* gallery is nonlinear, allowing visitors to immediately engage with things that spark their interest.

Figure 5. Night time in the Late Cretaceous as an evening sky is projected on the domed ceiling above the "*T. rex* Attacks" scene.

Here a *Gorgosaurus* eats its fill of a *Maiasaura* carcass, while small *Bambiraptors* watch warily nearby. Is this gorgosaur a hunter or a scavenger? Even a casual glance at the *Gorgosaurus* skeleton reveals pathologic evidence—the bones of this animal reveal testimony of both injury and disease. Some of the documented pathologies include: a broken and poorly healed fibula, broken femur, crushed caudal vertebrae, broken gastralia, osteomyletis of the dentary, and a shattered scapula with a large growth formed to stabilize a nonunion fracture. Researchers have also observed what appears to be a small bone-forming tumor within the braincase of this dinosaur (Natterson-Horowitz and Bowers, 2012). So how did these injuries affect the dinosaur's behavior? Was it still able to catch prey, or did it primarily scavenge for meals? Evidence suggests that theropods traveled in groups; so did the Gorgosaur receive help from pack members?

Figure 7. In the *Dinosphere* exhibit, it is important to clearly identify actual fossils and casts, such as those in this "Watering Hole" scene with *Hypacrosaurus* and *Prenoceratops*.

"Watering Hole"

It's early morning and a group of Cretaceous creatures gathers at a watering hole—a dangerous place for most animals. An adult and juvenile *Hypacrosaurus* are thirsty. They've separated from the herd to find water. Nearby, two small hypacrosaurs playfully chase a dragonfly. There's a crunching noise in some low-lying bushes by the water. One *Prenoceratops* snips and swallows leaves and twigs, while another slowly backs into a shallow depression to watch for predators (Fig. 7).

In this scene, there is no flesh-eating theropod stalking prey, but the dynamics of this group of dinosaurs still pose many questions. How did dinosaurs interact with one another? How do we know that some dinosaurs traveled in herds? How did some dinosaurs, such as these crested duck-billed dinosaurs, communicate with one another?

Children's museums provide families the opportunities to interact with exhibits and objects, but also with each other. We rely heavily on members of the interpretation staff to interact with the public, and these members of staff are a key element in the success of *Dinosphere*. Whether talking one-on-one with a visitor or leading large programs, these educators are adept at keeping audiences highly engaged. As facilitators, they are able to translate content from scholars and specialists to provide multiple levels and points of entry for visitors to a content-rich gallery. They help to ensure that the science behind dinosaurs is interesting and accessible to a wide audience (Fig. 8).

Dracorex

In 2004, during the gallery opening for *Dinosphere,* while guests were mingling and exploring the new exhibit, several

Figure 6. Museums are well equipped to provide tangible encounters with real fossils, such as the *Gorgosaurus* and *Maiasaura* in this "Predator or Scavenger" scene.

Figure 8. A staff interpreter draws a crowd of interested visitors with a cast jaw of the *Triceratops* Kelsey.

people had gathered in a back room. They were looking at a fossil skull covered with bumps, spikes, and knobs—an apparent member of the family of dinosaurs called Pachycephalosaurs also known as the dome-headed dinosaurs. However, this fossil was notably different—instead of having a dome, this creature sported two very large holes, called supratemporal fenestra, in the top of the skull. The fossil was still partially encased in rock and wrapped in a protective plaster jacket, but to all the observers it was clearly something new to science.

Standing with this group was Brian Buckmeier, along with brothers Steve and Pat Saulsbury. These skilled fossil collectors had discovered the dinosaur during a trip to the 66-m.y.-old Hell Creek Formation of South Dakota, and they generously donated this new find to the museum. The fossil was in remarkably good condition but still needed careful conservation. It had been found eroding out of the rock matrix—and erosion is seldom kind to fossils.

Luckily, museum staff member Victor Porter was well known for his skills in fossil preparation. He was able to repair areas of the skull that were fractured and piece together the four cervical, or neck vertebrae, that had been shattered by erosion long before the specimen was discovered. While Porter carefully worked on this fossil, Dr. Robert Bakker took the lead on researching this dinosaur.

"It looks like a dragon" was an exclamation that Porter heard daily from enthusiastic children. Usually scientifically significant specimens are prepared for research and exhibition in a back room, well away from any distractions. Porter, however, chose to work on the fossil skull in full view of the public and while answering their many questions. It was Porter's then 9-year-old daughter Victoria, who heard the visitors' comments and first suggested naming the animal dragon king. After being placed in the proper Latin, it became *Dracorex.*

After hearing this, Dr. Bakker was thrilled. He knew the value of a good dinosaur name. Bakker then suggested "*hogwartsia*" as the species, in honor of J.K. Rowling and her series of Harry Potter books. After a quick discussion among museum staff, donors, and paleontologists, the dinosaur would then be known as *Dracorex hogwartsia,* or Dragon King of Hogwarts (Fig. 9).

When hearing of this honor to her work, J.K. Rowling commented,

> The naming of *Dracorex hogwartsia* is easily the most unexpected honor to have come my way since the publication of the Harry Potter books! I am absolutely thrilled to think that Hogwarts has made a small (claw?) mark upon the fascinating world of dinosaurs. I happen to know more on the subject of paleontology than many might credit, because my eldest daughter was *Utahraptor*-obsessed, and I am now living with a passionate *Tyrannosaurus rex*-lover, aged three. My credibility has soared within my science-loving family, and I am very much looking forward to reading Dr. Bakker's paper describing "my" dinosaur, which I can't help visualising as a slightly less pyromaniac Hungarian Horntail. (Malvern, 2006, p. 7)

Years later, the exact status of this dinosaur is a current topic of debate in paleontology. Some researchers see *Dracorex* as a distinct species, while others suggest it's a juvenile form of the more well-known *Pachycephalosaurus wyomingensis.* Surprisingly this is something that aids museum staff and offers starting points for visitor conversations, such as how species are identified and named, the importance of sample size in science, and ontogeny.

Leonardo

In the spring of 2012, with the cooperation of the Great Plains Dinosaur Museum in Malta, Montana, a new exhibit element opened in *Dinosphere.* This joint effort between museums created the exhibition of *Leonardo the Mummified Dinosaur.* On display are the nearly complete remains of a *Brachylophosaurus canadensis,* a species of duck-billed dinosaur excavated from the 77-m.y.-old rocks of the Judith River Formation. While much of the skeleton is preserved and articulated, it is contained within an envelope of skin, scales, and muscle tissue. Even unlikely elements, such as stomach contents, are preserved in this remarkable specimen (Fig. 10).

Oftentimes when dinosaur material is displayed, despite the best efforts of exhibit designers, interpreters, and curators, it can still be difficult for visitors to see fossil remains and visualize them as once living and dynamic organisms. Art, however, is particularly good at communicating this across age, education, and language barriers.

The Lanzendorf Collection

Located on the upper level of *Dinosphere* is the Mann Properties Gallery, which features the Lanzendorf Collection of Dinosaur Imagery. This space is dedicated to the exhibition of artwork acquired by John Lanzendorf, a well-known Chicago hairstylist and art collector who assembled one of the world's largest collections of dinosaur themed art (Fig. 11).

Lanzendorf first became interested in collecting dinosaur items as a child, when he discovered a plastic dinosaur as a prize in a cereal box (Moffett, 2000). However, it wasn't until much later in life, after the slow recovery from a cancer diagnosis that he decided to pursue his passion for collecting paleoart (Daniels, 2001).

Figure 9. The pachycephalosaurid *Dracorex hogwartsia,* a holotype specimen. The skull was discovered in the Hell Creek Formation of South Dakota and is 51 cm in length.

Figure 10. Dr. Robert Bakker leading a special *Dinosphere* program on the discovery and science behind the mummified hadrosaur Leonardo.

Figure 11. John Lanzendorf in his downtown Chicago apartment surrounded by paleoart in his collection.

An intelligent and savvy collector, John was soon doing more than acquiring art; he began connecting with noted paleontologists and paleoartists. He went on excavations with Phil Currie in Alberta, was soon working with Paul Sereno at The University of Chicago, and became a valued member of the Society for Vertebrate Paleontology (SVP). While he collected works from established paleoartists, he would also seek out talented new artists, offering encouragement and becoming a well-known patron for dinosaur imagery. His active support for both artists and scientists has been recognized by the SVP through the creation of the John J. Lanzendorf PaleoArt Prize that recognizes outstanding achievement in scientific illustration of paleontologic and naturalistic art. These are the only awards chosen directly by scientists to support and acknowledge the artistry that helps bring their paleontological discoveries to life.

In 2000, Academic Press released a coffee-table book entitled, *Dinosaur Imagery: The Science of Lost Worlds and Jurassic Art: The Lanzendorf Collection.* That same year, Lanzendorf's preference for theropod images was apparent as nearly 80 pieces from the collection were displayed in conjunction with the grand opening of the *T. rex* "Sue" exhibit at the Field Museum in Chicago (Debus and Debus, 2002). All these activities brought John and his collection to the attention of staff at TCMI.

After brief negotiations in 2001, his collection of nearly 400 works of art was acquired by the museum. It contains drawings, paintings, and sculptures, with works including the vivid mixed media illustrations of Luis Rey; classic oil paintings of Michael Skrepnick; dynamic bronze sculptures by Michael Trcic; and illustrations by James Gurney, the author of the *Dinotopia* books. There is also art by Czerkas, Cooley, Krentz, Staab, Braginetz, and many others. This collection would become a core element in the development of *Dinosphere* (Figs. 12–14).

Art and science both involve observation, inquiry, critical thinking, and experimentation, but they are often taught separately (Cornett, 2011; Steele and Ashworth, 2013). This space, the art gallery within a dinosaur gallery, helps show the integration of the arts and sciences and how art is crucial in communicating

Figure 12. *Ornithomimus Family Portrait* by Donna Braginetz. Part of the John Lanzendorf Collection at The Children's Museum of Indianapolis.

Figure 13. A bronze sculpture *Daspletosaurus torosus* by artist Michael Trcic. Part of the John Lanzendorf Collection at The Children's Museum of Indianapolis.

new discoveries in paleontology. Visitors share the excitement of artistic and scientific discovery through interactive elements, fossils, and art. In this space, staff and volunteers act as the art gallery host—greeting children and their families and conducting informal interactions based on the visitor's interest. The gallery is a hands-on space for children to draw and sculpt their own dinosaur art.

Staff members lead programs to demonstrate the collaborative process that occurs between the arts and sciences when creating illustrations of dinosaurs or other extinct creatures. In one program entitled "Sketch the Unknown," visitors are presented with cast skulls of modern animals (such as a wolf eel or tapir) and try to recreate the animal's appearance based on observation of the skeletal anatomy. Another program called "Dissect a Rex" has staff members taking apart a soft sculpture *T. rex.* Here they learn about soft tissue preservation and the artistic interpretation of prehistoric life and environments based on the meager fossil record (Fig. 15).

Figure 14. A painting, T. rex *Attacks,* commissioned by Michael Skrepnick to illustrate one of the dinosaur scenes in the *Dinosphere* gallery. Part of the John Lanzendorf Collection at The Children's Museum of Indianapolis.

The size and scope of the Lanzendorf collection enables curators to easily rotate artwork on display. Museum staff members oversee its conservation and exhibition as well as its further development. To remain vital, a paleoart collection must expand and incorporate new discoveries, techniques, and even technologies.

Since the collection was first acquired in 2001, new species have been described, and compelling new technologies have developed to more closely examine the color, locomotion, and plumage of the dinosaurs. Advances in digital art software have changed how these ancient creatures can be represented. Museum staff members realize that it's vital that the Lanzendorf collection remain vibrant and useful, and they look to add to the collection.

This hands-on approach to the arts and sciences is also reflected in laboratory space within *Dinosphere.*

The Paleo Prep Lab

Prominently positioned in *Dinosphere* is the Polly Horton Hix Paleo Prep Lab. Here, museum staff, students, and volunteers skillfully work to remove rock matrix from dinosaur bones, consolidate and stabilize fossils, and prepare materials for research and display. As far as labs go, it's an active and somewhat messy place (Fig. 16).

It's not the first public prep lab for the museum. Two years prior to the opening of *Dinosphere,* a temporary lab was created as a test—to see what it takes to run such a space, to generate excitement for the *Dinosphere* project, and to conduct front-end evaluations to explore visitor perceptions and expectations for the upcoming gallery. While museum visitors could peer into the windows of the temporary lab, those windows generally remained closed, only open for a scheduled two hours a day for interpretation.

Figure 15. A museum staff member engages with families in the Mann Properties Gallery, which features the Lanzendorf Collection of Dinosaur Imagery.

A lot was learned in those first few months of lab operation about design, equipment needs, staffing and training, as well as public perception. The space must look scientific but clearly inviting as well. Staff must be trained in the detailed work of fossil prep but also in how to engage and interact with visitors of all ages. Visitors expressed a keen desire to see real fossils and research. Visitors also wanted to participate and not passively view. It became important to bring paleontology research out of the back rooms and make it available to the public and to provide the means for visitors to interact and ask questions about this research.

Today, the windows of the lab are kept open, and families are encouraged to ask questions and interact with staff. It is a space that provides tangible encounters with fossils and researchers. However, this open-window policy offers benefits for staff too—it provides a way to gauge the interest and knowledge of our visitors, creates an opportunity to identify geologic items brought in by young collectors, and allows us to provide additional community resources to those who would like to pursue their interests in the geosciences (Fig. 17).

The lab is meant to be a creative and inspiring space. It's brightly colored and highlighted with artwork from the Lanzendorf collection. Visitors peer into the lab and see fossil preparation, but also illustration, sculpting, molding, and casting. In the lab, staff members proudly discuss the interaction of the arts and sciences in paleontology.

There are, of course, challenges to operating a public lab space, particularly one in which interaction is rigorously encouraged. Even for the most ardent extrovert, working in a "fish bowl" can be difficult; therefore, staff need down time and quiet areas in which they can get detailed work done. The operation of louder tools such as pneumatics or cast-cutting saws must be used with care and well away from the windows.

Contributing to the operation of the lab are over 40 volunteers who assist with prep, exhibition, and education. Youth volunteers also work in the lab, which sends the important message to children that science is at its best when it is participatory—anyone can actively contribute to the geosciences.

Visiting scientists come to the prep lab to study objects within the natural history collection and in turn provide resources for staff, help create compelling narratives for exhibits, and offer unique interactions with visitors. Depending on the available time and comfort level of the researchers, lab staff can facilitate public interaction and explain the research being conducted, or the researchers themselves can directly discuss their current work with visitors. The museum values these special collaborations that provide great opportunities to engage ever-broader audiences and to promote scientific research.

It was in this lab that the pachycephalosaurid known as *Dracorex hogwartsia* was first introduced to the public and the holotype prepared in full view of museum visitors.

Since new fossil discoveries provide fresh educational and research opportunities as well as greatly contributing to visitor engagement, museum staff members are also active in the field—searching for some of the most iconic dinosaur faunas of the Late Cretaceous.

Fieldwork with Families

Whether it's a small child who laughingly roars at the mention of *T. rex* or an adult who has always been fascinated by ancient creatures, a few days working at a real dinosaur dig site can have a profound impact on a person's interest in science—and can also make for a pretty cool family vacation.

In a season, we may have had as many as 100 visitors dig with museum staff, with over 1200 participants over the life of the dig program. Children as young as eight years old can participate and dig alongside family members. Individual attention from museum staff makes the paleontology experience more fulfilling; therefore, space is limited to only 20 people digging at a time.

Figure 16. An opportunity to see science in action as staff in the Paleo Prep Lab staff prepare fossil specimens for research and exhibit.

Figure 17. Visitors interact with a museum volunteer at the Paleo Prep Lab window and are given an opportunity to touch a tibia of a *Tyrannosaurus rex*.

For families, the dig is a two-day experience, but they can extend their time in the field by signing up for an additional two, four, or six days. Those who sign up for these dinosaur digs run the spectrum from well-experienced outdoors people to those who have never before ventured outside a major metropolitan area. What brings this diverse group together is the desire for adventure and the love of dinosaurs.

Logistics for the dig are purposely kept simple. Participants meet museum staff in the town of Faith, South Dakota, the night before excavations begin. Meeting in town takes the burden away from museum staff to arrange national and sometimes international travel, but it also helps to cast the dig in light of a family adventure. Many families make the dinosaur excavation a part of a grand vacation to the American West.

Staff members transport guests from town to the site. Since Ruth Mason Quarry is located off-road on a remote cattle ranch, in an area with no cell phone reception, it gives everyone the chance to focus on the work at hand. Instruction always begins with essential information about safety and security. While these issues are well covered during orientation back at the museum, they need to be reinforced at the site. While the landscape is beautiful, there are specific hazards, such as summer heat, local wildlife, and easily turned ankles. When the public is at the quarry, there is always a locally contracted emergency medical technician (EMT) on site with medical kit, oxygen, and an emergency radio. Luckily, other than blisters and a few errant cactus spines, the services of the EMT have never been required.

Families who participate learn about the dinosaurs of the Hell Creek Formation, natural history of the area, scientific excavation techniques, and detailed mapping and cataloguing in the field. They learn the tools, techniques, and documentation needed to make the excavation a successful one. At the quarry, there is really no division of labor; everyone gets to participate equally. Visitors learn the process of excavation from start to finish. Working with children requires a basic fairness—no one is excluded. If a child discovers a significant fossil, they are never replaced by a more experienced digger. Staff members work closely with visitors to carefully excavate each fossil (Figs. 18–20).

In addition to family dinosaur digs, the museum hosts specific excavations for adults. For educators, this dig becomes part

Figure 18. Family dig participants uncover an ischium and femur from an *Edmontosaurus*, a duck-billed dinosaur.

Figure 19. Learning to carefully map each discovery is an integral part of the scientific process.

Figure 20. Creating a plaster field jacket to protect fragile dinosaur bones for transport back to the museum.

of a professional development program in which they earn continuing education credits. For others, it's often a "bucket list" item, the chance to explore.

If an educator discovers a particularly interesting dinosaur bone, a mold can be made in the Paleo Prep Lab. While the original fossil stays with the museum in the permanent collection, the rubber mold can go with the teacher into the classroom setting. The teacher can pour quick-drying plaster into the mold, and during the course of a day, make enough casts for an entire class. This makes for a great introduction to paleontology and becomes a cool, "Look what I did over summer vacation" program for the teacher. One educator can easily instill a love of the geosciences in hundreds of students.

Some of the most ardent supporters of the museum paleontology program are the teachers who participate in fieldwork and in the lab. They become invested in the program. Many have attended dinosaur excavations for well over a decade, coming back year after year to help dig duck-billed dinosaurs. In turn, to advocate the teaching of geosciences in schools, the museum offers professional development opportunities where teachers can work on graduate credits, license renewals, and other certifications in a one-of-a-kind environment. We offer in-school dinosaur programs, exclusive lab tours, and even the opportunity to display recent finds in the schools.

One of the most important aspects of these dino digs is the sense of community that develops. Of course there is a shared feeling of camaraderie for those who have been on a dig, but those who participate develop a greater shared sense of how science works. Their interest in paleontology becomes more than memorization of dinosaur names or superlative arguments about which dinosaur is bigger, faster, meaner; ... they begin to understand the concept of deep time and see paleontology as complex and dynamic science. Dig site participants collaborate on a discovery, they document their finds, and they see the fossils go back to a public institution to be studied, conserved, and carefully prepared in the Paleo Prep Lab for all to see (Figs. 21 and 22).

It takes time and effort to create an outstanding field experience for visitors. Much of the work and planning occurs at the museum, but once at the site museum, staff members also need to coordinate efforts with local police, fire and ambulance departments, as well as businesses. Since we operate out of a small town of less than 500 people, we realize that we impact the local infrastructure and economy and plan accordingly. While working

Figure 21. Dino dig participants use a small knife and brush to delicately excavate a dinosaur skull element.

Figure 22. A dinosaur element, an *Edmontosaurus* jugal, or cheek bone, after being prepared in the Paleo Prep Lab, placed on display, and then made available as an image free of copyright restrictions on Wikipedia.

in the field, it's easy to understand the importance of community in the geosciences (Fig. 23).

The success of *Dinosphere* and the excavations in South Dakota has led the museum to search for additional field sites and opportunities.

CONCLUSIONS

Museum exhibits and programs on the geosciences can be transformative in the lives of children and their families. Research indicates that playful interaction with science during childhood influences interest in science later in life (Bulunuz and Jarrett, 2010). Or as paleontologist Robert Bakker aptly states, "Dinosaurs are the jumper cables to the human mind. Kids can't curb their enthusiasm when they're in a hall of dinosaurs and mammoths and mammoth hunters and trilobites and giant fish that could chomp up a shark. These natural objects in motion and context make kids want to read, you can't stop them from reading and thinking" (Plushnick-Masti, 2012, p. 12).

Figure 23. Experiences such as these can be transformative in the lives of children and their families, instilling a greater interest in the geosciences.

While museums excel at audience evaluation, it can be difficult to get the big picture of the institution's long-range impact on geoscience education. Understanding this influence is increasingly relevant in a political system where every voter is in essence a policymaker—scientific literacy becomes fundamentally important.

Yet sometimes circumstances provide a glimpse of how far-reaching the impact of science programs at a children's museum can be. More than 70 years after participating in the 1938 Prairie Trek, author Kurt Vonnegut was asked about the program and its leader, Hillis Howie. Vonnegut, who was then well into his eighties, recalled, "The value system under which I try to operate relative to animals and plants and the earth and persons with cultures different from mine is one that I learned from him. There are thousands of us who were lucky enough to come under his influence, and my guess is that we are more at home on this planet, and more respectful of it, than most of our neighbors are" (Shields, 2011, p. 15).

ACKNOWLEDGMENTS

All photos courtesy of The Children's Museum of Indianapolis.

REFERENCES CITED

Boddenhammer, D., and Barrows, R., eds., 1994, The Encyclopedia of Indianapolis: Bloomington, Indiana, Indiana University Press, p. 385.

Bulunuz, M., and Jarrett, O., 2010, Developing an interest in science: Background experiences of preservice elementary teachers: International Journal of Environmental and Science Education, v. 5, no. 1, p. 65–84.

Christians, J., 1992, Taphonomy and Sedimentology of the Mason Quarry, Hell Creek Formation (Upper Cretaceous), South Dakota [M.S. thesis]: Madison, University of Wisconsin, 91 p.

Cornett, C., 2011, Creating Meaning through Literature and the Arts: Arts Integration for Classroom Teachers (4th ed.): Boston, Massachusetts, Allyn & Bacon/Pearson, 472 p.

Counts, C., 2009, Spectacular design in museum exhibitions: New York, Curator, v. 52, no. 3, p. 273–288.

Daniels, M., 2001 (14 May), Lost world found; It's in this extreme collector's Gold Coast apartment where period decorating takes on a whole new meaning: Chicago Tribune, Home and Garden, p. 1.

Debus, A.A., and Debus, D.E., 2002, Dinosaur Memories: Dino-trekking for Beasts of Thunder, Fantastic Saurians, 'Paleo-People,' 'Dinosaurabilia,' and Other 'Prehistoria': San Jose, California, Authors Choice Press, 572 p.

Dewey, J., 1910, How We Think: Boston, D.C. Heath & Company, 240 p.

Din, H., 1999, An investigation of children's museums in the United States—Their past, present and future: A proposed study: Marilyn Zurmuehlin Working Papers in Art Education, v. 15, p. 63–69.

Fiffer, S., 2000, *Tyrannosaurus* Sue: The Extraordinary Saga of the Largest, Most Fought over *T. rex* Ever Found: New York, W.H. Freeman and Company, 248 p.

Kriplen, N., 1982, Keep an Eye on That Mummy: A History of the Children's Museum of Indianapolis: The Children's Museum of Indianapolis, 217 p.

Larson, P., and Donnan, K., 2002, Rex Appeal: The Amazing Story of Sue, the Dinosaur That Changed Science, the Law, and My Life: Montpelier, Vermont, Invisible Cities Press, 424 p.

Malvern, J., 2006 (24 May), Dinosaur is dead ringer for Potter dragon: A new species discovered in the US has been named *Dracorex hogwartsia,* much to the delight of the author J. K. Rowling: London, The Times, p. 7.

Moffett, N., 2000 (9 April), Field to show dino-mite art: Chicago Sun-Times, p. 19.

Natterson-Horowitz, B., and Bowers, K., 2012, Zoobiquity: The Astonishing Connection between Human and Animal Health: Toronto, Vintage Books (Random House), 416 p.

Nuwer, H., 1987, A skull session with Kurt Vonnegut: South Central Review, v. 9, no. 2, p. 2–23.
Plushnick-Masti, R., 2012 (22 May), Houston museum unveils $85 million dinosaur hall: The San Diego Union-Tribune, Associated Press, p. 22.
Radu, L., 2011, John Dewey and Progressivism in American Education: Bulletin of the Transilvania University of Braşov, Series VII: Social Sciences, v. 4 (53), no. 2, p. 85–90.
Rebbert, C., 2007, The role of museums in geoscience education: A perspective: The Leading Edge, v. 26, no. 10, p. 1322–1325, https://doi.org/10.1190/1.2794398.
Shields, C.J., 2011, And So It Goes: Kurt Vonnegut: A Life: New York, Henry Holt and Company, 528 p.
Steele, A., and Ashworth, E.L., 2013, Walking the integration talk: An ArtSci project: The Canadian Journal for the Scholarship of Teaching and Learning, v. 4, no. 2, p. 6.
Vonnegut, K., 1985, Galápagos: New York, Dell Publishing, 295 p.
Winslow, D.R., ed., 1982, Necrology 43 Hillis Langdon Howie: Proceedings of the Indiana Academy of Science: Bloomington, Indiana, Indiana University, v. 92, 544 p.

Manuscript Accepted by the Society 3 January 2018
Manuscript Published Online 10 July 2018

Printed in the USA

The Geological Society of America
Special Paper 535

Live science in the Valley of the Last Dinosaurs: *A public window into the world of paleontology*

John Hankla
Denver Museum of Nature & Science, Department of Earth Sciences, 2001 Colorado Boulevard, Denver, Colorado 80205, USA

Samantha Sands
Denver Museum of Nature & Science, Museum Programs Department, 2001 Colorado Boulevard, Denver, Colorado 80205, USA

Megan Sims
University of Kansas Biodiversity Institute and Natural History Museum, Department of Vertebrate Paleontology, 1345 Jayhawk Boulevard, Lawrence, Kansas 66045, USA

Jeremy Wyman
Denver Museum of Nature & Science, Department of Earth Sciences, 2001 Colorado Boulevard, Denver, Colorado 80205, USA

ABSTRACT

LiveSci in the Valley of the Last Dinosaurs **(http://lastdinos.livesci.org/) was a website and social media presence that provided the global online community with unprecedented access to the exciting paleontological research happening in the remote badlands of North Dakota and Montana in the summer of 2016. A collaborative team of researchers, students, and citizen scientists from around the world excavated some of the last dinosaurs that ever walked the Earth, mapped the K/Pg boundary in high resolution, and uncovered fossils that show us how life recovered after the extinction of the dinosaurs. To engage the public in the ongoing process of scientific discovery, dedicated project staff and participating researchers posted videos, photos, blog entries, and social media content nearly every day during the seven-week field season.**

Researchers and science educators from the Howard Hughes Medical Institute and Denver Museum of Nature & Science, along with collaborators from Brooklyn College, the Smithsonian National Museum of Natural History, Yale Peabody Museum, the Royal Ontario Museum, and the Marmarth Research Foundation, were joined by young scientists and citizen scientist volunteers of all ages. The production team consisted of high school and college interns, public science outreach professionals, and research scientists. To expand the reach of the project, a bilingual intern maintained a parallel Spanish website.

Hankla, J., Sands, S., Sims, M., and Wyman, J., 2018, Live science in the *Valley of the Last Dinosaurs:* A public window into the world of paleontology, *in* Rosenberg, G.D., and Clary, R.M., eds., Museums at the Forefront of the History and Philosophy of Geology: History Made, History in the Making: Geological Society of America Special Paper 535, p. 289–299, https://doi.org/10.1130/2018.2535(19).

Hundreds of thousands of online viewers watched, contributed, and shared these authentic experiences with their communities during the live portion of the project, and many more continue to access the archived website and social media content. This project exemplifies how social media and real-time interaction with scientists have the potential to connect the public to science as it unfolds, removing myths and stereotypes about how science happens and who scientists are. Initiatives such as this one help to create citizens who are more connected to the process of science and who can use that understanding in their lives through more informed decision making.

INTRODUCTION

In June and July of 2016, Tangled Bank Studios, a production company dedicated to the creation of original science documentaries, provided support to earth science researchers and educators at the Denver Museum of Nature & Science (DMNS) to produce educational content about paleontological research and fieldwork. The result was an interactive website and social media presence titled *LiveSci in the Valley of the Last Dinosaurs,* which chronicled the daily life and work of a multi-institutional, multidisciplinary group of paleontologists as they worked in a remote field area along the border of North Dakota and Montana. The researchers and volunteers from various institutions collaborated by sharing data, equipment, laboratory space, and a field kitchen while they conducted their fieldwork pursuing the various lines of evidence that inform our understanding of the extinction event that killed the dinosaurs and the recovery of life on Earth that followed (Lyson et al., 2011). Throughout the seven-week field season, LiveSci produced and posted online video content for social media and maintained a project website (www.lastdinos.livesci.org) to host biographical content, written reflections, pictures, and videos from the scientists, students, and volunteers conducting this fieldwork and research. Unlike Tangled Bank's typical documentary content produced for broadcast or theatrical distribution such as the PBS series, *Your Inner Fish,* the NOVA special, *Vaccines—Calling the Shots,* and *Mass Extinction: Life at the Brink,* which aired on the Smithsonian Channel, this digital project made use of online platforms to connect its audience with developments from the field as they happened while providing an archive of the field season that can be easily accessed for years to come.

The website, video content, and social media pages were created, produced, and maintained by DMNS research associate John Hankla and high school intern Jeremy Wyman. In addition to the main website, the project also included a Facebook page (https://www.facebook.com/LiveSci/), a Twitter feed (https://twitter.com/LiveSci_HHMI), and an Instagram account (https://www.instagram.com/livesci_hhmi/) to introduce members of the general public to participants within the field of paleontology and to create ways for the two groups of people to communicate.

BACKGROUND AND OBJECTIVE

Tangled Bank Studios was created by the Howard Hughes Medical Institute (HHMI) in 2012 to "produce programs that capture compelling stories of discovery" (http://www.tangledbankstudios.org/about-us/about-tangled-bank-studios). The vision for their body of work involves creating productions that will have a lasting impact while being entertaining, including new science, and inspiring wonder. "The Denver Museum of Nature & Science envisions an empowered community that loves, understands, and protects our natural world" (http://www.dmns.org/about-us/mission-and-vision/). The Department of Earth Sciences at DMNS embodies this vision through their focus on specimen-based research, which they share with the public through the museum's programs and exhibits. Dr. Tyler Lyson, Curator of Vertebrate Paleontology at DMNS and Director of the Marmarth Research Foundation, collects and studies fossils from the Hell Creek Formation to make discoveries about dinosaur extinction (Lyson et al., 2011). Lyson's fossil localities produce incredible dinosaurs, turtles, and many other Late Cretaceous flora and fauna that provide details as to what life was like before and leading up to the Cretaceous–Paleogene (K/Pg) boundary horizon. His work attracted the attention of HHMI's "BioInteractive" educators, and a collaborative relationship between DMNS and HHMI was formed that resulted in the production of a film, *The Day the Mesozoic Died* (2012).

As mentioned on the HHMI website for the film, *The Day the Mesozoic Died,* HHMI and Tangled Bank Studio were attracted to the work of Lyson and his colleagues researching the end-cretaceous mass extinction because it "represents a rare instance in which many different disciplines—geology, physics, biology, chemistry, and paleontology—contributed to a revolutionary theory" (https://www.hhmi.org/biointeractive/day-mesozoic-died). The website also cites the popularity of the subject of dinosaur mass extinction with a quote from a Colorado teacher, "Who does not love dinosaurs and huge explosions!?" In the College Board's current Advanced Placement (AP) Biology curriculum framework, it is considered "Essential Knowledge" that "scientific evidence of biological evolution is supported by evidence from many disciplines." With this in mind, LiveSci chose to highlight the interdisciplinary nature of the project, developing content with scientists from multiple institutions, working in sedimentary geology, vertebrate paleontology, paleobotany, physics, computer science, and museum studies. Lyson's research questions about the changes in vertebrate communities before, during, and immediately after the mass extinction that is recorded in the rocks of the Hell Creek field area cannot be answered by solely observing the vertebrate fossil record. Dr. Antoine Bercovici's detailed understanding of paleobotanical data, such as the fossil record of pollen spores, resolves the timing of the changes in the vertebrate record relative to the event that caused so much drastic change to these vertebrate communities. Using sophisticated GIS modeling software, Gabriella Rosetto uses the finely resolved data from both the vertebrate and plant fossil records to create a visual model that takes advantage of one of the most comprehensive terrestrial data sets ever made for this extinction event. During the 2016 field season, LiveSci provided each of these collaborators and many more participating scientists a chance to interact directly with the public through online platforms. LiveSci's goal in introducing the public to the people behind these individual research objectives was to show how real-world networks help strengthen scholarly collaborations and allow researchers to answer bigger questions than could be addressed individually.

In addition to providing scientific expertise and access to research scientists, DMNS also has experience using the Internet to connect scientists with the public. Since 2001, DMNS distance learning programs have been using new technologies to give students a window into the world of scientists. In 2007, DMNS enhanced the interactivity of its distance learning program with the launch of Scientists in Action (SIA). This program uses digital satellite connections from real scientists' field or laboratory sites to the Museum's auditorium, where hundreds of students are gathered. From the auditorium, the students can see and hear a live demonstration and lecture with a talk-back circuit available for student questions. Drawing upon the production experience and funding of Tangled Bank Studios as well as the scientific and programming expertise of DMNS, the idea for *LiveSci in the Valley of the Last Dinosaurs* was born.

The LiveSci collaboration was developed to communicate science by establishing an online community and providing a social space where dialog could take place between working paleontologists and interested members of the general public. Unlike previous programming efforts such as SIA that check in with researchers only once, the vision for LiveSci was that the public could follow and interact with a group of working scientists through an entire season of fieldwork. By creating this online community, LiveSci aspired to increase public engagement in paleontology.

The idea for employing community-based learning environments is not new in the field of education and outreach. In the mid-1970s, Albert Bandura pioneered the concept of social learning (Bandura, 1977). Now a widely accepted pillar of modern educational pedagogy, his "social cognitive theory" proposes that people learn observationally through the process of modeling in a social setting. Because much of the process of modern paleontology takes place in remote field areas or behind the scenes in museums and universities, it can be challenging for the general public to find ways to model the behaviors of major contributors to this field. With recent advances in technology, social arenas can now be created in the form of online communities that provide access for the exchange of ideas and dialog between individuals, regardless of their background or geographic location. People who feel a sense of belonging to a community will exhibit more motivation, which leads to increasing engagement with the subject (Silius et al., 2011). Student engagement has been defined as "the amount of physical and psychological energy that a person devotes to the academic experience" (Astin, 1984). In their 2004 study on student engagement in higher education, Zhao and Kuh (2004) showed that participation in a learning community that includes social interactions with experts is directly related to personal and social development, improved academic performance, and engagement in their education. By taking advantage of web-based technologies such as social media, LiveSci was able to create an inclusive community for social learning that encompasses research scientists and members of the general public. According to Silius et al. (2011), "social media enhanced learning systems allow students to participate in educational online communities," giving people a sense of belonging that helps to assist in the learning process. Cerda and Planas (2011) found that social media sites, such as Facebook (www.facebook.com), provide a collaborative learning experience by fostering a "virtual community culture" and opportunities for social learning. Modern learners have been shown to "absorb information quickly in images and video" (Duffy, 2008); therefore, the online community developed by *LiveSci in the Valley of the Last Dinosaurs* created regular, episodic video content and posted it to the project website and social media pages. Videos introduced the research scientists who drive the field of paleontology, explored the daily realities of their scientific fieldwork, and provided biographical information, contact links, and comment threads to facilitate dialog between the scientists and the public.

The effectiveness of social educational communities is not novel, and the benefits of technology to facilitate these communities have been demonstrated and documented in educational pedagogy. Nevertheless, rapid advances in technology are giving us new ways to approach social learning each year. It is hard to imagine that pioneers in the field of social learning could have predicted the powerful set of tools that is considered commonplace today. A 2015 Pew Research Internet survey shows that two-thirds of Americans own smartphones. These common devices can facilitate the creation of content, including high-quality video, and are capable of posting content to social media platforms even from remote locations (Rainie, 2012). *LiveSci in the Valley of the Last Dinosaurs* demonstrates that an effective online community can be created to engage the public with paleontological research at minimal expense in terms of equipment and staff. The production team included one paid educator and one high school intern. Content was created and posted using their personal smartphones without the need for purchasing expensive devices.

METHODS

Format and Schedule

Video posts and written reflections were regularly posted to Facebook, Twitter, and the project's website. The general weekly schedule of content was outlined prior to the start of field season by producer John Hankla and the lead researcher, Lyson. This general outline accounted for the specific field schedule of faculty from each of the collaborating institutions while also allowing room for more spontaneous content.

Each week, two videos and several written blog entries were produced and posted revolving around a particular theme that was relevant to that week's research. These seven weekly themes detailed the range of fieldwork being conducted and how the work of various researchers involved in the collaborative project contributes to the understanding of Earth's last major extinction event as it is recorded in the rocks of North America.

During the seven-week field season, LiveSci recorded video conversations with research teams from DMNS, Yale University, Brooklyn College, the Smithsonian Institution, the Pioneer Trails Regional Museum, the Marmarth Research Foundation, Mansoura University, and the Royal Ontario Museum. The scientists, students, and volunteers who work in this field area and represent each of these varied institutions share tools, equipment, and other necessary resources with one another during the field season, and they also share data. The results of their individual research frequently overlap in publications, and these collaborators become coauthors as in the case of *Dinosaur Extinction: Closing the '3 m Gap,'* which was published in 2011 by Tyler Lyson, Antoine Bercovici, Stephen Chester, Eric Sargis, Dean Pearson, and Walter Joyce (Lyson et al., 2011). This paper featured scientists representing seven different departments at six institutions, all of whom were still working together during the 2016 field season. Researching a complex paleontological concept such as the end-Cretaceous mass extinction and subsequent recovery of life on Earth requires input from many different subdisciplines of paleontology and relies on the research of multiple scientists. Including the collaborative nature of this ongoing project was a priority of LiveSci.

The format of each video included a 10-second, eye-catching, and often humorous introduction, followed by a 7-second custom LiveSci musical theme, 2–5 minutes of conversation with a scientist or team, and a 7-second musical conclusion. Special attention was given to the framing of each video. The visual appeal and "mood" of each video was considered almost as important as the scientific content because the team sought to capture both the scientifically relevant and the beautiful aspects of individual field localities.

These video conversations delivered stories and ideas from the field researchers in their own words, conveying a sense of authenticity without looking over-produced or heavily edited. For this reason, LiveSci went with a "vlog" (video-blog) style approach when filming to get 2–5 succinct minutes of footage in as few takes as possible. In addition to appearing genuine, this approach has the added benefit of using less of the researcher's valuable time and attention while they are busy with the subject of their fieldwork. Many of the collaborating institutions are only able to spend a few days or weeks in the field area each year and simply do not have entire days to devote to outreach. Creating this kind of content in a few takes, usually 30 minutes or less, without using too much of the researcher's precious time and attention was often a challenge. With a heavily edited video piece, several hours of footage can be put together to make one succinct point in a few seconds to minutes of video time. Making simple, clear, and informative points can be a difficult task when filming time is limited to only a few minutes, cameras are scarce, and time does not allow the opportunity to spend hours poring over footage to edit out mistakes. Much of the work necessary to overcome these challenges fell upon the host of the episode. Before filming each episode, the host needed to conduct casual conversations with the subjects (often while lending a hand at whatever task the subject was working on) to decide what would be the most interesting talking points. Once a basic outline was constructed casually and a location for filming decided upon, the cameras and microphones were turned on and a quick conversation could be captured in less than 10 minutes.

Tools and Equipment

LiveSci video content resulted in a considerable reach despite being created with minimal expense in terms of equipment and expertise. By choosing to use minimal equipment, LiveSci avoided transporting large, cumbersome devices into field sites in remote locations and managed to record candid, casual conversations in some field sites where bringing traditional film equipment would be significantly more difficult. An added advantage of this minimalist approach was that LiveSci reduced the anxiety that can sometimes make capturing candid content difficult when talking to scientists who have various levels of experience on camera.

Video content was created using an Apple iPhone 6 in a simple mounting case that provided stability and protection. The "ALM mCAM" (http://www.actionlifemedia.com/mcam) mounting case included a wide-angle lens attachment and a threaded attachment point for tripods and grips. An "Opteka x-grip" (http://opteka.com/xgrip.aspx) was used as an unobtrusive way to increase stabilization during filming. Occasionally, in sites where no one was available to hold the camera, a small tripod was used.

Audio was recorded using the built-in microphone included in the standard "ear bud" headphones that come with the iPhone and the "voice recorder" that is built-in to the iOS software on the phone. An iPhone recorded audio in the pocket of each person featured in a video with the ear bud microphone tucked out of plain sight under clothing, near the chest of the individual. Audio files from the various iPhones used for recording sound were sent to the iPhone that recorded video via the built-in AirDrop function. This simple, cordless, and fast file transfer function proved to be very helpful in remote areas without phone carrier signal or WiFi. Using AirDrop to share the files on site was a considerable asset because most of the time, the LiveSci production team was relying on iPhones belonging to various scientists working in each field site. This one-touch file sharing allowed LiveSci to limit time spent using borrowed equipment. Content was edited using the free iMovie program (https://www.apple.com/imovie/) that comes pre-installed on a MacBook Air laptop computer. Audio and video files were synchronized and mixed with this easy-to-use program, often on the tailgate or inside the cab of a field vehicle.

This relatively small set of user-friendly equipment allowed LiveSci to capture video and show viewers actual paleontological fieldwork, even when it was conducted miles from any roads and hours away from town. Videos were regularly posted to the web from a high-speed Internet connection at the Marmarth Research Foundation's field laboratory in the town of Marmarth, North Dakota.

Release forms for each on-camera participant were printed and signed before filming, and an iPhone scanning application (https://itunes.apple.com/us/app/tiny-scanner-pdf-scanner-to-scan-document-receipt-fax/id595563753?mt = 8) was used to upload the documents into an organized folder where they could be accessed by Tangled Bank Studios staff when needed.

The project website was created using the Squarespace website-building platform (https://www.squarespace.com). A website template was modified to include pages for biographical information, a "blog" of notes from the various scientists and collaborators, a video blog page for posting video content, and multiple one-click forms for submitting audience questions.

At the end of the field season, viewership data from social media sites were gathered and used to provide insight into the online activity of the project. Statistical analyses were performed in Stata v11.2 (College Station, Texas). All data are reported as medians unless otherwise stated. Data were analyzed using Kruskal-Wallis test, and statistical significance was assigned at $p < 0.05$.

RESULTS

The results of video content posted to the social media outlets and the project website could be easily measured in real time and compared using analytical data built into the various platforms. The Facebook page was by far the most active. More than 206,000 unique users viewed the project's Facebook content. The Facebook platform refers to this metric as "reach." The project website (www.lastdinos.livesci.org) received 6874 page views, during the field season, with 49% of those visits coming as direct traffic from links in emails and documents. The other 22% of visits came from the Facebook platform itself. Video content was posted to Twitter as well and was delivered to the feed of 23,800 unique users. The top-performing Tweet reached 6468 people and was retweeted nine times. This top Tweet featured the collaboration between the LiveSci team and the paleontology team from the Royal Ontario Museum in Toronto, a large institution with a very active Twitter presence (more than 6000 followers) and no Facebook account. The Instagram platform was not used frequently or effectively to post video content and only generated 38 followers. This infrequent use was not due to any technical limitations having to do with the Instagram software application, but rather a failure on the part of the present author to grasp the popularity and staying power of this very important social media platform at the time of production.

The video content posted to Facebook fit into three broad categories: science personalities (SP), audience questions (AQ), and methods and fieldwork (MF). Descriptions and viewership metrics for this video content are shown in Table 1.

Nine videos were in the SP category, seven videos were in the AQ category, and seven videos were in the MF category. In the SP category, the "most viewed" video featured Smithsonian scientist Bercovici and his research on fossil pollen. The AQ video with the highest reach featured a question from a kindergarten student and the research of Yale University colleague Jakob Vinther. The highest reach in the MF video category was a video showing the teams from Brooklyn College, Yale University, and DMNS having fun and playing practical jokes while working in their quarry.

Median reaction rates per 100 views (our chosen measurement to understand the true interaction of viewers with the videos through Facebook's reaction buttons) were not statistically significantly different among the three video types (Fig. 1; $p = 0.263$).

Regarding all median viewership metrics (i.e., lifetime reach, views, clicks, and reactions), there were no statistically significant differences among the three content types (Fig. 2).

Lifetime reach is the number of unique users who received the post in their Facebook feed. A "view" on Facebook is registered when an individual watches the video content. "Clicks" is a metric that measures the number of individuals who choose to follow the link in a Facebook post or click somewhere on the video for more content. Facebook "reactions" refer to the action taken by a user to click the "like" button, leave a comment, or use one of the other built-in "reaction images." The Smithsonian video featuring Bercovici's paleobotanical fieldwork performed very well on Facebook and was viewed 19,766 times. In addition, more than 1000 people commented on or shared the video with their networks. Despite the large raw number of views, comments, and shares associated with Bercovici's paleobotany video, median lifetime reach (MLR) was not significantly different between the videos from each collaborating institution (Fig. 3; $p = 0.457$).

Making sense of why one video performs better than another online is difficult given the many variables that affect a video's production and distribution. Bercovici's conversation in his most-watched video was insightful, and the content was intellectually stimulating. It was filmed in beautiful evening light, becoming one of the project's most aesthetically pleasing videos. Bercovici was featured in multiple videos, as was his adorable dog, attracting viewers to their endearing personalities. Though nearly every video delivered solid scientific content in a visually interesting setting, only this video approached 20,000 views. The massive median lifetime reach of Bercovici's video probably has more to do with the influence of the museum he represents, the

TABLE 1. VIDEO DESCRIPTIONS WITH VIEWERSHIP METRICS

Video stars (Content)	Content type	Collaborating institutions	Lifetime reach	Views	Clicks	Reactions
Dr. Antoine Bercovici (Paleobotany)	SP	Smithsonian	56,107	19,766	3219	1111
Dr. Tyler Lyson (Question Tuesday)	AQ		14,269	6719	1317	438
Samantha Richards (DMNS)	SP	Denver	14,204	3538	852	334
Dr. Tyler Lyson (First Fossil Find)	SP	Denver	13,784	5465	1647	464
Dr. Kirk Johnson (Question Tuesday)	AQ	Smithsonian	13,494	3821	599	185
Dr. Ian Miller (Extinction)	SP	Denver	9621	2518	768	260
Shindig LIVE!	AQ	Multiple	7657	3158	1021	489
Shovel Wars	MF	Brooklyn and Yale	5485	2299	353	133
Pioneer Trails Museum	SP	Pioneer Trails	5284	1553	444	156
Jake Percival and Jeremy Wyman (Prospecting)	MF		5274	2487	239	86
Dr. Hesham Sallam (Question Tuesday)	AQ	Mansoura	4880	2529	683	279
Storm Video	MF	Multiple	4848	2585	495	141
Independence Day	MF		4417	2351	277	60
Dr. Stephen Chester (Camel Butte)	SP	Brooklyn and Yale	4349	1686	243	101
Erica Evans (Geotechnical Data)	SP	Denver	3907	1504	345	103
Mike Getty (Triceratops Quarry)	MF	Denver	3847	1047	221	91
Matt Hess (Question Tuesday)	AQ	Denver	3739	1482	312	91
Royal Ontario Museum	SP	Royal Ontario Museum	2634	597	202	42
John Hankla and Jeremy Wyman (Question Tuesday)	AQ		2497	664	139	54
Dr. Tyler Lyson, Matt Hess, Erica Evans (High Dinosaurs)	MF		2328	509	120	37
Gabi Rosetto (GIS Model)	SP	Denver	2170	837	267	112
John Hankla (Question Tuesday)	AQ		1833	294	59	12
Dr. Stephen Chester (Lab Students)	MF	Brooklyn and Yale	1117	434	128	42

Note: AQ—audience questions; MF—methods and fieldwork; SP—science personalities; Brooklyn—Brooklyn College of the City University of New York, Brooklyn, New York; Denver—Denver Museum of Nature & Science (DMNS), Denver, Colorado; Mansoura—Mansoura University, Al Manṣūrah, Egypt; Pioneer Trails—Pioneer Trails Regional Museum, Bowman, North Dakota; Royal Ontario Museum, Toronto, Ontario, Canada; Smithsonian—Smithsonian National Museum of Natural History; Washington, D.C.; Yale—Yale University, New Haven, Connecticut.

Smithsonian Institution's National Museum of Natural History, and the connections that Howard Hughes Medical Institute has with their social media team. By focusing on a large, multidisciplinary, multi-institutional research area such as *The Valley of the Last Dinosaurs,* LiveSci took advantage of the large, preexisting social media networks of each collaborating institution. Table 1 shows higher "lifetime reach" among the videos that featured Smithsonian Institution scientists.

By coalescing the social networks of each of the collaborating programs on the various platforms, the project grew its audience with each post, from a small group of participants, family, and friends to include members of the general public around the world.

The Facebook platform also allows page administrators to view data about the variety of individual users who view Facebook content. Choosing to "follow" a Facebook page means choosing to see posts from the page in the news feed of your Facebook page. On average, LiveSci Facebook followers were divided nearly equally among male and female viewers (Fig. 4).

On 30 July 2016, the LiveSci team used Facebook's "Facebook Live" feature to bring a panel of 11 world-class scientists together to answer audience questions in real time from the Carter County Museum in the tiny town of Ekalaka, Montana. This small museum happens to be one of the hubs of fieldwork in this region of the country. A total of 889 viewers watched during the 23 min stream, and as many as 50 people watched at a time. Cumulatively, 3443 viewers saw the video, which picked up thousands of views when each of the scientists from the 11 different institutions was tagged the following day. Despite the low-tech equipment (an iPhone camera with no external microphones)

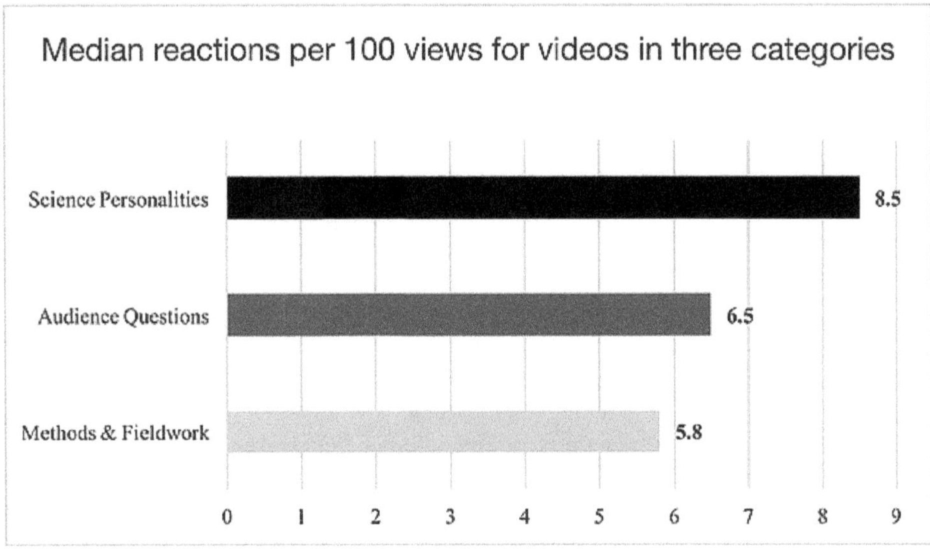

Figure 1. Reactions per 100 views among video content categories: AQ—audience questions; MF—methods and fieldwork; and SP—science personalities.

and less-than-ideal WiFi signal in the remote Carter County Museum, the audio and video quality were surprisingly good, and the signal was only minimally interrupted for buffering. The team chose to use museum curator, Nathan Carroll, as the "master of ceremonies" for the live event. Carroll helped to keep the ever-changing audience engaged by frequently explaining who the scientists were and that they were available to answer questions live. The Facebook Live video achieved the fourth highest number of direct engagements with the audience. The live function on Facebook's platform was nearly brand new at the time of production, and this video exceeded expectations for how much attention the new function could attract.

Considering the potential of this technology to reach such a large audience in real time, a few ideas came up during the Facebook Live "experiment" that could increase the peak live viewership when used again for a similar question-and-answer session. In terms of promotion, the event should be announced several times on various social media accounts and noted on the website. Only one promotion was posted to Facebook in the days before the event. Reminders and "plugs" could be included in other videos released throughout the duration of the project. Direct communications could be sent out via e-mail to key influencers, such as the media teams at the Smithsonian, DMNS, and the other collaborating institutions and media outlets to promote awareness of a scheduled event such as this. A higher number of viewers with high-quality questions might be assembled by tagging each of the participants in promotional posts ahead of time, as well as on the Facebook Live post as soon as it goes up. The highest number of viewers watching at the same time came during the final minutes of the stream, indicating that it takes time to build an audience and that the stream could be longer than 23 min in future iterations. Regarding content and layout, the panel of

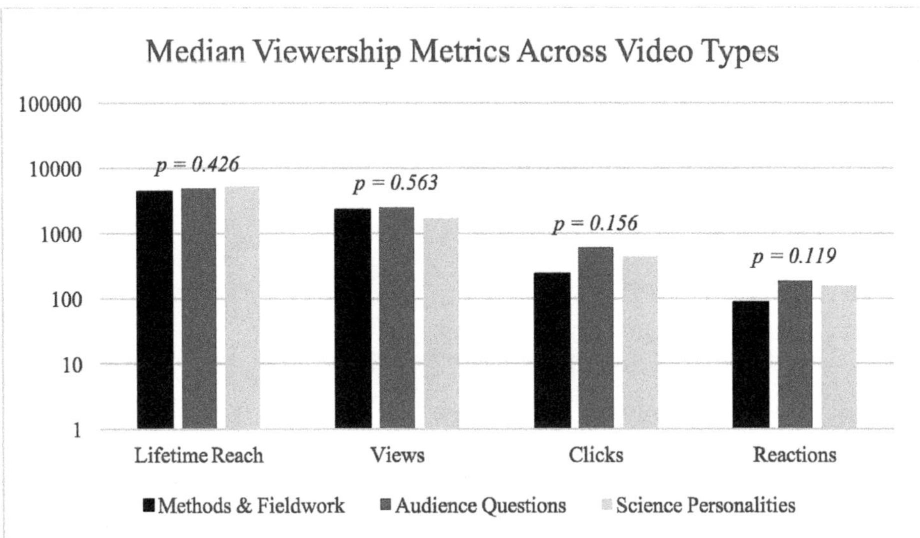

Figure 2. Comparison of median viewership metrics across video types.

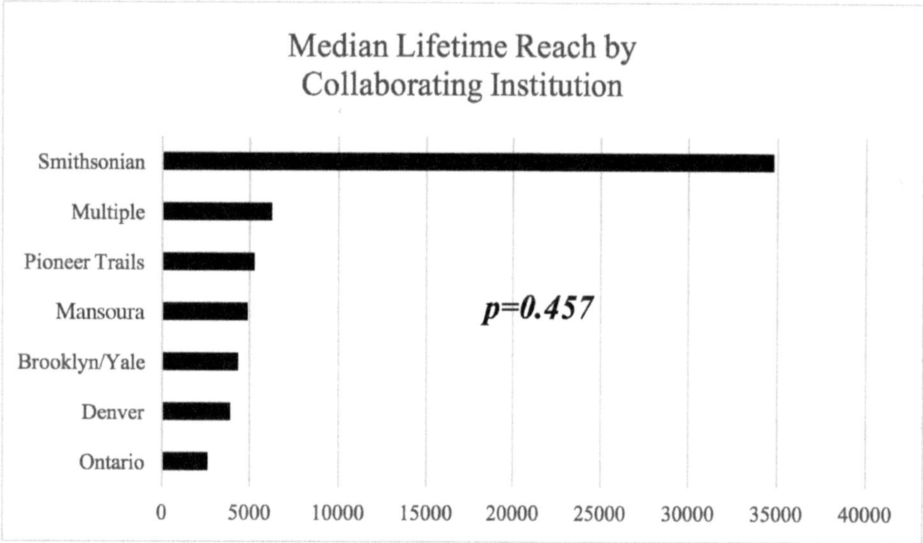

Figure 3. Median lifetime reach (MLR) on Facebook for videos by collaborating institutions. The value $p = 0.457$ indicates no significant difference between non-Smithsonian institutions.

scientists could wear prominent name tags displaying their institutional affiliations and titles to help the audience become better acquainted with the participants. As with much of the content that this project produced, the success of the video can be attributed to the high number of participating scientists involved, a key point to keep in mind when making improvements. In general, collaboration is a key function of this research that makes it suitable for creating successful online content.

Featuring institutions with giant social media networks proved to be a great way to take the LiveSci content beyond our average expectations in terms of Facebook views and overall reach. At the same time, featuring the input of smaller institutions can pay off as well. The Pioneer Trails Regional Museum (PTRM) in Bowman, North Dakota, has a tiny social media presence compared with that of the Smithsonian or DMNS. The PTRM is located only a few minutes from the remote field study area, but the network of local residents was quite interested in the activity of this small museum. Residents of Bowman shared the video in this region. As a result, the LiveSci Facebook page picked up 30 fans from one of the least populated regions in the United States. The value of these 30 Facebook fans may be overshadowed when compared with the thousands of fans from other regions, but the following testimonial submitted to the website's comments section by one such resident may offer an insight into why featuring the work of multiple, diverse institutions on projects such as these should be taken seriously:

> Watching the LiveSci videos on Facebook was the first time I ever really understood just what in the world all you folks were doing out here. I've seen you coming and going on the BLM lands and on The Hoff's Place but all I saw was strangers tearing up and down our roads.... After watching the [LiveSci] videos I'm starting to understand what's important about this research and how vital this land is in terms of the fossil resources here to be discovered. I'd like to invite you to come look on our place, we share a fence line with the ranch in your videos and we'd be happy to take you out. Maybe even ride a

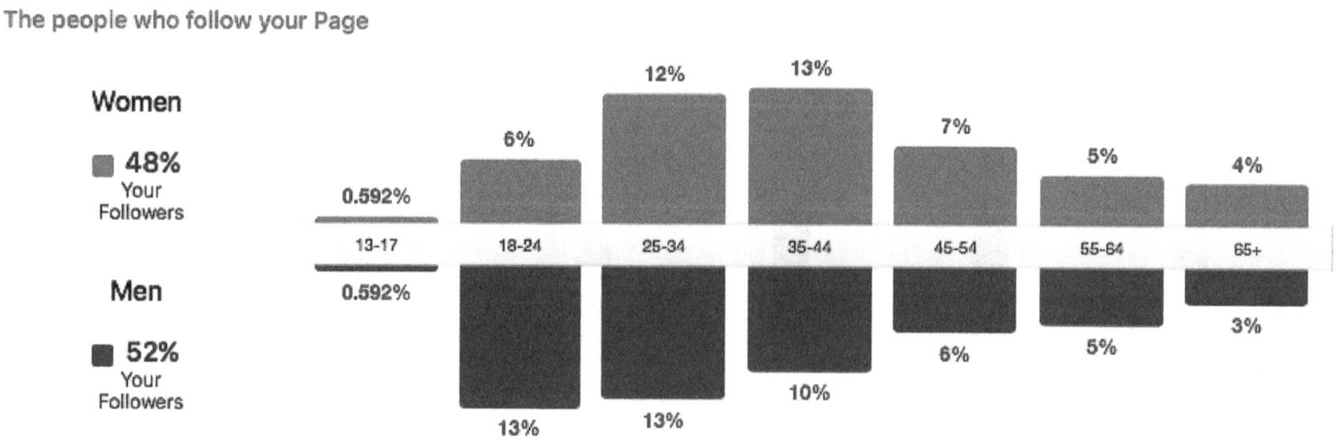

Figure 4. Facebook analytics showing male versus female followers and the ages provided in their Facebook profiles.

horse if you can manage? Keep the videos coming and let's make one about the Wang Ranch next. (Cindy, age 64, Baker, Montana)

The success of an educational video like the ones produced in the Valley of the Last Dinosaurs is not always measured in numbers. Reaching people with science content, as any educator will tell you, is about individual connections. Focusing on multi-institutional projects with diverse collaborators paid off. Land access is also important in the world of field paleontology, and the connection made through the online videos could be an added bonus that will have a positive effect on future work in the area.

Blog

In addition, the LiveSci written blog content received meaningful feedback. More than a month after the research project had wrapped up in the field area, an interesting story unfolded that highlights the value of archiving field experiences on the web for viewers to find even after the field season ends. A student pursuing a career in paleontology had stumbled upon a particular blog post written by a summer intern and LiveSci contributor, Megan Sims. Megan wrote an impassioned letter in the "field notes" section of the LiveSci website, expressing the all-too-common phenomenon known as "imposter syndrome" that many students face when pursuing their careers in science (http://lastdinos.livesci.org/field-notebook/7/7/an-open-letter-to-science-imposters). In her honest and open blog post, she describes sometimes feeling lost in academia and how her season of volunteer fieldwork with the collaborative team at the Marmarth Research Foundation helped her orient herself and meet academic mentors from institutions across the country. The post was read on 16 September 2016 by a paleontology student, who shared it to his Facebook page using the website's built-in share buttons. This student emphasized that it was a good read for students discouraged in their career path and that it had already helped him a lot. Less than an hour later, this student's Facebook post was read and shared yet again, this time by the Smithsonian Institution's National Museum of Natural History.

On 22 September 2016, using the form submission e-mail button on Megan's "bio" page on the LiveSci website, the student reached out directly to Megan Sims with a heartfelt thank you for posting this meaningful content during the field season:

Hi Megan, Last week I was browsing the Valley of the Last Dinosaurs website and I stumbled upon your post "An open letter to science imposters". Curious about the title, I read it, and I'm really glad I did. Like you, I've wondered if I was on the wrong path to paleontology, but upon reading about your experience, I feel much better… When I read your piece, it helped me realize that I shouldn't compare my educational path to that of others and should instead do what is best for me. I'm not exaggerating when I say it may very well be one of the most important things I've read regarding my education… Sincerely, Ian. (Garfalo, 2016)

LESSONS LEARNED

"Vlog" style scientific video content produced from the field is still relatively rare on the internet. Looking back at the seven weeks of filming, a few salient points stand out and provide insight for future projects that endeavor to record authentic science stories in remote locations with minimal equipment and budget.

The platform that worked best for LiveSci was Facebook. Videos that were uploaded directly to this platform made use of a high-resolution, large-format online video player on mobile devices and computers. The format was easily shareable across multiple preexisting social networks of collaborators, viewers, and host institutions.

Multi-institutional, collaborative projects such as the K/Pg extinction research featured in *The Valley of the Last Dinosaurs* work well for online science outreach because they make use of multiple social networks. This was most obvious with our videos that featured Smithsonian faculty; these videos were shared thousands of times on Facebook. Likewise, the shares that we got locally from the small Pioneer Trails Regional History Museum were also important in terms of connecting our researchers to the residents living in and around the field area.

Regarding video production, LiveSci benefited from meticulous schedule planning, but also benefited from having the flexibility to throw the schedule out and improvise on some occasions. The crew tried for nearly two weeks to film a video about the "camp life" of the DMNS team thriving in tents at their remote dinosaur quarry. However, bad weather made the typically exuberant camp look sad and depressing. It also kept some key characters in the camp from being able to physically make it to the site on the washed-out roads when the group was scheduled for filming. After a frustrating run of boring footage, the weather eventually became so extreme that it evolved into its own story, and a successful video about the camp living through a tornado touchdown with golf-ball–sized hail captured a unique view of scientific fieldwork that is not often shared.

Framing and camera work were an important part of the filming process. A beautiful and interesting scene will make or break a great conversation. The handheld "x-grip" mount for the camera provided a stable, yet moving frame that was as authentic in appearance as the conversation was casually informative. This setup depended upon having, at each location, an extra person who could hold the camera. In instances where extra hands were scarce, we relied on a simple and compact tripod, although the effect achieved using this method seems much less spontaneous and slightly more produced than the moveable, handheld set-up, which conveys the more authentic, informal nature of the project. The LiveSci crew also learned that there is more to "holding a camera" than simply "holding a camera," and they would have benefited greatly from the on-the-ground talents and insights of a skilled videographer. A dedicated videographer with an understanding of basic framing techniques would have allowed the featured scientists and hosts to concentrate more on video content without worrying about video quality.

Keeping the one-take conversations 100% scientifically accurate while also being casual was often challenging. Some researchers and volunteers had more experience with informal teaching and "in front of the camera" skills than others. In some cases, outlines were scripted to highlight key phrases and ensure that major talking points were addressed. In a few instances, a teleprompter app for the iPad was used (http://promptsmart.com), although this had a noticeable effect on the authenticity of the video content and was used only in extreme cases of camera shyness as can be seen in the video posted on 28 June 2016, titled "Question Tuesday #1" (http://lastdinos.livesci.org/blog/6/29/question-tuesday-1-kt-vs-kpg).

The typical field season for a paleontology project is approximately six weeks. Building a significant social media audience can often take several months, even with active posting several times each week. Excitement could be built in the months or weeks leading up to the field season, teasing with footage from previous years, interviews with experts as they are planning their work, and hints about the fossils that might be excavated that season. The seven-week duration of the LiveSci online video project in *The Valley of the Last Dinosaurs* was an excellent proof of concept that shows how two people equipped with little more than their iPhones and a computer can create an online community for engaging the public with science content. In just seven weeks, the Facebook page alone had grown from a few participants, family, and friends to reach the news feed of over 207,000 people in more than 50 countries around the world. Although the *Valley of the Last Dinosaurs* stands alone as a succinct body of work, it can also be considered a "season" of episodes in an enduring web-video series of field content for LiveSci at Tangled Bank Studios.

CONCLUSIONS

The 2016 project *LiveSci in the Valley of the Last Dinosaurs* shows that when the unique perspectives of scientists and audiences are carefully considered, museum professionals can capture engaging narrative stories from remote areas using easily accessible Internet technology and minimal equipment. The project also demonstrates that by mirroring the real-world networks of collaborators in large-scale research projects, science communicators from museums can expand their audiences and create meaningful interactions with large numbers of individuals around the world. The modern social media landscape provides an abundance of tools that can be easily employed to achieve museum outreach objectives, helping to raise awareness and spark interest in scientific work. These powerful online tools allow museums to share their mission-driven research with the world by moderating the direct intellectual exchanges between museum audiences and museum scientists in both directions. The benefit of this kind of science outreach to individual members of the public, and indeed to society at large, is obvious in a world where "public skepticism and concern are increasingly directed at scientific issues that appear to conflict with core human values and religious beliefs or that pose conflicts with political or economic expediency" (Leshner, 2007, p. 161). What can be less obvious is the tangible benefit to research scientists. In their critical evaluation of science outreach via social media, McClain and Neely (2015, p. 3) point out that "many of the arguments made in support of science outreach via social media … are simply special applications of arguments for outreach in general." It has been pointed out that certain barriers exist that hinder the research scientist from expending time and effort on public outreach. In a recent study of how academic biologists and physicists view science outreach, Eklund et al. (2012) found that 31% of scientists agree that, within the research university system that seems to value research productivity over all else, scientists have little time to engage in activities that are not directly connected to their academic pursuits. A study published in the *Journal of Geoscience Education* from 2005 demonstrates that many scientists view outreach as a form of volunteer work that is auxiliary to their other responsibilities (Andrews et al., 2005). There can exist indirect benefits to the careers of scientists who utilize social media for outreach, such as increasing overall communication skills, finding students, advertising a project to potential collaborators, appeasing funding agencies, and fostering the next generation of science enthusiasts who will help carry the mantle of science in years to come. However, indirect effects such as these can be difficult to quantify and assess. Numerous voices are arguing that U.S. scientists should be doing more to connect with the general public, including Dr. Alan Leshner, who was CEO of the American Association for the Advancement of Science and Executive Editor of *Science* magazine from 2001 to 2015. Leshner states that, "If science is going to fully serve its societal mission in the future, we need to both encourage and equip the next generation of scientists to effectively engage with the broader society in which we work and live" (Leshner, 2007). Yet, despite the obvious need for outreach, there is currently little incentive for scientists to participate. The scientific reward system does not currently do enough to support scientists who take the time and expense to interact with the public regarding their work. "Academic institutions need to reward faculty who fulfill commitments to such work. That will entail putting public outreach efforts among the metrics used to decide promotion and tenure" (Leshner, 2007). Currently, we are seeing a few universities begin to prioritize outreach. McClain and Neely (2015, p. 5) cite a change to University of Wisconsin–Madison, Division of Biological Sciences tenure guidelines, which call for "synergistic combinations of research, teaching and outreach or service … and the dissemination of information derived from scholarly inquiry for the benefit of society." The mid-1990s "Broader Impacts" directive in the National Science Foundation (NSF) grant application requires a description of how a proposed research project will affect society through outreach. This NSF directive is another illustration of how attitudes toward outreach are changing (National Science Foundation, 2002). While examples such as the directives from Alan Leshner and the changes taking place within NSF and the University of Wisconsin are encouraging, it is obvious that for the time being, the task of interpreting scientific research will continue to fall upon the science communicators such as HHMI's Tangled Bank Studios and other non-academic organizations. By working with these independently funded institutions on projects such as *LiveSci in the Valley of the Last Dinosaurs,* scientists can minimize the cost required to disseminate their work in terms of time and financial expense.

ACKNOWLEDGMENTS

The authors extend our appreciation to our many collaborators who helped make our seven weeks in the field such a joy and to make our project a success. Special thanks to Dr. Tyler Lyson and the Department of Earth Sciences at the DMNS for the countless hours of planning and support provided throughout the project in the planning stages and all along the way.

We thank Brisa Gonzales for laboring night and day to bring our work and the work of our many collaborators to the attention of Spanish-speaking audiences in Denver and beyond.

Sincere appreciation also goes out to Dr. Lou Taylor, Barb Benty, and the rest of the team at the Marmarth Research Foundation for providing us with amazing support, including an Internet connection and a dry place to sleep.

Our project depended on the many research paleontologists from various collaborating institutions who helped us create engaging content and share that content with their audiences. We thank Dave Evans from the Royal Ontario Museum; Ian Miller, Mike Getty, Matt Hess, Erica Evans, and Gabi Rossetto from DMNS; Kirk Johnson and Antoine Bercovici from the Smithsonian Institution's National Museum of Natural History; Stephen Chester from CUNY and Brooklyn College; Eric Sargis of the Yale Peabody Museum; and Dean Pearson of the Pioneer Trails Regional History Museum in Bowman, North Dakota.

REFERENCES CITED

Andrews, E., Weaver, A., Hanley, D., Shamatha, J., and Melton, G., 2005, Scientists and public outreach: Participation, motivations, and impediments: Journal of Geoscience Education, v. 53, no. 3, p. 281–293, https://doi.org/10.5408/1089-9995-53.3.281.

Astin, A.W., 1984, Student involvement: A developmental theory for higher education: Journal of College Student Personnel, v. 25, p. 297–308.

Bandura, A., 1977, Social Learning Theory: Englewood Cliffs, New Jersey, Prentice-Hall, 14 p.

Cerda, F.L., and Planas, N.C., 2011, Facebook's potential for collaborative e-learning: Revista de Universidad y Sociedad del Conocimiento, v. 8, no. 2, p. 197–210.

Duffy, P., 2008, Engaging the YouTube Google-eyed generation: Strategies for using web 2.0 in teaching and learning. The Electronic Journal of e-Learning, v. 6, no. 2, p. 119–130.

Ecklund, E.H., James, S.A., and Lincoln, A.E., 2012, How academic biologists and physicists view science outreach: PLoS One, v. 7, no. 2, p. 1–5.

Garfalo, I., 2016, "Just want to say thanks": Message to Megan Elayne Sims, 22 September 2016, e-mail.

Leshner, A., 2007, Outreach training needed: Science, v. 315, p. 161, https://doi.org/10.1126/science.1138712.

Lyson, T.R., Bercovici, A., Chester, S.G.B., Sargis, E.J., Pearson, D., and Joyce, W.G., 2011, Dinosaur extinction: Closing the '3 m gap': The Royal Society Publishing, Biology Letters, v. 7, p. 925–928, https://doi.org/10.1098/rsbl.2011.0470.

McClain, C., and Neeley, L., 2015, A critical evaluation of science outreach via social media: Its role and impact on scientists: F1000 Research, v. 3, no. 300, p. 1–12, https://doi.org/10.12688/f1000research.5918.1.

National Science Foundation, 2002, Merit review broader impacts criterion: Representative activities: National Science Foundation: https://www.nsf.gov/pubs/2002/nsf022/bicexamples.pdf (accessed September 2016).

Rainie, L., Smith, A., Schlozman, K., Brady, H., and Verba, S., 2012, Social media and political engagement: http://www.pewinternet.org/~/media/Files/Reports/2012/PIP_SocialMediaAndPoliticalEngagement_PDR.pdf (accessed September 2017).

Silius, K., Kailanto, M., and Tervakari, A., 2011, Evaluating the quality of social media in an educational context: International Journal of Emerging Technologies in Learning, v. 6, no. 3, p. 505–510.

Zhao, C., and Kuh, G., 2004, Learning communities and student engagement: Research in Higher Education, v. 45, no. 2, p. 115–138, https://doi.org/10.1023/B:RIHE.0000015692.88534.de.

Manuscript Accepted by the Society 30 January 2018
Manuscript Published Online 25 May 2018

The Geological Society of America
Special Paper 535

Digitization reveals and remediates challenges to research on dispersed museum collections from Florissant fossil beds, Colorado

Gwen S. Antell
National Park Service, Florissant Fossil Beds National Monument, P.O. Box 185, Florissant, Colorado 80816, USA, and *Department of Earth Sciences, University of Oxford, South Parks Road, Oxford OX1 3AN, UK*

ABSTRACT

Florissant fossil beds ranks among the best documented Cenozoic fossil deposits in the world in number of scientific publications and named species. The history of geoscience research on the Upper Eocene Florissant Formation spans nearly one and a half centuries. New excavations and transfers of historic collections have spread Florissant fossils to nearly 30 natural history museums during that period. The history of acquisition, conservation, and taxonomic study of each museum's collection is unique, so Florissant collections provide examples of how taxonomic diversity, physical conservation, and public exhibition of collections vary with provenance.

Dispersal of fossils among museums, including separation of type specimen parts and counterparts, has led to a variety of challenges for research on Florissant fossils. First, an exploratory, quantitative analysis of taxonomic diversity in four collections of fossil insects from Florissant uncovers a pattern of identification bias. Some taxonomists preferentially identify common taxa or consistently misidentify rare taxa, for instance. In light of this result, it is recommended that researchers vet any set of identifications made by multiple researchers or, ideally, identify specimens anew. Second, observations of Florissant specimens at different museums show that a large number of fossils have been lost, damaged, or destroyed due to actions such as travel on loan, display in exhibits, or application of non-archival conservation techniques. Through the digitization process, including cataloging and imaging specimens, curatorial staffs have discovered the extent of uncatalogued or missing material.

Digitization has mitigated some of the challenges associated with dispersion of specimens. Collaborative projects across museums have led to rediscovery of lost specimens or discovery for the first time of parts and counterparts that correspond to the same fossil but are housed at different institutions. Online databases that serve specimen images allow researchers to assign new taxonomic determinations, controlling for bias from earlier researchers, or to examine fossils remotely from photographs, reducing the need to handle and ship fragile material for loans. Moreover, providing public access to museum specimen records through collaborative digitization projects expands the opportunities to exhibit and develop specimen-based educational curricula.

Antell, G.S., 2018, Digitization reveals and remediates challenges to research on dispersed museum collections from Florissant fossil beds, Colorado, *in* Rosenberg, G.D., and Clary, R.M., eds., Museums at the Forefront of the History and Philosophy of Geology: History Made, History in the Making: Geological Society of America Special Paper 535, p. 301–309, https://doi.org/10.1130/2018.2535(20).

INTRODUCTION

Museums safeguard and control access to paleontological data. As institutions unleash species occurrence data from drawers and ledger books into digital platforms, research is shifting away from taxonomic descriptions of individual fossils and toward integrative studies on ecological, evolutionary, and environmental trends. Development of databases specific to paleontology has lagged behind similar efforts for neontology, but "paleoinformatics" is growing quickly (for review, see Benton, 1999; Uhen et al., 2013). There are many approaches to digitizing paleontology data. For instance, iDigBio (Integrated Digitized Biocollections; www.idigbio.org), the data hub of the National Science Foundation (NSF) Advancing Digitization of Biological Collections (ADBC) program, hosts modern and fossil specimen data uploaded by repository participants in Thematic Collections Networks (TCN). By centralizing specimen information from many collections, regions, and time periods into public databases, museums facilitate paleontology research on large temporal, spatial, and taxonomic scales.

Studies that pool data from multiple institutions assume ecological equivalency among different collections from the same or similar sites. Unfortunately, large-scale macroecology or macroevolution studies that acquire species occurrences from collaborative databases risk introducing additional noise or false signals. There may be biological differences between collection sites within a single formation, epoch, or other large-scale unit of analysis, as well as researcher bias in collecting, conserving, or identifying specimens in each collection.

Collections from the Upper Eocene Florissant Formation, Colorado (Fig. 1) offer a basis for evaluating differences caused by provenance, because shale fossils from the beds are equivalent in geologic age, elevation, and latitude but have been collected, curated, and identified by a unique combination of staff at different museums. Similarly, comparison of Florissant fossil condition across museums might indicate ways that conservation techniques or exhibit methods specific to institutions affect specimen stability. An astonishing number of fossil specimens have been described from Florissant, because more than 150 scientists have published articles related to the formation (Meyer, 2003). With ~1800 species names currently valid, the site has furnished more than one-third of all name-bearing specimens from National Park Service lands (Meyer, 2003; Tweet et al., 2016). Florissant Fossil Beds National Monument (FLFO) houses all fossils excavated (~10,000 specimens) after the establishment of the national park unit. An estimated 75%–80% of Florissant fossils, perhaps as many as 40,000 specimens, are part of historic collections housed at other museums (Meyer et al., 2008).

This work compares the taxonomic diversity and physical integrity of several Florissant Formation collections, identifies problems that resulted from differences in provenance, and provides examples of solutions to those problems. First, quantitative comparisons between museum holdings investigate possible sources of bias in collections. Second, observations assess the integrity of Florissant material among collections with the intention of validating, disputing, or proposing new methods of fossil conservation or exhibition.

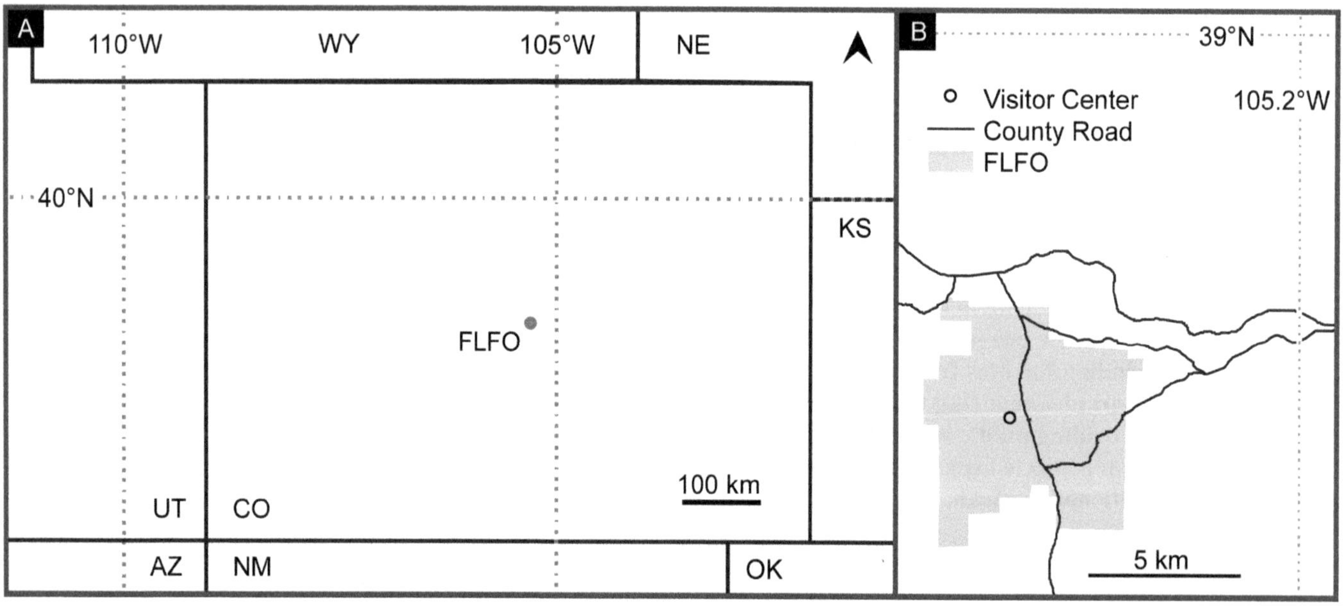

Figure 1. Map showing the location of Florissant Fossil Beds National Monument (FLFO), Teller County, Colorado. (A) The monument lies in the Rocky Mountains in central Colorado, within the Intermountain Region of the National Park Service. (B) FLFO encompasses over 24 km^2 of montane forest and meadows and operates a visitor center with exhibits and a paleontology lab space that includes a collection storage room. AZ—Arizona; CO—Colorado; NM—New Mexico; UT—Utah.

BACKGROUND

Geoscience at Florissant

Plant, insect, and occasionally vertebrate fossils are preserved in the Florissant fossil beds as organic compressions in lake shales. The upper part of the formation has been dated with $^{40}Ar/^{39}Ar$ isotopes at 34.07 Ma (Evanoff et al., 2001). Other fossils at the site include pollen, bones, petrified trees, and trace fossils. When shale is split open to reveal a compression fossil, the component slabs are called a part and counterpart. People have known of shale fossils at Florissant for centuries.

The native Ute peoples from Florissant, Colorado, likely knew of the fossils there generations before European settlers arrived. In 1873, the federally funded Hayden Survey of the American West explored the area, and expedition member Albert C. Peale described petrified wood and fossiliferous shale near Florissant (Peale, 1874). Notable collectors and taxonomists followed in the effort to document these resources. Samuel Scudder, an entomologist with the Survey who was later affiliated with Harvard, first visited and acquired specimens from the site in August 1877. One month earlier, the Princeton Scientific Expedition of 1877 traveled to the site. Princeton's disorganized party acquired most of its fossils not by splitting shale but by arranging transfers of substantial collections made by homesteader Charlotte Hill (Leopold and Meyer, 2012). T.D.A. Cockerell, Harry MacGinitie, and Henry Wickham made extensive new collections in the twentieth century, and private landowners ran commercial fossil collecting and tourism operations at Florissant until the mid-twentieth century (for review, see Veatch and Meyer, 2008). FLFO was established in 1969 after a precedent-setting environmental law case, described at length in *Saved in Time* (Leopold and Meyer, 2012). The monument supports a paleontology division that coordinates research at the site, develops materials to educate visitors about local geoscience, and maintains on-site historical archives and museum collections.

Disposition of Specimens

Nearly 30 different museums house Florissant specimens that were collected before the fossil beds were designated a national monument. Some museums acquired collections from associated researchers who visited Florissant. For instance, MacGinitie's extensive fossil plant collection resides almost entirely at the University of California Museum of Paleontology, the fossil repository for MacGinitie's home institution, the University of California, Berkeley. University museums also received collections resulting from expeditions that the school had funded. Princeton initially housed the fossils brought east by its Scientific Expedition of 1877, for example. The Milwaukee Public Museum acquired ~1500 paleobotany specimens from a six-week field trip in 1936 led by Assistant Curator of Geology E.R. Nelson (letters from Diane Gabriel, Milwaukee Public Museum, dated 3 January 1978; available at FLFO archive). Waynesburg University amassed ~3000 specimens collected during field trips to Florissant between 1940 and 1970 under former president Paul Stewart (Meyer, 2003).

Separation of collections among institutions allowed more scientists to access Florissant fossils at the time. Likely as part of an exchange, Dr. W.L. Fry gifted a small number of Florissant plant fossils from University of California Museum of Paleontology to Cornell University, Ithaca, New York; Ohio Wesleyan University, Delaware, Ohio; and University of Witwatersrand, Johannesburg, South Africa (letter from Howard Schorn, University of California, Berkeley, dated 3 January 1979; available at FLFO archive). Other researchers, notably Cockerell, sent fossils of rare taxa to institutions where specialists on those groups could describe the specimens. In this way, "the first really recognizable moss ever found fossil in America [...] was appropriately transmitted for study to Mrs. Gertrude Britton of the New York Botanical Garden, the best authority on American mosses. It may now be seen in the great Botanical Museum in Bronx Park" (Cockerell, 1908, p. 125–126). The New York Botanic Garden accessioned the specimen, and Britton formally described it, becoming the first woman to publish a scientific article about Florissant (Britton and Hollick, 1907; Meyer and Buskirk, 2016). Within a few years of sending the moss to Britton, Cockerell also sent Florissant wasps to Charles Brues, who was composing a monograph of parasitic Hymenoptera from the formation. Brues' Florissant type specimens in turn were reposited at several museums, including the American Museum of Natural History and the Milwaukee Public Museum (Brues, 1910). Florissant fossils transferred not only between researchers but also between museums.

Not every museum has maintained a commitment to house or conserve its accessions, and so some Florissant specimens have passed from one museum collection to another. The American Museum of Natural History transferred some non-type specimens to the National Museum of Natural History collections (USNM), Smithsonian Institution, for instance (letter from James Ferrigno, USNM, dated 5 January 1979; available at FLFO archive). Princeton University transferred many plant type specimens to the USNM and the rest of its Florissant collection to the Yale Peabody Museum and other institutions. The New York Botanic Garden type plant specimens—including Britton's moss—moved to the Yale Peabody Museum as well. Other collections have remained at their original repository but have been neglected in storage and never cataloged.

Digitization of Collections

FLFO anticipated the need for publicly accessible collections data by building a dedicated database (planning.nps.gov/flfo) that compiles photographs and citations for published specimens from the formation (Meyer et al., 2008). This open database predated many internal databases at institutions housing Florissant fossils. Subsequently, several museums, including the American Museum of Natural History, Natural History Museum in London, Harvard Museum of Comparative Zoology (MCZ),

TABLE 1. PROPERTIES OF COLLECTIONS

Museum collection	Locality*	Primary collector	Taxonomist
Florissant Fossil Beds National Monument	A	Thoene Henning	Antell
Harvard Museum of Comparative Zoology	A	Scudder	Perez de la Fuente
University of Colorado Museum of Natural History	B and C	Cockerell	Curatorial staff
Smithsonian National Museum of Natural History	B and C	Cockerell	Antell

*Letter codes indicate collections obtained at the same localities rather than site codes. Locality coordinates are available upon request to the National Park Service.

and University of Colorado Museum of Natural History (UCM), have relied on the Florissant database in order to create, verify, or add to their own records (Meyer et al., 2008). Now many individual institutions are uploading their internal database records to collaborative public databases.

Specimen records from Florissant are entering large searchable databases through the Fossil Insect Collaborative, an ADBC TCN (NSF award 1305066) that aims to make public the digital records of all major fossil insect collections in the United States. As of March 2018, the ten institutions that compose the group had contributed more than 95% of the ~12,700 Florissant fossil records in iDigBio (querying by "Florissant" in all fields and "FossilSpecimen" as the basis of record). The national monument is a participating institution in the TCN and at present is preparing to upload its specimen records to central data hubs.

MATERIALS AND METHODS

Data Acquisition

This study compiled specimen records for fossil insects from the Florissant Formation held at FLFO, MCZ, UCM, and USNM. The collection site(s), primary collector, and researcher(s) who identified the specimens for each collection are summarized in Table 1. FLFO records included in this analysis were exported from the Interior Collections Management System database and then filtered to fossils from excavation site 9 of Thoene Henning et al. (2012). Specimens from this site were collected using methodology intended to minimize collector bias. MCZ data were downloaded from MCZbase; the collection is largely the result of Scudder expedition excavations near to where specimens from the FLFO collection were excavated. The USNM houses collections from multiple expeditions to Florissant. Only fossil insects from the Cockerell collection were identified for this study. UCM records of invertebrate paleontology specimens collected at Florissant were exported from the internal museum database and then subset to specimens collected by Cockerell expeditions.

Additional information about collection provenance and data acquisition is available in the GSA Data Repository[1]. Excavation coordinates within the boundaries of the national monument are confidential under the Paleontological Resources Preservation Act of 2009 and may be obtained by request to Florissant Fossil Beds National Monument. Repositories holding type specimens were recorded from *The Fossils of Florissant* (Meyer, 2003).

Only specimens assigned to groups of Insecta traditionally ranked at a family or lower taxonomic level were included in analysis. This rank represents a trade-off between maximizing the number of specimens identified to taxa of a given rank and maximizing the number of taxa that exist at that given rank. Restricting specimen records to those identified at the "genus" level eliminates entire museum collections from analysis, and restricting to the "order" level decreases the number of taxa in every collection by an order of magnitude. It is well known that groups of equivalent rank are not necessarily equivalent degree of interspecific similarity, number of species included, or any other regard, because no standardized guidelines exist for delimiting higher taxa. Therefore, diversity measures calculated here at the "family" level are not intended to estimate ß diversity for collection sites or α diversity of the entire formation but instead are used for exploratory comparison of collections that may be high-graded due to researcher bias.

Diversity Analysis

Multiple metrics assessed the taxonomic diversity of each collection; all metrics were based on identification of specimens at the level of traditionally recognized "families." Each method assesses richness (the number of taxa present in a sample), evenness (the similarity in abundance of taxa), or a standard measure of diversity (a weighted combination of richness and evenness). There is no single or uncontested best diversity metric for evaluating community composition, and so multiple metrics are calculated for each Florissant collection. The diversity indices used here are widely employed, readily calculable without specialized software, and easily compared to one another.

The three diversity indices calculated for each collection are the Shannon-Wiener index, Simpson index, and estimated rarefied richness. Indices do not have any associated confidence intervals or significance thresholds; they are calculated as weighted sums of abundances over all species in a sample. Rarefied species richness accounts for unequal sample sizes by calculating richness on random subsamples. The size of each subsample was equal to the minimum number of specimens in

[1]GSA Data Repository Item 2018190, Elaboration of data sources and methods, is available at www.geosociety.org/datarepository/2018/, or on request from editing@geosociety.org or Documents Secretary, GSA, P.O. Box 9140, Boulder, CO 80301-9140, USA.

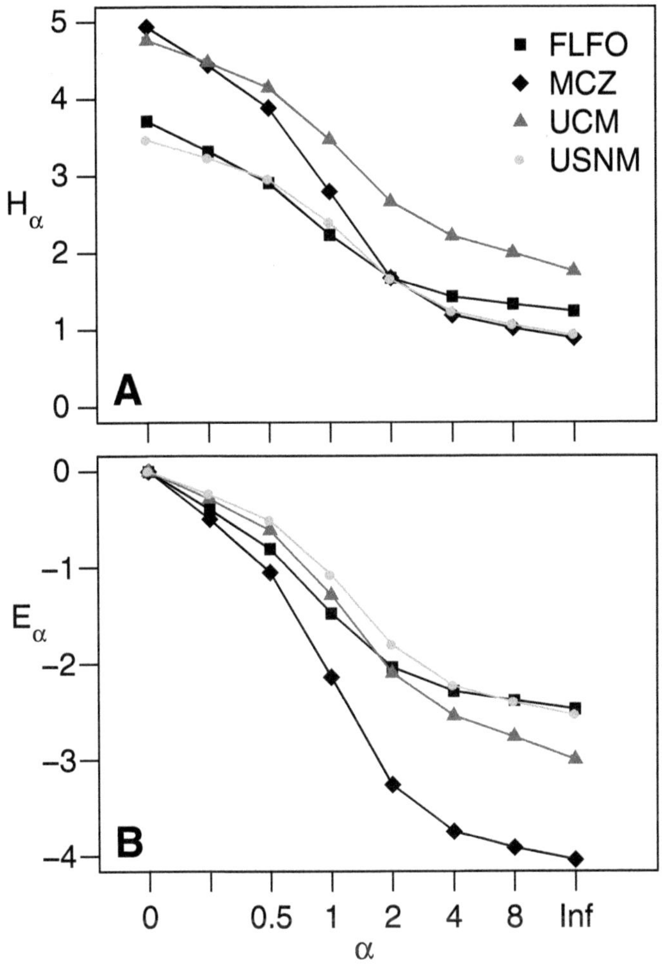

Figure 2. (A) Rényi diversity (H_α) and (B) Rényi evenness (E_α) profiles for all orders of α. A profile that is everywhere higher represents a sample that is more diverse or even. Profiles were calculated for fossil insect specimens from Florissant Fossil Beds National Monument (FLFO), Harvard Museum of Comparative Zoology (MCZ), University of Colorado Museum of Natural History (UCM), and Smithsonian National Museum of Natural History (USNM). Specimens were identified to taxa traditionally ranked as families.

any single collection. Two diversity ordering techniques, Rényi diversity (richness) and Rényi evenness profile comparisons, were performed on each collection. Profiles are functions calculated over orders (values from zero to infinity) of α; a richness or evenness profile that lies above another profile for all orders of α represents a sample (in this case, a collection) with greater richness or evenness, respectively. The numeric value of a profile at any point is less important for interpretation than is the shape of the profile relative to others. Additional information on diversity metrics, including equations and citations, is available in the GSA Data Repository (see footnote 1).

RESULTS

Quantitative Results

The Rényi diversity (H_α) of the UCM collection was higher than that of FLFO or USNM for every order α (Fig. 2A), and the UCM specimens were the most diverse by the Shannon-Wiener and Simpson indices (Table 2). The MCZ collection had the highest total richness, as shown by the greatest Rényi diversity for α of order zero, H_0 (Fig. 2A). Rarefied richness, however, which accounts for unequal n values among collections, was greatest for UCM specimens (Table 2). The MCZ collection had the lowest evenness, as shown in Figure 2B by an evenness profile lower than that of all other collections for all nonzero orders of α. Rényi evenness of the USNM collection was greater than that of the UCM collection for all nonzero orders of α (Fig. 2B). The ranking of FLFO, MCZ, and USNM collections in order of diversity depends on the index used (Table 2). The nearly overlapping Rényi profiles for FLFO and USNM specimens indicate that they were the most similar of all collections assessed, both in richness and in evenness (Fig. 2).

Observations

Each historic collection of Florissant fossils has been subjected to different curatorial practices that affect specimen stability. Past curatorial staff in at least two museums coated Florissant specimens with lacquer-like substances in efforts to enhance contrast between fossil and matrix. Similarly, preparators immersed some flaking or cracking specimens in consolidants. Most of the compounds applied to Florissant fossils have degraded over time. East Texas State University professor Joan Echols noted the fragility of the coating on one of the fossils in the university collection, for instance:

> [T]he big compound leaf that I wrote you about from the Florissant material here is about to break up. It was covered with shellac when

TABLE 2. TAXONOMIC DIVERSITY OF COLLECTIONS

Museum collection	Specimens identified*	Shannon-Wiener index	Simpson index	Rarefied richness† (± SE)
Florissant Fossil Beds National Monument	674	2.24	0.813	24 ± 2
Harvard Museum of Comparative Zoology	4663	2.80	0.813	39 ± 4
University of Colorado Museum of Natural History	852	3.48	0.930	51 ± 4
Smithsonian National Museum of Natural History	170	2.39	0.810	32 ± 0

*The number of specimens identified was the n value used in calculating diversity indices.
†Rarefied richness was calculated as the mean number of taxa occurring in random subsamples containing 170 specimens from a collection. SE—standard error.

it was collected … I can see that the shellac is drying out. … Drat! If I leave the shellac on it will pull the leaf off, and if I remove it, we'll lose it anyway. … It is a real beauty and I will hate for anything to happen to it. (letter dated 13 November 1979, available at FLFO archive)

Sometimes additional or more extreme conservation methods were used on type specimens compared to non-type specimens. In Scudder's collection at the MCZ, for instance, a stronger solvent is needed to remove the coating on type specimens than on other material; type specimens may have been coated with a more durable substance (Ricardo Perez de la Fuente, 2017, personal commun.). At the USNM, some Florissant fossil slabs were mounted onto larger pieces of shale. At least seven of Henry Wickham's beetle type specimens crumbled "into irretrievable bits" after the adhesive failed and the fossiliferous paper shale laminae peeled off the larger slabs (F.M. Brown, unpublished notes; available at FLFO archive). Furthermore, the USNM moved Florissant type specimens onto exhibit at least until 1979, and one Scudder holotype specimen there "was long on exhibit and has darkened very much" (F.M. Brown, 1979 unpublished notes; available at FLFO archive).

In addition to exposure of type specimens to chemical treatments or ultraviolet radiation, it was once common practice to sell, discard, or gift counterparts, or to deposit them in a different repository from that of the parts. The parts and counterparts of at least 97 different Florissant insect holotype specimens are housed at different museums. Of these, 39 are Hayden Survey fossils described by Scudder; the MCZ and USNM hold one component of each specimen. One fossil was described as a holotype after the establishment of the national monument, and the name-bearing part and counterpart for this species, *Oligodonta florissantensis* (Lepidoptera: Pieridae), are deposited at University of Florida Museum of Natural History and Waynesburg University. Cockerell described or collected the other 52 specimens, of which component pieces were sent to at least seven institutions (American Museum of Natural History, MCZ, Milwaukee Public Museum, Natural History Museum in London, UCM, USNM, and Yale University Peabody Museum of Natural History).

Parts and counterparts housed at different museums frequently receive different degrees of conservation and thus incur unequal amounts of damage, as exemplified by Cockerell's type specimen of *Lithosmylus columbianus* (Neuroptera: Osmylidae). One part shows wing pigmentation patterning but is missing wing venation around a repaired crack (Fig. 3A). The other component of the specimen is housed at a separate institution and remains intact, perhaps because of a coating applied to the specimen. The long-term stability of the coating was unknown at the time it was used; decades later it is discolored and crazing, and the underlying pigmentation is faint to absent (Fig. 3B). Only by combining information from part and counterpart is it possible to reconstruct in entirety both the venation and pigmentation of the type specimen.

DISCUSSION

Museum Collection Biases

One explanation for bias in community composition is that some taxonomists misidentify or preferentially identify specimens. If this problem was present in databased Florissant insect collections, one would expect that FLFO and USNM specimens would be the most similar to one another because both were identified by the same person (Table 1). The results agree with this prediction; moreover, MCZ and UCM specimens were identified by different researchers and were distinct from FLFO and from USNM specimens (Tables 1 and 2). Where identification bias exists, collections with lower diversity metrics are not necessarily less diverse. Rather, it is possible that some researchers preferentially identify the most common taxa, assigning proportionally fewer taxonomic identifications overall and causing the diversity of identified specimens to underrepresent the true diversity present in a collection. Two additional sources for researcher bias besides taxonomic determinations include collection locality and excavation methods.

A second hypothesized bias in community composition is that different depositional environments preferentially preserved different taxa. If this taphonomic process exerted a dominant

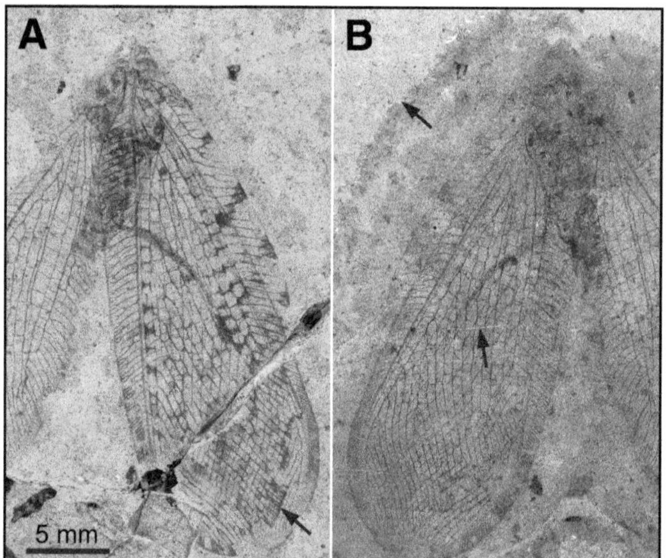

Figure 3. Part and counterpart of the type specimen of *Lithosmylus columbianus* (Neuroptera: Osmylidae). (A) University of Colorado Museum of Natural History (UCM) specimen 4103 is missing the areas of wing where the rock cracked. Original wing pigmentation is preserved, however, as indicated in one region by an arrow. (B) Harvard Museum of Comparative Zoology (MCZ) specimen 4820 is whole but coated in lacquer. The upper arrow shows the edge of this dark stain on the specimen. The lower arrow indicates one of the cracks in the coating. UCM-4103 photograph taken by David Zelagin (UCM). MCZ-4820 image courtesy of the MCZ, Harvard University. All original photography is © President and Fellows of Harvard College.

influence on the fossil record of insects at Florissant, one would expect that FLFO and MCZ collections and UCM and USNM collections would be similar to one another because specimens in these pairs were excavated at the same or nearby localities (Table 1). Instead, there was a clear disparity in both total and rarefied richness between FLFO and MCZ collections and in all diversity metrics between UCM and USNM specimens (Table 2; Fig. 2). It is possible that a collection locality bias was overshadowed by identification bias in the analysis. However, previous studies have found that Florissant fossils collected from sites or lithologies representing a variety of depositional environments were comparable in taxonomic composition, quality, abundance, and orientation (Smith and Moe-Hoffman, 2007; Thoene Henning et al., 2012). The negative evidence for a collection locality bias from the comparisons above can be interpreted as cautious support for a lack of bias associated with geographic locality.

Finally, the team of researchers who excavated any given collection could have selectively kept the rarest or most complete specimens they found. If community composition varied by collector, one would expect that UCM and USNM specimen diversities would be similar to one another and that FLFO specimen diversity would be distinct from that of all other collections. This is because both UCM and USNM collections resulted from Cockerell expeditions, while FLFO fossils were collected by a different team and with a protocol designed to eliminate collector bias (Table 1). The distinctly greater diversity of UCM collections and the similarity between FLFO and USNM collections counter these predictions (Table 2; Fig. 2). If there was a collection bias for higher-quality fossils, it did not cause detectable differences in diversity among the collections evaluated above. This result is subject to the same qualifications as the negative evidence for collection locality bias. Additionally, this study did not assess preservation quality, so it was not possible to determine whether field workers selectively collected specimens with a higher grade of preservation even if they did not selectively collect rarer taxa.

Specimen Loss and Damage

The dispersal of historic collections from Florissant among many institutions has caused practical difficulties for research involving physical examination of fossils. Specimens have been lost or destroyed through misplaced loans or non-archival conservation practices, such as coating fragile specimens with unknown chemicals or gluing them to other pieces of shale. Many parts and counterparts are housed in separate institutions, sometimes thousands of kilometers apart; Cockerell specimens are particularly widely scattered. In order to access all described Florissant specimens for a taxon of interest, a paleontologist must arrange travel to or loans from multiple museums.

When a paleoentomologist at the MCZ revised the fossil ants of North America, for instance, he acquired Florissant fossils from nine museums, on two continents (Carpenter, 1930). At that time collections staff at the British Museum (now the Natural History Museum in London) and Dublin Museum would have sent the specimens by ship; fossils from the UCM likely were delivered by train. Even after shipping specimens from around the world, Carpenter was unable to access all the Florissant material necessary for a complete revision of Formicidae, because one of Cockerell's holotype specimens was missing. Carpenter lamented "all the complications of the situation" resulting from this lost specimen (Carpenter, 1930, p. 42).

While storage of specimens in different institutions complicated research for Carpenter and many other paleontologists, it also ensured that some specimens were spared from non-archival conservation techniques practiced at particular institutions. The part and counterpart of Cockerell's type specimen of *Lithosmylus columbianus* preserve complementary morphological data due to differences in provenance, for instance (Fig. 3). In contrast, when component parts of other fossils were housed in the same museum and so were subject to the same conservation methods, some type specimens were lost or destroyed entirely—such was the fate of several of Wickham's fossil beetles.

Digital Solutions

Before collections were digitized, finding even published specimens across all the museums that housed them was no insignificant task, requiring copious correspondence with curators or hours of pulling drawers in museum cabinets. Only once cataloging and digitization began could researchers estimate the total number of type specimens from Florissant, count the number of museums that housed them, or discover the multitude of corresponding parts and counterparts in different museums (Meyer et al., 2008).

In addition to streamlining specimen acquisition for off-site researchers, online specimen catalogs with high-resolution images reduce the risk of loss or damage to fossils. Modern photography equipment and image stacking programs can enhance visible detail without applying damaging lacquers to a specimen. Hosting digitized specimen records online provides remote access to specimen data and so minimizes the need to ship and handle physical objects. Instead of requesting loans for material in bulk, scientists can study specimens from photographs or determine the specimens relevant for their studies and request loans for only those fossils. Exhibiting fossils through photographs can protect specimens as well. The University of Florida Museum of Natural History regularly displays images of the two FLFO holotype specimens it houses, and the objects themselves are not on exhibit (Roger Portal and Hongshan Wang, 2017, personal commun.). In this way, the museum showcases attractive and scientifically valuable fossils while preventing the damage incurred by Florissant type specimens that were physically exhibited at other museums. Whatever the intended purpose of digitization, modern software and hardware now automate many steps of imaging and cataloging specimens and of uploading data to online platforms, making these processes more efficient than ever before (Karim et al., 2016).

Online access to museum collections increases both the number of people who can view museum specimens and the number of museum specimens that people can view. The Fossil Insect Collaborative database iDigPaleo (idigpaleo.org), for instance, allows visitors to see images of Florissant insects from the American Museum of Natural History, MCZ, and USNM, institutions with substantial holdings of Florissant fossils that are not physically displayed. Database-driven outreach helps students and members of the public understand processes of scientific research and the enduring value of natural history museum collections. Digitization democratizes education as well, since school classes that are geographically or financially unable to visit museums where physical objects are on exhibit can view, measure, and download images online.

CONCLUSIONS

A quantitative analysis of taxonomic diversity among museum collections of fossil insects from the Florissant Formation indicates that taxonomists selectively identified taxa, substantially influencing the overall taxonomic richness and evenness of the determined fossils. Researchers should exercise caution when using identifications from previous workers. Where digitized specimen records are available, paleontologists could remediate or at least standardize researcher bias by reexamining and identifying specimens themselves from database images.

Digitization not only increases the accessibility of specimens to researchers, it also mitigates problems with physical conservation and furthers public engagement. Many Florissant type specimen parts and counterparts were separated between repositories, a fact discovered and documented largely during the process of databasing those specimens. Providing digital access to specimens reduces the need to loan physical objects and so minimizes the chance of specimen loss or degradation. Similarly, displaying images of fragile shale fossils protects those specimens from exposure to damaging environmental conditions outside of storage rooms. Simultaneously, virtual exhibition avoids the space constrains of floor plans, allowing the number of displayed specimens to increase by orders of magnitude. FLFO was one of the first institutions to create a searchable public database of fossil specimens. The monument is now part of an NSF program that uploads digitized fossil insect records into central data hubs, which are accessible to researchers, educators, and the public.

One cautionary note should follow any endorsement of virtual specimens: the growing usefulness of digital records in no way diminishes the value of the physical objects that those records represent. The best photogrammetry model could never capture the internal ridges of a gastropod shell, nor could the best 3D scan ever capture its isotopic signatures. The scientific history of Florissant fossils includes many examples of the irreversible damage and loss that occurs when repositories falter in long-term commitments to stabilize, catalog, and house paleontological specimens. Electronic databasing cannot replace more traditional methods of curation. Instead, digitization complements object-based curation, democratizes access to museum resources, and empowers professional, avocational, and aspiring paleontologists the world over to collaborate in large-scale geoscience research.

ACKNOWLEDGMENTS

The author thanks Herb Meyer (National Park Service) for feedback on drafts and discussions on the history of science at Florissant fossil beds. Talia Karim (University of Colorado Museum of Natural History), Ricardo Perez de la Fuente (Harvard University), and Finnegan Marsh (Smithsonian National Museum of Natural History) provided access to museum collections. Two anonymous reviewers suggested prudent additions and corrections to the manuscript. All archived documents referenced in the text belong to the F. Martin Brown collection at Florissant Fossil Beds National Monument, and Conni O'Connor (National Park Service) assisted with archive searches. Environmental Stewards, the Geological Society of America, and the National Park Service provided support through the Geoscientists-in-the-Parks internship program. The image of the figured University of Colorado Museum of Natural History specimen was generated by the Fossil Insect Collaborative Thematic Collections Network (National Science Foundation award 1305066). An Ernst Mayr Travel Grant in Animal Systematics from the Museum of Comparative Zoology funded travel that allowed examination of specimens at the Smithsonian National Museum of Natural History.

REFERENCES CITED

Benton, M., 1999, The history of life: Large databases in palaeontology, *in* Harper, D., ed., Numerical Palaeobiology: Chichester, UK, John Wiley & Sons Ltd., p. 249–283.

Britton, E.G., and Hollick, A., 1907, American fossil mosses, with description of a new species from Florissant, Colorado: Bulletin of the Torrey Botanical Club, v. 34, no. 3, p. 139–142, https://doi.org/10.2307/2479149.

Brues, C.T., 1910, The parasitic Hymenoptera of the Tertiary of Florissant, Colorado: Bulletin of the Museum of Comparative Zoology at Harvard College, v. 54, no. 1, p. 1–126.

Carpenter, F.M., 1930, The fossil ants of North America: Bulletin of the Museum of Comparative Zoology, v. 70, p. 3–66.

Cockerell, T.D.A., 1908, Florissant: A Miocene Pompeii: Popular: The Scientific Monthly, v. 74, p. 112–126.

Evanoff, E., McIntosh, W.C., and Murphey, P.C., 2001, Stratigraphic summary and $^{40}Ar/^{39}Ar$ geochronology of the Florissant Formation, Colorado, *in* Evanoff, E., Gregory-Wodzicki, K.M., and Johnson, K.R., eds., Fossil Flora and Stratigraphy of the Florissant Formation, Colorado: Proceedings of the Denver Museum of Nature & Science, ser. 4, no. 1, p. 1–16.

Karim, T.S., Burkhalter, R., Farrell, Ú.C., Molineux, A., Nelson, G., Utrup, J., and Butts, S.H., 2016, Digitization workflows for paleontology collections: Palaeontologia Electronica, v. 19, no. 3, p. 1–14.

Leopold, E.B., and Meyer, H.W., 2012, Saved in Time: The Fight to Establish Florissant Fossil Beds National Monument, Colorado: Albuquerque, New Mexico, University of New Mexico Press, 139 p.

Meyer, H.W., 2003, The Fossils of Florissant: Washington, D.C., Smithsonian Books, 258 p.

Meyer, H.W., and Buskirk, B.L., 2016, The significant involvement of women in science and conservation at the Florissant fossil beds, Colorado: Geological Society of America Abstracts with Programs, v. 48, no. 7, https://gsa.confex.com/gsa/2016AM/webprogram/Paper282011.html.

Meyer, H.W., Wasson, M.S., and Frakes, B.J., 2008, Development of an integrated paleontological database and web site of Florissant collections, taxonomy, and publications, *in* Meyer, H.M., and Smith, D.M., eds., Paleontology of the Upper Eocene Florissant Formation, Colorado: Geological Society of America Special Paper 435, p. 159–177, https://doi.org/10.1130/2008.2435(11).

Peale, A.C., 1874, Report of A.C. Peale, M.D., geologist of the South Park division: Annual Report of the U.S. Geological and Geographical Survey of the Territories, v. 7, p. 193–273, 20 pl.

Smith, D.M., and Moe-Hoffman, A.P., 2007, Taphonomy of Diptera in lacustrine environments: A case study from Florissant fossil beds, Colorado: Palaios, v. 22, no. 6, p. 623–629, https://doi.org/10.2110/palo.2006.p06-119r.

Thoene Henning, J., Smith, D.M., Nufio, C.R., and Meyer, H.W., 2012, Depositional setting and fossil insect preservation: A study of the late Eocene Florissant Formation, Colorado: Palaios, v. 27, no. 7, p. 481–488, https://doi.org/10.2110/palo.2011.p11-084r.

Tweet, J.S., Santucci, V.L., and McDonald, H.G., 2016, Name-bearing fossil type specimens and taxa named from National Park Service areas: New Mexico Museum of Natural History and Science Bulletin, v. 73, p. 277–288.

Uhen, M.D., Barnosky, A.D., Bills, B., Blois, J., Carrano, M.T., Carrasco, M.A., Erickson, G.M., Eronen, J.T., Fortelius, M., and Graham, R.W., 2013, From card catalogs to computers: Databases in vertebrate paleontology: Journal of Vertebrate Paleontology, v. 33, no. 1, p. 13–28, https://doi.org/10.1080/02724634.2012.716114.

Veatch, S.W., and Meyer, H.W., 2008, History of paleontology at the Florissant fossil beds, Colorado, *in* Meyer, H.M., and Smith, D.M., eds., Paleontology of the Upper Eocene Florissant Formation, Colorado: Geological Society of America Special Paper 435, p. 1–18, https://doi.org/10.1130/2008.2435(01).

Manuscript Accepted by the Society 21 November 2017
Manuscript Published Online 6 June 2018

Printed in the USA

The Geological Society of America
Special Paper 535

From public lands to museums: The foundation of U.S. paleontology, the early history of federal public lands and museums, and the developing role of the U.S. Department of the Interior

Gregory A. Liggett
Bureau of Land Management, Montana State Office, 5001 Southgate Drive, Billings, Montana 59101, USA

S. Terry Childs
Department of the Interior, Museum Program, 1849 C Street NW, Washington, D.C., 20240, USA

Nicholas A. Famoso
National Park Service, John Day Fossil Beds National Monument, 32651 Highway 19, Kimberly, Oregon 97848, USA

H. Gregory McDonald
Bureau of Land Management, Utah State Office, 440 West 200 South, Suite 500, Salt Lake City, Utah 84101, USA

Alan L. Titus
Bureau of Land Management, Grand Staircase–Escalante National Monument, 669 South Highway 89A, Kanab, Utah 84741, USA

Elizabeth Varner
Department of the Interior, Museum Program, 1849 C Street NW, Washington, D.C., 20240, USA

Cameron L. Liggett
Bureau of Land Management, Montana State Office, 5001 Southgate Drive, Billings, Montana 59101, USA

ABSTRACT

Today, the United States Department of the Interior manages 500 million acres of surface land, about one-fifth of the land in the United States. Since enactment of the Antiquities Act in 1906, historic and scientific resources collected on public land have remained government property, held in trust for the people of the United States. As a result, the Department of the Interior manages nearly 204 million museum objects. Some of these objects are in federally managed repositories; others are in the repositories of partner institutions.

The establishment of the United States as a nation corresponded with the development of paleontology as a science. For example, mastodon fossils discovered at or near present-day Big Bone Lick State Historic Site, Kentucky, found their way to

Liggett, G.A., Childs, S.T., Famoso, N.A., McDonald, H.G., Titus, A.L., Varner, E., and Liggett, C.L., 2018, From public lands to museums: The foundation of U.S. paleontology, the early history of federal public lands and museums, and the developing role of the U.S. Department of the Interior, *in* Rosenberg, G.D., and Clary, R.M., eds., Museums at the Forefront of the History and Philosophy of Geology: History Made, History in the Making: Geological Society of America Special Paper 535, p. 311–334, https://doi.org/10.1130/2018.2535(21).

notable scientists both in the United States and in Europe by the mid-eighteenth century and were instrumental in establishing the reality of extinction.

Public land policies were often contentious, but generally they encouraged settlement and use, which resulted in the modern pattern of federal public lands. Continued investigation for fossils from public land filled the nation's early museums, and those fossils became the centerpieces of many museum exhibitions.

Case studies of the management of fossils found in Fossil Cycad National Monument, the John Day fossil beds, the Charles M. Russell National Wildlife Refuge and surrounding areas of public land, the American Falls Reservoir, and Grand Staircase–Escalante National Monument are outlined. These examples provide a sense of the scope of fossils on federal public land, highlight how their management can be a challenge, and show that public land is vital for continued scientific collection and research.

INTRODUCTION

How has the United States Department of the Interior (DOI) come to hold a collection of nearly 204 million museum objects (Table 1)? The story is intimately connected with the development of paleontology as a science, the expansion of the nation across the continent and the consequent creation of public lands, and the rise of modern museums and research collections. Initially the federal government took little notice of the museum collections (archaeological and paleontological objects in particular but also any object of scientific or historical interest) made from its lands. That changed in 1906, when the federal government asserted ownership of historic and scientific resources collected from lands it managed. With the assertion of ownership in trust for the American people, the collection of museum objects to be managed by the DOI grew significantly, creating a museum collection that is now one of the largest in the United States.

Following its successful war for independence, the newly established United States of America immediately began to document the lands and resources it had acquired and to determine how those resources could be used to the country's advantage. From the very beginning, scientific inquiry, based on collection, was a key component of the study of the newly acquired lands. The vast spaces of the nation held significant historical and scientific assets including archaeological, paleontological, geological, and biological resources. The development of the young science of paleontology was intimately intertwined with the expansion of the young nation, because fossils from the West would significantly add to the scientific dialogue.

As an historical science, paleontology posits that we cannot understand the present without understanding the past. Likewise, to place in context present-day struggles related to public land, it is informative to explore the history of public lands. As the United States added to its land holdings and stretched across the continent, numerous policies were designed to encourage

TABLE 1. LIST OF BUREAUS AND OFFICES WITHIN THE DOI AND THE NUMBER OF MUSEUM OBJECTS CURRENTLY IDENTIFIED UNDER THEIR MANAGEMENT IN FISCAL YEAR 2016

Bureau	Objects (millions)	Linear (ft)	Volume (cubic ft)
BIA	3.736	450.0	6585
BLM	11.824	2541.5	5868
BOR	8.742	0	6086
FWS	4.473	0	0
IACB	0.011	0	0
IM	0.008	0	0
NPS	175.060	0	0
USGS	0.056	0	0
Totals	203.910	2991.5	18,539

Note: Several bureaus or offices with very small collections are not shown. Numbers given in millions. Numbers of objects are rounded. BIA—Bureau of Indian Affairs; BLM—Bureau of Land Management; BOR—Bureau of Reclamation; FWS—Fish and Wildlife Service; IACB—Indian Arts and Crafts Board; IM—Department of Interior Museum; NPS—National Park Service; USGS—United States Geological Survey.

development, use, and settlement. Early on, little regard was given to notions of preservation, and this was reflected in the lack of federal focus on the huge numbers of fossils from federal public land filling the nation's early museums. However, in the late nineteenth and early twentieth centuries, that began to change as a conservation ethic emerged. Concern over the obvious loss of resources, such as forested lands, and a desire to set areas aside slowly turned the course of the government away from promoting exploitation of resources. A landmark law for museums and federal public land was the Antiquities Act of 1906, in which for the first time the government laid claim to historic and scientific resources coming from the public land.

This paper provides brief histories of early fossil finds from the wilds of the continent, the early development of paleontology, and the rise of the modern museum. Also examined are the story of the United States public land (Fig. 1) and the laws and policies that directed its development. This will illuminate the DOI's role as trustee for a tremendous national treasure and will show that in this role, the DOI faces significant challenges that will last into the future. Included throughout the paper are illustrative examples, historic and recent, of the collection and management of fossil resources on public land. (The terms "public land" and "public domain" refer to federal public land and are used synonymously throughout this paper. Also, the terms "paleontological resources" and "fossils" can be thought of as meaning fossils holding scientific interest.)

FIRST FOSSILS FROM NORTH AMERICA

In 1739, long before the birth of the United States, a French military group came upon a site in what is today northern Kentucky. Fossil bones from an elephant-like creature from the site were shipped down the Mississippi River to New Orleans and transported to Paris, where they were studied by the first practitioners of paleontology, Georges Cuvier and Georges-Louis Leclerc, Comte de Buffon. Certainly, people had taken notice of fossils in North America for millennia (Simpson, 1943), but this event is one of the earliest that influenced science. The locality that the bones most likely came from is today called Big Bone Lick State Historic Site (see Simpson [1942] and [1943] for discussions). It was frequented by frontiersmen and Native Americans of the area due to the saline springs, but the visitors also marveled at the bones. Over the succeeding decades, fossils from Big Bone Lick were sent to Benjamin Franklin, George Washington, and other colonial leaders and scientists. Bones, teeth, and tusks of large proboscideans captivated the scientists of Europe and North America. This animal was called a "mammoth" at that time, although we now identify it as a mastodon. Questions about the bones' origins and their taxonomic affinities were subjects of great debate. Some thinkers favored the notion that these beasts must have come to Kentucky by way of a connection between North America and Asia. Others questioned how elephant-like animals (which live only in the tropics today) could have survived in temperate Kentucky. Still others argued that the remains represented more than one type of animal, and were a mixture of proboscidean (as evidenced by the tusks) and hippopotamus (as evidenced by the teeth) (Turner, 1799; Semonin, 2000).

On 21 January 1796, the French scientist Georges Cuvier read his paper, "On the species of living and fossil elephants," for the National Institute of Sciences and Arts in Paris. He pointed out that the species in the New World and those known from Siberia were not the same as the modern Asian and African elephants (Cuvier, 1799). With careful anatomical work and reasoning, Cuvier articulated the best evidence for extinction presented to date.

Extinction was a radical idea. It had been contemplated long before, but the world was vast, and the possibility always existed that an organism known only from fossils still lived in some remote, unstudied place of the globe. To some people, extinction implied that there was a flaw in nature's design, which was incompatible with an orderly divine creation. At the very least, the implications of extinction and deep geologic time forced scientists to question long-held understandings of the natural world (Rudwick, 1972). Fossils from the American West added to a vigorous debate of radical new scientific ideas. The contributions of American scientists would soon rival those of their European colleagues (Jaffe, 2000).

Thomas Jefferson took great interest in the "mammoth" bones of the West. The wondrous beast was sometimes referred to as the American *incognitum* (Semonin, 2000). Jefferson sought to acquire the fossils for his own research, and engaged George Rogers Clark to send some to him in 1782 (Semonin, 2000; Hedeen, 2008). After William Clark returned from the western expedition with Meriwether Lewis, he was financed by Jefferson to return to Big Bone Lick in 1807 to collect mostly head and foot bones, which were missing from the American Philosophical Society's "mammoth" skeleton (see the section titled "Peale's Museum" for more on this specimen). Clark's dig was an immense success, recovering over 300 bones of various species, including the coveted "mammoth" cranium. Clark sent the bones to Jefferson at his house. The sheer volume of the collection gave Jefferson the idea to send duplicate specimens to the National Institute of France, and the bones are still curated today at the Muséum National d'Histoire Naturelle in Paris (Semonin, 2000).

In 1797, Jefferson submitted a note to the American Philosophical Society on other bones recently uncovered in a Virginia (today West Virginia) cave. What most impressed Jefferson about these bones were the large claws of the beast, which he at first imagined to be lion-like in appearance and habit. By the time he published his paper on the animal, Jefferson had learned of an extinct giant sloth from South America with similar claws, so he did not specify that the animal was a cat (Boyd, 1958). He dubbed the beast *Megalonyx,* or "giant claw" (Jefferson, 1799).

Jefferson was not a supporter of the new idea of extinction. He subscribed to a notion known as the "Great Chain of Being," a traditional concept with roots in early science and Christian doctrine. This idea imagines a hierarchy that ranks everything in God's creation, inanimate and living, with humans at the top

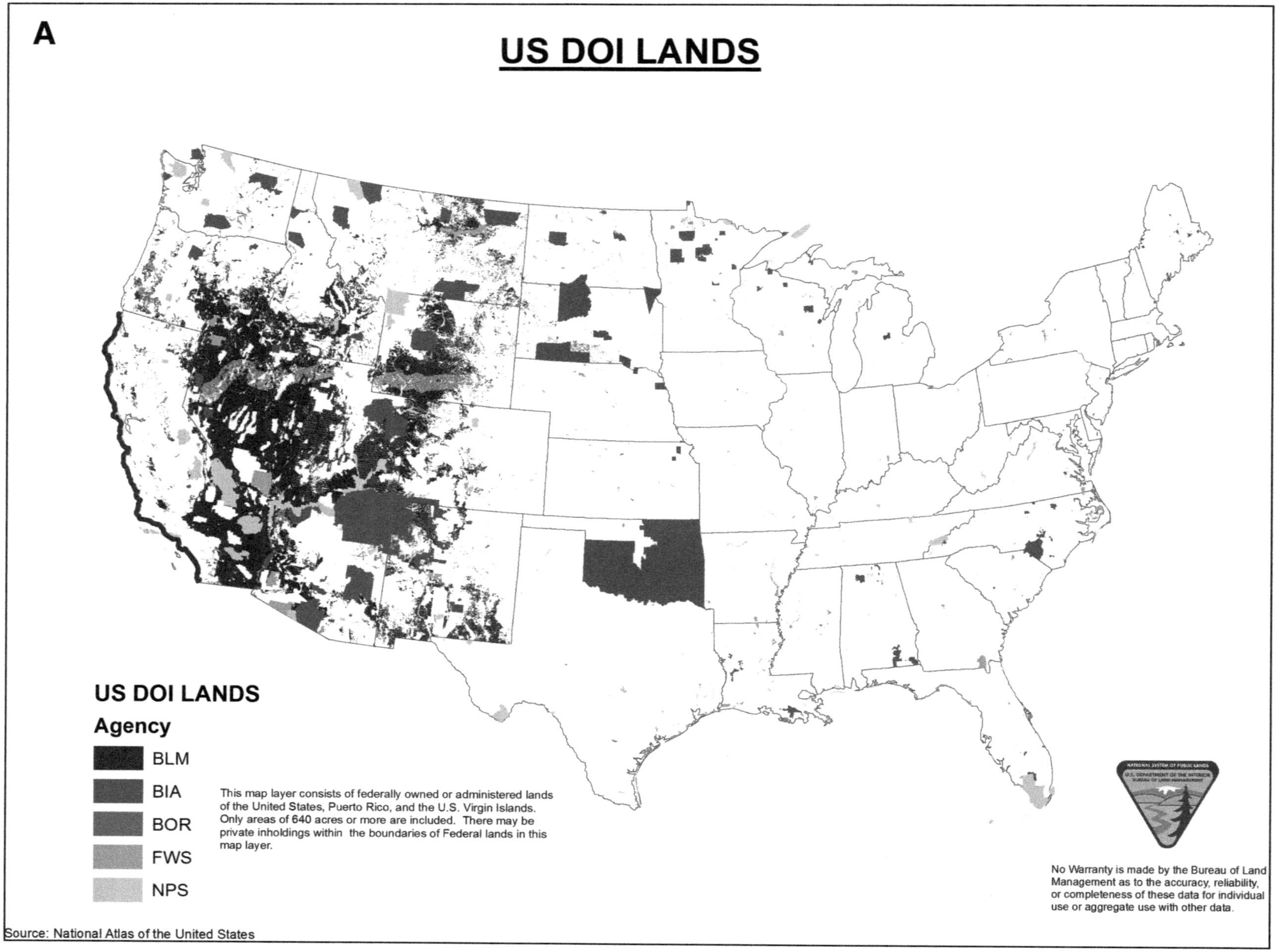

Figure 1 (*Continued on facing page*). Current United States Department of the Interior–managed lands: (A) in the continental United States; and (B) in Alaska. BLM—Bureau of Land Management; BIA—Bureau of Indian Affairs; BOR—Bureau of Reclamation; FWS—Fish and Wildlife Service; and NPS—National Park Service.

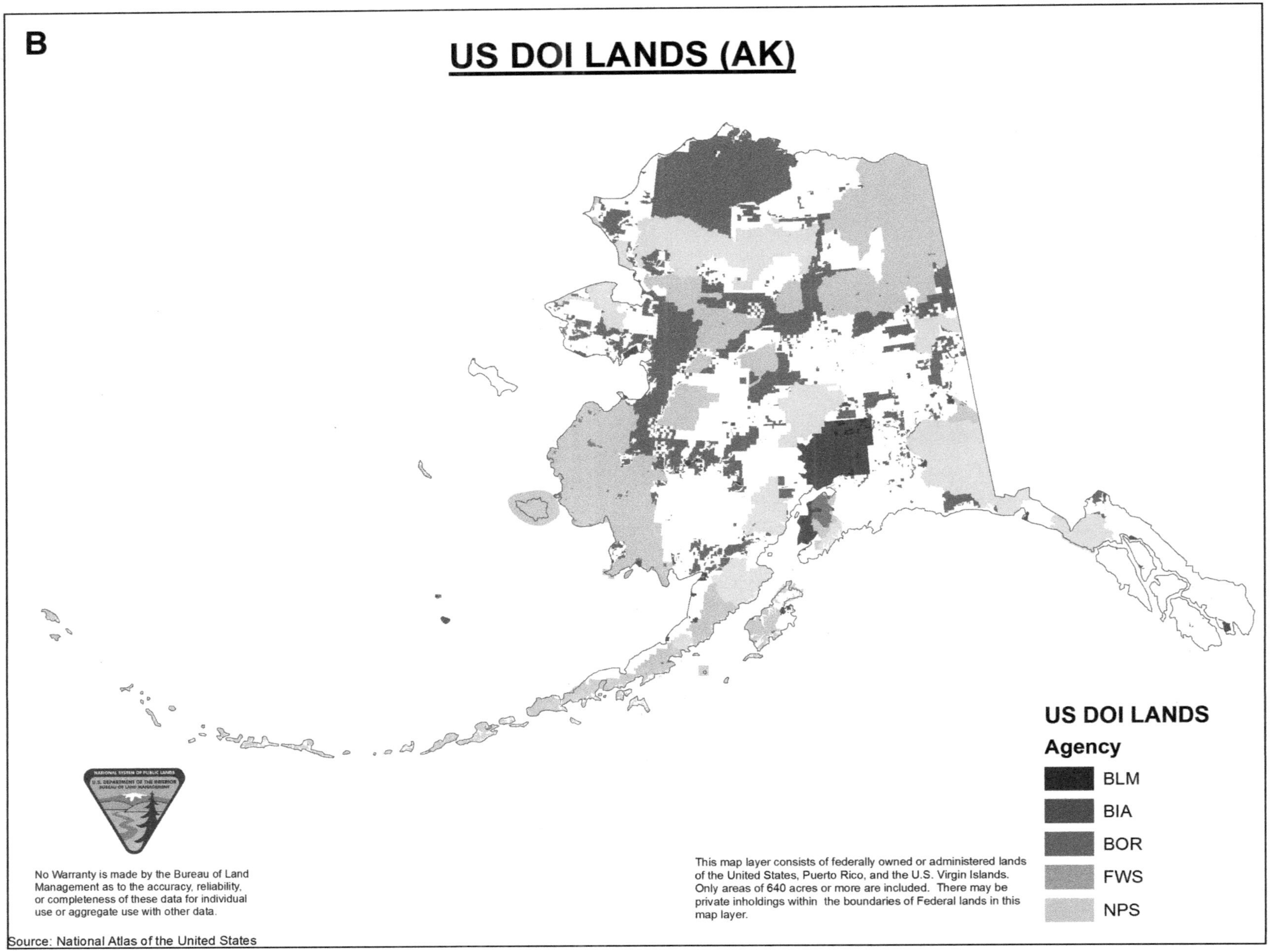

Figure 1 (*Continued*). AK—Alaska.

of the biological world (Semonin, 2000). Taking the notion further, some held that above humans were the moon, stars, angels, and then God. Like a physical chain, the linkages of all of creation could only endure if every link remained intact. If one link, or one species, were lost, it was thought that the entire system would collapse. "It may be asked," wrote Jefferson, "why I insert the Mammoth [as living in North America], as if it still existed? I ask in return, why I should omit it, as if it did not exist? Such is the economy of nature, that no instance can be produced of her having permitted any one race of her animals to become extinct; of her having formed any link in her great work so weak as to be broken" (Jefferson, 1788).

DEVELOPMENT OF THE PUBLIC DOMAIN

The territories west of the Appalachians and the issue of public domain were matters of early interest as the thirteen colonies joined in the War for Independence. Six states had been given charters from England granting them land beyond the Appalachian Mountains: Virginia, Massachusetts, Connecticut, North Carolina, South Carolina, and Georgia. New York also claimed land to the west from a concession by the Iroquois Nation (Muhn and Stuart, 1988). However, the other colonies not holding similar claims to western land worried that their economic and political power would be diminished compared to those with lands in which to expand. In November, 1776, Maryland contended "... that the back lands claimed by the British crown, if secured by the blood and treasure of all, ought, in reason, justice, and policy, to be considered as a common stock ..." (Scharf, 1879, p. 276) and threatened not to sign the Articles of Confederation that sutured the young nation together. Gradually, the land-claim states ceded their interests in the more western lands to the federal government, and Maryland joined in the struggle for independence.

The public domain again grew substantially after the purchase of the Louisiana Territory from France in 1803 (Fig. 2). President Thomas Jefferson authorized several exploratory and scientific expeditions of the continent, with Lewis and Clark's Corps of Discovery (1803–1806) being the most well-known. Paleontology was again not far from the minds of statesmen. Jefferson fully expected that Lewis and Clark would discover

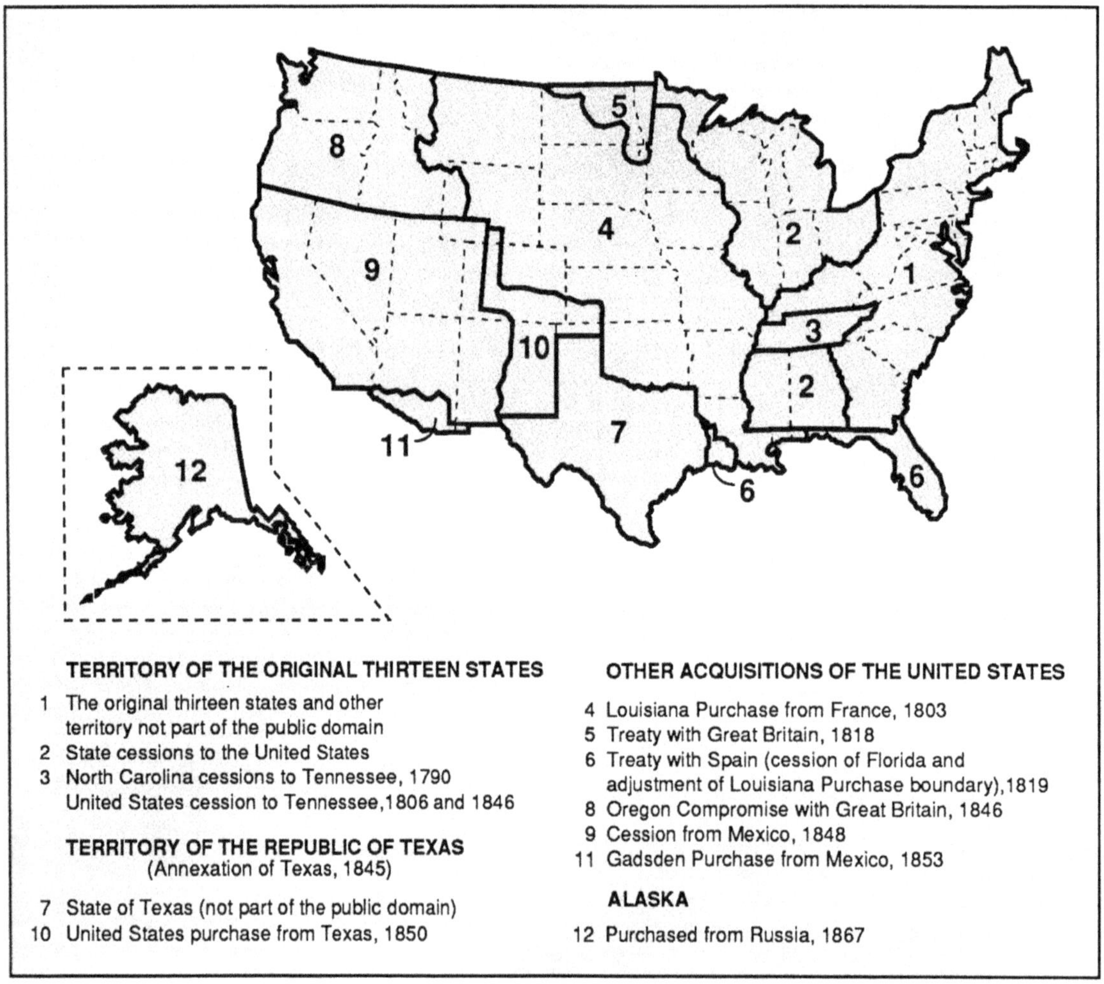

Figure 2. The acquisition of the U.S. lands occurred piecemeal and by a variety of actions. From Muhn and Stuart (1988).

both the "mammoth" and the *Megalonyx* living somewhere in the West, given his conviction that extinction was impossible (Ambrose, 1997; Semonin, 2000).

Whereas Lewis and Clark found many important species new to science, and sent evidence of them back east, they did not find a *Megalonyx* or a "mammoth" as Jefferson hoped. They did, however, find a few fossils, including a marine reptile in South Dakota and a fish fossil from Iowa or Nebraska (Chaky, 2015). Today the fish is in the collection of the Academy of Natural Sciences in Philadelphia, and is the only fossil known from the expedition still preserved (Simpson, 1942). One other fossil of note mentioned in their journals was a bone Clark described from along the banks of the Yellowstone River, several miles downstream from the landmark he named "Pompys Tower," now called Pompeys Pillar National Monument. The bone, which Clark described as a rib some 3 inches (7.6 cm) in diameter and 3 feet long (0.91 m), is likely the first notice made of a dinosaur fossil in the United States, foreshadowing the rich finds of fossils to be made in the West (Simpson, 1942). The first dinosaur fossils from America would not be recognized again and collected until 1855 when Ferdinand Hayden collected them from the banks of the Judith River in Montana (Leidy, 1856). That area is now part of the Upper Missouri River Breaks National Monument. Both Pompeys Pillar and the Upper Missouri River Breaks are managed today by the Bureau of Land Management (BLM).

Other early explorations authorized to study the continent included the Red River Expedition of 1806, which investigated the southwestern part of the continent, and Zebulon Pike's 1806–1807 expedition to examine the northern parts of Spain's territorial claims in what is now Colorado and New Mexico.

During the 1840s, the nation undertook another major growth spurt—annexing Texas in 1845; resolving the disputed boundary of Oregon with Britain in 1846; and with the end of the Mexican-American War, adding California, half of New Mexico, most of Arizona, all of Nevada and Utah, and parts of Wyoming and Colorado to the nation in 1848 (Fig. 2) (Utley and Mackintosh, 1989). The need for a consolidated department to manage domestic affairs was clear, and the DOI was created in 1849. The new department took on a wide variety of homeland tasks, including conducting a census of freed slaves moved to Haiti; exploring the West; overseeing the District of Columbia (D.C.) jail; regulating territorial governments; constructing the D.C. water system; managing hospitals and universities; managing parks and other public land; and taking responsibility for Native Americans. It is easy to see why the DOI has sometimes been called the "department of everything else" (Utley and Mackintosh, 1989).

Following the creation of the DOI, several independent surveys of the interior were authorized, including expeditions by John Wesley Powell (1869), Clarence King (1867–1872), Ferdinand Hayden (1869–1871), and George Wheeler (1872–1879). All these surveys of the West worked more or less in isolation from each other. Each contributed to the scientific understanding of the flora, fauna, geology, and paleontology of the United States. After Hayden's discovery of dinosaurs in Montana, many fossil remains were found throughout the West. These contributed to a colorful period in history sometimes called the "Great Dinosaur Rush," a period from ~1877–1892. This lust for fossils also forever linked the names of two rival scientists, Edward Drinker (E.D.) Cope and Othniel Charles (O.C.) Marsh. Each tried to outdo the other by possessing the best fossils in their own collections and rushing to publish new species faster than the other. The frontiers of the West were open for prospectors to roam freely, collecting fossils for wealthy scientists or for museums that could afford to fund their excavations or to outright purchase them. In this freewheeling period, thousands of fossils by count, and tons by weight, filled the museums of the nation, particularly in the East, mostly coming from public lands.

Paleontology played a central role in all the various surveys. Fossils are important indicators of geologic age, helping stratigraphers arrange the rocks they map into chronological sequence. Cope served as chief paleontologist during the Wheeler survey of the territories in 1874. There was no pay for the position, but it allowed Cope to claim he was on official government business and to procure supplies from the army (Jaffe, 2000). Political reformers came to argue that all of the independent government-funded survey efforts were inefficient. With political maneuvers, Marsh and his allies worked two items into legislation: (1) merging the surveys; and (2) providing that material collected under the surveys would become national property and be deposited in the Smithsonian National Museum of Natural History (Jaffe, 2000). This last provision was aimed directly at making Cope turn over any collections he had made under official government business. In 1879, the new United States Geological Survey (USGS) was formed (Muhn and Stuart, 1988). In 1882, Marsh himself was appointed as chief paleontologist with the USGS, which made him the nation's first paid government paleontologist. These two brilliant scientists apparently felt that, vast as it was, North America was not big enough for them to coexist, and the Cope and Marsh feud is now legendary (Lanham, 1973; Colbert, 1984; Jaffe, 2000).

Land Policy and Legislation

The history of land policy, the various related laws, and the many competing interests of the past provide a basis for understanding modern aspects of public land (see Gates, 1976). At the beginning, the United States was "land rich and cash poor," and the nation repeatedly turned to its land assets. For example, soldiers in the Continental Army were promised land as payment, as were soldiers in several subsequent national wars. There was also much debate about the best land policy. The two earliest broad notions that arose were either to have the government manage the selling of the land to individuals in an orderly fashion, or to wholesale the land to wealthy speculators with deep pockets who would then develop the land. The first view was espoused by Thomas Jefferson, and the second by Alexander Hamilton. The Public Land Act of 1796 took Hamilton's approach, and small farmers had to buy their farms from larger real estate interests.

However, this angered farmers, which brought about the Land Act of 1800. It reduced the minimum size of parcels that could be purchased and created a system to extend government credit to purchasers (Muhn and Stuart, 1988).

An essential development was a system to survey and divide land. Early on, claims to land were disorderly, leading to multiple disputes over land boundaries that burdened the court system. What was needed was an unambiguous system for subdividing land into unique parcels, and what emerged was the Rectangular Survey System, or the Public Land Survey System. Once a baseline was established for a region, rectangles of 36 square miles, 6 miles to a side, were measured out, and called townships. Each township contained 36 square parcels, or sections, which were numbered in an orderly fashion. Subparts, or aliquot parts (such as "northeast quarter"), could be further defined, creating a unique reference to any parcel of land within the system. This greatly improved land description and disposal. The task of mapping and recording the land ultimately fell to the General Land Office (GLO) of the federal government (White, 1991).

Ultimately, after a variety of other schemes were tried for selling land to citizens, the Homestead Act of 1862 opened public land to settlement without cash purchase. Settlers were required to apply for a patent to the land, improve the land through their own labor for five years, and pay fees. Although there was no cost for the land itself, there were still costs to build homes and develop crops, as well as sometimes to deal with the harsh and arid environments of the West. Of the more than 1.3 million homestead entries filed before 1900, only about half went to patent, in which the new owners gained title (Muhn and Stuart, 1988). (See Edwards [2008] for a discussion on the difficulty of compiling exact homestead numbers.) At least one homestead claim was made to secure the protection of fossils (see the example of Fossil Cycad National Monument).

In addition to encouraging settlement, the nation was eager to develop laws and policies to encourage exploitation of its mineral wealth. However, at the time of the discovery of gold in California in 1848 and the subsequent gold rush, the nation was not adequately prepared to oversee mining. This led to ad hoc rules, titles, and claims being devised by the miners themselves. After several initial attempts to get a handle on the issue, Congress passed the General Mining Law of 1872, which established that public lands were to be "free and open to exploration and purchase" (Muhn and Stuart, 1988, p. 19). Individuals and companies could apply for mining claims on public land, which made the development of minerals a priority over other uses (Muhn and Stuart, 1988). Mining claims were also tried as a tool to secure fossil collecting rights (see the example of *Diplodocus carnegii* in a following section).

Another use of public lands that helped encourage national development involved land grants made by Congress to encourage public education. In 1862, the Morrill Act granted to each state having federal public land 30,000 acres of that land from within its boundaries. This land was used to create land grant college campuses and research farms and for generating funds through such means as mining and grazing, or the land was sold outright to raise funds to support education. States without public lands were granted scrip to land elsewhere, issued in 160-acre increments that could be sold to raise funds for state colleges. In addition, one section of every township, section 16, was given to the states to help support education. After 1848, states received two sections of every township, and with the admission of Utah, Arizona, and New Mexico, the number of sections increased to four (Muhn and Stuart, 1988).

Land grants were also made to railroad companies to encourage transcontinental rail development, making travel and commerce easier across the nation, which now spanned the continent. Alternating mile-square sections of public land on either side of the railroad right-of-way were given to the railroad companies, as much as 40 sections of land for each mile across the territories. Even today, the public land has a noticeable "checkerboard" pattern of public and private ownership along the course of rail lands in western states.

Public projects were undertaken to harness the rivers in the arid West, both to generate electrical power and to provide for irrigation to increase the productivity of otherwise marginal lands. Federal lands were also set aside for wildlife preservation and for hunting, and to preserve ancient Native American ruins.

Policy related to public land evolved over the first two centuries of the nation's history. Public land served many roles. It was used early on to help pay national debt, encourage westward expansion, provide for private land ownership, and encourage the development of mining and mineral production, as well as to fund major infrastructure projects such as the railroads. However, the policy of land disposal was revised with the passage of the Federal Land Policy and Management Act (FLPMA) of 1976. That act consolidated several federal bureaus, including the GLO and the General Grazing Service, into the BLM. It also set a policy for public lands to be retained by the federal government rather than disposed of. The lands were to be managed for multiple use and sustainable yield to serve present and future generations. Today, the BLM administers more surface land (248.3 million acres) and subsurface minerals (700 million acres) than any other agency of the government (Bureau of Land Management, 2015, 2016).

DEVELOPMENT OF THE MODERN MUSEUM

Peale's Museum

Charles Willson Peale founded one of America's first museums, the Peale Museum in Philadelphia, which opened to the public in 1786 (Rainger, 2001). Peale's vision for the museum was as a public resource that would improve visitors' moral character through lessons in science. The exhibits of mammals, birds, and plants were arranged following the *scala naturae,* or the great chain of being. One of his prize exhibits was the skeleton of a mastodon discovered in 1799 on the property of John Masten outside of Newburgh, New York. Peale visited Masten with the hope of securing the fossil for his museum. After the two reached an agreement, Peale returned to Masten's farm with a crew and $500 in funding from the American Philosophical Society. Parts of the mastodon were missing, so Peale's workers explored a few other nearby farms that summer, including the farms of Captain J. Barber and Peter Millspaw.

Once all the mastodon bones were collected, they were transported to Philadelphia. There was enough material to attempt two mounts (Peale, 1803). Mounting the skeletons was done by Peale's son, Rembrandt, and Moses Williams, an enslaved man who worked for the Peales. Articulating the skeleton took three months. Initiating a practice that continues today for most mounted fossil skeletons, a composite of several individuals was made, and Rembrandt filled in missing parts of the skeleton (the top of the cranium and the tail) with sculpted elements. Wooden discs were placed between vertebrae to simulate the missing cartilage but their thickness slightly exaggerated the animal's length. The first completed mastodon mount was initially shown to the public in 1802, in the main hall of the American Philosophical Society in Philadelphia, because they had provided the funding for the excavation. The second mount was eventually displayed in the Peale Museum in Baltimore (Simpson and Tobien, 1954).

Lewis and Clark's Corps of Discovery was the first government-sponsored expedition for making scientific collections, beginning a long-standing tradition of federal museum property management. Acting as perhaps the earliest non-federal museum partner to the federal government, Peale's museum took in many of the American Indian artifacts and natural history specimens from the Lewis and Clark expedition because no one else could take them (Semonin, 2000). However, after Peale's death in 1827, his museum floundered. Rembrandt and his brother, Rubens Peale, struggled for years seeking support for the museum. They opened museums in New York and Baltimore trying to expand their audience and attendance, always hoping for some public support. In many ways, the Peale museums' issues were not dissimilar to those of modern museums, struggling with the balance between education and entertainment, and always short of funds. The Peale museums were eventually reduced from educational institutions to cheap sideshows when Phineas Taylor (P.T.) Barnum acquired the contents of Peale's New York museum in 1842, leased the Baltimore museum in 1845 and 1846, and purchased a portion of the contents of the Philadelphia museum in 1850 (Alderson, 1992). The Peale mastodons were put up for auction.

This illustrates another modern museum problem, that of federal collections orphaned when a museum or repository can no longer care for its collections. Orphaned collections are a problem for museums, and when federal collections are involved, for the DOI as well. Orphaned collections can languish without care, and deteriorate. When DOI collections are involved, the collections need to be moved, sometimes at great expense, to another repository that can properly care for them. Finding the funds to move and rehouse orphaned collections can be a problem. Some of the Peale collection, such as the fish fossil from the Lewis and Clark expedition, was transferred to the Academy of Natural Sciences in Philadelphia. The Academy, founded in 1812, also served an early non-federal repository for the various other government surveys until the establishment of the Smithsonian Institution on 10 August 1846. Fortunately, Peale's mastodons do survive to the present. The Philadelphia mastodon, the more complete of the two, was shopped around museums in Europe by speculators. The specimen was eventually purchased by Johann Jakob Kaup for the Dessischen Landesmuseum in Durmstadt, Germany (Simpson and Tobien, 1954). The Baltimore mastodon was purchased by John C. Warren (Warren, 1852), and is today at the American Museum of Natural History, disassembled (Simpson and Tobien, 1954).

Diplodocus carnegii

Many of the nation's most prestigious museums were stocked with paleontological material from public land during the "Great Dinosaur Rush." These included the Smithsonian National Museum of Natural History, the Yale Peabody Museum, the American Museum of Natural History, the Carnegie Museum of Natural History, and the Academy of Natural Sciences in Philadelphia. Peale's Philadelphia mastodon may have been the first mount, but the mounting of skeletons of ancient life was repeated many times by other museums. The first mounted dinosaur skeleton in the United States was a hadrosaur, which was erected in 1868 at the Academy of Natural Sciences in Philadelphia (Warren, 1998). Perhaps a bit ironically, the dinosaur mount was so popular that in the summer of 1870 the museum initiated an admission charge to reduce public attendance (Rainger, 2001). As more dinosaurs were discovered in the West, the public presentation of articulated, mounted skeletons gradually became more common.

One illustrative example is the story of *Diplodocus carnegii* (Rea, 2001). In 1898, the *New York Journal and Advertiser* ran a full page of stories and photographs under the headline "Most colossal animal ever on Earth just found out west" (Library of Congress, 2017) (Fig. 3). Andrew Carnegie wanted such a dinosaur for his namesake museum in Pittsburgh, and he dispatched the museum's director, William Jacob Holland, to obtain one. Obtaining enough fossilized skeletal elements for a complete mount was not a simple endeavor. Holland met with many setbacks and struggled with the strong personalities of Wyoming politicians, dinosaur hunters, and researchers. One of the issues that Holland and his collectors dealt with was the fact that their productive bone quarries were on public land, and it was feared that someone else might swoop in and collect from them first. Holland and his compatriots contemplated filing formal mineral claims for the area of their quarries in order to have a legal claim on the fossils. However, for whatever reason, they did not carry through with that while in Wyoming.

Eventually, Holland met the order for a complete dinosaur. By 1901, John Bell Hatcher put together a composite skeleton at the museum, mostly from the wild lands of Wyoming. He published his description of the animal with an illustration of

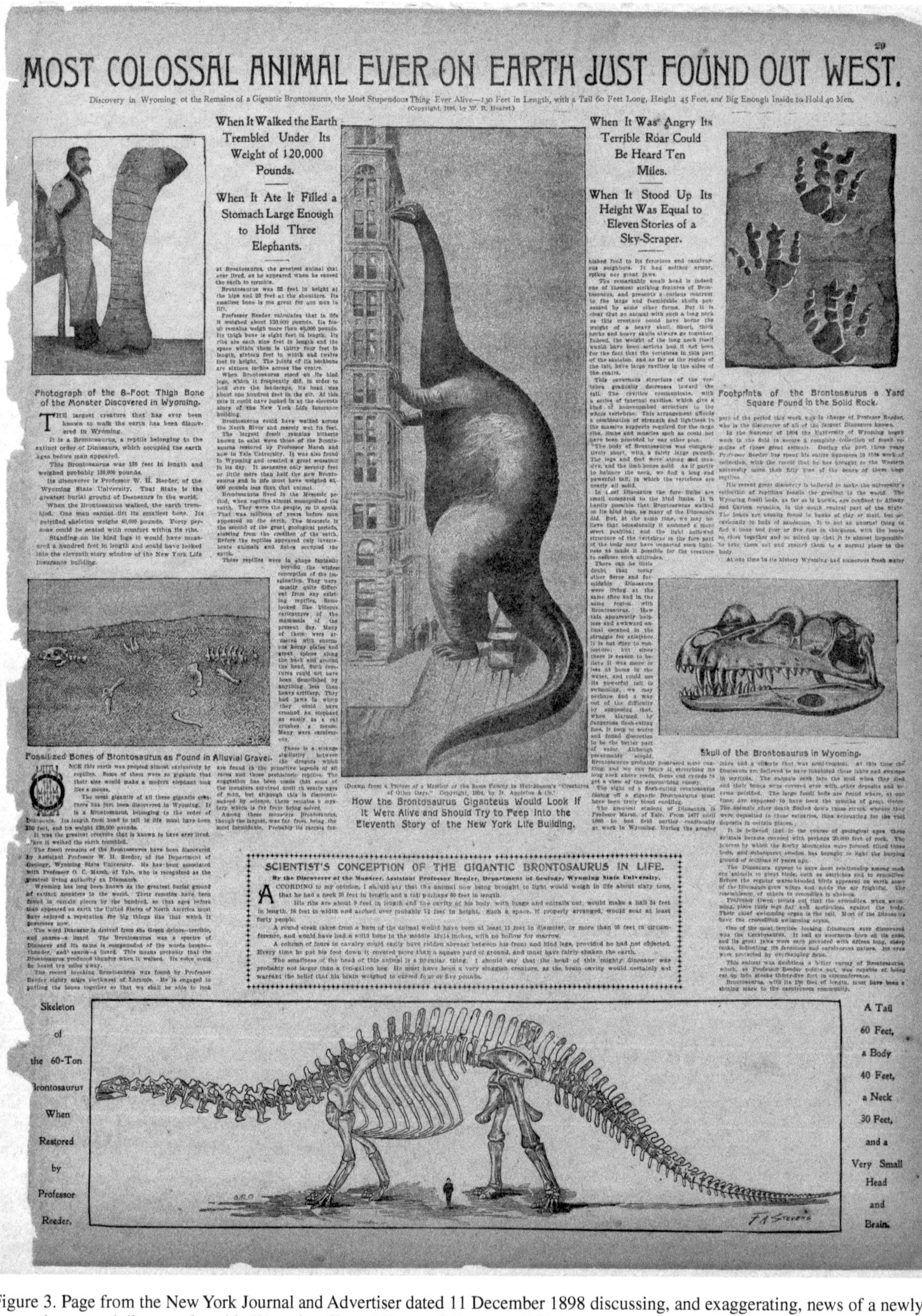

29

MOST COLOSSAL ANIMAL EVER ON EARTH JUST FOUND OUT WEST.

Discovery in Wyoming of the Remains of a Gigantic Brontosaurus, the Most Stupendous Thing Ever Alive—130 Feet in Length, with a Tail 60 Feet Long, Height 45 Feet, and Big Enough Inside to Hold 40 Men.

(Copyright, 1898, by W. R. Hearst.)

When It Walked the Earth Trembled Under Its Weight of 120,000 Pounds.

When It Ate It Filled a Stomach Large Enough to Hold Three Elephants.

When It Was Angry Its Terrible Roar Could Be Heard Ten Miles.

When It Stood Up Its Height Was Equal to Eleven Stories of a Sky-Scraper.

Photograph of the 8-Foot Thigh Bone of the Monster Discovered in Wyoming.

Footprints of the Brontosaurus a Yard Square Found in the Solid Rock.

Fossilized Bones of Brontosaurus as Found in Alluvial Gravel.

How the Brontosaurus Giganteus Would Look If It Were Alive and Should Try to Peep Into the Eleventh Story of the New York Life Building.

Skull of the Brontosaurus in Wyoming.

SCIENTIST'S CONCEPTION OF THE GIGANTIC BRONTOSAURUS IN LIFE.

By the Discoverer of the Monster, Assistant Professor Reeder, Department of Geology, Wyoming State University.

Skeleton of the 60-Ton Brontosaurus When Restored by Professor Reeder.

A Tail 60 Feet, a Body 40 Feet, a Neck 30 Feet, and a Very Small Head and Brain.

Figure 3. Page from the New York Journal and Advertiser dated 11 December 1898 discussing, and exaggerating, news of a newly discovered sauropod dinosaur found in Wyoming. From Library of Congress (2017).

the articulated skeleton (Hatcher, 1901) and named it *Diplodocus carnegii* for the museum's benefactor. The illustration was framed and sent to Carnegie at his home in Scotland. The illustration caught the eye of King Edward VII while visiting Carnegie, and the King requested a similar specimen for The Natural History Museum. This royal request thrilled Carnegie, who financed the casting and mounting of another full skeleton.

On 12 May 1905, the newly installed skeleton was dedicated in the halls of the British Museum, complete with speeches by Carnegie, Holland, the museum director, and British nobility. This was the first of many gifts by Carnegie to museums around the world. He was eager to show off "his" dinosaur[1], and with a sense of great national pride, to promote the wealth of the American fossil beds. All of these early museums, with their obligatory mounts of extinct skeletons, created the iconic impression of what a museum is in the public mind: a sanctuary of knowledge, public education, and research.

Throughout the period of the "Great Dinosaur Rush," the United States government seemed to take little notice of those riches coming from the public land. Another telling incident can be related on this point. Earl Douglass, another paleontologist working for the Carnegie Museum in 1913, did file a mineral claim for fossils. The claim was at what would become Dinosaur National Monument in Utah, today managed by the National Park Service (NPS). Attempting the ploy contemplated by Holland before him in Wyoming, Douglass filed his mineral claim to secure the area for the museum. The case was reviewed by the DOI, and it was ruled in 1915 that "[f]ossil remains of dinosaurs and other prehistoric animals are not mineral within the meaning of the United States mining laws, and lands containing such remains are not subject to entry under such laws" (Department of the Interior, 1916, p. 325). With this ruling, the use of mineral claims to reserve fossil rights was shut down. However, before the ink was dry on the DOI ruling, President Woodrow Wilson proclaimed the area as Dinosaur National Monument in 1915.

HOW THE DOI CAME TO MANAGE MUSEUM PROPERTY

A New Conservation Ethic

As Americans developed their vast country, little thought was initially given to conservation. The resources and extent of the land seemed limitless. Forests were something to clear to make room for grazing and agricultural production, and good new land was always just over the horizon. However, it was obvious that timberlands were disappearing by late in the nineteenth century. Many areas around the Great Lakes had been clear-cut. Fear arose that the nation would soon have no more forests, so calls began for conserving and managing forest lands (Muhn and Stuart, 1988; Williams, 2005).

As early as 1876, Congress appropriated funding to study the nation's forest health, which resulted in the creation of forest reserves, and ultimately the United States Department of Agriculture Forest Service (USFS) in 1905 (Williams, 2005). Another sign of the emerging conservation ethic was the establishment of the world's first national park with the creation of Yellowstone National Park in 1872 (Utley and Mackintosh, 1989). The remarkable scenery, wildlife, and geologic wonders such as geysers and thermal pools were set aside in perpetuity after Ferdinand Hayden reported the results of his survey to Congress.

It was significant to science and the burgeoning museum profession when the federal government finally took notice of the scientific resources on the land it managed. Fossils had been collected from public lands for decades, but the first assertion of ownership by the federal government was to claim the fossils collected by Cope while he was with the geological surveys. Other than that, little notice was taken of all the fossils collected from federal public land and mounted in the nation's museums. However, in 1906 that situation changed.

The Antiquities Act

With the growing conservation ethic, the desire to set aside national parks, and a desire to manage resources for the future, Congress became alarmed at the looting by private collectors of archaeological resources from the southwest. Congress passed the Antiquities Act of 1906 to punish anyone who "… shall appropriate, excavate, injure, or destroy any historic or prehistoric ruin or monument, or any object of antiquity …" on federal public land. Enactment of the Antiquities Act was a watershed moment for science and for resource management. For the first time, the government asserted a blanket claim to the historic and scientific resources on the public lands of the United States. The Antiquities Act also required that the government regulate, through a system of permits, who was authorized to collect, and directed what would happen to the collections made: Those resources would remain available to the public, under public ownership and control.

There was difficulty, however, in enforcing the Antiquities Act. For one thing, it was not entirely clear if the act applied to fossil resources—were they correctly seen as "objects of antiquity"? Legal opinions on that differed (Sakurai, 1994; Kuizon, 2006). Even for the archaeological resources to which the law more obviously applied, violations of the Antiquities Act were inconsistently prosecuted and often hard to win in court. In the case of the *United States v. Diaz*, argued before the Ninth Circuit Court of Appeals, it was ruled in 1974 that the law was vague, and therefore unconstitutional (United States v. Ben Diaz, 1974). However, the Tenth Circuit Court ruled in 1979 that the law was not vague, and that a person of ordinary intelligence would be able to ascertain that excavating a prehistoric Native American burial ground is prohibited (United States v. Smyer and May, 1979). Thus, this part of the Antiquities Act potentially faced a ruling in the United States Supreme Court. However, it was not

[1]Still making an impact today, the Carnegie specimen, now called "Dippy" by the museum, will tour the United Kingdom from 2018–2020.

appealed to that level, and that court did not get a chance to hear it. In effect, it remains law, but subsequent legislation has clarified the government's policy on historic and scientific resources from public land.

Fossil Cycad National Monument

This national monument was one of the first established under the authority of the Antiquities Act and can be viewed as a negative example of the management of fossil resources on public lands. It illustrates that, even with the best intentions of managing the fossil resource in perpetuity, maladroit scientific and political management can unravel those intents.

Along the flank of the Black Hills in South Dakota are exposures of the Lower Cretaceous Lakota Formation. The area came to the attention of scientists around 1890, when fossil plants called cycadeoids (Fig. 4) were found in abundance. The first scientific publication on the fossils was produced by a University of Iowa professor, Thomas MacBride (1893). The paleobotanist at the Smithsonian, Lester Ward, also obtained and published on specimens from the site (Ward, 1899). By 1896, O.C. Marsh had acquired a very large collection of the cycadeoids for the Yale Peabody Museum.

A paleontology student, George Wieland, who was working with Marsh at Yale, collected the majority of the cycadeoids for Marsh, and devoted his entire career to these fossil plants. He lobbied for the protection of the rich collecting site and dreamed of making it a public attraction, complete with a visitor center and in situ fossil displays (Wieland, 1937). In 1919, concerned because exposed fossils were being carried off, Wieland applied for a homestead patent to protect the parcel of land where he and others had excavated hundreds of fossil specimens. He happily renounced his patent claim when President Woodrow Wilson signed an executive order in 1920 to withdraw the land from homestead claims so the case for a national monument could be studied. Two years later, President Warren G. Harding signed a proclamation establishing Fossil Cycad National Monument, and the management of the monument was passed to the NPS. It had taken decades of lobbying and two presidential actions for Wieland's dream to come to fruition.

Figure 4. A fossil cycadiod, *Cycadeoidea furcata,* a species of uncertain status from the now-decommissioned Fossil Cycad National Monument in South Dakota. From Wieland (1916, p. 296).

Long after its establishment as a national monument, Wieland continued to lobby for the site's development. He led an excavation at the site in 1935, likely to prove to detractors that in situ fossils were still there, and reported collecting another ton of fossil material (Wieland, 1937). He publicly shamed the NPS for dragging its feet, and proposed additional congressional action to force it to act (Wieland, 1938, 1945). That strategy may not have furthered his cause. The NPS was not enthusiastic about the monument and its development, and years of fossil collecting by museums and fossil poaching by locals eventually resulted in all visible fossils being removed. The passage of time, the death of the monument's advocates, and the visible absence of the fossils for which the monument was established were all ammunition that the NPS used to lobby to have the monument decommissioned. In 1957, they got their wish. The monument returned to general federal land management under the BLM. The monument and dreams for its future faded. More detail on this interesting story can be found in Santucci and Ghist (2014).

Today, the NPS and the BLM are working in collaboration to preserve the historical record of the former monument. There is rich archival material, including publications, newspaper stories, photographs, and letters all related to the site. There also exists a wealth of fossils that were collected, many of which are preserved in the Yale Peabody Museum. Many also found their way around the nation and around the world to several other museums. The BLM recognizes the site with the designation of an Area of Critical Environmental Concern (ACEC) in management plans, ensuring that its unique history will be considered in future decisions.

Additional Preservation Laws

Given the legal uncertainty of whether the Antiquities Act actually protected antiquities and other scientific objects, other laws were passed by Congress to clarify the situation. For archaeological resources, clear protections were accomplished with the Archaeological Resources Protection Act (ARPA) of 1979. The National Environmental Policy Act (NEPA) of 1969 also outlined the policy of the nation to evaluate potential impacts of federal projects upon all natural resources and the human environment on public land.

Paleontological resources were generally managed by federal agencies under the authority of the Antiquities Act, even though that law's application to fossils was unclear. The Act was, until recently, the primary authority cited for requiring and issuing permits to collect fossils on public land. The professional paleontological community recognized the need for clear legislation related to the status of fossils on public land, as had been passed for archaeology with ARPA. An extensive record of legal and policy actions related to fossils on federal land was summarized by Kuizon (2006). Legislation was finally passed in the Omnibus Public Land Management Act of 2009 in a section that is cited as the Paleontological Resources Preservation Act (PRPA). Key provisions in PRPA include the following:

- Fossils from these lands shall be managed and protected using scientific principles and expertise.
- Casual collection of a reasonable amount of common invertebrate and plant fossils is allowed for personal noncommercial use on lands administered by the BLM, the Bureau of Reclamation (BOR), and the USFS.
- Paleontological resources may not be collected from the specified federal lands without a permit, except as allowed for casual collection.
- An applicant for a permit must be qualified to conduct the permitted activity.
- Fossils collected under permit from public land will remain the property of the United States, and specimens and related records must be preserved and made available to the scientific community.
- Certain acts are prohibited, and criminal and civil penalties are established for violations. Such acts include unauthorized collecting, damaging, defacing, or selling of fossils from federal land.
- Information about the locality and nature of fossil resources on or collected from the specified federal lands must be held as confidential to protect those resources, unless disclosure is judged by the government to cause no potential harm.

PRPA reasserts or establishes several key points. Prior to PRPA, most land management agencies by policy focused on vertebrate fossils as a management priority, and generally only required authorization (by permit) for collecting vertebrate fossils. This focus on vertebrate fossils, however, gave many people the impression that other groups of fossils, such as plants and non-vertebrates, could be collected on federal land without any restrictions. That notion was incorrect. While agencies may not have required a permit to collect plants and non-vertebrates, any ground disturbance on federal land has required assessment under NEPA since its enactment in 1969. This unfortunate misunderstanding in some cases led to rather large areas of ground-disturbing work with the excuse that "it is ok, we are not collecting vertebrates."

PRPA established clearly that it is the policy of the government to manage all paleontological resources, not just vertebrates. Whereas plant and non-vertebrate researchers collected without a permit in the past (but could not cause ground disturbances), they too now need to apply for authorization for collection. This allows land managing agencies to have a better handle on the range and extent of research across the taxonomic spectrum that originates on public land. And PRPA unambiguously establishes that fossil material collected from the public domain must be reposited in public museums.

Another point to be made is that requiring curation of material in museums demands responsible collection and acquisition policies. This is true across the board, including federal and non-federal researchers working on public land. Museum storage space is always at a premium, and additions to any collection should be purpose and mission driven. All of the provisions of PRPA, the authorization and tracking of research on public land, and best collection and museum management policies will result in a benefit for the American people. There will be a better understanding of the value of public land to science, better accountability for the resources that come from public lands, and assurance that those resources are used for the benefit of all people, through research and educational opportunities.

MANAGING LANDSCAPES AND FOSSILS

Millions of fossils have come from public lands, both before and after Congress expressed a desire to manage the resource. Whereas Section 1 of the Antiquities Act outlined resources to be protected and punishments for violations, Section 2 gave the President of the United States the authorization to designate national monuments. It allowed the creation of national monuments to protect "… historic landmarks, historic and prehistoric structures, and other objects of historic or scientific interest…." Almost immediately after the passage of the Act, paleontological resources were deemed significant when in 1906 President Theodore Roosevelt created one of the first monuments, Petrified Forest National Monument, later designated Petrified Forest National Park in 1962. Either through presidential proclamation or through legislation, at least 23 national parks and monuments have been created with the authorization explicitly mentioning their fossil resources (Table 2). In addition to those places, the NPS has documented fossils in more than 260 additional NPS-managed units (National Park Service, 2017).

One recent publication documented holotype specimens from within (or likely collected within) NPS areas. Including microfossils and fossil plants, invertebrates, vertebrates, and ichnotaxa, over 4850 holotype specimens were documented (Tweet et al., 2016). The NPS manages ~80 million acres of public land, about 16% of the total land managed by the DOI. Recognition that many DOI-managed public lands are rich in fossils creates opportunities for research and public education, as well as challenges for land and resource management. Following are some focused case studies highlighting different geologic eras, different agencies, and different management issues.

TABLE 2. LIST OF PARKS AND NATIONAL MONUMENTS CREATED AT LEAST IN PART FOR THEIR PALEONTOLOGICAL RESOURCES

Bureau	Unit	Description of fossils	Year	State
NPS	Petrified Forest National Park	Late Triassic trees and other fossils	Monument in 1906; Park in 1962	Arizona
NPS	Zion National Park	Triassic plant and vertebrate remains	Monument in 1909; Park in 1919	Utah
NPS	Dinosaur National Monument	Dinosaurs	1915	Utah
NPS	Fossil Cycad National Monument	Fossil plants	1922; subsequently Monument status revoked 1957	South Dakota
NPS	Death Valley National Park	Precambrian through Paleozoic fossils, Paleogene and Neogene body and traces	1933	California
NPS	Joshua Tree National Park	Paleontology resources	Monument in 1936; Park in 1994	California
NPS	Channel Islands National Park	Dwarf mammoths	1938	California
NPS	Agate Fossil Beds National Monument	Miocene mammals	Authorized in 1965; established 1997	Nebraska
NPS	Florissant Fossil Beds National Monument	Eocene to Oligocene plants, insects, and vertebrates	1969	Colorado
NPS	Fossil Butte National Monument	Eocene plants, insects, and vertebrates	1972	Wyoming
NPS	John Day Fossil Beds National Monument	Eocene to Miocene plants and animals	Authorized in 1974; opened in 1975	Oregon
NPS	Bering Land Bridge National Preserve	Pleistocene megafauna	1978	Alaska
NPS	Yukon Charley National Preserve	Paleontology resources	Monument in 1978; Preserve in 1980	Alaska
NPS	Hagerman Fossil Beds National Monument	Pliocene vertebrates	1988	Idaho
BLM	Grand Staircase–Escalante National Monument	Dinosaurs	1996	Utah
BLM and NPS	Grand Canyon–Parashant National Monument	Paleozoic invertebrates	2000	Nevada
FWS	Hanford Reach National Monument	Mammalian fossils from the Miocene to Pliocene Ringold Formation	2000	Washington
BLM	Prehistoric Trackways National Monument	Paleozoic reptile tracks	2009	New Mexico
NPS	Tule Springs Fossil Beds National Monument	Pleistocene fossils	2014	Nevada
BLM and USFS	Browns Canyon National Monument	Pennsylvanian non-vertebrates	2015	Colorado
NPS	Waco Mammoth National Monument	Mammoths	2015	Texas
NPS	Katahdin Woods and Waters National Monument	Paleozoic fossils	2016	Maine
BLM and USFS	Bears Ears National Monument	Paleozoic and Mesozoic marine and terrestrial fossils	2016	Utah

Note: BLM—Bureau of Land Management; FWS—Fish and Wildlife Service; NPS—National Park Service; USFS—United States Forest Service.

Grand Staircase–Escalante National Monument

When Grand Staircase–Escalante National Monument (GSENM) was created and assigned to the BLM to manage, the Late Cretaceous fossil record of the Kaiparowits Plateau region was listed as one of the primary resources worthy of protection (Clinton, 1996). The BLM assigned a full-time paleontologist to focus on the resource and to promote research.

The Kaiparowits Plateau, in south-central Utah, is underlain by ~2100 m of lower Cenomanian to upper Campanian (and possibly early Maastrichtian) strata (dated 100–70 Ma) that is mostly terrestrial, although marine beds do occur in the lower half of the section (Titus et al., 2013). Vertebrate fossils occur throughout the entire section, with a general trend toward better preservation and higher abundance up-section (Titus et al., 2017). The units yielding the highest numbers of terrestrial vertebrate fossils are the Wahweap and Kaiparowits formations, both of which occur at the top of the sequence. Largely paralic and marine units in the lower half (Naturita, Tropic Shale, and Straight Cliffs formations) have much lower fossil abundances overall, and macrovertebrate remains are much less common. However, locally spectacular exceptions are present, especially in the fully marine Tropic Shale, where articulated sharks, fishes, marine reptiles, and dinosaurs occur (Albright et al., 2013).

The significance of the monument's "world class" Late Cretaceous record was largely based on the study of the abundant vertebrate microfauna (Titus et al., 2017). With the monument's designation, interest in macrovertebrate remains began to build. Numerous macroskeletal sites had been previously documented during microvertebrate surveys, although very little diagnostic material had been collected as of 1996. In fact, only two individual dinosaurs had been described in the literature as identifiable to genus (Weishampel and Jensen, 1979; DeCourten and Russell, 1985). Between 1998 and 2000, the Utah Geological Survey was permitted by the BLM to conduct fieldwork to document known sites and provide a basis for planning decisions related to fossil resources (Foster et al., 2001). Relatively few new sites were documented, however, because the focus of this field effort was to verify and assess sites known from previous literature. The BLM's Monument Management Plan (MMP) was signed and implemented in 2000. The planning process did identify the Kaiparowits region as having tremendous potential for significant macrofauna even though almost nothing was known about it. Key decisions in the MMP (SCI-7 and PAL-1) provided the basis for establishing a program of fossil inventory, collection, and research to fill in this key knowledge gap.

Thus, every year since 2000, the monument's in-house paleontology program, along with a handful of other partner institutions (e.g., the Natural History Museum of Utah, the Utah Geological Survey, the Denver Museum of Nature & Science, and the Museum of Northern Arizona), have conducted inventories and excavations to broaden the scientific understanding of the Late Cretaceous fossil ecosystems and to preserve some of the priceless fossils. One of the key partnerships in this endeavor is with the Natural History Museum of Utah, which not only conducts fieldwork and collecting each year, but also permanently curates large Kaiparowits collections. The Natural History Museum of Utah currently houses ~20,000 Kaiparowits fossil objects, many of which are type and figured specimens of new dinosaurs and other vertebrate taxa. The Denver Museum of Nature & Science, the other major partner in the current inventory and collection effort, houses ~27,000 GSENM Cretaceous fossil objects, over half of which are paleobotanical specimens. From these and other collections, a whole new North American Late Cretaceous flora and fauna are being cataloged.

Since 2005, 12 new genera of Kaiparowits dinosaurs, such as *Nasutoceratops titusi* (Fig. 5), have been named (Titus et al., 2017). Numerous other taxa are currently under study for publication. The Kaiparowits dinosaur assemblage contains the oldest named members of the tyrannosaurid and lambeosaurine dinosaur clades, *Lythronax argestes* and *Adelolophus hutchisoni,* respectively (Loewen et al., 2013b; Gates et al., 2014), as well as the highest ceratopsid diversity (minimum of five genera) for a single faunal zone in the world (Loewen et al., 2013a). It has also yielded the most complete Late Cretaceous terrestrial bird fossil from North America (*Avisaurus* sp.) and numerous specimens of hadrosaurs, ceratopsids, and maniraptorans retaining evidence of soft tissue (Zanno and Sampson, 2005; Gates et al., 2013; Sampson et al., 2013; Zanno et al., 2013). The near total uniqueness of the late Campanian (ca. 75 Ma) Kaiparowits dinosaur assemblage, in concert with turtle, mammal, crocodyle, and squamate evidence, has made it clear that the Kaiparowits Plateau represents a previously unrecognized, higher-diversity biogeographic realm (Roberts et al., 2013) distinct from that which characterizes the northern Great Plains (Montana and Alberta, Canada). Overall, the region can claim to hold one of the most complete Late Cretaceous terrestrial vertebrate fossil successions in the world. With ~25% of the entire plateau region surveyed, additional important future discoveries are assured.

Charles M. Russell National Wildlife Refuge

The Charles M. Russell National Wildlife Refuge (CMRNWR) surrounds Fort Peck Reservoir in Montana and is the second-largest National Wildlife Refuge in the lower 48

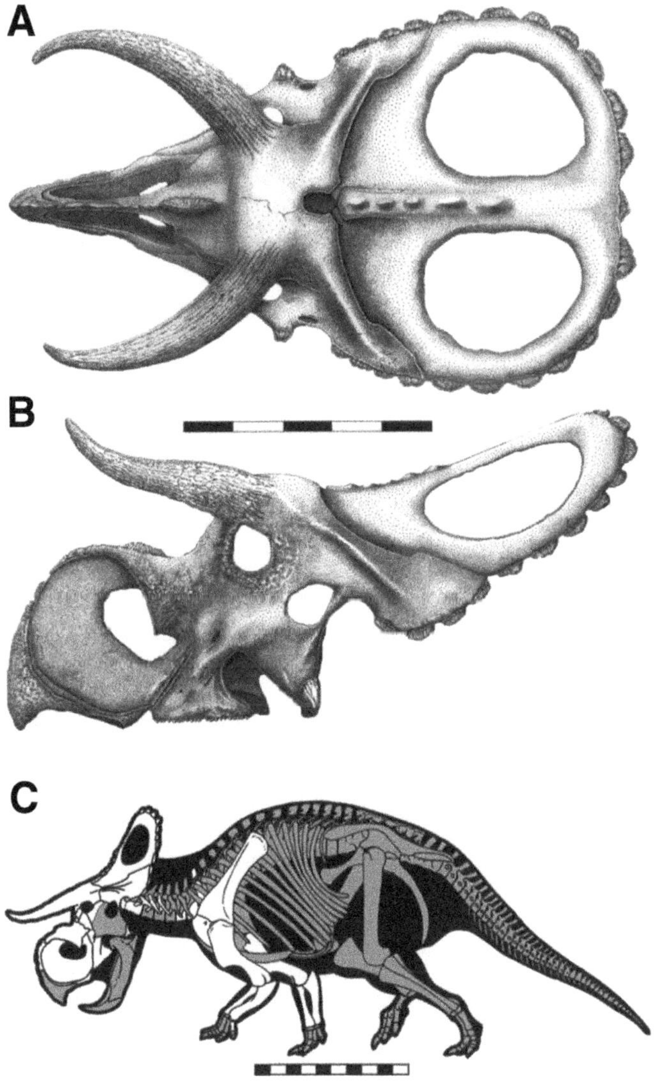

Figure 5. *Nasutoceratops titusi,* a newly identified ceratopsian species from Grand Staircase–Escalante National Monument. Scale bars in (A) and (B) are 50 cm, and in (C) 1 m. From Sampson et al. (2013).

states. The refuge, managed by the U.S. Fish and Wildlife Service (FWS), protects ~125 linear miles (not river miles) of the Missouri River ranging from the Fort Peck Dam west to the boundary of the Upper Missouri River Breaks National Monument (UMRBNM). The latter is managed by the BLM and protects another 80 plus linear miles of the river to the west. The CMRNWR was created in 1936 as a "range" and officially became a refuge in 1976. This area is significant for discoveries made in the early years of paleontology, as well as intensive ongoing modern research.

Fossils from this area came to light in 1901, when William T. Hornaday, then director of the New York Zoological Society, was exploring the area. He picked up three fragments of *Triceratops* nasal horn (Clemens and Hartman, 2014). He showed the bones to H.F. Osborn at the American Museum of Natural History, who dispatched Barnum Brown to the area to prospect the next year. Brown later recounted that soon after arriving in the area indicated by Hornaday, while the cook prepared supper, he walked up the drainage and located bones running into a hillside. This discovery would forever change paleontology and the imaginations of children everywhere—it became the type specimen of *Tyrannosaurus rex* (Osborn, 1905) (Fig. 6). The type locality today is just outside the modern boundaries of the refuge, and on private land. However, in 1902, the land was under federal management.

The formation in which the *Tyrannosaurus rex* and several *Triceratops* skulls were collected in 1902 now takes its name from a small drainage flowing north into the Missouri River—it is known as the Hell Creek Formation. From this early success, fossil hunters have been drawn to the Hell Creek badlands ever since. Paleontologists have amassed extensive floral and faunal lists. The depositional setting for the shales and sands of the Hell Creek Formation was a low-lying, low-relief margin of the Cretaceous Interior Seaway, with fluctuating fluvial channels, deltas, and peaty swamps (Clemens and Hartman, 2014). The general area of the Hell Creek Formation and the overlying Tullock Member of the Fort Union Formation is of intense interest because it is one of the few places in the world to preserve a terrestrial record of the Cretaceous–Paleogene (K–Pg) boundary.

During the 1970s, the discovery of anomalously high concentrations of iridium in marine formations at the K–Pg boundary led to the advancement of the asteroid extinction hypothesis (Alvarez et al., 1980). Soon thereafter, samples were taken at a site dubbed "Iridium Hill" (Clemens and Hartman, 2014) on BLM-managed land outside the preserve. The iridium layer represents an isochron that marks the end of the Cretaceous. Using lithological, geochemical, palynological, and geochronological analyses, the layer was located at other areas across Montana and North Dakota. Hell Creek fossils include invertebrates, diverse plant macrofossils, fishes, amphibians, lizards, dinosaurs, and mammals (Clemens and Hartman, 2014). The overlying Tullock Member also preserves flora and fauna. Together these units provide excellent information about the final years of the Cretaceous Period and the Paleogene recovery within the ecosystem after the end-Cretaceous extinction. Because of this, the formations have consistently attracted multidisciplinary researchers who work on both the CMRNWR and surrounding BLM land. Partner museums with significant Hell Creek and Tullock collections from public land include the Museum of the Rockies, the Burke

Figure 6. First published image of *Tyrannosaurus rex*, taken from Osborn's 1905 holotype description paper.

Museum, the Smithsonian National Museum of Natural History, and the Denver Museum of Nature & Science, among others.

American Falls Reservoir

The American Falls Reservoir was created when the Bureau of Reclamation (BOR) built the American Falls Dam on the Snake River in Idaho. The dam and reservoir were constructed as part of the Minidoka Project on the Snake River Plain to provide flood control, irrigation, and recreation. Construction of the original dam by the BOR began in 1925 and was completed in 1927. When a second dam was built in 1978, the original structure was demolished. Although the dam itself is located in Power County, the reservoir stretches northeastward into Bingham and Bannock counties. It is one of 348 reservoirs managed by the BOR (Stratton et al., 2014) and the largest in the upper Snake River Basin. The management of fossils along the shoreline of an active reservoir, combined with ongoing erosion, provides a challenge.

The American Falls fauna is recovered as float from the lower member of the American Falls Formation, which is exposed on the beach ringing the reservoir (Ridenour, 1969). The formation represents a floodplain of the ancestral Snake River and consists of sands, silty sands, silts, and a small proportion of clayey silt beds. Forming cliffs above the lower member are lacustrine sediments composed of clay, which are considered the upper member of the American Falls Formation. The upper member accumulated in a lake that formed when the Cedar Butte basalt flow impounded the Snake River. The two sedimentary units are separated by a distinct change in texture and an undulatory erosional surface. The upper member contains the Rainbow Beach fauna (McDonald and Anderson, 1975).

The first documented fossil vertebrate collected from the American Falls Formation was a partial cranium referred to as *Bison alaskensis,* collected in 1894 "from the banks of the Snake River, Idaho, about 9 miles above American Falls, in the gravel overlying the lava beds" (Lucas, 1899, p. 766). This general area is now inundated by the American Falls Reservoir. However, the composition and mineralogy of matrix adhering to the cranium is consistent with that of the basal sand and gravel of the lower member of the American Falls Formation. The skull was originally housed at Stanford University but was subsequently transferred to the Idaho Museum of Natural History on the Idaho State University campus in Pocatello (Pinsof, 1991).

The American Falls fauna is Sangamonian interglacial in age and the Rainbow Beach fauna is Wisconsinan. Both faunas have similar megafauna including the giant bison, *Bison latifrons* (Fig. 7), along with the horse *Equus scotti*; the extinct camel *Camelops hesternus*; two types of ground sloth, *Paramylodon harlani* and *Megalonyx jeffersonii*; the Columbian mammoth, *Mammuthus columbi*; and the American mastodon, *Mammut americanum* (McDonald and Anderson, 1975; Pinsof, 1998). The largest collection of fossil vertebrates of both the American Falls and the Rainbow Beach faunas is held in the John A. White Repository at the Idaho Museum of Natural History. The earliest collections made by the museum started in the 1930s and have continued uninterrupted to the present (Stratton et al., 2014). Over 17,000 vertebrate specimens from the American Falls fauna alone are recorded in the museum's database.

Figure 7. John White, former curator of vertebrate paleontology, Idaho Museum of Natural History, with skull of the giant bison, *Bison latifrons,* at American Falls Reservoir. The skull is part of a partial skeleton (IMNH 75002/27) collected in 1975. Photo courtesy of the John A. White Paleontological Repository, Idaho Museum of Natural History.

At the American Falls Reservoir, fluctuations in the water level occur both seasonally and in response to long-term changes in yearly precipitation in the region. The difference between the high and low stands in the reservoir is ~18 m, and the shoreline ranges from a minimum of 75 km at normal minimum elevation to 185 km at normal maximum elevation (Bureau of Reclamation, 1995). This results in constant erosion of both the lower member of the American Falls Formation along the beach and the cliff face of the upper member. Control of this erosion by the BOR through the installation of riprap rock barriers requires considering the impact to the fossil resources, and mitigation often involves their collection and documentation (Stratton et al., 2014). The continuous exposure of fossils also increases the potential for their illegal removal by individuals visiting American Falls Reservoir for recreation. Usually the fossils are found as float along the beach strandline, so articulated material is rare. Constant monitoring of the shoreline around the reservoir is needed. This is particularly important in the summer when the reservoir is heavily used for recreation and in the fall when the water level is at its lowest, which is when exposure of the fossil-bearing sediments is at its greatest. The ongoing partnership between the BOR and the Idaho Museum of Natural History has ensured that this paleontological resource is preserved for future generations and available for both scientific study and public interpretation in museum exhibits.

MANAGING MUSEUM PROPERTY

The DOI as Steward

The "treasure" of the nation is not only in the landscapes and resources utilized in some productive way, but also in the objects and specimens that have come from the land and are today curated as museum property. The museum property held by the DOI represents a significant prehistoric, historic, cultural, and scientific legacy for the nation. Collectively, it makes the DOI one of the largest managers of museum collections in the United States. The scope of the DOI collections includes objects from all scientific and cultural disciplines, including archaeology, archives, art, biology, ethnography, geology, history, and paleontology. Ten DOI bureaus and offices presently manage museum property (Table 3), and the total of all museum objects across all bureaus and offices and collection types is presently estimated at nearly 204 million objects. The largest single collection discipline is archives, representing over 131 million documents along with almost 3000 additional linear feet of archives (Table 4). The NPS, which has hundreds more museum staff than all other agencies combined (Table 5), is the largest collection holder in the DOI and manages an estimated 175 million objects (Table 1). (Readers interested in scientific collections held across the federal government beyond the DOI should refer to the Interagency Working Group on Scientific Collections, 2009).

When dealing with collections of this size and scope, challenges are inevitable. In 2009, the DOI Office of Inspector General (OIG) conducted an audit of the overall museum program and correctly noted a number of important deficiencies (Office of Inspector General, 2009). In broad terms, the audit found that the DOI lacked sufficient oversight of the bureau and office collections; had a very large backlog of objects that were not accessioned and/or cataloged; was not caring for some objects properly; was not conducting regular inventories for accountability purposes; and did not know where all its collections were housed, including those in non-federal museums and repositories. The OIG issued recommendations for the DOI to implement, which are being addressed. Each year one or more of the recommendations is completed.

The DOI provides oversight, policy direction, technical support, and training for all the bureaus and offices that manage museum collections. Department staff work closely with national curators of the bureaus and offices to write policy, address current issues and challenges, and cooperate on projects. One such project was a recent survey of ~200 non-federal repositories to find the DOI collections they possibly housed. The survey included questions about collections by discipline, so the hope was to

TABLE 3. BUREAUS WITHIN DEPARTMENT OF THE INTERIOR (DOI) RESPONSIBLE FOR MUSEUM PROPERTY AND A BRIEF OVERVIEW OF WHAT EACH BUREAU MANAGES

Bureau	Abbreviation	Acres or types of resources managed
Bureau of Indian Affairs	BIA	55.7 million acres of land held in trust for Native Americans
Bureau of Land Management	BLM	248 million surface, 700 million subsurface acres
Bureau of Reclamation	BOR	476 dams and 338 reservoirs
Bureau of Safety and Environmental Enforcement	BSEE	1.7 billion acres on the U.S. Outer Continental Shelf
Fish and Wildlife Service	FWS	150 million acres
Indian Arts and Crafts Board	IACB	Operates three regional museums
Department of the Interior Museum	IM	Houses a collection related to DOI and curates exhibits at the Main Interior Building in Washington, D.C., and online
National Park Service	NPS	80 million acres, including 56 national parks and 417 other units
Office of the Special Trustee for American Indians	OST	Manages assets for trust beneficiaries that the federal government holds in trust
U.S. Geological Survey	USGS	Conducts scientific research on a wide range of natural systems of national importance. Much of the resulting museum collections are curated by the Smithsonian Institution

TABLE 4. NUMBERS OF MUSEUM OBJECTS CURRENTLY IDENTIFIED UNDER DEPARTMENT OF THE INTERIOR MUSEUM PROGRAM MANAGEMENT IN FISCAL YEAR 2016

Discipline	Museum objects managed (millions)	Cubic feet	Linear feet
Archaeology	63.452	18,109	0
Archives	131.855	0	2992
Art	0.104	0	0
Biology	3.396	0	0
Ethnography	0.034	0	0
Geology	0.081	0	0
History	3.472	360	0
Paleontology	1.515	105	0
Totals	203.909	18,574	2992

TABLE 5. MUSEUM STAFF BY BUREAU FOR FISCAL YEARS 2014–2016

Bureau	2014	2015	2016
BIA	3	3	3
BLM	5	4	4
BOR	6	55	
BSEE	2	2	2
FWS	3	33	
IACB	5	5	5
IM	4	4	4
NPS	660	588	646
OST	1	11	
USGS	1	1	1

Note: BIA—Bureau of Indian Affairs; BLM—Bureau of Land Management; BOR—Bureau of Reclamation; BSEE—Bureau of Safety and Environmental Enforcement; FWS—Fish and Wildlife Service; IACB—Indian Arts and Crafts Board; IM—Department of the Interior Museum; NPS—National Park Service; OST—Office of the Special Trustee for American Indians; USGS—United States Geological Survey.

find fossils and other objects and specimens in repositories that bureaus did not know about. Another goal of the survey was to forge partnerships with the newly discovered repositories and strengthen relationships with known partner repositories.

DOI Repositories

Most bureaus and offices have few full-time staff dedicated to museum collection care (Table 5). The NPS is the exception to this rule with 646 museum staff members. The NPS operates 1469 museum repository facilities located at 329 units, where a unit is a park, monument, historic site, or historic house. Some of those units have multiple repositories that house collections, such as exhibits at visitor centers, museums, historic structures, and administrative or storage spaces. The other bureaus have far fewer bureau-run repositories. For example, the BLM has three repositories, whose scope of collections is limited to archaeological and historical objects. As a result, the BLM has 158 non-federal partner facilities housing the remainder of its museum objects. All DOI facilities must comply with DOI policy and procedures for the physical security and environmental conditions of museum collections to ensure their proper care. The DOI encourages non-federal partner facilities to also comply with DOI collection standards, which reflect best museum practices as promoted by the American Alliance of Museums.

John Day Fossil Beds National Monument

An example of a DOI repository can be found at the John Day Fossil Beds National Monument, the Thomas Condon Paleontology Center. Covering ~10,000 square miles, the John Day fossil beds contain one of the most complete terrestrial Cenozoic stratigraphic sequences in the world. Agency boundaries and private lands are comingled in this fossil-rich landscape, lending a perfect opportunity for collaborative, interagency management (Fremd et al., 2009).

The fossils of the John Day Basin have been of interest to researchers for over 150 years, beginning with Thomas Condon, O.C. Marsh, and E.D. Cope and later followed by J.C. Merriam, Ralph Chaney, and many others. It was clear that the vertebrate and plant records in the region were a valuable resource to the scientific community. J.C. Merriam was one of the early proponents of protecting the fossil beds as public land, and by the 1930s, three state parks had been created: Thomas Condon–John Day Fossil Beds State Park, Clarno State Park, and Painted Hills State Park. With the support of the governor and legislature of the state of Oregon, the state parks were turned over to the NPS to manage (Enabling Legislation of 1974, Public Law 93-486). The modern 14,000 acre monument was authorized when the enabling legislation adopted by Congress was signed into law on 26 October 1974. The monument consists of three separate and distinct management units, corresponding to the original three state parks, as well as other federal land; however, the monument could not encompass all the important fossil sites within the John Day River valley. Agencies that also manage significant fossil localities in the region include NPS, BLM, FWS, Bureau of Indian Affairs (BIA), USFS, and the state of Oregon. In particular, the NPS and the BLM have formally been partners in the John Day region since at least 1987, and these agencies have worked collaboratively to preserve the paleontological resources of the region. BLM specimens comprise ~26% of the collections at the Paleontology Center.

The geologic layers at the John Day fossil beds represent ~40 million years, or nearly two-thirds of the Cenozoic Era (Fig. 8). The fossiliferous units are the Clarno, John Day, Mascall, and Rattlesnake formations. The history of evolution and changing climates is well preserved within rocks exposed on these public lands. The Eocene Clarno Formation (54–39 Ma) is ~1800 m thick and includes both volcanic and volcanogenic sedimentary rocks deposited by lava flows, ashfalls, lahars, and fluvial mechanisms (Bestland et al., 1999; Dillhoff et al., 2009; McClaughry et al., 2009). The most captivating NPS-administered sites from the Clarno Formation are the Hancock Mammal Quarry and the Clarno Nutbeds outcrops, representing a diverse flora and fauna.

The John Day Formation (39–18 Ma) is subdivided into seven members: Big Basin, Turtle Cove, Kimberly, Haystack Valley, Balm Creek, Johnson Canyon, and Rose Creek (Retallack et al., 2000; Hunt and Stepleton, 2004; Albright et al., 2008). A rich fossil record of plants and animals is preserved, as well as paleosols. Thirteen of the volcanic layers in the John Day

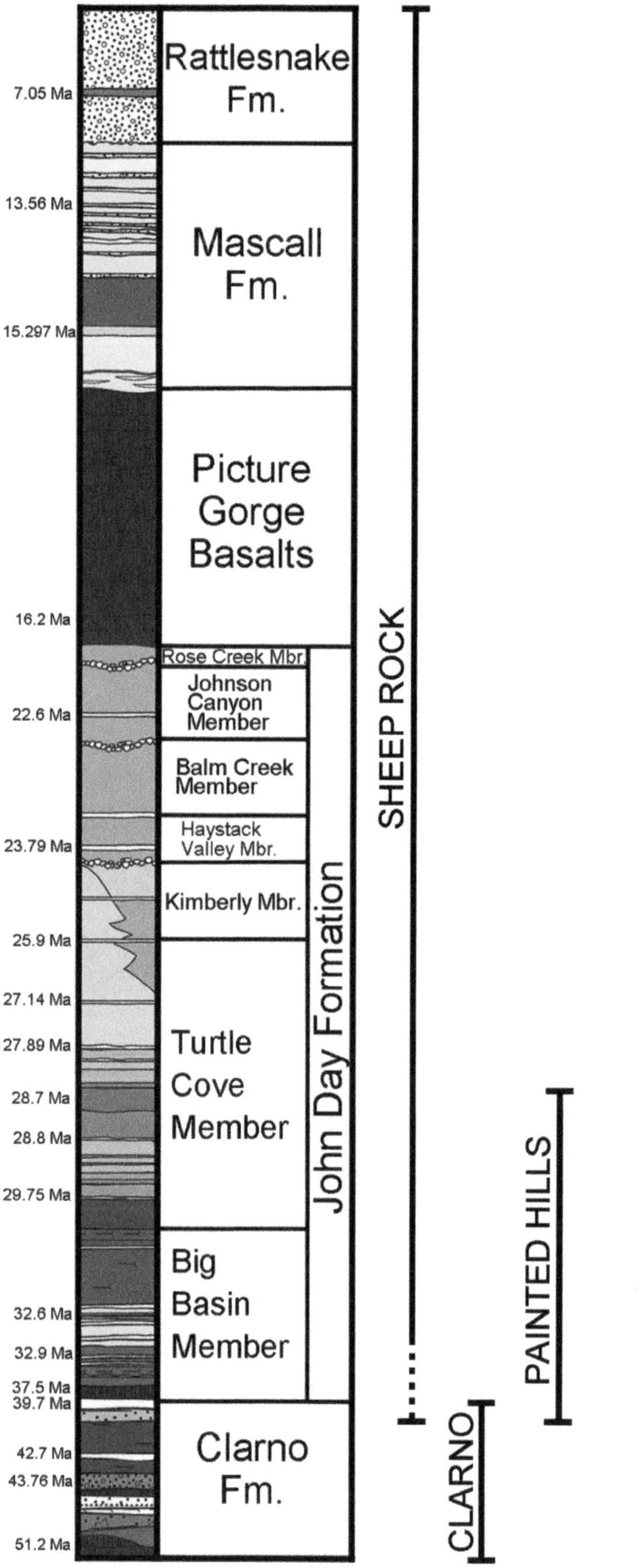

Figure 8. Composite stratigraphy of John Day Fossil Beds National Monument. The approximate stratigraphic ranges of the three units of the monument are indicated by black lines on the right. Courtesy of the National Park Service.

Formation are named and many of them have been radiometrically dated. One of the largest volcanic deposits is the Picture Gorge ignimbrite (28.7 Ma) in the Turtle Cove Member, which represents a large caldera eruption possibly associated with the Yellowstone hotspot (Seligman et al., 2014). The John Day Formation includes some rare and interesting taxa including one of the last North American primates before the arrival of humans, *Ekgmowechashala zancanellai* (Samuels et al., 2015).

The Mascall Formation (17–13 Ma) is composed of volcaniclastic sediment derived from the coeval volcanoes of Oregon, Washington, and Nevada (Rytuba and McKee, 1984; Bestland et al., 2008). Like the John Day Formation, several paleosols are preserved. A well-preserved floral assemblage (leaves and pollen) can be found in the lower diatomite layers, and there is a good record of extinct vertebrates throughout the entire Mascall Formation (Downs, 1956; Chaney, 1959).

The Rattlesnake Formation (7 Ma) is the stratigraphically highest geologic unit and comprises fluvial deposits (Enlows, 1976; Martin, 1983; Martin and Fremd, 2001). It is dissected by the Rattlesnake Ash Flow Tuff (7.05 Ma). Plant fossils are rare in the Rattlesnake Formation (Dillhoff et al., 2009), but there are numerous vertebrate fossils that were first described in 1925 and include the first occurrence of true beavers and fishers (Merriam et al., 1925; Samuels and Zancanella, 2011; Samuels and Cavin, 2013).

One fascinating story that is preserved at the John Day fossils beds is the evolution of the horse. Fossils of 24 species of horse can be found within the section. They chronicle the evolution of the family Equidae in the face of an environment changing from a closed forest to an open grassland (Fremd, 2010). Recently, the horse fossil record at the John Day fossil beds has been utilized in studies of the evolution of tooth morphology (Famoso et al., 2013; Famoso and Davis, 2014).

With 197 confirmed type specimens (Tweet et al., 2016), a relatively continuous section, abundant plant and animal fossils, and numerous volcanic layers (ashes, ignimbrites, and lava flows), the John Day fossil beds are a unique resource. The long history of collecting in the area has meant that fossil material can be found in many repositories across North America. The partnership of all the various public lands in the John Day region ensures that the rich 40-million-year record of plant and animal evolution and climate change is preserved for future generations.

Non-Federal Facilities

Whereas properly managing museum property within DOI facilities is challenging in itself, an entirely different set of issues arises when trying to manage and be accountable for federal property held in non-federal facilities. For example, unlike the NPS, which houses many fossil collections on-site at park units, almost all of the BLM's fossil resources are housed in non-federal partner repositories. The other bureaus are in a similar situation. The non-federal partners range from large public or private museums to small county museums, or even laboratories within university departments. More than 25 million museum objects managed by the DOI bureaus are housed in almost 900 non-federal partner facilities.

The DOI partnerships with all the non-federal facilities are an essential part of museum object care, and the DOI is grateful to its partner facilities for the work they do caring for public material in trust. Many of these collections were made by researchers at their own behest under permit, and some resulted in sizeable collections at their institutions. The non-federal facilities have taken on the task of documenting, curating, and housing these public materials in trust, preserving them for public benefit. All of the museum collections are available for exhibition and educational programming, and countless visitors to federal and non-federal museums are enriched by these objects.

Because the objects and specimens remain property of the federal government, each of the bureaus within the DOI is responsible to the American people to account for them. Sometimes records at both the bureaus and the repositories are inadequate (or otherwise unprocessed) to identify which objects are federally held. Backlogs of repository accessioning and cataloging; failure of the stewarding institution to note which objects were collected from public land; and old, lost, or incomplete records of permit authorizations by the bureaus (or cases where authorization was never sought) all contribute to the challenges of knowing what the DOI is responsible for, and where those collections are held.

In addition to discoveries of legacy collections (collections made historically, and newly discovered to be federal material), new collections are actively being made. This is especially true in paleontology. On BLM land, for example, the number of permits issued each year continues to rise. The BLM issued 440 paleontology research permits in fiscal year (FY) 2015, and 465 permits in FY2016, to researchers outside the federal government. Thousands of fossils are added annually to repository collections through researcher-driven efforts. This has led to many discoveries of scientific interest, including new species and other unique occurrences, but it also contributes to collection growth, backlogs, and more responsibilities to the public and researchers for access and use.

Despite these challenges, there has been considerable progress. Each year recommendations in the OIG report are completed. In recognition of the need to help support collections in non-federal partner facilities, the DOI has been able to offer some support for collection management improvements through project funding and internships. This support has successfully improved collection conditions and records across many non-federal partners. Strides have been made in identifying non-federal partners and developing working relationships and partnerships with repository staff. Outreach about the roles and responsibilities of the DOI in museum collection management is a continuing effort, of which this paper is a prime example.

The DOI systems for documenting and tracking collections are continuously being improved. A notable advancement being planned for the coming years is a complete upgrade to the DOI-wide computerized collection management system. The goal is to have an enterprise-wide solution for collection management that will dramatically improve access to information about DOI collections to better serve researchers and the American public. This will augment all the work the bureaus are doing to develop online exhibits and share other information about the public's collections.

CONCLUSIONS

The young U.S. government documented its natural resources and utilized its wealth in many ways. Among the resources on the public land were fossils, which ignited scientific curiosity and facilitated addressing significant questions such as the reality of extinction and evolution, and helped to propel paleontology into its modern form. Land settlement was encouraged with a variety of schemes to transfer land into private ownership, making more than a million ordinary people into landowners. A conservation ethic slowly emerged in the late nineteenth century, particularly with the obvious loss of forest lands. The federal government established the USFS and the first national park. The federal government also exerted ownership of scientific and cultural resources on the public lands by the turn of the twentieth century, which is when the DOI's accountability to the public began for the museum objects that it managed. Systems for permits were developed to provide a mechanism for tracking public material into either federal or non-federal museums charged with holding the material in trust for the American public. The permits also allow land managers to account for the amount and scope of research being conducted on public land. More recent efforts to account for and track the nation's historic and scientific treasure from public lands have led to the recognition that the DOI is responsible for one of the largest collections of museum property in the nation. Land-managing bureaus and offices are charged with overseeing historic and scientific resources while they are in the field, inventorying them, and mitigating potential damage to resources not yet collected. The DOI museum staff is charged with accounting for and preserving museum collections in both DOI and non-federal facilities. There are several hurdles to overcome to be successful in these daunting tasks, and recent successes and improvements make the DOI better positioned than ever to do so.

Responsible collecting by researchers on public lands, proper issuance and recordkeeping of bureaus' authorizations to researchers to collect from public lands, and the cooperation of research professionals and non-federal museums are essential. Both DOI museums and non-federal museums have taken on a significant trust responsibility to care for the nation's heritage. The partnership between the DOI and non-federal facilities is critical, and wherever possible the DOI is offering resources to help with the burden. The natural and cultural heritage under DOI management is immense, and the DOI is striving to manage this treasure to protect it for future generations.

ACKNOWLEDGMENTS

Many people contributed toward this paper. Justin Tweet was helpful with information on fossils in the NPS. Robert Marcell from the NPS pointed out useful reference material on homesteads. The BLM paleontology team of Scott Foss, Phil Gensler, Brent Breithaupt, Harley Armstrong, Rebecca Hunt-Foster, and Neffra Mathews is always supportive. The USFS paleontologists, Bruce Schumacher and Barbara Beasley, provided insights about their agency. BOR staff Kathleen Hogue, Lynne MacDonald, George Shannon, Zac Nelson, and Bill Chada assisted with information about fossil resources on BOR-administered land. BLM colleagues Cyndi Eide and Debbie Sorg assisted with research related to land patents, and Jeff Erwin produced Figure 1. Gary Smith, also of the BLM, was extremely supportive of this effort as well. Mary Thompson from the Idaho Museum of Natural History provided the photo in Figure 7. Paula Gouse provided information about fossils on FWS-managed land. Shelley Hall and Chris Schierup from John Day Fossil Beds National Monument also provided support. The editors and reviewers of the paper provided valuable comments for improvement.

REFERENCES CITED

Albright, L.B., III, Woodburne, M.O., Fremd, T.J., Swisher, C.C., III, MacFadden, B.J., and Scott, G.R., 2008, Revised chronostratigraphy and biostratigraphy of the John Day Formation (Turtle Cove and Kimberly Members), Oregon, with implications for updated calibration of the Arikareean North American Land Mammal Age: The Journal of Geology, v. 116, p. 211–237, https://doi.org/10.1086/587650.

Albright, L.B., III, Gillette, D.D., and Titus, A.L., 2013, Fossil vertebrates from the Tropic Shale (Upper Cretaceous), southern Utah, *in* Titus, A.L., and Loewen, M.A., eds., At the Top of the Grand Staircase: The Late Cretaceous of Southern Utah: Bloomington, Indiana University Press, p. 536–562.

Alderson, W.T., 1992, Mermaids, Mummies, and Mastodons: The Emergence of the American Museum: Washington, D.C., American Association of Museums, 104 p.

Alvarez, L.W., Alvarez, W., Asaro, F., and Michel, H.V., 1980, Extraterrestrial cause for the Cretaceous–Tertiary extinction: Science, v. 208, p. 1095–1108, https://doi.org/10.1126/science.208.4448.1095.

Ambrose, S.E., 1997, Undaunted Courage: Meriwether Lewis, Thomas Jefferson, and the Opening of the American West: New York, Simon and Schuster, 528 p.

Bestland, E.A., Hammond, P.E., Blackwell, D.L.S., Kays, M.A., Retallack, G.J., and Stimac, J., 1999, Geologic framework of the Clarno Unit, John Day Fossil Beds National Monument: Central Oregon: Oregon Geology, v. 61, p. 3–19.

Bestland, E.A., Forbes, M.S., Krull, E.S., Retallack, G.J., and Fremd, T.J., 2008, Stratigraphy, paleopedology, and geochemistry of the middle Miocene Mascall Formation (Type area, central Oregon, USA): PaleoBios, v. 28, p. 41–61.

Boyd, J.P., 1958, The *Megalonyx*, the *Megatherium*, and Thomas Jefferson's lapse of memory: Proceedings of the American Philosophical Society, v. 103, no. 5, p. 420–435.

Bureau of Land Management, 2015, Public Land Statistics: Lakewood, Colorado, Government Printing Office, 261 p.

Bureau of Land Management, 2016, The Federal Land Policy and Management Act of 1976 as amended: Washington, D.C., U.S. Department of the Interior, Bureau of Land Management, 106 p.

Bureau of Reclamation, 1995, Bureau of Reclamation American Falls Resource Management Plan, Bureau of Reclamation, Pacific Northwest Region Boise, Idaho and Snake River Area Office Minidoka Project, 254 p.

Chaky, D., 2015, Fossils and the fur trade: The Chouteaus as patrons of paleontology: We Proceeded On: Lewis and Clark Trail Heritage Foundation, v. 41, no. 2, p. 12–23.

Chaney, R.W., 1959, Miocene floras of the Columbia Plateau. Part I, Composition and interpretation: Washington, D.C., Carnegie Institution of Publication, v. 617, p. 1–134.

Clemens, W.A., and Hartman, J.H., 2014, From *Tyrannosaurus rex* to asteroid impact: Early studies (1901–1980) of the Hell Creek Formation in its type area, *in* Wilson, G.P., Clemens, W.A., Horner, J.R., and Hartman, J.H., eds., Through the End of the Cretaceous in the Type Locality of the Hell Creek Formation in Montana and Adjacent Areas: Geological Society of America Special Paper 503, p. 1–87, https://doi.org/10.1130/2014.2503(01).

Clinton, W.J., 1996, Proclamation 6920—Establishment of the Grand Staircase–Escalante National Monument, *in* Peters, G., and Woolley, J.T., The American Presidency Project: Santa Barbara, California, http://www.presidency.ucsb.edu/ws/?pid=51948.

Colbert, E.H., 1984, The Great Dinosaur Hunters and Their Discoveries: New York, Dover Publications, 283 p.

Cuvier, G., 1799, Mémoire sur les espèces d'éléphans vivantes et fossiles: Mémoires de l'Institut National des Sciences et Arts, Sciences: Mathematics and Physics, v. 2, p. 1–22.

DeCourten, F., and Russell, D.A., 1985, A specimen of *Ornithomimus velox* (Theropoda, Ornithomimidae) from the terminal Cretaceous Kaiparowits Formation of southern Utah: Journal of Paleontology, v. 59, p. 1091–1099.

Department of the Interior, 1916, Earl Douglass, *in* Hesselman, G.J., ed., Decisions of the Department of the Interior in Cases Relating to the Public Lands, Volume 44: Washington, D.C., Department of the Interior, p. 325–326.

Dillhoff, R.M., Dillhoff, T.A., Dunn, R.E., Myers, J.A., and Strömberg, C.A.E., 2009, Cenozoic paleobotany of the John Day Basin, central Oregon, *in* O'Connor, J.E., Dorsey, R.J., and Madin, I.P., eds., Volcanoes to Vineyards: Geologic Field Trips through the Dynamic Landscape of the Pacific Northwest: Geological Society of America Field Guide 15, p. 135–164, https://doi.org/10.1130/2009.fld015(07).

Downs, T., 1956, The Mascall fauna from the Miocene of Oregon: University of California Publications in Geological Sciences, v. 31, p. 199–354.

Edwards, R., 2008, Why the homesteading data are so poor (and what can be done about it): Great Plains Quarterly, v. 28, p. 181–190.

Enlows, H.E., 1976, Petrography of the Rattlesnake Formation at the type area, central Oregon: Department of Geology and Mineral Industries, State of Oregon: Oil and Gas Investigations, v. 25, p. 1–34.

Famoso, N.A., and Davis, E.B., 2014, Occlusal Enamel Complexity in middle Miocene to Holocene Equids (Equidae: Perissodactyla) of North America: PLoS One, v. 9, no. 2, e90184, 11 p., https://doi.org/10.1371/journal.pone.0090184.

Famoso, N.A., Feranec, R.S., and Davis, E.B., 2013, Occlusal enamel complexity and its implications for lophodonty, hypsodony, body mass, and diet in extinct and extant ungulates: Palaeogeography, Palaeoclimatology, Palaeoecology, v. 387, p. 211–216, https://doi.org/10.1016/j.palaeo.2013.07.006.

Foster, J.R., Titus, A.L., Winterfeld, G.F., Hayden, M.C., and Hamblin, A.H., 2001, Paleontological survey of the Grand Staircase–Escalante National Monument, Garfield and Kane counties, Utah: Utah Geological Survey Special Study, v. 97, p. 1–98.

Fremd, T.J., 2010, Guidebook: SVP Field Symposium 2010. John Day Basin Field Conference: Society of Vertebrate Paleontology, 153 p., https://scholarsbank.uoregon.edu/xmlui/handle/1794/12193.

Fremd, T.J., Foss, S.E., and Zancanella, J., 2009, Interagency management of fossil resources, *in* Foss, S.E., Cavin, J.L., Brown, T., and Kirkland, J.I., eds., Proceedings of the Eighth Conference on Fossil Resources: Salt Lake City, Bureau of Land Management, p. 29–31.

Gates, P.W., 1976, An overview of American land policy: Agricultural History, v. 50, p. 213–229.

Gates, T.A., Lund, E.K., Boyd, C.A., DeBlieux, D.D., Titus, A.L., Evans, D.C., Getty, M.A., Kirkland, J.I., and Easton, J.G., 2013, Ornithopod dinosaurs from the Grand Staircase–Escalante National Monument region, Utah, and their role in paleobiogeographic and macroevolutionary studies, *in* Titus, A.L., and Loewen, M.A., eds., At the Top of the Grand Staircase: The Late Cretaceous of southern Utah: Bloomington, Indiana University Press, p. 237–272.

Gates, T.A., Jinnah, Z., Levitt, C., and Getty, M.A., 2014, New hadrosaurid specimens from the lower middle Campanian Wahweap Formation of Utah, *in* Eberth, D.A., and Evans, D.C., eds., Hadrosaurs: Proceedings of the International Hadrosaur Symposium: Bloomington, Indiana University Press, p. 156–173.

Hatcher, J.B., 1901, *Diplodocus* (Marsh): Its osteology, taxonomy, and probable habits, with a restoration of the skeleton: Memoirs of the Carnegie Museum, v. 1, no. 1, p. 1–63.

Hedeen, S., 2008, Big Bone Lick: The Cradle of American Paleontology: Lexington, University Press of Kentucky, 182 p.

Hunt, R.M., and Stepleton, E., 2004, Geology and paleontology of the upper John Day beds, John Day River Valley, Oregon: Lithostratigraphic and biochronologic revision in the Haystack Valley and Kimberly areas (Kimberly and Mt. Misery quadrangles): Bulletin of the American Museum of Natural History, no. 282, p. 1–90, https://doi.org/10.1206/0003-0090(2004)282<0001:GAPOTU>2.0.CO;2.

Interagency Working Group on Scientific Collections, 2009, Scientific Collections: Mission-Critical Infrastructure of Federal Science Agencies: Washington, D.C., Office of Science and Technology Policy, 49 p.

Jaffe, M., 2000, The Gilded Dinosaur: The Fossil War between E.D. Cope and O.C. Marsh and the Rise of American Science: New York, Crown Publishers, 424 p.

Jefferson, T., 1788, Notes on the State of Virginia: Philadelphia, Prichard and Hall, 244 p.

Jefferson, T., 1799, A memoir on the discovery of certain bones of a quadruped of the clawed kind in the western parts of Virginia: Transactions of the American Philosophical Society, v. 4, p. 246–260, https://doi.org/10.2307/1005103.

Kuizon, L., 2006, Legislative and regulatory history of paleontological resources: New Mexico Museum of Natural History and Science Bulletin, v. 34, p. 102–108.

Lanham, U., 1973, The Bone Hunters: The Heroic Age of Paleontology in the American West: New York, Dover Publications, 285 p.

Leidy, J., 1856, Notices of remains of extinct reptiles and fishes, discovered by Dr. F.V. Hayden in the bad lands of the Judith River, Nebraska Territory: American Journal of Science and Arts, v. 22, p. 118–120.

Library of Congress, 2017, Most colossal animal ever on Earth just found out west: https://www.loc.gov/resource/sn83030180/1898-12-11/ed-1/?sp=33 (accessed February 2017).

Loewen, M.A., Farke, A.A., Sampson, S.D., Getty, M.A., Lund, E.K., and O'Connor, P.M., 2013a, Ceratopsid dinosaurs from the Grand Staircase of southern Utah, *in* Titus, A.L., and Loewen, M.A., eds., At the Top of the Grand Staircase: The Late Cretaceous of Southern Utah: Bloomington, Indiana University Press, p. 488–503.

Loewen, M.A., Irmis, R.B., Sertich, J.J.W., Currie, P.J., and Sampson, S.D., 2013b, Tyrant dinosaur evolution tracks the rise and fall of Late Cretaceous oceans: PLoS One, v. 8, no. 11, p. e79420, https://doi.org/10.1371/journal.pone.0079420.

Lucas, F.A., 1899, The fossil bison of North America: Proceedings of the United States National Museum, v. 21, p. 755–771, https://doi.org/10.5479/si.00963801.21-1172.755.

MacBride, T.H., 1893, A new cycad: Bulletin from the Laboratories of Natural History of the State University of Iowa, v. 2, p. 391–393.

Martin, J.E., 1983, Additions to the early Hemphillian (Miocene) Rattlesnake Fauna from central Oregon: Proceedings of the South Dakota Academy of Science, v. 62, p. 23–33.

Martin, J.E., and Fremd, T.J., 2001, Revision of the lithostratigraphy of the Hemphillian Rattlesnake units of central Oregon: PaleoBios, v. 21, p. 89.

McClaughry, J.D., Ferns, M.L., Streck, M.J., Patridge, K.A., and Gordon, C.L., 2009, Paleogene calderas of central and eastern Oregon: Eruptive sources of widespread tuffs in the John Day and Clarno Formations, *in* O'Connor, J.E., Dorsey, R.J., and Madin, I.P., eds., Volcanoes to Vineyards: Geologic Field Trips through the Dynamic Landscape of the Pacific Northwest: Geological Society of America Field Guide 15, p. 407–434, https://doi.org/10.1130/2009.fld015(20).

McDonald, H.G., and Anderson, E., 1975, A late Pleistocene vertebrate fauna from southeastern Idaho: Tebiwa, v. 18, no. 1, p. 19–37.

Merriam, J.C., Stock, C., and Moody, C.L., 1925, The Pliocene Rattlesnake Formation and fauna of eastern Oregon, with notes on the geology of the Rattlesnake and Mascall deposits: Carnegie Institution of Washington: Contributions to Palaeontology, v. 347, p. 43–92.

Muhn, J., and Stuart, H.R., 1988, Opportunity and Challenge: The Story of BLM: Washington, D.C., U.S. Department of the Interior, Bureau of Land Management, 301 p.

National Park Service, 2017, Preserving our rich fossil heritage: https://www.nps.gov/subjects/fossils/index.htm (accessed February 2017).

Office of Inspector General, 2009, Department of the Interior Museum Collections: Accountability and Preservation: Washington, D.C., Office of Inspector General, 20 p.

Osborn, H.F., 1905, *Tyrannosaurus* and other Cretaceous carnivorous dinosaurs: Bulletin of the American Museum of Natural History, v. 21, p. 259–265.

Peale, R., 1803, An Historical Disquisition on the Mammoth, or Great American Incognitum, an Extinct, Immense, Carnivorous Animal, Whose Fossil Remains Have Been Found in North America: London, E. Lawrence, 91 p.

Pinsof, J.D., 1991, A cranium of *Bison alaskensis* (Mammalia: Artiodactyla: Bovidae) and comments on fossil *Bison* diversity in the American Falls area, southeastern Idaho: Journal of Vertebrate Paleontology, v. 11, no. 4, p. 509–514, https://doi.org/10.1080/02724634.1991.10011418.

Pinsof, J.D., 1998, The American Falls local fauna: Late Pleistocene (Sangamonian) vertebrates from southeastern Idaho, *in* Akersten, W.A., McDonald, H.G., Meldrum, D.J., and Flint, M.E.T., eds., And Whereas…Papers on the Vertebrate Paleontology of Idaho Honoring John A. White, Volume 1, Idaho Museum of Natural History Occasional Paper 36, p. 121–145.

Rainger, R., 2001, The museum in America: Historical and cultural perspectives, *in* Mares, M.A., ed., The Sam Noble Oklahoma Museum of Natural History: A University Natural History Museum for the New Millennium: Norman, Oklahoma, Sam Noble Oklahoma Museum of Natural History, p. 9–26.

Rea, T., 2001, Bone Wars: The Excavation and Celebrity of Andrew Carnegie's Dinosaur: Pittsburgh, Pennsylvania, University of Pittsburgh Press, 276 p.

Retallack, G.J., Bestland, E.A., and Fremd, T.J., 2000, Eocene and Oligocene Paleosols of Central Oregon: Geological Society of America Special Paper 344, 196 p.

Ridenour, J., 1969, Depositional environments of the late Pleistocene American Falls Formation, southeastern Idaho [M.S. thesis]: Pocatello, Idaho State University, 82 p.

Roberts, E.M., Sampson, S.D., Deino, A., and Bowring, S., 2013, The Kaiparowits Formation: A remarkable record of Late Cretaceous terrestrial environments, ecosystems and evolution in western North America, *in* Titus, A.L., and Loewen, M.A., eds., At the Top of the Grand Staircase: The Late Cretaceous of Southern Utah: Bloomington, Indiana University Press, p. 85–106.

Rudwick, M.J.S., 1972, The Meaning of Fossils: Episodes in the History of Paleontology: Chicago, University of Chicago Press, 287 p.

Rytuba, J.J., and McKee, E.H., 1984, Peralkaline ash flow tuffs and calderas of the McDermitt: Journal of Geophysical Research, v. 89, p. 8616–8628, https://doi.org/10.1029/JB089iB10p08616.

Sakurai, D.S., 1994, Animal, mineral, or cultural antiquity: The management and protection of paleontological resources: Loyola of Los Angeles International and Comparative Law Review, v. 17, p. 197–243.

Sampson, S.D., Lund, E.K., Loewen, M.A., Farke, A.A., and Clayton, K.E., 2013, A remarkable short-snouted horned dinosaur from the Late Cretaceous (late Campanian) of southern Laramidia: Proceedings of the Royal Society: Biological Sciences, v. 280, https://doi.org/10.1098/rspb.2013.1186.

Samuels, J.X., and Cavin, J.L., 2013, The earliest known fisher (Mustelidae), a new species from the Rattlesnake Formation of Oregon: Journal of Vertebrate Paleontology, v. 33, no. 2, p. 448–454, https://doi.org/10.1080/02724634.2013.722155.

Samuels, J.X., and Zancanella, J., 2011, An early Hemphillian occurrence of *Castor* (Castoridae) from the Rattlesnake Formation of Oregon: Journal of Paleontology, v. 85, p. 930–935, https://doi.org/10.1666/11-016.1.

Samuels, J.X., Albright, L.B., III, and Fremd, T.J., 2015, The last fossil primate in North America, new material of the enigmatic *Ekgmowechashala* from the Arikareean of Oregon: American Journal of Physical Anthropology, v. 158, p. 43–54, https://doi.org/10.1002/ajpa.22769.

Santucci, V.L., and Ghist, J.M., 2014, Fossil Cycad National Monument: A history from discovery to deauthorization: Dakoterra, v. 6, p. 82–93.

Scharf, J.T., 1879, History of Maryland from the Earliest Period to the Present Day: Hatboro, Pennsylvania, Tradition Press, 635 p.

Seligman, A.N., Bindeman, I.N., McClaughry, J., Stern, R.A., and Fisher, C., 2014, The earliest low and high $\delta^{18}O$ caldera-forming eruptions of the Yellowstone plume: Implications for the 30–40 Ma Oregon calderas and speculations on plume-triggered delaminations: Frontiers in Earth Science, v. 2, p. 1–9, https://doi.org/10.3389/feart.2014.00034.

Semonin, P., 2000, American Monster: How the Nation's First Prehistoric Creature Became a Symbol of National Identity: New York, New York University Press, 482 p.

Simpson, G.G., 1942, The beginnings of vertebrate paleontology in North America: Proceedings of the American Philosophical Society, v. 86, no. 1, p. 130–188.

Simpson, G.G., 1943, The discovery of fossil vertebrates in North America: Journal of Paleontology, v. 17, no. 1, p. 26–38.

Simpson, G.G., and Tobien, H., 1954, The rediscovery of Peale's Mastodon: Proceedings of the American Philosophical Society, v. 98, no. 4, p. 279–281.

Stratton, A.E., Tapanila, L., and Stratton, J.H., 2014, Collecting history of vertebrate fossils at American Falls, Idaho: A reservoir of data to inform land-use policy: Palaios, v. 29, p. 393–400, https://doi.org/10.2110/palo.2013.131.

Titus, A.L., Roberts, E.M., and Albright, L.B., III, 2013, Geologic overview, *in* Titus, A.L., and Loewen, M.A., eds., At the Top of the Grand Staircase: The Late Cretaceous of Southern Utah: Bloomington, Indiana University Press, p. 13–41.

Titus, A.L., Eaton, J.G., and Sertich, J., 2017, Late Cretaceous stratigraphy and vertebrate faunas of the Markagunt, Paunsaugunt, and Kaiparowits plateaus, southern Utah: Geology of the Intermountain West, v. 3, p. 229–291.

Turner, G., 1799, Memoir on the extraneous fossils, denominated mammoth bones: Principally designed to shew, that they are the remains of more than one species of non-descript animal: Transactions of the American Philosophical Society, v. 4, p. 510–518, https://doi.org/10.2307/1005126.

Tweet, J.S., Santucci, V.L., and McDonald, H.G., 2016, Name-bearing fossil type specimens and taxa named from National Park Service areas: New Mexico Museum of Natural History and Science Bulletin, v. 73, p. 277–288.

United States v. Ben Diaz, 1974, United States Ninth Circuit Court of Appeals: 499 F2d 113 74-1177.

United States v. William R. Smyer and Byron R. May, 1979, United States Tenth Circuit Court of Appeals: 596 F2d 939 78-1134, 78-1135.

Utley, R.M., and Mackintosh, B., 1989, The Department of Everything Else: Highlights of Interior History: Washington, D.C., United States Department of the Interior, 57 p.

Ward, L.F., 1899, Cretaceous formation of the Black Hills as indicated by the fossil plants: U.S. Geological Survey Annual Report, v. 19, p. 521–958.

Warren, J.C., 1852, The Mastodon Giganteus of North America: Boston, John Wilson and Son, 219 p., https://doi.org/10.5962/bhl.title.62016.

Warren, L., 1998, Joseph Leidy: The Last Man Who Knew Everything: New Haven, Connecticut, Yale University Press, 303 p.

Weishampel, D.B., and Jensen, J.A., 1979, *Parasaurolophus* (Reptilia: Hadrosauridae) from Utah: Journal of Paleontology, v. 53, no. 6, p. 1422–1427.

White, C.A., 1991, A History of the Rectangular Survey System: Washington, D.C., Bureau of Land Management, 774 p.

Wieland, G.R., 1916, American Fossil Cycads: Volume 2 Taxonomy: Carnegie Institute Publication, v. 34, 407 p.

Wieland, G.R., 1937, Fossil Cycad National Monument: Science, v. 85, no. 2203, p. 287–289, https://doi.org/10.1126/science.85.2203.287.

Wieland, G.R., 1938, Our national monuments: Science, v. 87, no. 2246, p. 37–39, https://doi.org/10.1126/science.87.2246.37.

Wieland, G.R., 1945, The Yale cycadeoids: Science, v. 102, no. 2655, p. 513–514, https://doi.org/10.1126/science.102.2655.513-a.

Williams, G.W., 2005, The USDA Forest Service—The First Century: Washington, D.C., U.S. Department of Agriculture Forest Service Office of Communication, 156 p.

Zanno, L.E., and Sampson, S.D., 2005, A new oviraptorosaur (Theropoda, Maniraptora) from the Late Cretaceous (Campanian) of Utah: Journal of Vertebrate Paleontology, v. 25, no. 4, p. 897–904, https://doi.org/10.1671/0272-4634(2005)025[0897:ANOTMF]2.0.CO;2.

Zanno, L.E., Loewen, M.A., Farke, A.A., Kim, G., Claessens, L.P.A.M., and McGarrity, C.T., 2013, Late Cretaceous theropod dinosaurs of southern Utah, *in* Titus, A.L., and Loewen, M.A., eds., At the Top of the Grand Staircase: The Late Cretaceous of Southern Utah: Bloomington, Indiana University Press, p. 504–525.

Manuscript Accepted by the Society 21 November 2017
Manuscript Published Online 6 July 2018

The Geological Society of America
Special Paper 535

Museums, paleontology, and a biodiversity science–based approach

Bruce S. Lieberman
Department of Ecology and Evolutionary Biology and Biodiversity Institute, University of Kansas, 1345 Jayhawk Boulevard, Dyche Hall, Lawrence, Kansas 66045, USA

Julien Kimmig
Biodiversity Institute, University of Kansas, 1345 Jayhawk Boulevard, Dyche Hall, Lawrence, Kansas 66045, USA

ABSTRACT

Museum collections provide a tremendous wealth of data bearing on biogeography, the field that focuses on the study of the distribution of organisms in space and time. Biogeography is a discipline that played a fundamental role in the development of ideas on evolution in the nineteenth century, and it still is a vibrant research area today. One way that biogeography has remained vibrant is through the burgeoning area of biodiversity science. There are many aspects of biodiversity science relevant to paleontology, running the gamut from conservation paleobiology to ecosystem observations, etc. Our especial focus here is on biodiversity science approaches involving the analysis of museum specimen records using mapping and analytical approaches, such as the geographic information system (GIS) and ecological niche modeling (ENM), to quantify how climate change has caused (and will continue to cause) species to move across the face of the globe through time. Initial efforts considered extant taxa, but now analyses of extinct taxa are becoming more commonplace. These analyses of fossil taxa offer extensive opportunities to gain increased insight into biogeography and also macroevolution. This contribution focuses specifically on approaches using fossil taxa and their associated museum specimen data. Such approaches have shown how invasive species have contributed to ancient biodiversity crises, how species niches largely remain stable over geological time scales, how it is predominately abiotic factors, as opposed to competition, that influence species distributions and determine species survival in the long term, and finally how extant species that have been present in marine ecosystems for millions of years are now in grave peril due to impending climate changes projected to occur in the near term. Each of these discoveries will be highlighted in order to help show the value that museum collections of fossils continue to have in the twenty-first century.

Lieberman, B.S., and Kimmig, J., 2018, Museums, paleontology, and a biodiversity science–based approach, *in* Rosenberg, G.D., and Clary, R.M., eds., Museums at the Forefront of the History and Philosophy of Geology: History Made, History in the Making: Geological Society of America Special Paper 535, p. 335–348, https://doi.org/10.1130/2018.2535(22).

INTRODUCTION

Today scientists live in the age of "big data": where analyses of large data sets fuel new scientific discoveries and allow refined testing of old hypotheses. It is in this present-day context that the field of paleontology needs to be considered, for paleontologists have been amassing copious amounts of data in the United States for roughly two centuries. These data are now housed in museum collections. Truly, these paleontological museum collections are our great data repositories, and the fossils they comprise are our data. One of the express purposes why these fossils were gathered was to aid future generations of researchers, though likely at the time little did the collectors realize the powerful scientific approaches that would be available to scientists in the twenty-first century. For the existence of these collections, today's scientists clearly owe a great debt of gratitude to these paleontologists. The collections they made are literally irreplaceable from a financial and time burden sense, and in some cases were derived from localities no longer accessible. Without them paleontologists would literally be stranded in an age of big data, without any data to analyze. A challenge does exist, though, because some of these data are actually hidden in museum cabinets and are not available in a manner that makes them easily analyzed. The express focus of this paper is to consider how fossils housed in collections have been used in a particular type of biodiversity science perspective, one based on the use of computer mapping and analysis technologies, to advance our understanding of how past environmental changes have shaped species distributions through time and influenced macroevolution. (There are many types of biodiversity science perspectives [Dirzo and Loreau, 2005]. Among the most important types on the geological side are those in the area of conservation paleobiology [e.g., Kowalewski, 2009; Dietl and Flessa, 2011; Casey et al., 2014].) The present-day scientific emphasis on data is why we should always remember that it is "the actual collector … The (hu)man who, in whatever station of life, devotes (their) time, by night and by day … to the positive capture and preservation of those specimens which serve as the object for all our observations: (s)he is the real labourer in the field, and if we would keep the lamp of our science constantly burning, it is (them) alone that we can look for fuel to feed its flame" (Newman, 1854, p. 144–145).

For the purposes of this paper, what is meant by paleontology museum collections data are individual specimens and the associated collection record: a taxon name, preferably a species name, as well as a geographic occurrence in space, and a stratigraphic occurrence in time. These data, once input into a relational database structure, can be readily shared via the World Wide Web. The geographic occurrence information can further, via the process of georeferencing (see Wieczorek et al., 2004, for an early discussion), be converted into a latitude and longitude, so that a species can be precisely located on a map, which maximizes the utility of the collections data for scientific research in biogeography and macroevolution. Knowing the present-day latitude and longitude of a fossil collection site is of course valuable, but what is of even greater value is knowing the latitude and longitude during the time period when that fossil was actually alive (Rode and Lieberman, 2004; Hendricks et al., 2008; Myers and Lieberman, 2011). Thankfully, several different approaches (*PaleoGIS,* now *PaleoWeb,* is one such approach) can be used to rotate present-day latitudes and longitudes back to an ancient time period to visualize the hypothesized distribution of a specimen when it was extant. Some of the most efficacious and powerful approaches available to paleontologists utilize geographic information systems (GIS) because these provide a means of visualizing and analyzing data. In a GIS framework, the next analytical step that can be taken is to use the geographic occurrence records of species to produce polygons that reflect reconstructed species ranges. These polygons make it possible to produce precise estimates of geographic ranges. Geographic range values are especially pertinent in paleontology because they are related to several key macroevolutionary parameters such as speciation and extinction rates (Eldredge, 1979; Stanley, 1979; Vrba, 1980; Jablonski, 1986; Lieberman, 2000). There is also evidence that geographic ranges are heritable (Jablonski, 1987; Hunt et al., 2005) and thus of relevance to phenomena like species selection (Lieberman and Vrba, 1995). Indeed, macroevolution is the field devoted to the study of the patterns and processes pertaining to the birth, death, and persistence of species (Lieberman and Eldredge, 2014), and geographic factors play a fundamental role in motivating species birth, death, and persistence (see Lieberman, 2000, 2008, and Eldredge et al., 2005, for review). Thus, being able to track geographic range in individual species through time, which is afforded via GIS, is extremely beneficial. Several studies of how speciation and extinction rates wax and wane relative to geographic range values in deep time using fossil collections data will be discussed herein.

Geographic range values not only have relevance to macroevolution, but they are associated with various key research questions in paleoecology and paleobiogeography. GIS is valuable in this context because it makes it possible to see if there is any evidence for interaction or correlation between the ranges of different species through time (Myers and Lieberman, 2011). This affords the opportunity to consider the role that competition might play in macroevolution (Gould and Calloway, 1980; Benton, 1996). GIS also has the added benefit of making it possible to input several other data layers in addition to geographic range data, such that the correlations between range data and other variables can be considered. One set of data layers frequently constructed and utilized consists of various environmental variables. The connection between environmental variables and geographic ranges can be considered using a technique known as ecological niche modeling (ENM), an approach described in detail in Peterson et al. (2011), Saupe et al. (2012), Myers et al. (2015) and the references therein. This is particularly valuable in macroevolutionary, paleoecological, and paleobiogeographic research because it makes it possible to consider which abiotic factors are most influencing species distribution, survival, and diversification through time. Therefore, several studies that have applied an

ENM approach to the fossil record, i.e., paleoENM approaches sensu Myers et al. (2015), will also be discussed herein.

EXTENDING AND IMPROVING COLLECTIONS DATA FOR RESEARCH

In recent years the National Science Foundation (NSF) and its sponsored programs have significantly increased funding for the digitization of paleontological collections (Karim et al., 2016). This funding has supported data generation and entry at museums, and then these data are sent to data repositories such as *iDigBio* (Integrated Digitized Biocollections, https://www.idibio.org). Data are made publicly available, at no cost, and also standardized to Darwin Core (see Wieczorek et al., 2012). This facilitates direct comparison of specimens from different collections without having to manually manipulate individual data sets from different institutional databases. It also enhances scientists' ability to analyze big data sets including specimens from several museums, standardized in one database, which in turn allows for more detailed and quicker paleobiogeographic and macroevolutionary studies.

The University of Kansas, Biodiversity Institute, Division of Invertebrate Paleontology (KUMIP) collections have been the subject of digitization efforts for several years, and work is continuing with the ultimate goal of making information on all KUMIP specimens available online to researchers and the general public. KUMIP specimen data exist in a *Specify* (http://specifyx.specifysoftware.org) database and include geographic and stratigraphic data. The data are also shared with *iDigBio* and GBIF (Global Biodiversity Information Facility, http://www.gbif.org). The KUMIP collection houses over 900,000 specimens, and at present, core digitization efforts concentrate on Pennsylvanian and Cretaceous fossil specimens and their corresponding data (geographic location where collected, chronostratigraphy, lithostratigraphy, date collected, collector, etc.). Two NSF Advancing the Digitization of Biological Collections (ADBC) grants are currently funding these projects at the KUMIP, and so far, more than 350,000 specimens have been entered into the database and more than 12,000 localities have been georeferenced via the point-radius method (Wieczorek et al., 2004). In addition to entering specimen data, data are also screened for accuracy, to maximize their utility in research. Screening includes correcting taxonomic names and synonymies, georeferencing localities to the smallest possible area and providing error bounds, and updating the stratigraphic information provided. To make these data even more useful to researchers and the general public, type specimens and well-preserved non-type specimens are photographed, and those photographs are included with the specimen data entries.

In an effort to use the generated data for public outreach, the information is also shared via the Digital Atlas of Ancient Life (http://www.digitalatlasofancientlife.org/), a Web site developed and managed by Jonathan Hendricks at the Paleontological Research Institution. The Web site is aimed at informing interested individuals from the general public, and avocational and professional paleontologists, about more than 1000 species of fossils that can be found in different parts of the United States.

CHALLENGES TO DATA QUALITY AND ACCESS

Data quality is an important issue when it comes to reconstructing the geographic ranges of fossil and modern species using museum collections data. Of course, the more detailed the locality information provided with the specimen, the easier georeferencing is. When one works with fossil museum collections, the quality of geographic, taxonomic, or stratigraphic information can vary from specimen to specimen. In a good-case scenario, the individual georeferencing might be given a specimen identified to the species level with a well-characterized locality, perhaps including township, section, and range data, or to county level, along with the name of the member and formation it was recovered in, and even the specimen's relative position in a stratigraphic section. In other instances, the specimen might not have a species or genus identification, and the locality information is poorly resolved both geographically (i.e., only to state, province, or country) and stratigraphically (i.e., only to period).

Georeferencing protocols are important to establish early on in the digitization process, and faculty, staff, and students should undergo rigorous training before starting to georeference on their own. It is recommended that well-established guidelines (e.g., Chapman and Wieczorek, 2006) be used. These guidelines will help the digitization efforts, because the data will be comparable. It is also recommended that a common language protocol be followed—one that identifies which software and methods were used, as well as how the digitizer dealt with occurrence issues. In fact, with many specimen database programs, the method used to georeference can be included in the specimen record. Common problems encountered while digitizing and georeferencing are misspelled labels, inverted numbers, and other inaccuracies regarding the locality. In general, while the digitizer might be confident that they have successfully identified and resolved problems, it is always prudent to leave a note in the digitization remarks regarding these resolutions. A valuable aspect of georeferencing is that uncertainties can be included. This allows the end user of the data to exclude data that have uncertainties outside the scope pertinent regarding the research being performed.

In addition to potential issues with geographic location, chronostratigraphic and lithostratigraphic information can also have their own unique set of issues. These include the fact that over time there may be changes to the name of lithologic units or time stages and series, changes regarding the reconstructed stratigraphic age of lithologic units, and regional differences in the use of names, just to name a few. Thanks to the World Wide Web, many of these issues can be resolved, but this can be time consuming. Similar to the issues that crop up regarding resolving localities, it is again recommended that remarks be added regarding what changes have been made to the data by the digitizer.

One thing worth considering in this context is the recent policy developments in the United States at the Department of the Interior; these might create challenges and restrict access to locality information, and even to new specimens. In particular, the proposed Paleontological Resources Preservation Act (PRPA) (Department of the Interior et al., 2016) would implement a provision that would exempt information pertaining to the specific location of paleontological sites from the Freedom of Information Act, unless "the authorized officer determines that disclosure would: (1) further the purposes of PRPA; (2) not create risk of harm to, or theft or destruction of, the resource or site containing the resource; and (3) be in accordance with other applicable laws" (Department of the Interior, 2016, p. 88,176). This provision could create challenges in the future for projects aiming to build large digital resources, such as georeferenced localities, from museum collections, since it might not be possible to share or publish the locality data, at least in much detail, for some fossils. This could potentially make future scientific studies more difficult to reproduce.

Another challenge created by the proposed PRPA would be restrictions on casual collecting. The PRPA would establish casual collecting limits on federal lands governed by the agencies covered in the PRPA to 100 pounds per year per collector. This could have a significant effect on donations of invertebrate fossils to museum collections. Considering the fact that often important specimens can be derived from such donations, this might not only constrain the discovery of new species. This rule might also in the future restrict the ability of scientists to accurately reconstruct the geographic distributions of fossil species because apparent geographic ranges will be smaller due to the lack of specimens in museum collections.

APPLYING GIS TO THE STUDY OF BIOGEOGRAPHY AND MACROEVOLUTION DURING KEY TIME PERIODS IN THE HISTORY OF LIFE

The Ice Ages

The FAUNMAP project (Graham et al., 1996) was probably the first study to use GIS in conjunction with museum collections data to study biogeographic patterns in the fossil record. They used Quaternary fossil vertebrates from North America to look at how climate change during and at the end of the Ice Ages influenced the distributions of organisms. The use of GIS to study not only biogeography, but specifically the effects of climate change on geographic range, speciation, and extinction, became a key facet of subsequent biodiversity science studies that used paleontological collections.

The Late Devonian Biodiversity Crisis

Studies by Rode and Lieberman (2004, 2005) were the first to apply GIS to the invertebrate fossil record. They focused on the Late Devonian interval, a key time period in the history of life traditionally treated as representing one of the "Big Five" mass extinctions. They used information derived from the distribution of hundreds of species of brachiopods (Rode and Lieberman, 2004) and dozens of species of phyllocarid crustaceans (Rode and Lieberman, 2005) in space and time to investigate what happened to species' geographic ranges and rates of speciation and extinction before, during, and after this interval. Museum specimens housed at various places, including the Yale Peabody Museum of Natural History (YPM), were considered. What Rode and Lieberman (2004, 2005) found was that the Late Devonian interval per se did not actually represent a time period of elevated extinction; instead, extinction rates were fairly constant and moderately high throughout much of the Devonian. However, during the Late Devonian, speciation rates fell precipitously (Rode and Lieberman, 2004, 2005; Stigall, 2012a). This was associated with climate change (a profound episode of global warming) that led to sea-level rise and dramatic geographic range expansions, with concomitant multiple episodes of species invasions, and the homogenization of the global biota. The expansion in species ranges associated with the rise in sea level that eliminated geographic barriers to the free movement of marine organisms led to a decline in opportunities for allopatric speciation, the primary motor of speciation (Wiley and Lieberman, 2011). Given relatively constant but moderate extinction rates and a long-term drop in speciation rate, overall diversity will fall over the course of several hundreds of thousands to millions of years. The net decline in diversity might look like a mass extinction, but in actuality, it is better termed a biodiversity crisis because declining speciation was the cause. The patterns they discovered matched the patterns predicted and identified by McGhee (1996), Lieberman (2000), and Bambach et al. (2004). These analyses were usefully reviewed by Svenning et al. (2011), Stigall (2012a), and Stigall et al. (2017) (see also Stigall and Lieberman, 2006).

The studies by Rode and Lieberman (2004, 2005) also provided insight into the present-day biodiversity crisis. In particular, today we know that there are significant elevations in extinction rate associated with the habitat degradation our own species is causing. However, another prominent phenomenon, associated with the present-day biodiversity crisis causing extinction and driven by humans, is the accidental or purposeful introduction of new (invasive) species into different regions far beyond their ancestral ranges. The patterns during the Late Devonian biodiversity crisis suggest that the invasive species phenomenon can also drive down speciation rates (Rode and Lieberman, 2004, 2005). This, in conjunction with the present-day increase in extinction rate, suggests a one-two punch for the planet's biota. Not only are we eliminating species, but we are also thwarting the means for biodiversity recovery.

The museum collections data were obviously central to being able to conduct meaningful tests of biogeographic and macroevolutionary hypotheses. In most cases, these data had not been digitized prior to the beginning of the project; therefore, all information was obtained from specimen labels. The data were then corrected when necessary and the localities were georeferenced. Further, it also bears mentioning that of course the fossil record and the collections available in museums are never complete. However, in this respect, GIS also proves highly beneficial. First, the available collections distributional data can be viewed in a GIS and compared with an outcrop layer for the relevant time intervals also created in GIS. This suggested clear targets for further needed fieldwork (see Rode and Lieberman, 2004, 2005), which was then conducted to fill the gaps in sampling that could be readily identified. (Walls and Stigall [2012] also used a paleoENM approach to do something similar, in a predictive way, described more fully below.) The new material collected could in turn then be added to collections; in this way, collections can be built using the knowledge obtained by considering preexisting collections. The GIS approach also provides a number of means for testing the quality and completeness of the fossil record, including, for instance, comparisons relating outcrop area to species' distributional ranges to see if the two are strongly correlated (Myers and Lieberman, 2011). In the case of the studies of the Late Devonian biodiversity crisis, it did not appear that biases in the fossil record were governing the distributional patterns observed (Rode and Lieberman, 2004).

The Biogeography of Burgess Shale–Type Biotas

Cambrian soft-bodied fossil deposits provide a key window on the history of life, by revealing a panoply of ancient life forms not typically recovered in the fossil record (Whittington, 1985; Gould, 1989; Briggs and Fortey, 2005; Kimmig and Pratt, 2015). The Burgess Shale of Canada (Briggs et al., 1992) and the Chengjiang Biota of China (Hou et al., 2017) are likely the most well-known and intensely studied soft-bodied deposits, but in fact a plethora of such deposits exist from many other localities circumscribing the globe (Muscente et al., 2017). Further, several hypotheses have been put forward about the distinctive evolutionary and ecological nature of the non-skeletonized taxa that occur in these deposits (see Lieberman, 2001, and Smith and Lieberman, 1999, for further discussion). Hendricks et al. (2008) were interested in determining what might be distinctive about the soft-bodied arthropods, from a biogeographic perspective, relative to other types of organisms (Fig. 1). They used GIS to study what happened to the geographic range of soft-bodied species through time and compared these patterns with those involving the ranges of co-occurring trilobite taxa. In turn, they considered how geographic range was related to patterns of long-term species survival. The analyses incorporated direct observations on museum specimens housed in the KUMIP. They found that in fact the soft-bodied arthropods found in deposits such as the Burgess Shale showed fairly typical biogeographic patterns relative to many marine invertebrates, including modern marine arthropods. By contrast, it was the trilobites that showed the distinctive biogeographic patterns; in particular, they had relatively narrow geographic distributions and displayed a greater predilection to go extinct than species in other groups. Hendricks (2013) extended this work, again using GIS, to consider not just the soft-bodied arthropods, but also a much broader swath of the soft-bodied biota. His work reiterated the results found by Hendricks et al. (2008).

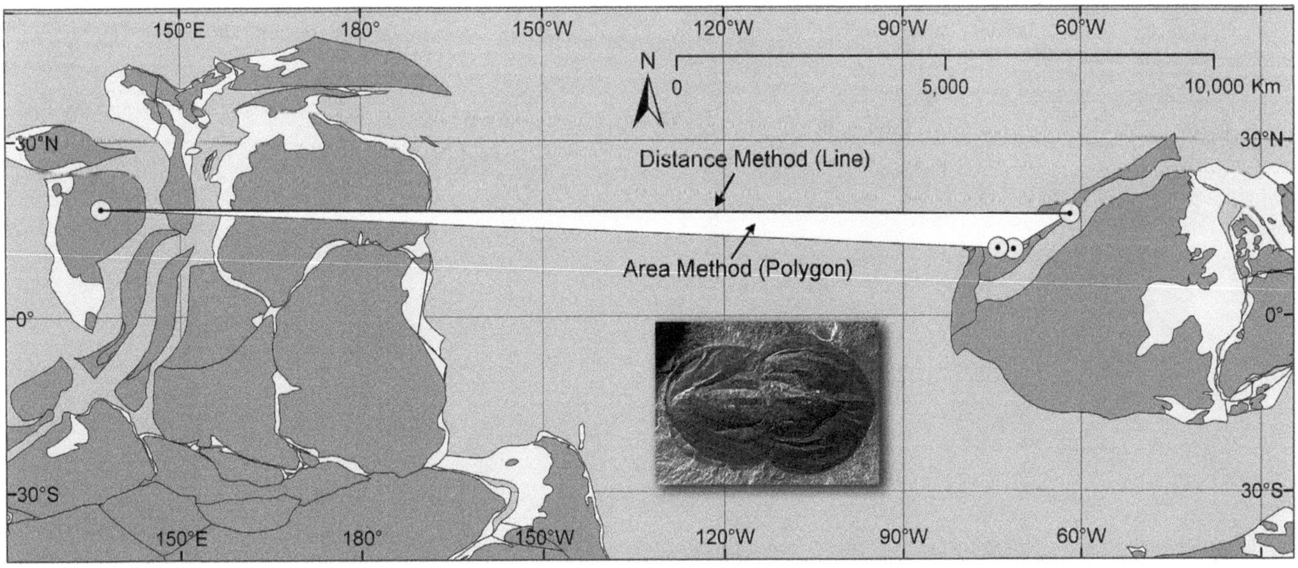

Figure 1. Species occurrence records of the most widely distributed species of Cambrian Burgess Shale–type arthropod, *Naraoia compacta,* plotted as points. The continents are shown in the approximate position they occupied during the middle Cambrian. The geographic range reconstruction for the species is shown diagrammatically and was reconstructed in two ways (each labeled): (1) as the maximum linear distance between the most spatially separated occurrence records; and (2) as a polygon circumscribing the different occurrence records. Reprinted from Hendricks et al. (2008) with permission from Elsevier.

Competition in the Cretaceous Western Interior Seaway (WIS) of North America

The fossil record, museum collections data, and GIS also provide a useful means to test the role that competition plays on species over macroevolutionary time scales. One area that has been investigated in detail involves the phenomenon of competitive replacement, whereby a new species enters the geographic region occupied by a preexisting species; if these species are ecologically similar, they directly compete; through time, the new species slowly yet inexorably drives the original species to extinction (Benton, 1996). There is in fact a long tradition in the history of evolutionary biology of ascribing an important role to various biotic factors, especially competitive replacement, in shaping macroevolutionary patterns, especially the long-term survival and extinction of species. However, there is evidence that some of this emphasis on biotic factors may have been overstated (Gould and Calloway, 1980; Vrba, 1980; Eldredge, 1989; Benton, 1996). To consider this issue, Myers and Lieberman (2011) used GIS and museum collections from a variety of repositories, including the University of Kansas, Biodiversity Institute, Division of Vertebrate Paleontology (KUVP). They focused on fossils from the Late Cretaceous WIS, a place and time period that housed (or in some cases lay beneath) the preferred habitat of a panoply of spectacular life forms including giant sharks, mosasaurs, and pterosaurs. The fossil record from this time period is exceptionally complete and well-studied (Myers and Lieberman, 2011). The value of this type of biodiversity science–based approach is that it allows direct quantification of geographic range in general and range overlaps in particular, so that it provides an excellent means of testing the evidence for competitive replacement. Myers and Lieberman (2011) found no evidence whatsoever for competitive replacement in the fossil record of the Cretaceous WIS (Fig. 2). In actuality, this is perhaps not surprising since the

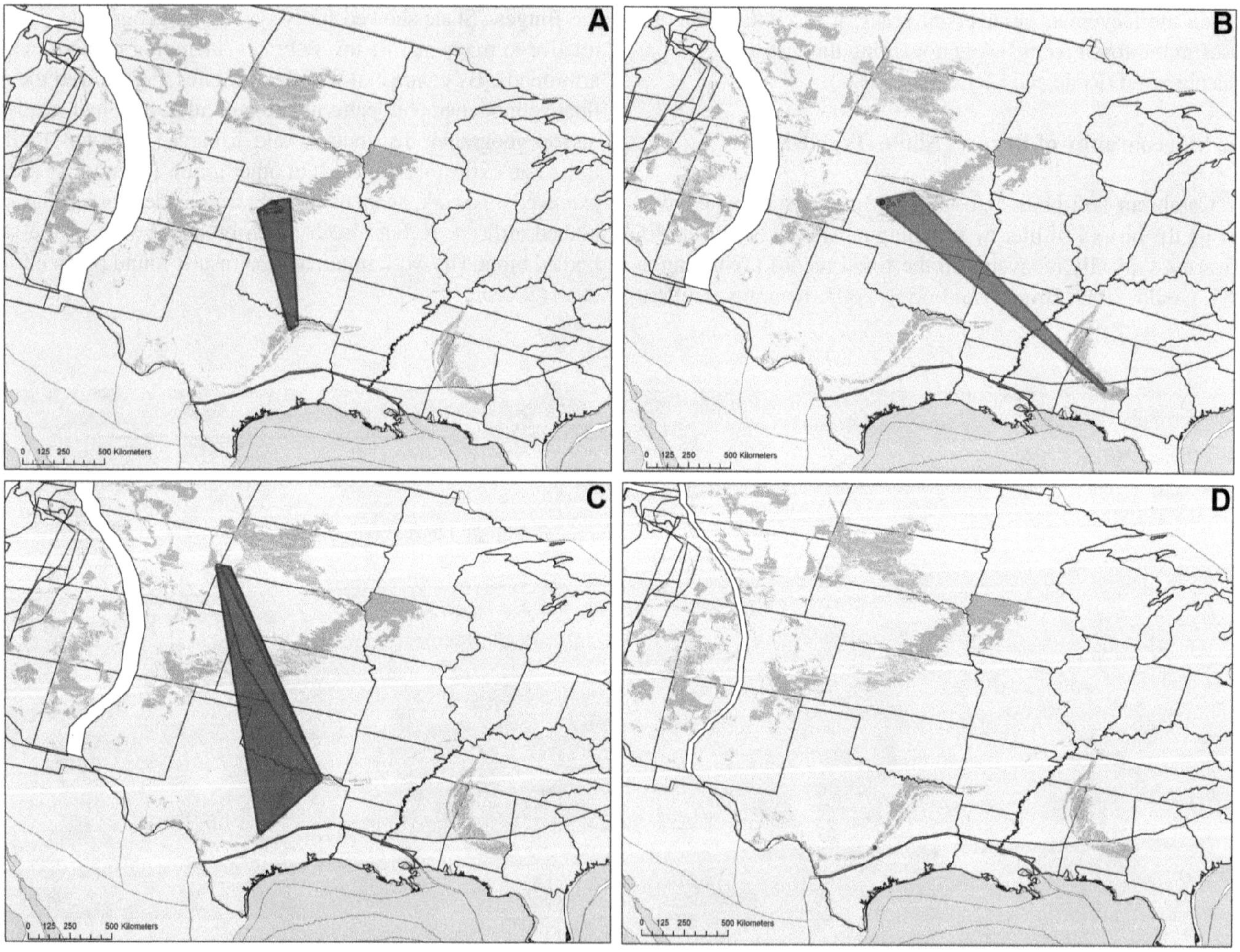

Figure 2. Range reconstructions for two different species of mosasaur over four different time intervals (stages) in the Late Cretaceous period: (A) Coniacian; (B) Santonian; (C) Campanian; and (D) Maastrichtian. Each species changed its distribution through time, but there is no biogeographic evidence for competitive replacement or competitive interactions between the two. Scale bars represent 500 km. From Myers and Lieberman (2011); used with permission.

time period over which we might expect competitive interactions to play out and lead to extinction should comprise hundreds or thousands of years. These are of course intervals too short in duration to typically be preserved in the fossil record. Competitive replacement may happen, but even if it does, it should not be expected to occur gradually over hundreds of thousands to millions of years. This is because the selection pressures such competition would entail are infinitesimally low. This is in fact analogous to one of the reasons why we should not expect to observe directional gradual evolution in species lineages over long time scales in the fossil record (see Gould and Eldredge, 1977 and Lieberman and Dudgeon, 1996). In short, competitive replacement is not a phenomenon we should expect to be manifested in the fossil record.

When and Why Does Geographic Range Size Not Always Confer Increased Extinction Resistance?

In the fossil record, there is a typically identified pattern whereby species with larger geographic ranges tend to have reduced extinction rates (Jablonski, 1986). However, there are times where that relationship breaks down. Dunill and Wills (2015) employed a biodiversity science approach to consider this topic in greater detail. They focused on patterns of survival and geographic range in terrestrial vertebrates before and during the end-Triassic mass extinction. Dunill and Wills (2015) found that, before the mass extinction, geographic range size was positively associated with survival, but intriguingly, during the mass extinction event itself, there was no longer an association between geographic range and survival (Fig. 3). It is almost as if during times of precipitous change, all bets are off, and the standard rules determining which species survive get thrown out of the proverbial window. This indicates that within clades, macroevolutionary dynamics can vary across time periods as they encounter different ranges of environmental variation. Dunill and Wills' (2015) study also nicely elucidated how GIS can be used to consider whether the quality of the fossil record might be biasing their results, and they notably concluded that this was not the case (see Rode and Lieberman, 2004, and Myers and Lieberman, 2011, for additional information about how the quality of the fossil record can be assessed using GIS and for further evidence that the record in general does not appear to be too biased to study biogeographic patterns). Each of these findings is very important and speaks to the broader utility of applying biodiversity science approaches to study macroevolution in general and the terrestrial fossil record in particular.

Notably, Myers et al. (2013) also uncovered evidence that geographic range size is not always associated with increased extinction resistance, especially during times of extreme global warmth or greenhouse conditions. Myers et al.'s (2013) analyses focused on species of mollusks (gastropods, cephalopods, and bivalves) from the Cretaceous WIS and could have especial relevance given the predicted effects of warming climate today. In particular, predicting which species might go extinct in the

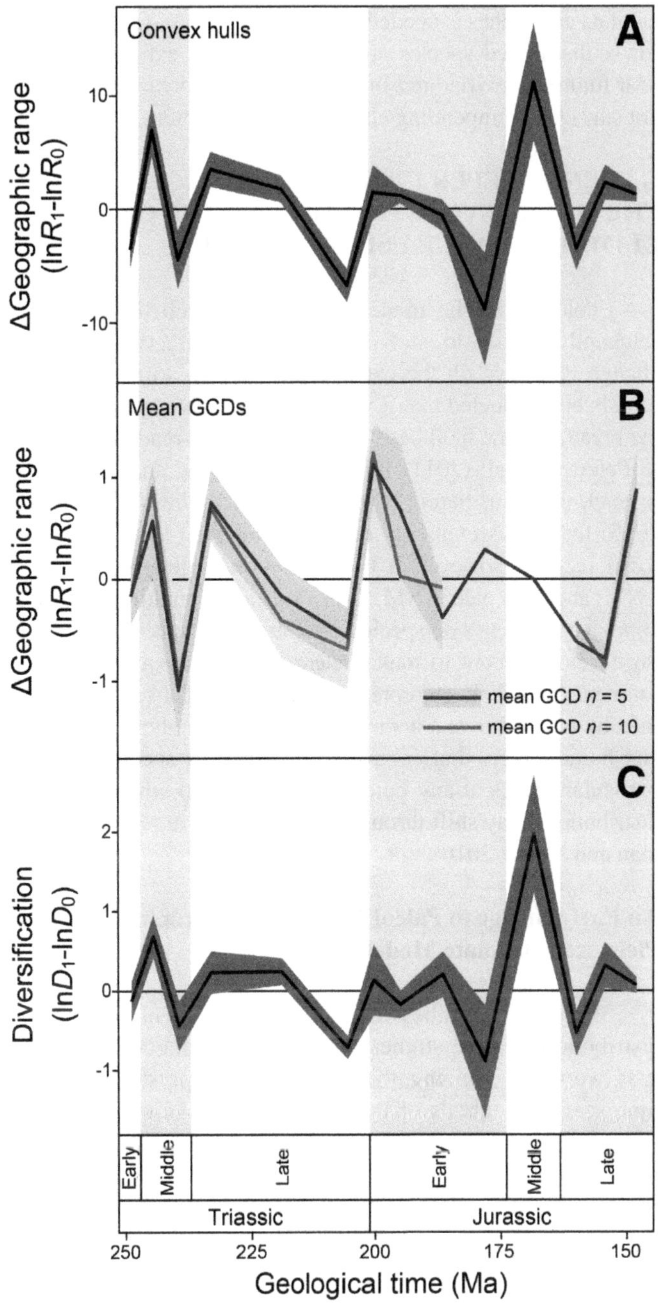

Figure 3. The mean rate of change in geographic range size (A and B) and diversification rates (C) for terrestrial vertebrates from before, during, and after the end-Triassic Mass Extinction. (A) Geographic range values calculated using occurrence records with convex hulls in GIS, labeled "convex hulls." (B) Standardized mean great circle distances, which are corrected for sample number, labeled "GCD." Note the paucity of change in calculated mean geographic ranges during the mass extinction interval; diversification rates declined during this interval. From Dunill and Wills (2015); reprinted by permission of Macmillan Publishers Ltd: *Nature Communications* © 2015.

future simply on the basis of their current geographic ranges may be problematic, and additional information about their preferred habitats and niches is needed (Myers et al., 2013). Certainly, narrowly distributed species are threatened with extinction in the near future for myriad reasons, but even widespread species may not survive the impending changes facing the biota.

GAINING DEEPER INSIGHTS INTO THE MACROEVOLUTIONARY EFFECTS OF CLIMATE CHANGE USING PaleoENM

Ecological niche modeling is an approach that has been frequently applied to study and predict the effects that climate change will have on the extant biota. In fact, so many studies have been conducted that it would be impossible to do justice to the breadth of the field here, and the interested reader is referred to Peterson et al. (2011) for a detailed review and discussion. Instead, the focus herein is on paleoENM studies (Myers et al., 2015) that focus explicitly on niche modeling that utilizes the fossil record. However, it bears mentioning that the value of ENM, and thus paleoENM, is that it is a quantitative, statistical approach that relies on species distributions in space and according to environment to make determinations about which environments are likely to represent the habitat of a species. It can thereby be used to test a variety of hypotheses about the abiotic and biotic factors that control where and why species occur in particular places at any point in time and also where and why distributions may shift through time (Myers et al., 2015; Lieberman and Saupe, 2016).

An Early Bridge to PaleoENM: Phylogeography and Pleistocene Climate Models

Many ENM studies focus solely on present-day species distributions. These studies are of course quite valuable, but it is worth recognizing that many species persist over long time scales (hundreds of thousands to millions of years) (see Eldredge et al., 2005, for a comprehensive review). Thus, when one examines present-day distributions, one is only seeing a snap shot of a distribution that has moved, and expanded and contracted, across geographic space over time (see Lieberman, 2000, for a detailed review). Several studies have recognized the importance of trying to connect current species distributions to longer evolutionary and climatic histories. These studies have utilized climate models that exist for some windows of time during the Pleistocene Ice Ages. The first studies to take this approach were those of Hugall et al. (2002), Peterson et al. (2004), Peterson and Nyári (2007), and Waltari et al. (2007) (see also Martínez-Meyer et al., 2004): these used ENM to attempt to align modern phylogeographic breaks within individual species lineages to the predicted distributions of these species during a part of the Ice Ages using a late Pleistocene General Circulation Model (GCM) (a mathematical model of climate that incorporates the respective influences of the oceans and the atmosphere). They found evidence that in many cases the signatures of Pleistocene climate change were preserved within extant species, indicating that many of these species were long lived; further, these species retained the history of some of the biogeographic changes they had experienced throughout their existence (see Peterson and Lieberman, 2012, for a review). Ultimately, this type of work, and many of the important current research topics in historical biogeography and evolutionary biology, build on and connect to the ground-breaking publication by Wiley and Mayden (1985) that used phylogenetic and biogeographic patterns to examine the role that climate change during the Pleistocene, with concomitant change in river basin geometry, played in structuring the evolution of diverse freshwater biota.

The First PaleoENM Studies

Stigall Rode and Lieberman (2005) represented the first paleoENM study. Their analysis focused on brachiopods extant during the Late Devonian biodiversity crisis, and they considered how environmental conditions interacted to determine species ranges during this time interval. One complication of working with such ancient taxa is that detailed climatic models, in particular, GCMs, unfortunately do not exist for the Late Devonian. Thus, Stigall Rode and Lieberman (2005) created one using the rocks these fossils occur in as proxies for the environment (see Stigall Rode and Lieberman, 2005, for more detailed discussion). The results of this study have been considered and discussed in several subsequent papers, including Stigall and Lieberman (2006) and Stigall (2012b). Maguire and Stigall (2009) and Nogués-Bravo et al. (2008) extended paleoENM approaches to the vertebrate fossil record. (Nogués-Bravo et al. [2008] is discussed in the next section.) Again, because GCM climate models were lacking for the time intervals they analyzed, Maguire and Stigall (2009) used a series of proxies to develop a means of relating climatic factors to geographic distribution (see Maguire and Stigall, 2009, for more detailed discussion). They focused on the horse lineage and considered how environmental change played a critical role in driving the evolution of this clade. This was important because it provided a quantitative means to consider long-standing issues in the history of evolutionary biology extending at least as far back as Matthew (1926) and Simpson (1944) and relating to the relevance of the horse clade for understanding macroevolution. (In particular, how are evolutionary changes in morphology related to climatic changes?) Maguire and Stigall (2009) was also reviewed by Varela et al. (2011) and Stigall (2012b). In the case of Stigall Rode and Lieberman (2005) and Maguire and Stigall (2009), museum collections data proved crucial in making it possible to use paleoENM. The lead authors of Stigall Rode and Lieberman (2005) and Maguire and Stigall (2009) also played a fundamental role in building the georeferenced data sets that existed for each group, using museum collections. Furthermore, the relevant digital collections resources available for fossil invertebrates and fossil horses continue to grow (see Hendricks et al., 2015, and MacFadden and Guralnick, 2017, respectively).

PaleoENM and Biotic Invasions during the Ordovician

The Ordovician is a key time period in the history of life and the invertebrate fossil record of North America in general, and the Cincinnatian region in particular, has been the subject of a variety of paleoENM studies by Alycia Stigall and her students. Part of the impetus for these studies and what made them possible were the databases and georeferencing for several significant museum collections housing Ordovician fossils (Stigall et al., 2014). One important topic that was considered using these newly generated collections data in conjunction with paleoENM was how to make predictions about where new fossil localities and geographic occurrence records for species should be found (Walls and Stigall, 2012). This is valuable because it shows how museum collections can be used in a predictive, quantitative manner to identify sites from which new collections can be derived (see Walls and Stigall, 2012, for more discussion of this topic). Another important topic considered using paleoENM of Cincinnatian fossils has been the analysis of whether individual species niches stay conserved over long periods of time. As mentioned above, most species show long-term stability in morphology through time. One of the important questions pertaining to punctuated equilibria (Eldredge and Gould, 1972) is what role ecology, in general, and niche characteristics, in particular, might play in motivating long-term species stasis. Stigall (2011), Malizia and Stigall (2011), and Walls and Stigall (2011) (see also Stigall, 2014, for additional discussion) subjected a variety of brachiopod taxa to paleoENM and found that although on the whole and over millions of years species niches were stable, there were some cases in which statistically significant changes in niches were found to occur, even in the absence of morphological change. (Note that niches are not being defined in this case by species morphology, of course, but rather by the results from the ENM.) This could suggest that niche stability in and of itself is not one of the primary factors causing stasis. (Saupe et al. [2014a] also considered this topic, and their work is discussed more fully below.) What was also very useful is that Malizia and Stigall (2011) and Walls and Stigall (2011) considered the phenomenon of niche stability in the context of species invasions. Building on this work, Stigall and Brame (2014) considered the topic in greater detail and focused especially on the role that biotic as opposed to abiotic factors might play in motivating the episodes of niche stability and change. In particular, they found that niche change was most pronounced during times of biotic invasion, suggesting that taxa were adjusting their niches in the face of new competition. By contrast, abiotic changes did not seem to induce shifts in apparent niches (Stigall and Brame, 2014). For additional discussion of each of these studies, see also the comprehensive reviews by Stigall (2012b) and Stigall et al. (2017).

The First PaleoENM Studies to Incorporate Climate Data Derived from GCMs

An approach related to both the aforementioned phylogeographic and paleoENM studies was employed by Guralnick (2007) and Waltari and Guralnick (2009). They looked at the long-term effects of climate change on modern species. However, they did not use phylogeographic patterns, but rather Pleistocene fossil distributional records, and further they were able to base their environmental data on a Pleistocene GCM (rather than considering other proxies for climate). This was among the first, if not the first, study to connect the skein between fossil and still-living species lineages using paleoENM. They found that the intervals of major climate change during this time period profoundly influenced the distribution of species throughout western North America. Further, they found that the actual peregrinations of various species across the region and through different corridors reflected individualistic environmental tolerances. The study by Di Febbraro et al. (2017) is another recent example of this type of approach.

Nogués-Bravo et al.'s (2008) work on the woolly mammoth is directly related to this. Although they could not connect fossil with extant distributional records, it was through no fault of their own: it was because their focal organism, the woolly mammoth, had been driven to extinction. However, they were able to use paleoENM in conjunction with environmental data derived from climate models for several intervals during the Quaternary to consider the extent to which humans or climate change played a role in delivering the final coup de grace to this species (arguing that it was the latter). Recent work on ancient DNA in this species (Rogers and Slatkin, 2017) suggests that it might at one point be possible to connect phylogeographic information to the study of this topic, in order to ascertain, for instance, the patterns of genetic relationship that existed among extinct mammoth populations, and whether predicted environmental discontinuities are correlated with reconstructed genetic distinctions.

Expanding the Macroevolutionary Purview of PaleoENM Studies

Another major development in the field of paleoENM involved extending the use of GCM-based climate data even further back in time, encompassing a broader swath of species' evolutionary history. In particular, Saupe et al. (2014a) focused on marine mollusks from the Atlantic and Gulf Coastal Plains; they considered species alive today that also had a fossil record extending back more than 3 million years. Using digital collections records of fossil and modern specimens, some of it preexisting and some of it captured during the course of the project (see Hendricks et al., 2015), they focused on what happened to species niches from the Pliocene (ca. 3 Ma) to the Last Interglacial (ca. 120 ka), and then through to the present day. Saupe et al. (2014a) found pronounced niche stability even in the face of significant environmental change over millions of years (Fig. 4). This

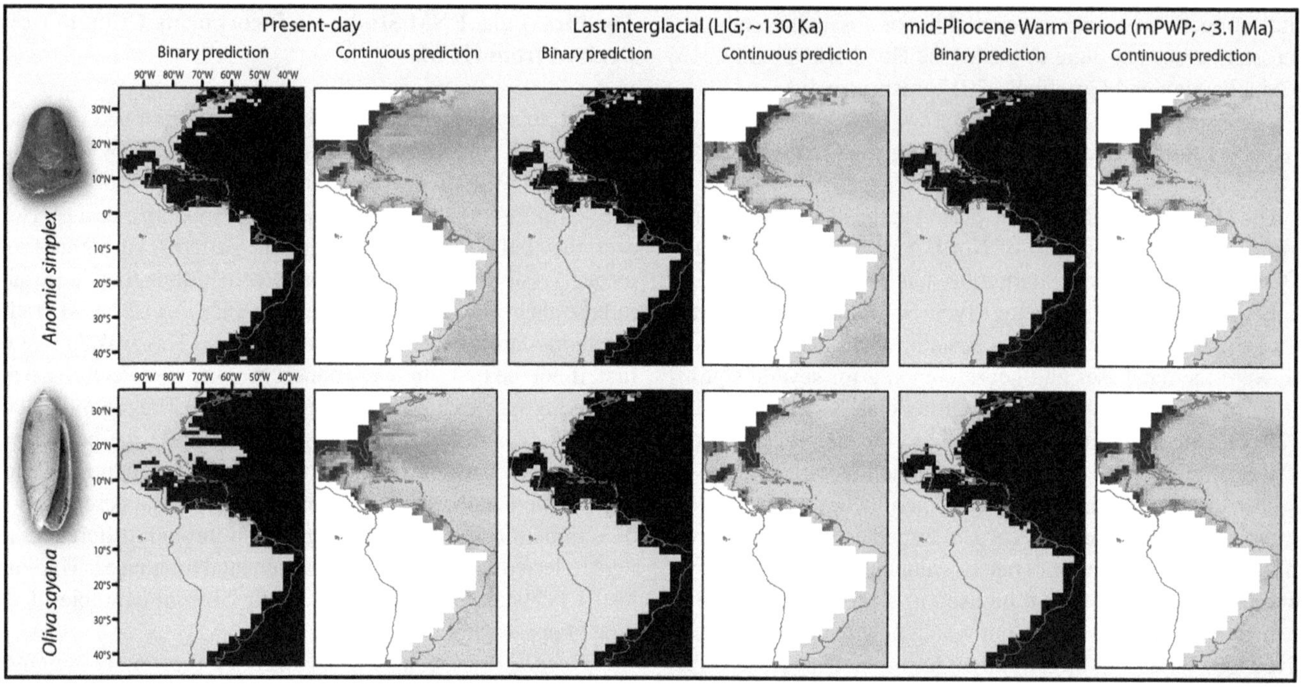

Figure 4. Ecological niche models for the present, the Last Interglacial (labeled LIG), and the mid-Pliocene Warm period (labeled mPWP) for the bivalve *Anomia simplex* and the gastropod *Oliva sayana*. The mPWP corresponds to conditions likely to be experienced by the end of the twenty-first century in the face of continued human-induced climatic warming. Both binary and continuous predictions are shown. In the case of binary predictions, light gray represents suitable habitat, and black represents unsuitable habitat; in the case of continuous predictions, darker denotes habitat that is more suitable for the species; lighter denotes habitat that is less suitable for the species. Results suggest that if by 2100, climate returns to conditions similar to those during the mPWP, these species will lose suitable habitat, spelling problems for the future of the marine biota. From Saupe et al. (2014a); used with permission.

may suggest critical challenges to the survival of these species in the future, especially as climate may warm beyond values experienced during the past 3 million years. If not faced with wholesale extinction, we should at least predict to see severe reductions in the geographic distribution of these marine mollusks and many other species occupying the marine realm. Indeed, research by Saupe et al. (2014b) predicted that during the coming 100 years, given the most conservative anticipated climate changes, many marine mollusk species will show significant range reductions (Fig. 5). Those climate models that predicted more severe climate changes concomitantly suggested that it would not be just range reductions but actual widespread extinction of the marine mollusk fauna that would result; by extension, many other species in the marine realm would also be expected to go extinct (Saupe et al., 2014b).

One other valuable aspect of paleoENM (and ENM) is that it makes it possible to assess aspects of both the fundamental and realized niches of species (Lieberman and Saupe, 2016). The former is "the full suite of environmental conditions in which a species can survive and maintain viable populations sensu Hutchinson (1957), whereas the latter is considered the set of environmental conditions currently occupied by a species (Peterson et al., 2011)" (Lieberman and Saupe, 2016, p. 145). Thus, researchers can differentiate between the relative roles of innate physiological tolerances, as opposed to the range of environments species experience in the face of competition, in determining geographic distributions (see Peterson et al., 2011, and Saupe et al., 2015). This is important in the context of macroevolutionary studies because factors such as fundamental and realized niche are often invoked as consequential determinants of the extinction rates of species over deep time. To consider the relative influences of geographic range, the fundamental niche, and the realized niche on long-term species survival, Saupe et al. (2015) conducted an analysis of more than 90 species of Neogene marine mollusks from the Gulf and Atlantic Coastal Plains. Some of these 90 species survive to the present day; others are no longer extant. Saupe et al. (2015) found that geographic range size was the strongest predictor of species survival. The realized niche also played a role; by contrast, the breadth of the fundamental niche occupied by a species seemed to have no effect when it comes to species survival. Further, they found that the geographic range occupied by a species was decoupled from its fundamental niche. Ultimately, it appears that the ability of a species to "fill their potential suitable area seems to provide extinction resistance, which is a function of both the dispersal ability of a species and biotic interactions" (Saupe et al., 2015, p. 7–8). However, the primary factor dictating survival is a broad geographic range, which enables decreased predisposition to extinction, and not simply

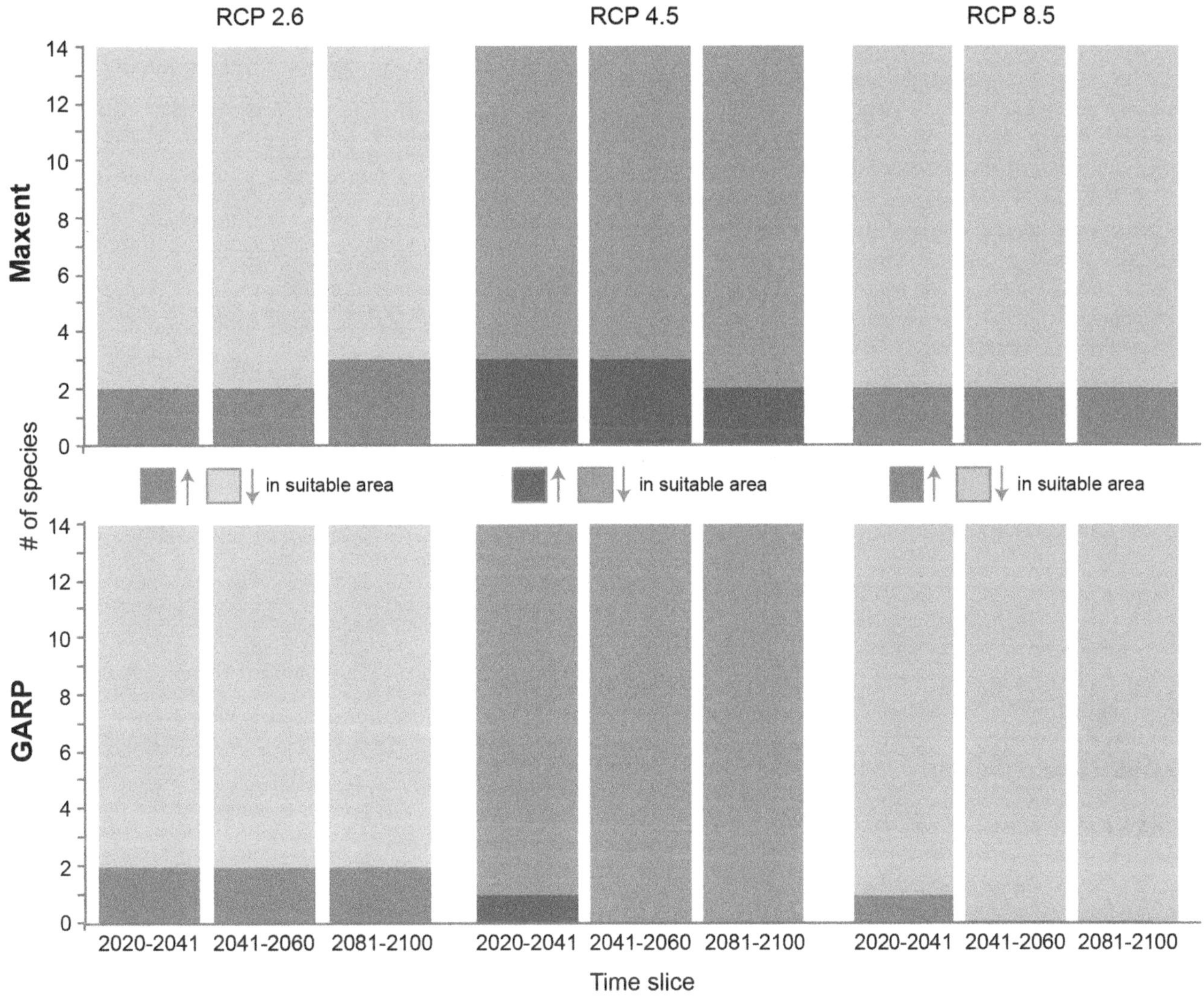

Figure 5. Histograms displaying the number of marine mollusk species that are predicted to gain or lose suitable area through the end of the twenty-first century. Analyses were conducted on western Atlantic species using two different ecological niche modeling (ENM) approaches (GARP and Maxent). Representative concentration pathway (RCP) represents different modeled scenarios for future anticipated climate change, with RCP 2.5 representing the most limited likely amount of global temperature increase, RCP 4.5 a more moderate amount of likely global temperature increase, and RCP 8.5 the most extreme anticipated likely global temperature increase. As can be seen, many more species will lose suitable area than will gain area, and the more extreme the climate change, the more this effect will be exacerbated. This indicates there should be significant concern for the future survival of important components of the present-day marine biota. From Saupe et al. (2014b); used with permission.

because species with broader ranges have broader environmental tolerances (Lieberman and Saupe, 2016).

CONCLUSIONS

Extensive opportunities for research and new advances in theoretical understanding are being facilitated by ongoing as well as new efforts to create data bases for fossil museum collections (see reviews in Hendricks et al., 2015, and MacFadden and Guralnick, 2017). These efforts are expanding the actual and perceived scientific value of paleontological collections, which is ultimately very positive for the fields of paleontology and museum science. At the same time as these digital data resources have been expanding, new analytical approaches have been developed that are enhancing the precision with which we can test hypotheses. The special emphasis here has been placed on data and research approaches most relevant to work in the area of biogeography and macroevolution, but many different aspects of paleontology are benefiting from these opportunities. Ultimately, the availability of new research approaches and data has been extremely valuable for paleontology, but also indicates where the remaining impediments to further research progress lie. In particular, what is crucially needed now is an expansion in the number of time periods throughout the history of life for which GCM climate models are available. Currently, models are lacking for many time periods, and this is a critical gap that needs to be addressed. Likely the next very important step for the field of macroevolution will involve expanding the number of such climate models. This will enable scientists to obtain a much more comprehensive picture of how the current biota have been shaped by long-term climate change and competition, which in turn will better integrate paleontology and neontology. Further, it will be a key component of developing an expanded macroevolutionary synthesis (Eldredge, 1985; Vrba, 1985; Myers and Saupe, 2013). In any event, it is clearly apparent that research involving fossil museum collections data is undergoing a renaissance, and new digital approaches are making it possible to consider longstanding questions of relevance to evolutionary biology.

ACKNOWLEDGMENTS

We thank Gary Rosenberg for graciously inviting us to participate in this volume. We thank him, Richard Davis, and two anonymous reviewers for comments on an earlier version of this paper. Thanks to A. Dunhill, J. Hendricks, C. Myers, E. Saupe, and M. Wills for generously sharing figures. This research was supported by National Science Foundation grants NSF-ADBC-1206757 and NSF-DBI-1602067.

REFERENCES CITED

Bambach, R.K., Knoll, A.H., and Wang, S., 2004, Origination, extinction, and mass depletions of marine diversity: Paleobiology, v. 30, p. 522–542.

Benton, M.J., 1996, Testing the roles of competition and expansion in tetrapod evolution: Proceedings of the Royal Society B: Biological Sciences, v. 263, no. 1370, p. 641–646, https://doi.org/10.1098/rspb.1996.0096.

Briggs, D.E.G., and Fortey, R.A., 2005, Wonderful strife: Systematics, stem groups, and the phylogenetic signal of the Cambrian radiation: Paleobiology, v. 31, sp5, p. 94–112, https://doi.org/10.1666/0094-8373(2005)031[0094:WSSSGA]2.0.CO;2.

Briggs, D.E.G., Erwin, D.H., and Collier, F.J., 1992, The Fossils of the Burgess Shale: Washington, D.C., Smithsonian Institution Press, 238 p.

Casey, M.M., Dietl, G.P., Post, D.M., and Briggs, D.E.G., 2014, The impact of eutrophication and commercial fishing on molluscan communities in Long Island Sound, USA: Biological Conservation, v. 170, p. 137–144, https://doi.org/10.1016/j.biocon.2013.12.037.

Chapman, A.D., and Wieczorek, J., 2006, Guide to Best Practices for Georeferencing: Copenhagen, Global Biodiversity Information Facility, 90 p.

Department of the Interior, 2016, Paleontological Resources Preservation: Proposed rule: Federal Register, v. 81, no. 235, p. 88,173–88,196.

Dietl, G.P., and Flessa, K.W., 2011, Conservation paleobiology: Putting the dead to work: Trends in Ecology & Evolution, v. 26, no. 1, p. 30–37, https://doi.org/10.1016/j.tree.2010.09.010.

Di Febbraro, M.D., Carotenuto, F., Castiglione, S., Russo, D., and Loy, A., 2017, Does the jack of all trades fare best?: Survival and niche width in Late Pleistocene megafauna: Journal of Biogeography, v. 44, p. 2828–2838, https://doi.org/10.1111/jbi.13078.

Dirzo, R., and Loreau, L., 2005, Biodiversity science evolves: Science, v. 310, p. 943, https://doi.org/10.1126/science.1119958.

Dunill, M.A., and Wills, M.A., 2015, Geographic range did not confer resilience to extinction in terrestrial vertebrates at the end-Triassic crisis: Nature Communications, v. 6, p. 7980, https://doi.org/10.1038/ncomms8980.

Eldredge, N., 1979, Alternative approaches to evolutionary theory: Bulletin of the Carnegie Museum of Natural History, v. 13, p. 7–19.

Eldredge, N., 1985, Unfinished Synthesis: New York, Oxford University Press, 237 p.

Eldredge, N., 1989, Macroevolutionary Dynamics: New York, McGraw-Hill, 226 p.

Eldredge, N., and Gould, S.J., 1972, Punctuated equilibrium: An alternative to phyletic gradualism, *in* Schopf, T.J., ed., Models in Paleobiology: San Francisco, W.H. Freeman, p. 82–115.

Eldredge, N., Thompson, J., Brakefield, P., Gavrilets, S., Jablonski, D., Jackson, J., Lenski, R., Lieberman, B.S., McPeek, M., and Miller, W., III, 2005, The dynamics of evolutionary stasis: Paleobiology, v. 31, p. 133–145, https://doi.org/10.1666/0094-8373(2005)031[0133:TDOES]2.0.CO;2.

Gould, S.J., 1989, Wonderful Life: New York, Norton, 347 p.

Gould, S.J., and Calloway, C.B., 1980, Clams and brachiopods: Ships that pass in the night: Paleobiology, v. 6, p. 383–396, https://doi.org/10.1017/S0094837300003572.

Gould, S.J., and Eldredge, N., 1977, Punctuated equilibria: The tempo and mode of evolution reconsidered: Paleobiology, v. 3, p. 115–151, https://doi.org/10.1017/S0094837300005224.

Graham, R.W., Ernest L. Lundelius, E.L., Jr., Graham, M.A., Schroeder, E.K., Toomey, R.S., III, Anderson, E., Barnosky, A.D., Burns, J.A., Churcher, C.S., Grayson, D.K., Guthrie, R.D., Harrington, C.R., Jefferson, G.T., Martin, L.D., McDonald, H.G., Morlan, R.E., Semken, H.A., Jr., Webb, S.D., Werdelin, L., and Wilson, M.C., 1996, Spatial response of mammals to late Quaternary environmental fluctuations: Science, v. 272, p. 1601–1606, https://doi.org/10.1126/science.272.5268.1601.

Guralnick, R., 2007, Differential effects of past climate warming on mountain and flatland species distribution: A multispecies North American mammal assessment: Global Ecology and Biogeography, v. 16, p. 14–23, https://doi.org/10.1111/j.1466-8238.2006.00260.x.

Hendricks, J.R., 2013, Global distributional dynamics of Cambrian clades as revealed by Burgess Shale-type deposits, *in* Harper, D.A.T., and Servais, T., eds., Early Palaeozoic Biogeography and Palaeogeography: Geological Society, London, Memoir 38, p. 35–43, https://doi.org/10.1144/M38.4.

Hendricks, J.R., Lieberman, B.S., and Stigall, A.L., 2008, Using GIS to study the paleobiogeographic and macroevolutionary patterns in soft-bodied Cambrian arthropods: Palaeogeography, Palaeoclimatology, Palaeoecology, v. 264, p. 163–175, https://doi.org/10.1016/j.palaeo.2008.04.014.

Hendricks, J.R., Stigall, A.L., and Lieberman, B.S., 2015, The Digital Atlas of Ancient Life: Delivering information on paleontology and biogeography via the web: Paleontologia Electronica, v. 18.2.3E, p. 1–9.

Hou, X.-G., Siveter, D.J., Siveter, D.J., Aldridge, R.J., Cong, P.-Y., Gabbott, S.E., Ma, X.-Y., Purnell, M.A., and Williams, M., 2017, The Cambrian Fossils of Chengjiang, China: Malden, Massachusetts, Blackwell Publishing, 316 p., https://doi.org/10.1002/9781118896372.

Hugall, A., Moritz, C., Moussalli, A., and Stanisic, J., 2002, Reconciling paleodistribution models and comparative phylogeography in the Wet Tropics rainforest land snail *Gnarosophia bellendenkerensis* (Brazier 1875): Proceedings of the National Academy of Sciences of the United States of America, v. 99, p. 6112–6117, https://doi.org/10.1073/pnas.092538699.

Hunt, G., Roy, K., and Jablonski, D., 2005, Species-level heritability reaffirmed: A comment on "On the heritability of geographic range sizes": American Naturalist, v. 166, p. 129–135, https://doi.org/10.1086/430722.

Hutchinson, G.E., 1957, Concluding remarks: Cold Spring Harbor Symposia on Quantitative Biology, v. 22, p. 415–427, https://doi.org/10.1101/SQB.1957.022.01.039.

Jablonski, D., 1986, Larval ecology and macroevolution in marine invertebrates: Bulletin of Marine Science, v. 39, p. 565–587.

Jablonski, D., 1987, Heritability at the species level: Analysis of geographic ranges of Cretaceous mollusks: Science, v. 238, p. 360–363, https://doi.org/10.1126/science.238.4825.360.

Karim, T.S., Burkhalter, R., Farrell, Ú.C., Molineux, A., Nelson, G., and Butts, S.H., 2016, Digitization workflows for paleontology collections: Palaeontologia Electronica, v. 19.3.4T, p. 1–14.

Kimmig, J., and Pratt, B.R., 2015, Soft-bodied biota from the middle Cambrian (Drumian) Rockslide Formation, Mackenzie Mountains, northwestern Canada: Journal of Paleontology, v. 89, p. 51–71, https://doi.org/10.1017/jpa.2014.5.

Kowalewski, M., 2009, The youngest fossil record and conservation biology: Holocene shells as eco-environmental recorders, *in* Dietl, G.P., and Flessa, K.W., eds., Conservation Paleobiology: Using the Past to Manage for the Future: The Paleontological Society Papers, v. 15, p. 1–23.

Lieberman, B.S., 2000, Paleobiogeography: New York, Kluwer Academic/Plenum Publishing, 208 p., https://doi.org/10.1007/978-1-4615-4161-5.

Lieberman, B.S., 2001, A probabilistic analysis of rates of speciation during the Cambrian radiation: Proceedings of the Royal Society B: Biological Sciences, v. 268, no. 1477, p. 1707–1714, https://doi.org/10.1098/rspb.2001.1712.

Lieberman, B.S., 2008, Emerging syntheses between palaeobiogeography and macroevolutionary theory: Proceedings of the Royal Society of Victoria, v. 120, p. 51–57.

Lieberman, B.S., and Dudgeon, S., 1996, An evaluation of stabilizing selection as a mechanism for stasis: Palaeogeography, Palaeoclimatology, Palaeoecology, v. 127, p. 229–238, https://doi.org/10.1016/S0031-0182(96)00097-1.

Lieberman, B.S., and Eldredge, N., 2014, What is punctuated equilibrium?: What is macroevolution: A response to Pennell et al.: Trends in Ecology & Evolution, v. 29, no. 4, p. 185–186, https://doi.org/10.1016/j.tree.2014.02.005.

Lieberman, B.S., and Saupe, E.E., 2016, Palaeoniches get stiches: Analyses of niches informing macroevolutionary theory: Lethaia, v. 49, p. 145–149, https://doi.org/10.1111/let.12170.

Lieberman, B.S., and Vrba, E.S., 1995, Hierarchy theory, selection, and sorting: A phylogenetic perspective: Bioscience, v. 45, p. 394–399, https://doi.org/10.2307/1312719.

MacFadden, B.J., and Guralnick, R.P., 2017, Horses in the cloud: Big data exploration and mining of fossil and extant *Equus* (Mammalia: Equidae): Paleobiology, v. 43, p. 1–14, https://doi.org/10.1017/pab.2016.42.

Maguire, K.C., and Stigall, A.L., 2009, Using ecological niche modeling for quantitative biogeographic analysis: A case study of Miocene and Pliocene Equinae in the Great Plains: Paleobiology, v. 35, p. 587–611, https://doi.org/10.1666/0094-8373-35.4.587.

Malizia, R.W., and Stigall, A.L., 2011, Niche stability in Late Ordovician articulated brachiopod species before, during, and after the Richmondian Invasion: Palaeogeography, Palaeoclimatology, Palaeoecology, v. 311, p. 154–170, https://doi.org/10.1016/j.palaeo.2011.08.017.

Martínez-Meyer, E., Peterson, A.T., and Hargrove, W.W., 2004, Ecological niches as stable distributional constraints on mammal species, with implications for Pleistocene extinctions and climate change projections for biodiversity: Global Ecology and Biogeography, v. 13, p. 305–314, https://doi.org/10.1111/j.1466-822X.2004.00107.x.

Matthew, W.D., 1926, The evolution of the horse: A record and its interpretation: The Quarterly Review of Biology, v. 1, p. 139–185, https://doi.org/10.1086/394242.

McGhee, G.R., 1996, The Late Devonian Mass Extinction: New York, Columbia University Press, 303 p.

Muscente, A.D., Schiffbauer, J.D., Broce, J., Laflamme, M., O'Donnell, K., Boag, T.H., Meyer, M., Hawkins, A.D., Huntley, J.W., McNamara, M., MacKenzie, L.A., Stanley, G.D., Jr., Hinman, N.W., Hofmann, M.H., and Xiao, S., 2017, Exceptionally preserved fossil assemblages through geologic time and space: Gondwana Research, v. 48, p. 164–188, https://doi.org/10.1016/j.gr.2017.04.020.

Myers, C., and Lieberman, B.S., 2011, Sharks that pass in the night: Using GIS to investigate competition in the Cretaceous Western Interior Seaway: Proceedings of the Royal Society, Series B. v. 278, p. 681–689, https://doi.org/10.1098/rspb.2010.1617.

Myers, C.E., and Saupe, E.E., 2013, A macroevolutionary expansion of the Modern Synthesis and the importance of extrinsic abiotic factors: Palaeontology, v. 56, p. 1179–1198.

Myers, C.E., MacKenzie, R.A., III, and Lieberman, B.S., 2013, A GIS approach to studying greenhouse biogeography: Invasion and extinction in the Cretaceous Western Interior Seaway: Paleobiology, v. 39, p. 135–148, https://doi.org/10.1666/0094-8373-39.1.135.

Myers, C.E., Stigall, A.L., and Lieberman, B.S., 2015, PaleoENM: Applying ecological niche modeling to the fossil record: Paleobiology, v. 41, p. 226–244, https://doi.org/10.1017/pab.2014.19.

Newman, E., 1854, President's address for 1853: Transactions of the Entomological Society of London, new series, v. 2, p. 142–154.

Nogués-Bravo, D., Rodríguez, J., Hortal, J., Batra, P., and Araújo, M.B., 2008, Climate change, humans, and the extinction of the woolly mammoth: PLoS Biology, v. 6, e79, p. 685–692.

Peterson, A.T., and Lieberman, B.S., 2012, Species' geographic distributions through time: Playing catchup with changing climates: Evolution Education and Outreach, v. 5, p. 569–581, https://doi.org/10.1007/s12052-012-0385-2.

Peterson, A.T., and Nyári, Á., 2007, Ecological niche conservatism and Pleistocene refugia in the Thrush-like Mourner, *Schiffornis* sp., in the Neotropics: Evolution: International Journal of Organic Evolution, v. 62, p. 173–183.

Peterson, A.T., Martínez-Meyer, E., and Gonzalez-Sálazar, C., 2004, Reconstructing the Pleistocene geography of the *Aphelocoma* jays (Corvidae): Diversity and Distributions, v. 10, p. 237–246, https://doi.org/10.1111/j.1366-9516.2004.00097.x.

Peterson, A.T., Soberón, J., Pearson, R.G., Anderson, R.P., Martínez-Meyer, E., Nakamura, M., and Araújo, M.B., 2011, Ecological niches and geographic distributions: Princeton, New Jersey, Princeton University Press, 314 p.

Rode, A.L., and Lieberman, B.S., 2004, Using GIS to unlock the interactions between biogeography, environment, and evolution in Middle and Late Devonian brachiopods and bivalves: Palaeogeography, Palaeoclimatology, Palaeogeography, v. 211, p. 345–359, https://doi.org/10.1016/j.palaeo.2004.05.013.

Rode, A.L., and Lieberman, B.S., 2005, Integrating biogeography and evolution using phylogenetics and PaleoGIS: A case study involving Devonian crustaceans: Journal of Paleontology, v. 79, p. 267–276, https://doi.org/10.1666/0022-3360(2005)079<0267:IEABAC>2.0.CO;2.

Rogers, R.L., and Slatkin, M., 2017, Excess of genomic defects in a woolly mammoth on Wrangel island: PLoS One, v. 13, e1006601, p. 1–16.

Saupe, E.E., Barve, V., Myers, C.E., Soberón, J., Barve, N., Hensz, C.M., Peterson, A.T., Owens, H.L., and Lira-Noriega, A., 2012, Variation in niche and distribution model performance: The need for a priori assessment of key causal factors: Ecological Modelling, v. 237–238, p. 11–22, https://doi.org/10.1016/j.ecolmodel.2012.04.001.

Saupe, E.E., Hendricks, J.R., Portell, R.W., Dowsett, H.J., Haywood, A., Hunter, S.J., and Lieberman, B.S., 2014a, Macroevolutionary consequences of climate change on niche evolution in marine mollusks over the past 3 million years: Proceedings of the Royal Society B: Biological Sciences, v. 281, no. 1795, p. 1995–2013, https://doi.org/10.1098/rspb.2014.1995.

Saupe, E.E., Hendricks, J.R., Peterson, A.T., and Lieberman, B.S., 2014b, Climate change and marine molluscs of the western North Atlantic: Future prospects and perils: Journal of Biogeography, v. 41, p. 1352–1366, https://doi.org/10.1111/jbi.12289.

Saupe, E.E., Huijie, Q., Hendricks, J.R., Portell, R.W., Hunter, S.J., Soberón, J., and Lieberman, B.S., 2015, Niche breadth and geographic range size as determinants of species survival on geological time scales: Global Ecology and Biogeography, v. 24, no. 10, p. 1159–1169, https://doi.org/10.1111/geb.12333.

Simpson, G.G., 1944, Tempo and Mode in Evolution: New York, Columbia University Press, 237 p.

Smith, L.H., and Lieberman, B.S., 1999, Disparity and constraint in olenelloid trilobites and the Cambrian radiation: Paleobiology, v. 25, p. 459–470, https://doi.org/10.1017/S0094837300020327.

Stanley, S.M., 1979, Macroevolution: San Francisco, California, W.H. Freeman, 332 p.

Stigall, A.L., 2011, Application of niche modeling to analyze biogeographic patterns in Palaeozoic brachiopods: Evaluating niche stability in deep time: Memoirs of the Association of Australasian Palaeontologists, v. 41, p. 229–255.

Stigall, A.L., 2012a, Speciation collapse and invasive species dynamics during the Late Devonian "Mass Extinction": GSA Today, v. 22, p. 4–9, https://doi.org/10.1130/G128A.1.

Stigall, A.L., 2012b, Using ecological niche modeling to evaluate niche stability in deep time: Journal of Biogeography, v. 39, p. 772–781, https://doi.org/10.1111/j.1365-2699.2011.02651.x.

Stigall, A.L., 2014, When and how do species achieve niche stability over long time scales?: Ecography, v. 37, p. 1123–1132.

Stigall, A.L., and Brame, H-M.R., 2014, Relating environmental change and species stability in Late Ordovician seas: GFF: A Scandinavia Journal of Earth Sciences, v. 136, p. 249–253.

Stigall, A.L., and Lieberman, B.S., 2006, Quantitative paleobiogeography: GIS, phylogenetic biogeographic analysis, and conservation insights: Journal of Biogeography, v. 33, p. 2051–2060, https://doi.org/10.1111/j.1365-2699.2006.01585.x.

Stigall, A.L., Bauer, J.E., and Brame, H.-M.R., 2014, The Digital Atlas of Ordovician Life: Digitizing and mobilizing data for paleontologists and the public: Estonian Journal of Earth Sciences, v. 63, p. 312–316, https://doi.org/10.3176/earth.2014.36.

Stigall, A.L., Bauer, J.E., Lam, A.L., and Wright, D.A., 2017, Biotic immigration events, speciation, and the accumulation of biodiversity in deep time: Global and Planetary Change, v. 148, p. 242–257, https://doi.org/10.1016/j.gloplacha.2016.12.008.

Stigall Rode, A.L., and Lieberman, B.S., 2005, Using environmental niche modelling to study the Late Devonian biodiversity crisis, *in* Over, D.J., Morrow, J.R., and Wignall, P.B., eds., Understanding Late Devonian and Permian–Triassic Biotic and Climatic Events: Towards an Integrated Approach. Developments in Palaeontology and Stratigraphy: Amsterdam, the Netherlands, Elsevier, p. 93–180.

Svenning, J.-C., Fløjgaard, C., Marske, K.A., Nógues-Bravo, D., and Normand, S., 2011, Applications of species distribution modeling to paleobiology: Quaternary Science Reviews, v. 30, p. 2930–2947, https://doi.org/10.1016/j.quascirev.2011.06.012.

Varela, S., Lobo, J.M., and Hortal, J., 2011, Using species distribution models in paleobiogeography: A matter of data, predictors and concepts: Palaeogeography, Palaeoclimatology, Palaeoecology, v. 310, p. 451–463, https://doi.org/10.1016/j.palaeo.2011.07.021.

Vrba, E.S., 1980, Evolution, species, and fossils: How does life evolve?: South African Journal of Science, v. 76, p. 61–84.

Vrba, E.S., 1985, Environment and evolution: Alternative causes of the temporal distribution of evolutionary events: South African Journal of Science, v. 81, p. 229–236.

Walls, B.J., and Stigall, A.L., 2011, Analyzing niche stability and biogeography of Late Ordovician brachiopod species using ecological niche modeling: Palaeogeography, Palaeoclimatology, Palaeoecology, v. 299, p. 15–29, https://doi.org/10.1016/j.palaeo.2010.10.024.

Walls, B.J., and Stigall, A.L., 2012, A field based analysis of the accuracy of niche models applied to the fossil record: University of Kansas Paleontological Contributions, v. 6, p. 1–12.

Waltari, E., and Guralnick, R., 2009, Ecological niche modelling of montane mammals in the Great Basin, North America: Examining past and present connectivity of species across basins and ranges: Journal of Biogeography, v. 36, p. 148–161, https://doi.org/10.1111/j.1365-2699.2008.01959.x.

Waltari, E., Perkins, S., Hijmans, R., Peterson, A.T., Nyári, Á., and Guralnick, R., 2007, Locating Pleistocene refugia: Comparing phylogeographic and ecological niche model predictions: PLoS One, v. 2, e563, 11 p., https://doi.org/10.1371/journal.pone.0000563.

Whittington, H.B., 1985, The Burgess Shale: New Haven, Connecticut, Yale University Press, 151 p.

Wieczorek, J., Guo, Q., and Hijmans, R.J., 2004, The point-radius method for georeferencing locality descriptions and calculating associated uncertainty: International Journal of Geographical Information Science, v. 18, p. 745–767, https://doi.org/10.1080/13658810412331280211.

Wieczorek, J., Bloom, D., Guralnick, R., Blum, S., Döring, M., Giovanni, R., Robertson, T., and Vieglais, D., 2012, Darwin Core: An Evolving Community-Developed Biodiversity Data Standard: PLoS One, v. 7, e29715, 8 p., https://doi.org/10.1371/journal.pone.0029715.

Wiley, E.O., and Lieberman, B.S., 2011, Phylogenetics [2nd ed.]: Hoboken, New Jersey, Wiley-Blackwell, 406 p., https://doi.org/10.1002/9781118017883.

Wiley, E.O., and Mayden, R.L., 1985, Species and speciation in phylogenetic systematics, with examples from the North American fish fauna: Annals of the Missouri Botanical Garden, v. 72, p. 596–635, https://doi.org/10.2307/2399217.

Manuscript Accepted by the Society 6 December 2017
Manuscript Published Online 6 June 2018